LARGE-SCALE CONVEX OPTIMIZATION

Starting from where a first course in convex optimization leaves off, this text presents a unified analysis of first-order optimization methods – including parallel-distributed algorithms – through the abstraction of monotone operators. With the increased computational power and availability of big data over the past decade, applied disciplines have demanded that larger and larger optimization problems be solved. This text covers the first-order convex optimization methods that are uniquely effective at solving these large-scale optimization problems. Readers will have the opportunity to construct and analyze many well-known classical and modern algorithms using monotone operators, and walk away with a solid understanding of the diverse optimization algorithms. Graduate students and researchers in mathematical optimization, operations research, electrical engineering, statistics, and computer science will appreciate this concise introduction to the theory of convex optimization algorithms.

Ernest K. Ryu is Assistant Professor of Mathematical Sciences at Seoul National University. He previously served as Assistant Adjunct Professor with the Department of Mathematics at the University of California, Los Angeles from 2016 to 2019, before joining Seoul National University in 2020. He received a BS with distinction in physics and electrical engineering from the California Institute of Technology in 2010; and then an MS in statistics and a PhD – with the Gene Golub Best Thesis Award – in computational mathematics at Stanford University in 2016. His current research focuses on mathematical optimization and machine learning.

Wotao Yin is Director of the Decision Intelligence Lab with Alibaba Group (US), DAMO Academy, and a former Professor of Mathematics at the University of California, Los Angeles. He received his PhD in operations research from Columbia University in 2006. His numerous accolades include an NSF CAREER Award in 2008, an Alfred P. Sloan Research Fellowship in 2009, a Morningside Gold Medal in 2016, and a DAMO Award and an Egon Balas Prize in 2021. He invented fast algorithms for sparse optimization, image processing, and large-scale distributed optimization problems, and is among the top 1 percent of cited researchers by Clarivate Analytics. His research interests include computational optimization and its applications in signal processing, machine learning, and other data science problems.

"Ryu and Yin's *Large-Scale Convex Optimization* does a great job of covering a field with a long history and much current interest. The book describes dozens of algorithms, from classic ones developed in the 1970s to some very recent ones, in unified and consistent notation, all organized around the basic concept and unifying theme of a monotone operator. I strongly recommend it to any mathematician, researcher, or engineer who uses, or has an interest in, convex optimization."

– Stephen Boyd, Stanford University

"This is an absolute must-read research monograph for signal processing, communications, and networking engineers, as well as researchers who wish to choose, design, and analyze splitting-based convex optimization methods best suited for their perplexed and challenging engineering tasks."

– Georgios B. Giannakis, University of Minnesota

"This is a very timely book. Monotone operator theory is fundamental to the development of modern algorithms for large-scale convex optimization. Ryu and Yin provide optimization students and researchers with a self-contained introduction to the elegant mathematical theory of monotone operators, and take their readers on a tour of cutting-edge applications, demonstrating the power and range of these essential tools."

– Lieven Vandenberghe, University of California, Los Angeles

"First-order methods are the mainstream optimization algorithms in the era of big data. This monograph provides a unique perspective on various first-order convex optimization algorithms via the monotone operator theory, with which the seemingly different and unrelated algorithms are actually deeply connected, and many proofs can be significantly simplified. The book is a beautiful example of the power of abstraction. Those who are interested in convex optimization theory should not miss this book."

– Zhouchen Lin, Peking University

"The book covers topics from the basics of optimization to modern techniques such as operator splitting, parallel and distributed optimization, and stochastic algorithms. It is the natural next step after Boyd and Vandenberghe's *Convex Optimization* for students studying optimization and machine learning. The authors are experts in this kind of optimization. Some of my graduate students took the course based on this book when Wotao Yin was at UCLA. They liked the course and found the materials very useful in their research."

– Stanley Osher, University of California, Los Angeles

LARGE-SCALE CONVEX OPTIMIZATION

Algorithms & Analyses via Monotone Operators

ERNEST K. RYU
Seoul National University

WOTAO YIN
University of California, Los Angeles
and DAMO Academy, Alibaba Group

CAMBRIDGE
UNIVERSITY PRESS

University Printing House, Cambridge CB2 8BS, United Kingdom

One Liberty Plaza, 20th Floor, New York, NY 10006, USA

477 Williamstown Road, Port Melbourne, VIC 3207, Australia

314–321, 3rd Floor, Plot 3, Splendor Forum, Jasola District Centre, New Delhi – 110025, India

103 Penang Road, #05-06/07, Visioncrest Commercial, Singapore 238467

Cambridge University Press is part of the University of Cambridge.

It furthers the University's mission by disseminating knowledge in the pursuit of education, learning, and research at the highest international levels of excellence.

www.cambridge.org
Information on this title: www.cambridge.org/9781009160858
DOI: 10.1017/9781009160865

First published 2023

A catalogue record for this publication is available from the British Library.

ISBN 978-1-009-16085-8 Hardback

Dedicated to our wives
Bora and Rui

Contents

Preface

We write this book to share an elegant perspective that provides powerful higher-level insight into first-order convex optimization methods. The study of first-order convex optimization methods, which are more effective at solving large-scale optimization problems, started in the 1960s and 1970s, but the field at the time was focused rather on second-order methods, which are more effective at solving smaller problems. It was in the 2000s that increased computation power and the availability of big data brought first-order optimization methods into the mainstream. During this modern era, the authors entered the field of optimization and discovered (but did not invent) the perspective mentioned above, and we wish to share it through this book.

> *Our goal is to present a unified analysis of convex optimization algorithms through the abstraction of monotone operators.*

The widespread modern use of first-order methods makes this perspective more relevant than ever for both researchers and users of optimization.

This book has a somewhat unconventional organization: the chapters are structured around the techniques for deriving and analyzing optimization methods, rather than around optimization methods themselves. Through this organization, we aim to provide structure to the theory and achieve intellectual economy in that we present and analyze many optimization methods with a handful of mathematical concepts. The result is, we hope, a book that serves as a concise introduction to the theory of convex optimization algorithms.

We should also explain what this book is not. This book is not a text on monotone operator theory. We use monotone operators as a means to the end of developing and analyzing optimization algorithms, but we do not focus on the study of monotone operators themselves. This book is not a comprehensive reference on the best convex optimization methods or the strongest convergence analyses. We utilize a handful of techniques to derive and analyze optimization methods, and we only present methods and results that fit this approach.

Audience

This book is meant for both mathematicians and engineers. We appeal to mathematicians by showing that the abstraction is elegant and, in some aspects, challenging

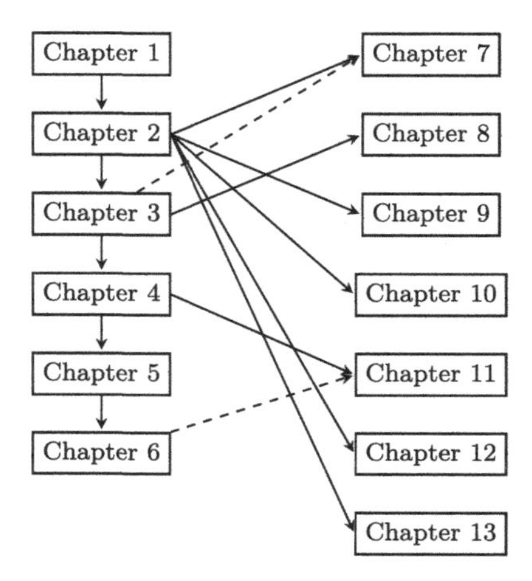

Figure 0.1 Chapter dependencies. Solid arrows denote hard dependence, while dashed arrows denote soft dependence. For example, Chapter 11 can be read after Chapter 4, but an understanding of the materials up to Chapter 6 is beneficial.

(interesting). We appeal to engineers, users of optimization, with the simplicity of the techniques and the diversity of the algorithms. In several instances, we have met engineers who know only gradient descent and ADMM, which, although powerful, are not universally feasible or best choices. This book empowers the reader to choose and even design the splitting methods best suited for any given problem.

The background required of the reader is a good knowledge of advanced calculus, linear algebra, basic probability, and basic notions of convex analysis on the topics of convex sets, convex functions, convex optimization problems, and convex duality at the level of chapters 2 through 5 of Boyd and Vandenberghe's *Convex Optimization*. Background in (mathematical) analysis and measure-theoretic probability theory is helpful but not necessary.

Informally, this book presupposes interest in convex optimization, and an appreciation of it as a useful tool. To keep the discussion concise, we focus on optimization algorithms without discussing the engineering and science origins of the optimization problems that the algorithms solve. Boyd and Vandenberghe's *Convex Optimization* is an excellent reference on the applications.

Note to Instructors

The material of this book can be taught in 15 weeks of a graduate or advanced undergraduate course. We have taught this book in an undergraduate course at SNU after covering the first five chapters of Boyd and Vandenberghe's *Convex Optimization* and in a graduate-level course at UCLA. The chapters of Part I should be taught in a linear order, while the chapters of Part II can be selected independently. Figure 0.1 illustrates the chapter dependencies. While the book does not delve deeply into the analysis of any single method, it covers many methods, as listed in Table 1. In our experience, many students appreciate the variety rather than the depth of the coverage.

This book contains almost no discussion of applications. Students without prior exposure to applications may find lectures solely on algorithms dry, so an instructor using this book may need to supplement the lectures with applications of interest to the audience.

Table 1 Optimization methods covered in each chapter

Chapters	**Methods**
Chapter 2	Gradient descent, dual ascent, proximal point method, method of multipliers, proximal method of multipliers, forward-backward splitting, Douglas–Rachford splitting, Davis–Yin splitting, proximal gradient method, iterative soft thresholding, consensus optimization, forward-Douglas–Rachford, variable metric proximal point, variable metric forward-backward splitting, backward-backward method, averaged alternating modified reflections, PPXA
Chapter 3	ADMM, alternating minimization algorithm (Tseng), PDHG (Chambolle–Pock), Condat–Vũ, proximal method of multipliers with function linearization, PAPC/PDFP^{2}O, linearized method of multipliers, PD3O, proximal ADMM, linearized ADMM, Chen–Teboulle, DYS 3-block ADMM, doubly-linearized method of multipliers.
Chapter 5	Coordinate gradient descent block-coordinate descent, coordinate proximal-gradient descent, stochastic dual coordinate ascent, MISO/Finito, coordinate updates on conic programs.
Chapter 6	ARock, asynchronous coordinate gradient descent, asynchronous ADMM.
Chapter 7	Stochastic forward-backward method, stochastic gradient descent, stochastic proximal gradient method, stochastic proximal simultaneous gradient method, stochastic Condat–Vũ.
Chapter 8	Function-linearized proximal ADMM, golden ratio ADMM, doubly-linearized ADMM, partial linearization, near-circulant splitting, Jacobi ADMM, 2-1-2 ADMM, Trip-ADMM, split Bregman method, four-block 2-1-2-4-3-4 ADMM.
Chapter 11	Distributed ADMM, decentralized ADMM, distributed gradient descent, method of diffusion, adapt-then-combine, PG-EXTRA, NIDS.
Chapter 12	Nesterov accelerated gradient method, FISTA, accelerated proximal point method.

For example, at SNU, we discussed engineering and machine learning applications from Boyd and Vandenberghe's *Convex Optimization*.

The textbook contains adequate homework exercises with varying levels of difficulties; some basic exercises complement the main exposition, while the difficult ones are designed to challenge the mathematically gifted students. We have also made public course material, including lecture slides and videos, on the website
`https://large-scale-book.mathopt.com/`
to help prospective instructors prepare for their lectures.

Acknowledgments

This book was greatly improved by the suggestions of Pontus Giselsson, Shuvomoy Das Gupta, Howard Heaton, Jongmin Lee, Daniel McKenzie, Chanwoo Park, Jisun Park, Ruoyu Sun, Jaewook Suh, Matthew Tam, and Taeho Yoon. We also thank the students in our courses who provided us with valuable feedback.

We also acknowledge Stephen P. Boyd, the Ph.D. advisor of Ernest Ryu. Boyd has a writing style of extreme clarity, and Ernest Ryu has strived to learn from and emulate it. Those familiar with Boyd's work may recognize his influence. In particular, Chapter 2 has much overlap with the review paper "Primer on Monotone Operator Methods," written by Ernest Ryu and Stephen Boyd in 2016 [RB16].

1 Introduction and Preliminaries

Monotone operator theory is an elegant and powerful tool for analyzing first-order convex optimization methods and, as such, plays a central role in convex analysis and convex optimization theory. In this book, we use this tool to provide a unified analysis of many classical and modern convex optimization methods.

This book is organized into two parts. Part I presents analysis of convex optimization methods via monotone operators, the core content. The content of Part I has sequential dependence, so the chapters should be read in a linear order. Part II presents additional auxiliary topics. The chapters can be read independently of each other. A diagram in the preface illustrates the dependency of the chapters.

1.1 FIRST-ORDER METHODS IN THE MODERN ERA

Many convex optimization methods can be classified into first or second-order methods. First-order methods can be described and analyzed with gradients and subgradients, while second-order methods use second-order derivatives or their approximations.

In the early days of convex optimization, the 1970s through the 1990s, researchers focused primarily on second-order methods, as they were more effective in solving the relatively smaller optimization problems of the era. Within the past decade, however, the demand to solve ever-larger problems grew, and so did the popularity of first-order methods.

Second-order methods require relatively fewer iterations to solve the optimization problem to high accuracy, even up to machine precision. However, the computational cost per iteration quickly becomes expensive as the problem size grows. In contrast, first-order methods have a much lower computational cost per iteration. For some large-scale optimization problems, running even a single iteration of a second-order method is infeasible, while first-order methods can solve such problems to acceptable accuracy.

Another advantage of first-order methods is that they are extremely simple; we can usually describe the entire method with two or three lines of equations. This is a significant advantage in practice, as simpler methods are easy for practitioners to implement and try out quickly, and the simplicity tends to make efficient parallelization easier.

The two classes of methods are usually not in competition. When a high-accuracy solution is needed, second-order methods should be used. In large-scale problems, one should use first-order methods and tolerate inaccuracy. After all, most engineering applications require only a few digits of accuracy in their solution. If the problem size is small, one should use second-order methods since there is little reason to forgo the high accuracy.

The total cost of a method is

$$(\text{cost per iteration}) \times (\text{number of iterations}).$$

We can analyze the cost per iteration by examining the computational cost of the individual components of the method. We can analyze the number of iterations required for convergence by analyzing the *rate of convergence*.

In convex optimization, arguments advocating one method over another are often based on the cost per iteration. In fact, we just made this very argument in comparing first-order and second-order methods. However, it is important to keep in mind that these arguments are incomplete since the cost per iteration is only half of the equation, literally. A method with a low cost per iteration has the potential, not a guarantee, to be efficient.

Nevertheless, primarily focusing on the cost per iteration of a method is still a useful simplification, so we adopt it in this book. With the exception of §12 and §13, this book almost entirely focuses on establishing convergence without paying much attention to the rate of convergence. We do prove convergence rates, but the rates are discussed infrequently.

1.2 LIMITATIONS OF MONOTONE OPERATOR THEORY

One of the main goals of this book is to provide streamlined and simple convergence proofs, and we only discuss results that fit this approach. Such results are simple but often not the strongest. The strongest results in convex optimization usually involve arguments that go beyond monotone operator theory.

Proofs based on monotone operator theory use monotonicity, rather than convexity, as the key property. This line of analysis does not lead to results involving function values. For example, the gradient method $x^{k+1} = x^k - \alpha\nabla f(x^k)$ converges, under suitable assumptions, with rate $\|\nabla f(x^k)\|^2 \leq O(1/k)$ and $f(x^k) - f(x^\star) \leq O(1/k)$. We can prove the first result with properties of monotone operators, but the second result requires properties of convex functions. Also, topics such as line searching, Frank–Wolfe, and second-order methods are not explained very well with monotone operator theory. Monotone operators do play a central role, but convex optimization theory does go beyond monotone operators.

1.3 PRELIMINARIES

In this section, we quickly review preliminary topics. We simply state, without proof, many of the results based on convex analysis and refer interested readers to standard references such as [Roc70d, Roc74, HL93, HL01, BV04, Nes04, BL06, NP06, Ber09, BV10, BC17a].

1.3.1 Sets

A set is empty when it contains no element. Let $\emptyset$ denote the empty set. When a set contains one element, we say it is a *singleton*.

A set S is *convex* if $x, y \in S$ implies $\theta x + (1-\theta)y \in S$ for all $\theta \in [0,1]$. The empty set, singletons, and $\mathbb{R}^n$ are also convex sets.

In this book, we overload the standard notation defined for points to sets. In particular, when $\alpha \in \mathbb{R}$, $x \in \mathbb{R}^n$, $A, B \subseteq \mathbb{R}^n$, and $M \in \mathbb{R}^{m \times n}$, we write

$$\begin{aligned} \alpha A &= \{\alpha a \mid a \in A\} \\ x + A &= \{x + a \mid a \in A\} \\ MA &= \{Ma \mid a \in A\} \\ A + B &= \{a + b \mid a \in A, \, b \in B\}. \end{aligned}$$

These operations preserve convexity; if A and B are convex, all of these sets are convex. The sum $A + B$ is called the *Minkowski sum*.

1.3.2 Linear Algebra

Write $\mathbb{R}^n$ for the n-dimensional Euclidean space. For any $x, y \in \mathbb{R}^n$, write

$$\langle x, y \rangle = x^\intercal y = \sum_{i=1}^{n} x_i y_i$$

for the standard inner product.

Given a matrix $A \in \mathbb{R}^{m \times n}$, write $\mathcal{R}(A)$ for the range of A and $\mathcal{N}(A)$ for the nullspace of A. If $A \in \mathbb{R}^{n \times n}$, we say A is a square matrix. If $A^\intercal = A$, which implies A is square, we say A is symmetric. If A is symmetric, the eigenvalues of A are real. Write $\lambda_{\max}(A)$ and $\lambda_{\min}(A)$ respectively for the largest and smallest eigenvalues of A, when A is symmetric.

If all eigenvalues of a symmetric matrix A are nonnegative, we say A is symmetric positive semidefinite and write $A \succeq 0$. If all eigenvalues of a symmetric matrix A are strictly positive, we say A is symmetric positive definite and write $A \succ 0$. We write $A \succeq B$ and $A \succ B$ if $A - B \succeq 0$ and $A - B \succ 0$, respectively.

Given $M \succeq 0$, write $M^{1/2}$ for the matrix square root, the unique symmetric positive semidefinite matrix that satisfies $(M^{1/2})^2 = M$. If $M \succ 0$, then $M^{1/2} \succ 0$, and we write $M^{-1/2} = (M^{1/2})^{-1}$.

Consider a symmetric matrix $X \in \mathbb{R}^{(m+n)\times(m+n)}$ partitioned as

$$X = \begin{bmatrix} A & B \\ B^\intercal & C \end{bmatrix},$$

where $A = A^\intercal \in \mathbb{R}^{m \times m}$, $B \in \mathbb{R}^{m \times n}$, and $C = C^\intercal \in \mathbb{R}^{n \times n}$. When A is invertible, we call the matrix

$$S = C - B^\intercal A^{-1} B$$

the *Schur complement* of A in X. Note that $S \in \mathbb{R}^{n \times n}$ is symmetric. Given $A \succ 0$, X is positive (semi)definite if and only if S is positive (semi)definite. Likewise, when C is invertible,

$$T = A - BC^{-1}B^{\intercal}$$

is the Schur complement of C in X. Given $C \succ 0$, X is positive (semi)definite if and only if T is positive (semi)definite. We use the Schur complement to assess whether a symmetric matrix is positive (semi)definite.

The 2-norm or the Euclidean norm is

$$\|x\| = \|x\|_2 = \sqrt{\langle x, x \rangle}.$$

In some cases, we will use the 1-norm and the ∞-norm respectively defined as

$$\|x\|_1 = \sum_{i=1}^{n} |x_i|, \qquad \|x\|_\infty = \max_{i=1,\ldots,n} |x_i|.$$

Given $A \succ 0$, define the A-norm as

$$\|x\|_A = \sqrt{x^{\intercal}Ax}.$$

Given $A \succeq 0$, define the A-seminorm as

$$\|x\|_A = \sqrt{x^{\intercal}Ax}.$$

Since this is a seminorm, the triangle inequality $\|x + y\|_A \leq \|x\|_A + \|y\|_A$ and absolute homogeneity $\|\alpha x\|_A = |\alpha|\|x\|_A$ hold, but $\|x\|_A = 0$ is possible when $x \neq 0$.

Given a matrix $A \in \mathbb{R}^{m \times n}$, write

$$\sigma_{\max}(A) = \sqrt{\lambda_{\max}(A^{\intercal}A)} = \max_{x \neq 0} \frac{\|Ax\|}{\|x\|}$$

for the maximum singular value of A and

$$\sigma_{\min}(A) = \sqrt{\lambda_{\min}(A^{\intercal}A)} = \min_{x \neq 0} \frac{\|Ax\|}{\|x\|}$$

for the minimum singular value of A. While a real eigenvalue can be negative, all singular values are nonnegative.

We say $V \subseteq \mathbb{R}^n$ is a (linear) subspace if $0 \in V$, $x, y \in V$ implies $x + y \in V$, and $x \in V$ implies $\alpha x \in V$ for any $\alpha \in \mathbb{R}$. Under this definition, $\{0\}$ and $\mathbb{R}^n$ are also subspaces. For any $A \in \mathbb{R}^{m \times n}$, $\mathcal{R}(A)$ and $\mathcal{N}(A)$ are subspaces.

1.3.3 Analysis

For $L > 0$, we say that a mapping $\mathbb{T}\colon \mathbb{R}^n \to \mathbb{R}^m$ is L-Lipschitz (continuous) if

$$\|\mathbb{T}(x) - \mathbb{T}(y)\| \leq L\|x - y\| \qquad \forall x, y \in \mathbb{R}^n.$$

We say $\mathbb{T}$ is Lipschitz (continuous) if $\mathbb{T}$ is L-Lipschitz for some unspecified $L \in (0, \infty)$. (One could say that a constant function is 0-Lipschitz, but we exclude this degenerate case from our definition, since we will later encounter quantities like $2/L$.)

If a mapping is Lipschitz, it is a continuous mapping. If $\mathbb{T}_1$ and $\mathbb{T}_2$ are respectively L_1- and L_2-Lipschitz, then $\mathbb{T}_1 \circ \mathbb{T}_2$ is L_1L_2-Lipschitz since

$$\|\mathbb{T}_1(\mathbb{T}_2(x)) - \mathbb{T}_1(\mathbb{T}_2(y))\| \le L_1\|\mathbb{T}_2(x) - \mathbb{T}_2(y)\| \le L_1L_2\|x - y\|.$$

If $\mathbb{T}_1$ and $\mathbb{T}_2$ are respectively L_1- and L_2-Lipschitz, then $\alpha_1\mathbb{T}_1 + \alpha_2\mathbb{T}_2$ is $(|\alpha_1|L_1 + |\alpha_2|L_2)$-Lipschitz.

A matrix $A \in \mathbb{R}^{m\times n}$ can be viewed as a mapping from x to Ax. Since

$$\|Ax\| \le \sigma_{\max}(A)\|x\|,$$

we can view A as a $\sigma_{\max}(A)$-Lipschitz mapping.

Write

$$B(x, r) = \{y \in \mathbb{R}^n \mid \|y - x\| \le r\}$$

for the closed ball of radius r centered at x. Define the interior of a set C as

$$\operatorname{int} C = \{x \in C \mid B(x, r) \subseteq C \text{ for some } r > 0\}.$$

Denote the closure of a set C as $\operatorname{cl} C$. Define the boundary of C as $\operatorname{cl} C \backslash \operatorname{int} C$.

An affine set A can be expressed as

$$A = x_0 + V,$$

where $x_0 \in \mathbb{R}^n$ and $V \subseteq \mathbb{R}^n$ is a subspace. The affine hull of C is defined as

$$\operatorname{aff} C = \{\theta_1x_1 + \cdots + \theta_kx_k \mid x_1, \ldots, x_k \in C,\ \theta_1 + \cdots + \theta_k = 1,\ k \ge 1\}.$$

The affine hull is the smallest affine set containing C; if $C \subseteq A$ and A is affine, then $\operatorname{aff} \operatorname{cl} C \subseteq A$.

Define the relative interior of a set C as

$$\operatorname{ri} C = \{x \in C \mid B(x, r) \cap \operatorname{aff} C \subseteq C \text{ for some } r > 0\}.$$

The relative interior of a nonempty convex set is nonempty. Under this definition, the relative interior of a singleton is the singleton itself. Define the relative boundary of C as $\operatorname{cl} C \backslash \operatorname{ri} C$. When we are dealing with low-dimensional sets placed in higher-dimensional spaces, the notion of relative interior is useful.

Example 1.1 Consider the line segment

$$S = \left\{(x, y) \in \mathbb{R}^2 \mid x \in [0.5, 1],\ y = 4x - 3\right\}.$$

The relative interior is the line segment with the end points excluded.

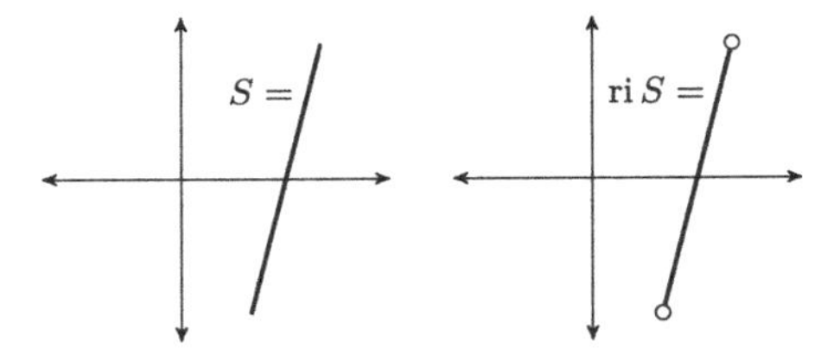

Define the distance of a point $x \in \mathbb{R}^n$ to a nonempty set $X \subseteq \mathbb{R}^n$ as

$$\operatorname{dist}(x, X) = \inf_{z \in X} \|z - x\|.$$

When X is nonempty and closed, the infimum is attained and $\operatorname{dist}(x, X) = 0$ if and only if $x \in X$. For notational convenience, write $\operatorname{dist}^2(x, X) = (\operatorname{dist}(x, X))^2$.

1.3.4 Functions

An *extended real-valued* function is a function that maps to the extended real line, $\mathbb{R} \cup \{\pm\infty\}$. Unless otherwise specified, functions in this book are extended real-valued. Write

$$\operatorname{dom} f = \{x \in \mathbb{R}^n \mid f(x) < \infty\}$$

for the (effective) domain of f. We use $\leq$, $<$, $\geq$, and $>$ for elements of the extended real line in the obvious way; for any finite α, we have $-\infty < \alpha < \infty$. We allow $\infty \leq \infty$ and $-\infty \leq -\infty$, but not $\infty < \infty$ or $-\infty < -\infty$.

A function f is *convex* if $\operatorname{dom} f$ is a convex set and

$$f(\theta x + (1-\theta)y) \leq \theta f(x) + (1-\theta) f(y), \quad \forall x, y \in \operatorname{dom} f,\ \theta \in (0,1). \tag{1.1}$$

A function f is *strictly convex* if the inequality (1.1) is strict when $x \neq y$. We say f is (strictly) concave if $-f$ is (strictly) convex.

The *epigraph* of a function is defined as

$$\operatorname{epi} f = \{(x, \alpha) \in \mathbb{R}^n \times \mathbb{R} \mid f(x) \leq \alpha\}.$$

A function f is convex if and only if $\operatorname{epi} f$ is convex. A function is *proper* if its value is never $-\infty$ and is finite somewhere. A proper function is *closed* if its epigraph is a closed set in $\mathbb{R}^{n+1}$. A proper function is closed if and only if it is lower semicontinuous. We say a function is CCP if it is closed, convex, and proper. As most convex functions of interest are closed and proper, we focus exclusively on CCP functions in this book. A function is CCP if and only if its epigraph is a nonempty closed convex set without a "vertical line," a line of the form $\{(x_0, t) \mid t \in \mathbb{R}\}$ for some $x_0 \in \mathbb{R}^n$.

Example 1.2 Whether a convex function f is closed is determined by f's behavior on the boundary of $\operatorname{dom} f$.

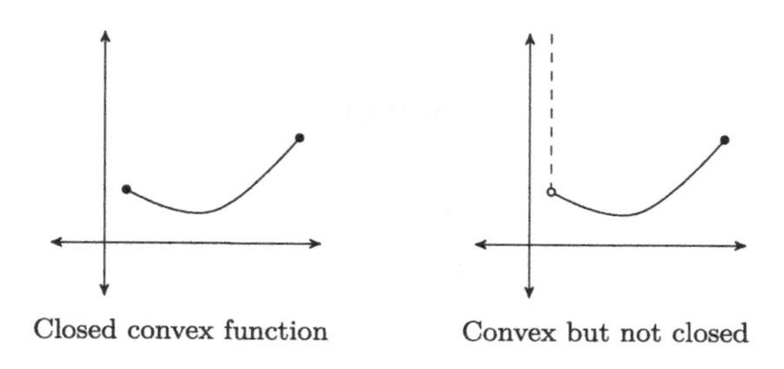

The dashed line denotes the function value of ∞.

Example 1.3 The epigraph of the CCP function $-\log$ is a nonempty closed convex set.

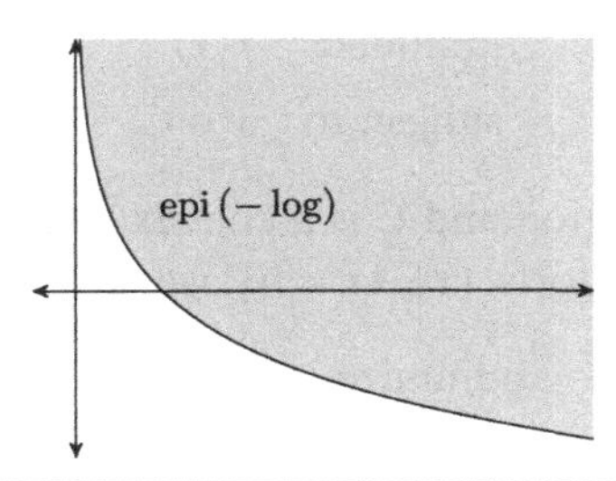

If f is a CCP function and $\alpha > 0$, then αf is CCP. If f and g are CCP functions and there is an x such that $f(x) + g(x) < \infty$, then $f + g$ is CCP. If f is a CCP function on $\mathbb{R}^n$, $A \in \mathbb{R}^{n\times m}$, and there is an $x \in \mathbb{R}^m$ such that $f(Ax) < \infty$, then $g(x) = f(Ax)$ is CCP.

We say $f\colon \mathbb{R}^n \to \mathbb{R}\cup\{\pm\infty\}$ is differentiable if $f\colon \mathbb{R}^n \to \mathbb{R}$ (so f is not extended real-valued), gradient $\nabla f(x) = [\frac{\partial f}{\partial x_1}(x), \dots, \frac{\partial f}{\partial x_n}(x)]^\intercal$ exists for all $x \in \mathbb{R}^n$, and

$$\lim_{h\to 0} \frac{f(x+h) - f(x) - \langle \nabla f(x), h\rangle}{\|h\|} = 0$$

for all $x \in \mathbb{R}^n$. A differentiable function f is convex if and only if

$$f(y) \geq f(x) + \langle \nabla f(x), y - x\rangle \qquad \forall x, y \in \mathbb{R}^n.$$

In other words, f is convex if its first-order Taylor expansion is a global lower bound of f. A twice continuously differentiable function f is convex if and only if $\nabla^2 f(x) \succeq 0$ for all $x \in \mathbb{R}^n$. (By the classic Schwarz's theorem, $\nabla^2 f(x) \in \mathbb{R}^{n\times n}$ is symmetric when f is twice continuously differentiable.) Intuitively speaking, $\nabla^2 f$ measures curvature, and f is convex if f is flat or has upward curvature everywhere. If f is a one-dimensional differentiable function, f is convex if and only if $f'(x)$ is monotonically nondecreasing. See the bibliographical notes for further discussion.

Write

$$\operatorname{argmin} f = \left\{ x \in \mathbb{R}^n \,\middle|\, f(x) = \inf_{z\in\mathbb{R}^n} f(z) \right\}$$

for the set of minimizers of f. When f is CCP, $\operatorname{argmin} f$ is a closed convex set, possibly empty. When f is strictly convex, $\operatorname{argmin} f$ has at most one point.

For $S \subseteq \mathbb{R}^n$, define the *indicator function*

$$\delta_S(x) = \begin{cases} 0 & \text{if } x \in S \\ \infty & \text{otherwise.} \end{cases}$$

If S is convex, closed, and nonempty, then δ_S is CCP.

1.3.5 Convex Optimization Problems

An *unconstrained optimization problem*

$$\underset{x\in\mathbb{R}^n}{\text{minimize}} \quad f(x)$$

is convex if f is a convex function. We call f the *objective function*. The *constrained optimization problem*

$$\begin{array}{ll} \underset{x\in\mathbb{R}^n}{\text{minimize}} & f(x) \\ \text{subject to} & x \in C \end{array}$$

is convex if f is a convex function and C is a convex set. We call $x \in C$ the *constraint*. When C is an affine set of the form $\{x \mid Ax = b\}$, we also write

$$\begin{array}{ll} \underset{x\in\mathbb{R}^n}{\text{minimize}} & f(x) \\ \text{subject to} & Ax = b. \end{array}$$

In these problems, $x \in \mathbb{R}^n$ is the *optimization variable*. If a solution to an optimization problem exists, write superscript $\star$ to denote a solution. So if x is the optimization variable, $x^\star$ denotes a solution. If u is the optimization variable, $u^\star$ denotes a solution.

Indicator functions allow us to move the constraint into the objective function and treat a constrained problem as an unconstrained problem:

$$\underset{x\in\mathbb{R}^n}{\text{minimize}} \quad f(x) + \delta_C(x).$$

This use of indicator functions and extended value functions greatly simplifies the notation.

1.3.6 Subgradient

We say $g \in \mathbb{R}^n$ is a *subgradient* of a convex function f at x if

$$f(y) \geq f(x) + \langle g, y - x\rangle \qquad \forall\, y \in \mathbb{R}^n. \tag{1.2}$$

In other words, a subgradient provides an global affine lower bound of f. We call (1.2) the *subgradient inequality*. The *subdifferential* of a convex function f at x is

$$\partial f(x) = \{g \in \mathbb{R}^n \mid f(y) \geq f(x) + \langle g, y - x\rangle, \forall\, y \in \mathbb{R}^n\}.$$

In other words, $\partial f(x)$ is the set of subgradients of f at x. It is straightforward to see that $\partial f(x)$ is a closed convex set, possibly empty. A convex function f is differentiable at x if and only if $\partial f(x)$ is a singleton.

By definition, $x^\star \in \operatorname{argmin} f$ if and only if $0 \in \partial f(x^\star)$. This fact, called Fermat's rule, illustrates why subgradients are central in convex optimization.

Example 1.4 The absolute value function is differentiable everywhere except at 0.

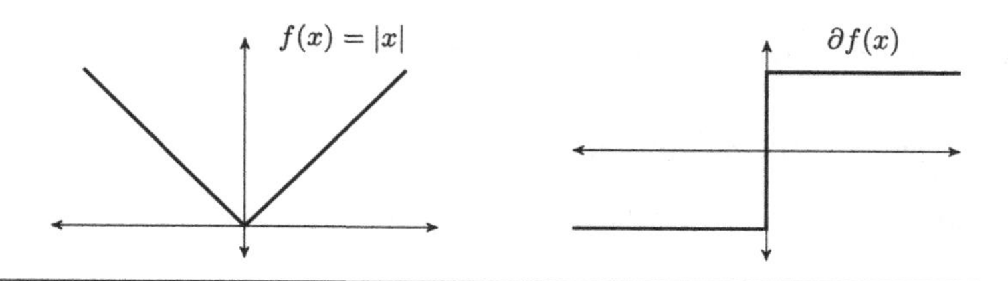

Example 1.5 At x_1 the convex function f is differentiable and $\partial f(x_1) = \{\nabla f(x_1)\}$. At x_2, f is not differentiable and has many subgradients.

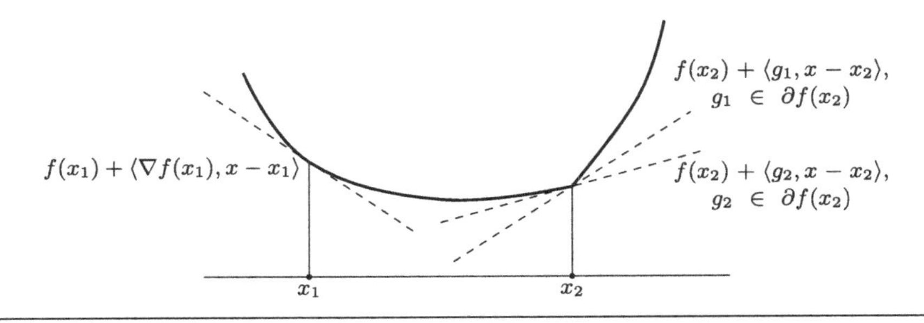

Example 1.6 Let $C \subseteq \mathbb{R}^n$ be a closed convex set. Then $\partial \delta_C(x) = \mathbb{N}_C(x)$, where

$$\mathbb{N}_C(x) = \begin{cases} \emptyset & \text{if } x \notin C \\ \{y \mid \langle y, z - x \rangle \leq 0 \, \forall z \in C\} & \text{if } x \in C \end{cases}$$

is the normal cone operator. For $x \in \operatorname{int} C$, $\mathbb{N}_C(x) = \{0\}$, and for $x \notin C$, $\mathbb{N}_C(x) = \emptyset$; $\mathbb{N}_C(x)$ is nontrivial only when x is on the boundary of C.

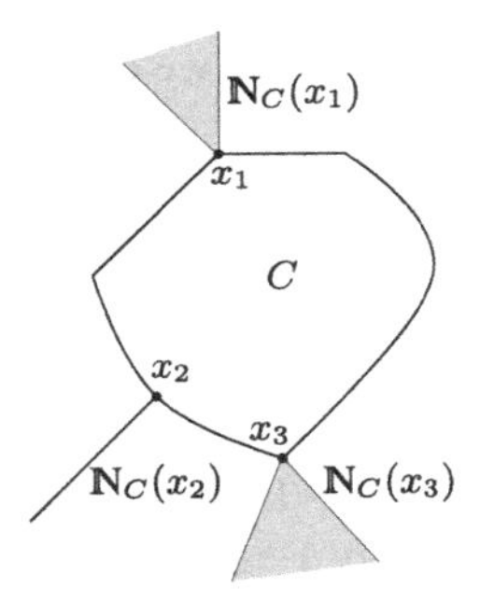

In this book, we will not pay too much attention to the meaning of $\mathbb{N}_C$. Rather, we use $\mathbb{N}_C$ as notational shorthand for $\partial \delta_C$.

We say a convex f is subdifferentiable at x if $\partial f(x) \neq \emptyset$. When f is convex and proper, $\partial f(x) = \emptyset$ where $f(x) = \infty$. When f is convex and proper, $\partial f(x) \neq \emptyset$ for any $x \in \operatorname{ri} \operatorname{dom} f$. So a convex and proper function is not subdifferentiable outside its domain, is subdifferentiable within the relative interior of its domain, and may or may not be subdifferentiable on the relative boundary of its domain.

Example 1.7 The CCP function f defined as

$$f(x) = \begin{cases} -\sqrt{x} & \text{for } x \geq 0 \\ \infty & \text{for } x < 0 \end{cases}$$

is not subdifferentiable at $x = 0$. The slope is $-\infty$, but we do not allow infinite gradients.

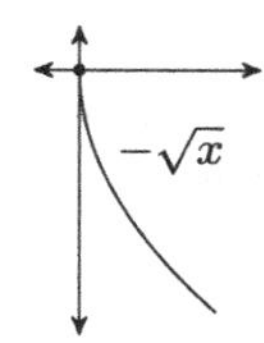

Several standard identities for gradients also hold for subdifferentials. Let f be CCP and $\alpha > 0$. Then

$$\partial(\alpha f)(x) = \alpha \partial f(x).$$

Let f be CCP and $\mathcal{R}(A) \cap \operatorname{ri} \operatorname{dom} f \neq \emptyset$. If $g(x) = f(Ax)$, then

$$\partial g(x) = A^\intercal \partial f(Ax). \tag{1.3}$$

Let f and g be CCP and $\operatorname{dom} f \cap \operatorname{int} \operatorname{dom} g \neq \emptyset$. Then

$$\partial(f+g)(x) = \partial f(x) + \partial g(x). \tag{1.4}$$

To clarify, $\partial f(x) + \partial g(x)$ is the Minkowski sum of the sets $\partial f(x)$ and $\partial g(x)$. Without the regularity conditions involving interiors, we can say

$$\partial g(x) \supseteq A^\intercal \partial f(Ax), \qquad \partial(f+g)(x) \supseteq \partial f(x) + \partial g(x).$$

Using the operator notation we define in §2, we can more concisely write

$$\partial \alpha f = \alpha \partial f, \qquad \partial g = A^\intercal \partial f A, \qquad \partial(f+g) = \partial f + \partial g,$$

provided the regularity conditions involving interiors hold.

1.3.7 Regularity Conditions

Say we have a mathematical statement "If P then Q". Then, if P "usually" holds, then Q "usually" holds. In this case, we say P is a *regularity condition*, since P is satisfied in the usual "regular" case. We just saw an example of this; if the regularity condition $\operatorname{dom} f \cap \operatorname{int} \operatorname{dom} g \neq \emptyset$ holds, then the identity $\partial(f+g) = \partial f + \partial g$ holds.

Statements in this book involving interiors and relative interiors can be considered regularity conditions. We keep track of these conditions, as they are necessary for a rigorous treatment of the subject. However, we do not focus on them.

1.3.8 Conjugate Function, Strong Convexity, and Smoothness

Define the conjugate function of f as

$$f^*(y) = \sup_{x \in \mathbb{R}^n} \{\langle y, x\rangle - f(x)\},$$

which is also known as the Fenchel conjugate or Legendre–Fenchel transform. When f is CCP, f^* is CCP and $f^{**} = f$; that is, the conjugate is CCP and the conjugate of the conjugate function is the original function. We call f^{**} the *biconjugate* of f. Note that we use the symbol $*$ for the notion of conjugate or dual, while we use the symbol $\star$ for the notion of optimality.

The conjugate function appears in optimization often because if f is CCP, then ∂f is an "inverse" of ∂f^* in the sense we define in §2.1. When f and f^* are both differentiable, then $(\nabla f)^{-1} = \nabla f^*$ as functions from $\mathbb{R}^n$ to $\mathbb{R}^n$.

We say a CCP f is μ-strongly convex if any of the following equivalent conditions are satisfied:

- $f(x) - (\mu/2)\|x\|^2$ is convex.
- $\langle \partial f(x) - \partial f(y), x - y\rangle \geq \mu\|x - y\|^2$ for all x, y.
- $\nabla^2 f(x) \succeq \mu I$ for all x if f is twice continuously differentiable.

The second condition is written with set-valued notation; the left-hand side is a subset of $\mathbb{R}$, so the inequality means the subset lies in $[\mu\|x - y\|^2, \infty)$. In the third condition, $I \in \mathbb{R}^{n\times n}$ denotes the identity matrix.

Strongly convex CCP functions have unique minimizers. If f is μ-strongly convex and g is convex, then $f+g$ is μ-strongly convex. Informally speaking, a function is μ-strongly convex if it has upward curvature of at least μ, and we can think of nondifferentiable points to be points with infinite curvature. To clarify, strong convexity does not imply differentiability.

Example 1.8 Informally speaking, μ-strongly convex functions have upward curvature of at least μ and L-smooth convex functions have upward curvature of no more than L.

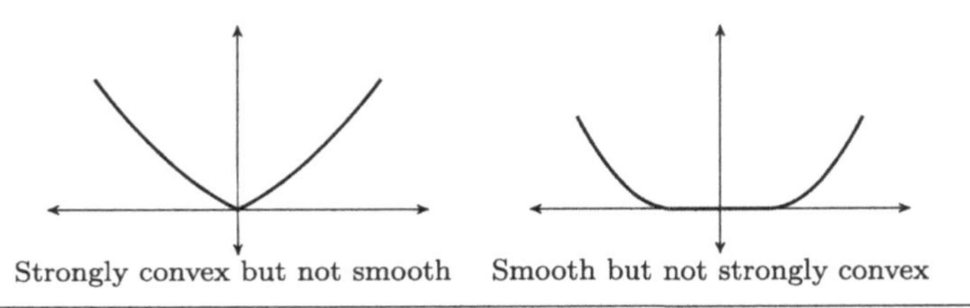

We say a CCP f is L-smooth if any of the following equivalent conditions are satisfied:

- $f(x) - (L/2)\|x\|^2$ is concave.
- f is differentiable and $\langle \nabla f(x) - \nabla f(y), x - y\rangle \geq (1/L)\|\nabla f(x) - \nabla f(y)\|^2$ for all x, y.
- f is differentiable and ∇f is L-Lipschitz.
- $\nabla^2 f(x) \preceq LI$ for all x if f is twice continuously differentiable.

(Remember, a function g is concave if $-g$ is convex.) The terminology "L-smoothness" is somewhat nonstandard; "smoothness" often means infinite differentiability in other fields of mathematics. Under our definition, L-smooth functions only need to be once-continuously differentiable.

Informally speaking, a convex function is L-strongly convex if it has upward curvature of at most L. Since non-differentiable points of convex functions can be thought of as points with infinite upward curvature, it is natural that smooth functions are differentiable.

If f is μ-strongly convex and L-smooth, then $\mu \leq L$. This follows from

$$\mu\|x - y\|^2 \leq \langle \nabla f(x) - \nabla f(y), x - y\rangle \leq \|\nabla f(x) - \nabla f(y)\|\|x - y\| \leq L\|x - y\|^2,$$

where we used the Cauchy–Schwartz inequality and the Lipschitz continuity of ∇f. Strong convexity and smoothness are dual properties; a CCP f is μ-strongly convex if and only if f^* is $(1/\mu)$-smooth. This follows from the fact that ∂f and ∂f^* are inverse operators, which we show in §2.1.

1.3.9 Convex Duality

In many introductory texts of convex optimization, one starts with a primal optimization problem and finds a corresponding dual problem. In this book, we take a slightly different viewpoint. We view the primal and dual problems as the two halves of a larger saddle point problem.

Let $\mathbf{L}\colon \mathbb{R}^n \times \mathbb{R}^m \to \mathbb{R} \cup \{\pm\infty\}$. We say $\mathbf{L}(x,u)$ is convex-concave if $\mathbf{L}$ is convex in x when u is fixed and concave in u when x is fixed. We say $(x^\star, u^\star)$ is a saddle point of $\mathbf{L}$ if

$$\mathbf{L}(x^\star, u) \le \mathbf{L}(x^\star, u^\star) \le \mathbf{L}(x, u^\star) \qquad \forall x \in \mathbb{R}^n,\ u \in \mathbb{R}^m.$$

We call

$$\underset{x\in\mathbb{R}^n}{\text{minimize}} \quad \sup_{u\in\mathbb{R}^m} \mathbf{L}(x,u)$$

the *primal problem* generated by $\mathbf{L}$ and write $p^\star = \inf_x \sup_u \mathbf{L}(x,u)$ for the primal optimal value. We call

$$\underset{u\in\mathbb{R}^m}{\text{maximize}} \quad \inf_{x\in\mathbb{R}^n} \mathbf{L}(x,u)$$

the *dual problem* generated by $\mathbf{L}$ and write $d^\star = \sup_u \inf_x \mathbf{L}(x,u)$ for the dual optimal value. In most engineering settings, one starts with an optimization problem, not a convex-concave saddle function. With this view of duality, the trick is to find a convex-concave saddle function that generates the primal problem of interest.

Example 1.9 Let f be a CCP function on $\mathbb{R}^n$, $A \in \mathbb{R}^{m\times n}$, and $b \in \mathbb{R}^m$. Consider the Lagrangian

$$\mathbf{L}(x,u) = f(x) + \langle u, Ax - b\rangle, \tag{1.5}$$

which generates the primal problem

$$\begin{array}{ll} \underset{x\in\mathbb{R}^n}{\text{minimize}} & f(x) \\ \text{subject to} & Ax = b \end{array} \tag{1.6}$$

and dual problem

$$\underset{u\in\mathbb{R}^m}{\text{maximize}} \quad -f^*(-A^\intercal u) - b^\intercal u. \tag{1.7}$$

The dual variable u is also called the Lagrange multipliers. If the constraint qualification

$$\{x \mid Ax = b\} \cap \operatorname{int} \operatorname{dom} f \neq \emptyset$$

holds, then $d^\star = p^\star$.

Example 1.10 Consider the Lagrangian

$$\mathbf{L}(x,u) = f(x) + \langle u, Ax\rangle - g^*(u), \tag{1.8}$$

which generates the primal problem

$$\underset{x\in\mathbb{R}^n}{\text{minimize}} \quad f(x) + g(Ax) \tag{1.9}$$

and dual problem

$$\underset{u\in\mathbb{R}^m}{\text{maximize}} \quad -f^*(-A^\intercal u) - g^*(u). \tag{1.10}$$

If the constraint qualification

$$A \operatorname{dom} f \cap \operatorname{int} \operatorname{dom} g \neq \emptyset$$

holds, then $d^\star = p^\star$. This primal-dual problem pair is sometimes called the Fenchel–Rockafellar dual.

Weak duality, which states $d^\star \le p^\star$, always holds. To prove this, note that for any x, u we have

$$\begin{aligned}
\inf_x \mathbf{L}(x,u) &\le \mathbf{L}(x,u)\\
\sup_u \inf_x \mathbf{L}(x,u) &\le \sup_u \mathbf{L}(x,u)\\
d^\star = \sup_u \inf_x \mathbf{L}(x,u) &\le \inf_x \sup_u \mathbf{L}(x,u) = p^\star.
\end{aligned}$$

Strong duality, which states $d^\star = p^\star$, holds often but not always in convex optimization. Regularity conditions that ensure strong duality are sometimes called constraint qualifications. The constraint qualifications for strong duality are similar to the regularity conditions for subgradient identities. Again, interested readers can refer to standard references such as [Roc74, Ber09, Boţ10] for a careful discussion of this subject.

Total duality states that a primal solution exists, a dual solution exists, and strong duality holds. Total duality holds if and only if $\mathbf{L}$ has a saddle point. Solving the primal and dual optimization problems is equivalent to finding a saddle point of the saddle function generating the primal and dual problems, provided that total duality holds. We will see in §2 and §3 that total duality is the regularity condition that ensures primal-dual methods converge.

Let us prove the equivalence. Assume $\mathbf{L}$ has a saddle point $(x^\star, u^\star)$. Then

$$\begin{aligned}
\mathbf{L}(x^\star,u^\star) &= \inf_x \mathbf{L}(x,u^\star)\\
&\le \sup_u \inf_x \mathbf{L}(x,u) = d^\star\\
&\le \inf_x \sup_u \mathbf{L}(x,u) = p^\star\\
&\le \sup_u \mathbf{L}(x^\star,u) = \mathbf{L}(x^\star,u^\star),
\end{aligned}$$

and equality holds throughout. Since $\inf_x \sup_u \mathbf{L}(x,u) = \sup_u \mathbf{L}(x^\star,u)$, $x^\star$ is a primal solution. Since $\inf_x \mathbf{L}(x,u^\star) = \sup_u \inf_x \mathbf{L}(x,u)$, $u^\star$ is a dual solution. Since $d^\star = \sup_u \inf_x \mathbf{L}(x,u) = \inf_x \sup_u \mathbf{L}(x,u) = p^\star$, strong duality holds.

On the other hand, assume total duality holds and $x^\star$ and $u^\star$ are primal and dual solutions. Then

$$\begin{aligned}\inf_x \mathbf{L}(x,u^\star) &= \sup_u \inf_x \mathbf{L}(x,u) = d^\star \\ &= \inf_x \sup_u \mathbf{L}(x,u) = p^\star \\ &= \sup_u \mathbf{L}(x^\star,u).\end{aligned}$$

Since

$$\mathbf{L}(x^\star,u^\star) \le \sup_u \mathbf{L}(x^\star,u) = \inf_x \mathbf{L}(x,u^\star) \le \mathbf{L}(x^\star,u^\star),$$

equality holds throughout and we conclude

$$\sup_u \mathbf{L}(x^\star,u) = \mathbf{L}(x^\star,u^\star) = \inf_x \mathbf{L}(x,u^\star),$$

that is, $(x^\star,u^\star)$ is a saddle point.

An *augmented Lagrangian* is a saddle function that has additional terms while sharing the same saddle points as its unaugmented counterpart.

Example 1.11 Consider the Lagrangian

$$\mathbf{L}(x,u) = f(x) + \langle u, Ax - b\rangle$$

with the associated primal problem

$$\begin{array}{ll}\underset{x\in\mathbb{R}^n}{\text{minimize}} & f(x) \\ \text{subject to} & Ax = b.\end{array}$$

We will often use the augmented Lagrangian

$$\mathbf{L}_\rho(x,u) = f(x) + \langle u, Ax - b\rangle + \frac{\rho}{2}\|Ax - b\|^2 \tag{1.11}$$

with $\rho > 0$. It is straightforward to show that (x,u) is a saddle point of $\mathbf{L}$ if and only if it is a saddle point of $\mathbf{L}_\rho$ for any $\rho > 0$.

Certain augmented Lagrangians arise naturally in monotone operator theory. In this book, we simply use these augmented Lagrangians without ascribing meaning to them.

1.3.10 Slater's Constraint Qualification

In the context of convex duality, regularity conditions that ensure strong duality are sometimes called *constraint qualifications*. The so-called Slater's constraint qualification is widely used, although not all constraint qualifications are due to Slater.

Consider the primal problem

$$\begin{array}{ll} \underset{x\in\mathbb{R}^n}{\text{minimize}} & f_0(x) \\ \text{subject to} & f_i(x) \leq 0 \quad \text{for } i = 1,\dots,m \\ & Ax = b, \end{array}$$

where $f_0, f_1, \dots, f_m$ are CCP functions, $A \in \mathbb{R}^{p\times n}$, and $b \in \mathbb{R}^p$, generated by the Lagrangian

$$\mathbf{L}(x,\lambda,\nu) = f_0(x) + \sum_{i=1}^{m} \lambda_i f_i(x) + \langle \nu, Ax - b\rangle - \delta_{\mathbb{R}^m_+}(\lambda),$$

where $\lambda \in \mathbb{R}^m$, $\nu \in \mathbb{R}^p$, and $\mathbb{R}^m_+ = \{(\lambda_1,\dots,\lambda_m) \mid \lambda_i \geq 0 \text{ for } i = 1,\dots,m\}$ is the nonnegative orthant.

Slater's constraint qualification states that if there exists an x such that

$$x \in \operatorname{ri} \bigcap_{i=0}^{m} \operatorname{dom} f_i, \quad f_i(x) < 0 \quad \text{for } i = 1,\dots,m, \quad Ax = b,$$

then strong duality holds (i.e., $d^\star = p^\star$), and if, furthermore, the optimal values are finite (i.e., $d^\star = p^\star > -\infty$), then a dual solution exists.

1.3.11 Proximal Operators

Let f be a CCP function on $\mathbb{R}^n$. Let $\alpha > 0$. We define the proximal operator with respect to αf as

$$\text{Prox}_{\alpha f}(y) = \underset{x\in\mathbb{R}^n}{\text{argmin}} \left\{ \alpha f(x) + \frac{1}{2}\|x - y\|^2 \right\}.$$

When $\alpha = 1$, we write Prox_f. If f is CCP, then $\text{Prox}_{\alpha f}$ is well defined, that is, the argmin uniquely exists.

Let us prove the well-definedness of $\text{Prox}_{\alpha f}$. Let $x_0 \in \operatorname{ri}\operatorname{dom} f$ and $g \in \partial f(x_0)$. (A CCP f has a nonempty domain, which is convex, the relative interior of a nonempty convex set is nonempty, and a CCP function is subdifferentiable on the relative interior of its domain.) Then, $f(x) \geq f(x_0) + \langle g, x - x_0\rangle$, and

$$\underbrace{\alpha f(x) + \frac{1}{2}\|x - y\|^2}_{=\tilde{f}(x)} \geq \underbrace{\alpha f(x_0) + \alpha\langle g, x - x_0\rangle + \frac{1}{2}\|x - y\|^2}_{=h(x)}.$$

Since $\lim_{\|x\|\to\infty} h(x) = \infty$ and $\tilde{f} \geq h$, we have $\lim_{\|x\|\to\infty} \tilde{f}(x) = \infty$. Therefore, $\tilde{f}(x^k) \to \inf_x \tilde{f}(x)$ implies $x^0, x^1, \dots$ is bounded. For any convergent subsequence $x^{k_j} \to \bar{x}$, lower semicontinuity of $\tilde{f}$ implies $\tilde{f}(\bar{x}) \leq \inf_x \tilde{f}(x)$. Thus $\tilde{f}(\bar{x}) = \inf_x \tilde{f}(x)$, that is, a solution exists. Finally, $\tilde{f}$ is strictly convex, so the minimizer is unique.

Example 1.12 The *soft-thresholding operator* $S(x;\kappa)$ for $x \in \mathbb{R}^n$ and $\kappa \geq 0$ is defined by

$$(S(x;\kappa))_i = \begin{cases} x_i - \kappa & \text{for } \kappa < x_i \\ 0 & \text{for } -\kappa \leq x_i \leq \kappa \\ x_i + \kappa & \text{for } x_i < -\kappa \end{cases}$$

for $i = 1,\dots,n$. This is the proximal operator with respect to ℓ_1 norm, that is, $S(x;\kappa) = \mathrm{Prox}_{\kappa\|\cdot\|_1}(x)$.

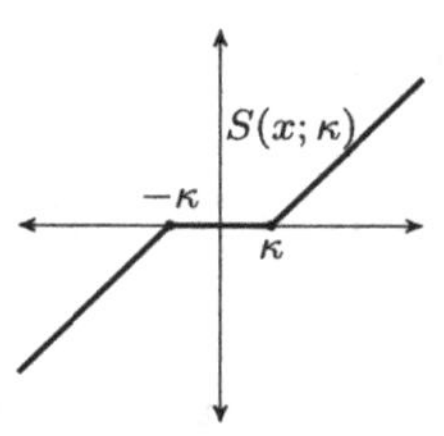

Example 1.13 Let C be a nonempty closed convex set. Define the projection onto C as

$$\Pi_C(y) = \operatorname*{argmin}_{x \in C} \|x - y\|.$$

It is straightforward to check that $\mathrm{Prox}_{\alpha\delta_C} = \mathrm{Prox}_{\delta_C} = \Pi_C$ for any $\alpha > 0$. In this sense, proximal operators generalize projections.

In general, evaluating a proximal operator is an optimization problem itself. For many interesting convex functions, however, the proximal operator has a closed-form solution and, if so, is suitable to use as a subroutine. We loosely say a function is *proximable* if its proximal operator is computationally efficient to evaluate. Several references such as [CP11b], [PB14b, Section 6], [BSS16, Section 3], and website [CCCP] catalog a list of proximable functions.

The field of monotone operator and splitting methods revolve around the idea of decomposing a given optimization problem (which is presumably not simple as a whole) into smaller, simpler pieces and operating on them separately. These simple pieces are functions for which we can easily evaluate the gradient or the proximal operators.

1.3.12 Asymptotic Notation

Write $f(x_1,\dots,x_r) = O(g(x_1,\dots,x_r))$ if

$$\limsup_{x_1,\dots,x_r\to\infty} \left|\frac{f(x_1,\dots,x_r)}{g(x_1,\dots,x_r)}\right| < \infty.$$

We call this the *O-notation* (and read it as "big O notation"). For example,

$$6n^2m + n^{3/2}m = O(n^2m).$$

Write $f(x_1,\dots,x_r) = o(g(x_1,\dots,x_r))$ if

$$\limsup_{x_1,\dots,x_r\to\infty} \left|\frac{f(x_1,\dots,x_r)}{g(x_1,\dots,x_r)}\right| = 0.$$

We call this the *o-notation* (and read it as "little o notation"). For example,

$$\frac{1}{k\log k} = o(1/k).$$

Write $f(x_1,\ldots,x_r) \sim g(x_1,\ldots,x_r)$ if

$$\limsup_{x_1,\ldots,x_r\to\infty} \frac{f(x_1,\ldots,x_r)}{g(x_1,\ldots,x_r)} = 1$$

and say f and g are *asymptotically equivalent*. For example,

$$2n^2m^3 + 3nm^3 \sim 2n^2m^3.$$

These are examples of *asymptotic notation*. Asymptotic notation is useful for identifying the limiting behavior of a function as the inputs tend toward a regime of interest. When discussing the convergence of methods, often the regime of interest is $k \to \infty$, where k is the iteration count, as we wish to know how the method eventually behaves. When discussing problem sizes, the regime of interest is $m,n \to \infty$, where m and n describe the problem size, because a method is judged by how well it can solve large (difficult) problems rather than small (easy) problems. That is not to say that non-asymptotic information is irrelevant. Sometimes we should ask at what iteration count or at what problem size the behavior described by the asymptotic notation becomes visible. Nevertheless, the asymptotic notation is a useful simplification.

BIBLIOGRAPHICAL NOTES

The 10-page lecture notes on subgradients by Boyd, Duchi, and Vandenberghe [BDV18] is a great resource to learn more about subgradients. Chapter 23 of Rockafellar's textbook [Roc70d] is another great resource providing a careful convex analytical treatment of subgradients.

The use of the conjugate function in convex analysis was pioneered by Fenchel in his unpublished lecture notes that were later distributed in mimeographed form [Fen53]. In particular, the result that $f = f^{**}$ when f is CCP is called the Fenchel–Moreau theorem and was first presented in [Fen49] and [Fen53, Theorem 37].

In careful treatments of calculus and analysis, the existence of partial derivatives, differentibility, and continuous differentiability are carefully distinguished. For convex functions, however, these notions coincide. By [Roc70d, Theorem 25.2], if f is a convex function and $x \in \mathbb{R}^n$ is a point such that $f(x) < \infty$, then f is differentiable at x if and only if

$$\frac{\partial f}{\partial x_i}(x) = \lim_{h\to 0} \frac{f(x+he_i)}{h}$$

exists and is finite for all $i = 1,\ldots,n$ (where e_i is the ith unit vector and the limit is two-sided). By [Roc70d, Corollary 25.5.1], if $f\colon \mathbb{R} \to \mathbb{R}$ is convex and differentiable, then f is necessarily *continuously* differentiable, that is, when f is convex, existence of $\nabla f(x)$ for all $x \in \mathbb{R}^n$ implies $\nabla f(x)$ is continuous.

Showing that the equivalent definitions for strong convexity and smoothness are indeed equivalent is a relatively straightforward exercise in vector calculus, when the function is twice continuously differentiable. Proofs in the general case can be found in references such as [Nes04]. The equivalence of the smoothness definitions is called the Baillon–Haddad theorem [BH77, Corollaire 10] [BC10].

There are multiple related but distinct viewpoints of convex duality. The view that primal-dual problem pairs are two halves of a larger saddle-point problem was developed in the mid 1960s by Dantzig, Eisenberg, and Cottle [DEC65], Stoer [Sto63, Sto64], and Mangasarian and Ponstein [MP65]. The presentation of this book closely follows Rockafellar's 1974 book [Roc74]. This 74-page book is still one of the best references on convex duality. Regularity conditions that ensure strong duality in optimization is an area with a large body of research. Slater's constraint qualification, the most widely used such condition, dates back to 1950 [Sla50]. Rockafellar's book [Roc74] provides a thorough discussion on this subject.

To expand on the discussion of §1.2, one can, in fact, establish an improved rate $\|\nabla f(x^k)\|^2 \le \mathcal{O}(1/k^2)$ for the gradient method using properties of convex functions [TB19, Theorem 3]; but this result cannot be established using only properties of monotone operators.

EXERCISES

1.1 Assume $\mathbb{T}_1 : \mathbb{R}^n \to \mathbb{R}^m$ is L_1-Lipschitz and $\mathbb{T}_2 : \mathbb{R}^n \to \mathbb{R}^m$ is L_2-Lipschitz. Show that $\alpha_1\mathbb{T}_1 + \alpha_2\mathbb{T}_2$ is $(|\alpha_1|L_1 + |\alpha_2|L_2)$-Lipschitz.

1.2 Let f be a convex function on $\mathbb{R}^n$. Show that $\partial f(x)$ is a closed convex set for all $x \in \mathbb{R}^n$.
Hint. Write $\partial f(x)$ as an intersection of closed half-spaces.
Remark. Remember that $\partial f(x)$ can be empty, but the empty set is a closed convex set.

1.3 Show that if f is a CCP function on $\mathbb{R}^m$, $A \in \mathbb{R}^{m\times n}$, and $g(x) = f(Ax)$, then

$$\partial g(x) \supseteq A^\intercal \partial f(Ax)$$

for all $x \in \mathbb{R}^n$. Also show that if f and g are CCP functions on $\mathbb{R}^n$, then

$$\partial(f + g)(x) \supseteq \partial f(x) + \partial g(x)$$

for all $x \in \mathbb{R}^n$.

1.4 Consider the function f: $\mathbb{R}^2 \to \mathbb{R} \cup \{\pm\infty\}$ defined as

$$f(x,y) = \begin{cases} x^2/y & \text{for } y > 0, \\ 0 & \text{for } x = y = 0, \\ \infty & \text{otherwise.} \end{cases}$$

Clearly f is proper, and it is possible to show that f is convex. Show that
(a) f is closed, and
(b) $f|_{\mathrm{dom} f} : \mathrm{dom} f \to \mathbb{R}$ is not continuous at $(0,0)$, that is, show that f restricted to where it is finite is not continuous at $(0,0)$.

Remark. This example demonstrates that a CCP function need not be continuous on its domain. In convex optimization, lower semi-continuity, not continuity, is the regularity condition of interest. However, a proper convex function is continuous on the relative interior of its domain.

1.5 *Existence of a minimizer with Slater.* Let f be a CCP function on $\mathbb{R}^n$ and $A \in \mathbb{R}^{m\times n}$. Assume $\mathcal{R}(A^\intercal) \cap \operatorname{ri}\operatorname{dom} f^* \neq \emptyset$. Consider the optimization problem

$$\begin{array}{ll}\underset{\mu\in\mathbb{R}^m,\,\nu\in\mathbb{R}^n}{\text{minimize}} & f^*(\nu) - \mu^\intercal y + \frac{1}{2}\|\mu\|^2 \\ \text{subject to} & A^\intercal\mu - \nu = 0\end{array}$$

generated by the Lagrangian

$$\mathbf{L}(\mu,\nu,x) = f^*(\nu) - \mu^\intercal y + \frac{1}{2}\|\mu\|^2 + \langle x, A^\intercal\mu - \nu\rangle.$$

Using Slater's constraint qualification, show

$$\underset{x\in\mathbb{R}^n}{\operatorname{argmin}} \left\{f(x) + (1/2)\|Ax - y\|^2\right\} \neq \emptyset.$$

1.6 *Saddle points of augmented Lagrangians.* Let f be a CCP function on $\mathbb{R}^n$, $A \in \mathbb{R}^{m\times n}$, and $b \in \mathbb{R}^m$. Show that the Lagrangian

$$\mathbf{L}(x,u) = f(x) + \langle u, Ax - b\rangle$$

and the augmented Lagrangian

$$\mathbf{L}_\alpha(x,u) = f(x) + \langle u, Ax - b\rangle + \frac{\alpha}{2}\|Ax - b\|^2,$$

where $\alpha > 0$, share the same set of saddle points.

1.7 Assume that a CCP function $f\colon \mathbb{R}^n \to \mathbb{R}\cup\{\pm\infty\}$ is proximable. Define $g\colon \mathbb{R}^n\times\mathbb{R}^n \to \mathbb{R}\cup\{\pm\infty\}$ as

$$g(x_1,x_2) = f(x_1 + x_2).$$

Show that

$$\operatorname{Prox}_g(x_1,x_2) = \frac{1}{2}\begin{bmatrix} x_1 - x_2 + \operatorname{Prox}_{2f}(x_1 + x_2) \\ x_2 - x_1 + \operatorname{Prox}_{2f}(x_1 + x_2)\end{bmatrix}.$$

Likewise, show that if

$$h(x_1,x_2) = f(x_1 - x_2),$$

then

$$\operatorname{Prox}_h(x_1,x_2) = \frac{1}{2}\begin{bmatrix} x_1 + x_2 + \operatorname{Prox}_{2f}(x_1 - x_2) \\ x_1 + x_2 + \operatorname{Prox}_{2f}(x_1 - x_2)\end{bmatrix}.$$

Hint. Note that $g = f \circ \begin{bmatrix} I & I\end{bmatrix}$ and show that $(y_1,y_2) = \operatorname{Prox}_g(x_1,x_2)$ if and only if there exists a $v \in \partial f(y_1 + y_2)$ such that

$$\begin{aligned} 0 &= v + (y_1 - x_1) \\ 0 &= v + (y_2 - x_2). \end{aligned}$$

1.8 Assume a CCP function $f\colon \mathbb{R}^n \to \mathbb{R}\cup\{\pm\infty\}$ is proximable. Assume $a = (a_1,\ldots,a_m) \in \mathbb{R}^m$ satisfies $a \neq 0$. Define $g\colon \mathbb{R}^{mn} \to \mathbb{R}\cup\{\pm\infty\}$ as

$$g(x_1,\ldots,x_m) = f(a_1x_1 + \cdots + a_mx_m).$$

Show that

$$v = \frac{1}{\|a\|^2}\left(a_1x_1 + \cdots + a_mx_m - \operatorname{Prox}_{\|a\|^2 f}(a_1x_1 + \cdots + a_mx_m)\right)$$

$$\mathrm{Prox}_g(x_1,\ldots,x_m) = \begin{bmatrix} x_1 - a_1\nu \\ \vdots \\ x_m - a_m\nu \end{bmatrix}.$$

1.9 *Basic normal cone example.* Let $\mathbb{R}^n_+ = \{(x_1,\ldots,x_n)\,|\,x_i \geq 0 \text{ for } i = 1,\ldots,n\}$ be the nonnegative orthant.

(i) Characterize $\mathbf{N}_{\mathbb{R}^n_+}$, that is, describe the set $\mathbf{N}_{\mathbb{R}^n_+}(x)$ for all $x \in \mathbb{R}^n$.

(ii) Let $f\colon \mathbb{R}^n \to \mathbb{R}^n$ be CCP and differentiable. Directly show, without using the subgradient identity $\partial(f+g) = \partial f + \partial g$, that x solves

$$\begin{array}{ll} \underset{x\in\mathbb{R}^n}{\text{minimize}} & f(x) \\ \text{subject to} & x \geq 0 \end{array}$$

if and only if $-\nabla f(x) \in \mathbf{N}_{\mathbb{R}^n_+}(x)$.

1.10 *Linear programming duality.* Consider the convex–concave saddle function

$$\mathbf{L}(x,\nu,\mu) = \langle c,x\rangle + \langle Ax+b,\nu\rangle - \langle x,\mu\rangle - \delta_{\mathbb{R}^m_+}(\nu) - \delta_{\mathbb{R}^n_+}(\mu),$$

convex in $x \in \mathbb{R}^n$ and concave in $(\nu,\mu) \in \mathbb{R}^m \times \mathbb{R}^n$. Here, $\mathbb{R}^m_+$ and $\mathbb{R}^n_+$ denote the m and n-dimensional nonnegative orthants. Remember that δ_C denotes the indicator function with respect to the set C.

Show that the saddle function $\mathbf{L}$ generates the primal problem

$$\begin{array}{ll} \underset{x\in\mathbb{R}^n}{\text{minimize}} & c^\intercal x \\ \text{subject to} & Ax + b \leq 0 \\ & x \geq 0. \end{array}$$

Here, the inequalities denote element-wise nonnegativity. Show that $\mathbf{L}$ generates a dual problem that is equivalent to

$$\begin{array}{ll} \underset{\nu\in\mathbb{R}^m}{\text{maximize}} & b^\intercal \nu \\ \text{subject to} & c + A^\intercal \nu \geq 0 \\ & \nu \geq 0. \end{array}$$

PART ONE Monotone Operator Methods

2 Monotone Operators and Base Splitting Schemes

In this chapter, we present the basic notion of monotone operators and the base splitting schemes. Throughout this book, we use this machinery to derive and analyze a wide variety of classical and modern algorithms in a unified and streamlined manner. The approach is to first pose the problem at hand as a monotone inclusion problem, then use one of the base splitting schemes to encode the solution as a fixed point of a related operator, and finally find the solution with a fixed-point iteration.

2.1 SET-VALUED OPERATORS

We say $\mathbb{T}$ is a *(set-valued) operator*, *point-to-set mapping*, *set-valued mapping*, *multi-valued function*, or *correspondence* on $\mathbb{R}^n$ if $\mathbb{T}$ maps a point in $\mathbb{R}^n$ to a (possibly empty) subset of $\mathbb{R}^n$. We denote this as $\mathbb{T}\colon \mathbb{R}^n \rightrightarrows \mathbb{R}^n$. So, $\mathbb{T}(x) \subseteq \mathbb{R}^n$ for all $x \in \mathbb{R}^n$. For notational simplicity, we write $\mathbb{T}x = \mathbb{T}(x)$.

If $\mathbb{T}x$ is a singleton or empty for all x, then $\mathbb{T}$ is a *function* or is *single-valued* with domain $\{x \mid \mathbb{T}(x) \neq \emptyset\}$. In this case, we mix function and operator notation and write $\mathbb{T}x = y$ (function notation) although $\mathbb{T}x = \{y\}$ (operator notation) would be strictly correct.

We define the graph of an operator as

$$\operatorname{Gra}\mathbb{T} = \{(x,u) \mid u \in \mathbb{T}x\} \subseteq \mathbb{R}^n \times \mathbb{R}^n.$$

An operator and its graph are mathematically equivalent. In other words, we can view $\mathbb{T}\colon \mathbb{R}^n \rightrightarrows \mathbb{R}^n$ as a point-to-set mapping and as a subset of $\mathbb{R}^n \times \mathbb{R}^n$. In this book, we will often not distinguish the operator itself and its graph; we will often write $\mathbb{T}$ when we really mean $\operatorname{Gra}\mathbb{T}$.

We extend many notions for functions to operators. For example, the domain and range of an operator $\mathbb{T}$ are defined as

$$\operatorname{dom}\mathbb{T} = \{x \mid \mathbb{T}x \neq \emptyset\}, \qquad \operatorname{range}\mathbb{T} = \{y \mid y \in \mathbb{T}x,\ x \in \mathbb{R}^n\}.$$

If $C \subseteq \mathbb{R}^n$, we write $\mathbb{T}(C) = \cup_{c\in C}\mathbb{T}(c)$ for the image of C under $\mathbb{T}$. If $\mathbb{T}$ and $\mathbb{S}$ are operators, we define the composition as

$$\mathbb{T} \circ \mathbb{S}x = \mathbb{T}\mathbb{S}x = \mathbb{T}(\mathbb{S}(x))$$

and the sum as

$$(\mathbb{T} + \mathbb{S})x = \mathbb{T}(x) + \mathbb{S}(x),$$

where $\mathbb{T}(x) + \mathbb{S}(x)$ is the Minkowski sum. Alternate equivalent definitions that use the graph are

$$\mathbb{T}\mathbb{S} = \{(x,z) \mid \exists y\ (x,y) \in \mathbb{S},\ (y,z) \in \mathbb{T}\},$$
$$\mathbb{T} + \mathbb{S} = \{(x, y+z) \mid (x,y) \in \mathbb{T},\ (x,z) \in \mathbb{S}\}.$$

We write $\mathbb{I}$ and $\mathbb{O}$ for the identity and zero operators

$$\mathbb{I} = \{(x,x) \mid x \in \mathbb{R}^n\}, \qquad \mathbb{O} = \{(x,0) \mid x \in \mathbb{R}^n\}.$$

So, for any operator $\mathbb{T}$, we have $\mathbb{T} + \mathbb{O} = \mathbb{T}$, $\mathbb{T}\mathbb{I} = \mathbb{T}$, and $\mathbb{I}\mathbb{T} = \mathbb{T}$.

For an $L > 0$, we say an operator $\mathbb{T}$ is L-Lipschitz if

$$\|u - v\| \le L\|x - y\| \qquad \forall (x,u), (y,v) \in \mathbb{T},$$

or, more concisely, if

$$\|\mathbb{T}x - \mathbb{T}y\| \le L\|x - y\| \qquad \forall x, y \in \operatorname{dom} \mathbb{T}.$$

If $\mathbb{T}$ is L-Lipschitz, it is single-valued; if $\mathbb{T}x$ is not a singleton, then we have a contradiction by setting $y = x$. (This generalizes the previous definition of §1, as it allows $\operatorname{dom} \mathbb{T} \neq \mathbb{R}^n$.)

The *inverse operator* of $\mathbb{T}$ is defined as

$$\mathbb{T}^{-1} = \{(y,x) \mid (x,y) \in \mathbb{T}\}.$$

Since $\mathbb{T}^{-1}$ can be multi-valued, it is always well defined. It is easy to see that $(\mathbb{T}^{-1})^{-1} = \mathbb{T}$ and $\operatorname{dom} \mathbb{T}^{-1} = \operatorname{range} \mathbb{T}$. As a note of caution, the inverse operator is not an inverse in the usual sense, as we can have $\mathbb{T}^{-1}\mathbb{T} \neq \mathbb{I}$. The zero operator is such an example. However, we do have $\mathbb{T}^{-1}\mathbb{T}x = x$ when $\mathbb{T}^{-1}$ is single-valued and $x \in \operatorname{dom} \mathbb{T}$. See Exercise 2.1.

Example 2.1 The inverse of an operator $\mathbb{T}$ always exists since we do not require it to be single-valued.

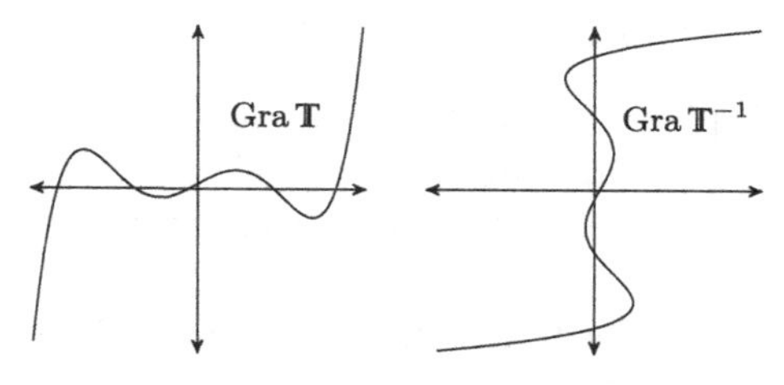

When $0 \in \mathbb{T}(x)$, we say that x is a *zero* of $\mathbb{T}$. We write the *zero set* of an operator $\mathbb{T}$ as

$$\operatorname{Zer} \mathbb{T} = \{x \mid 0 \in \mathbb{T}x\} = \mathbb{T}^{-1}(0).$$

We will see that many interesting problems can be posed as finding zeros of an operator.

Subdifferential

Let f be a convex function on $\mathbb{R}^n$. Then ∂f is a set-valued operator, and

$$\operatorname{argmin} f = \operatorname{Zer} \partial f,$$

that is, $0 \in \partial f(x)$ if and only if x minimizes f. When f is known or assumed to be differentiable, we write ∇f instead of ∂f. As an aside, $\operatorname{dom} \partial f \subseteq \operatorname{dom} f$, and it is possible to have $\operatorname{dom} \partial f \neq \operatorname{dom} f$. Example 1.7 is one such example.

When f is CCP, we have the elegant formula

$$(\partial f)^{-1} = \partial f^*, \tag{2.1}$$

which is known as *Fenchel's identity*. This follows from

$$\begin{aligned} u \in \partial f(x) \quad &\Leftrightarrow \quad 0 \in \partial f(x) - u \\ &\Leftrightarrow \quad x \in \underset{z}{\operatorname{argmin}} \{ f(z) - u^\intercal z \} \\ &\Leftrightarrow \quad -f(x) + u^\intercal x = f^*(u) \\ &\Leftrightarrow \quad f(x) + f^*(u) = u^\intercal x \\ &\Leftrightarrow \quad f^{**}(x) + f^*(u) = u^\intercal x \\ &\Leftrightarrow \quad x \in \partial f^*(u), \end{aligned}$$

where the second-to-last step uses the fact that $f^{**} = f$ when f is CCP, as discussed in §1.3.8, and the last step takes the whole argument backward.

Consider $g(y) = f^*(A^\intercal y)$, where f is CCP. If $\mathcal{R}(A^\intercal) \cap \operatorname{ri} \operatorname{dom} f^* \neq \emptyset$, we have

$$\begin{aligned} u \in \partial g(y) \quad &\Leftrightarrow \quad u \in A \partial f^*(A^\intercal y) \\ &\Leftrightarrow \quad u = Ax,\ x \in \partial f^*(A^\intercal y) \\ &\Leftrightarrow \quad u = Ax,\ \partial f(x) \ni A^\intercal y \\ &\Leftrightarrow \quad u = Ax,\ 0 \in \partial f(x) - A^\intercal y \\ &\Leftrightarrow \quad u = Ax,\ x \in \underset{z}{\operatorname{argmin}} \{ f(z) - \langle y, Az \rangle \}. \end{aligned} \tag{2.2}$$

This means we can find an element of ∂g by solving a minimization problem.

2.2 MONOTONE OPERATORS

An operator $\mathbb{T}$ on $\mathbb{R}^n$ is said to be *monotone* if

$$\langle u - v, x - y \rangle \geq 0 \qquad \forall (x, u),\ (y, v) \in \mathbb{T}.$$

Equivalently and more concisely, we can express monotonicity as

$$\langle \mathbb{T}x - \mathbb{T}y, x - y \rangle \geq 0 \qquad \forall x, y \in \mathbb{R}^n.$$

To clarify, $\langle \mathbb{T}x - \mathbb{T}y, x - y \rangle$ is a subset of $\mathbb{R}$ and the inequality means the subset is contained in $[0, \infty)$. When $x \notin \operatorname{dom} \mathbb{T}$ or $y \notin \operatorname{dom} \mathbb{T}$, then $\langle \mathbb{T}x - \mathbb{T}y, x - y \rangle = \emptyset$ and the inequality is vacuous.

An operator $\mathbb{T}$ is *maximal monotone* if there is no other monotone operator $\mathbb{S}$ such that $\operatorname{Gra}\mathbb{T} \subset \operatorname{Gra}\mathbb{S}$ properly. In other words, if the monotone operator $\mathbb{T}$ is not maximal, then there exists $(x,u) \notin \mathbb{T}$ such that $\mathbb{T} \cup \{(x,u)\}$ is still monotone. Maximality is a technical but fundamental detail.

Example 2.2 The Heaviside step function $u\colon \mathbb{R} \to \mathbb{R}$ defined as

$$u(x) = \begin{cases} 0 & \text{for } x \le 0 \\ 1 & \text{for } x > 0 \end{cases}$$

is monotone but not maximal. The operator $U\colon \mathbb{R} \rightrightarrows \mathbb{R}$ defined as

$$U(x) = \begin{cases} \{0\} & \text{for } x < 0 \\ [0,1] & \text{for } x = 0 \\ \{1\} & \text{for } x > 0 \end{cases}$$

is maximal monotone.

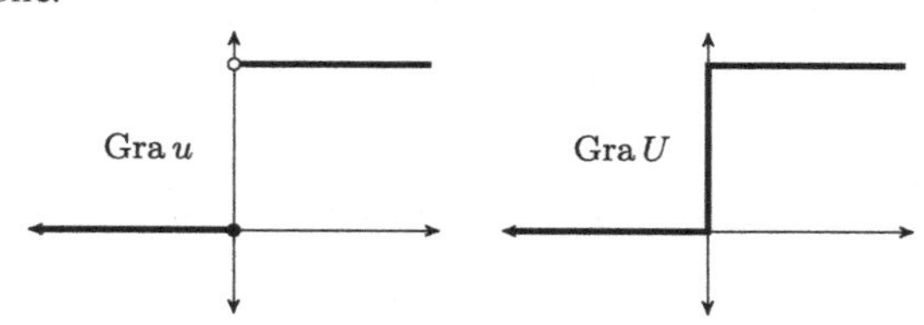

Subdifferential

If f is convex and proper, then ∂f is a monotone operator. If f is CCP, then ∂f is maximal monotone. To prove monotonicity, add the inequalities

$$f(y) \ge f(x) + \langle \partial f(x), y - x\rangle, \qquad f(x) \ge f(y) + \langle \partial f(y), x - y\rangle,$$

which hold by the definition of subdifferentials, to get

$$\langle \partial f(x) - \partial f(y), x - y\rangle \ge 0.$$

We prove maximality later in §10. See Exercise 2.2 for an example where ∂f is not maximal.

Not all maximal monotone operators are subdifferential operators. Subdifferential operators of CCP functions form a subclass of monotone operators that enjoy certain nice properties that general maximal monotone operators do not.

2.2.1 Stronger Monotonicity Properties

An operator $\mathbb{A} : \mathbb{R}^n \rightrightarrows \mathbb{R}^n$ is *μ-strongly monotone* or *μ-coercive* if $\mu > 0$ and

$$\langle u - v, x - y\rangle \ge \mu\|x - y\|^2 \qquad \forall\,(x,u),(y,v) \in \mathbb{A}.$$

We say $\mathbb{A}$ is strongly monotone if it is μ-strongly monotone for some unspecified $\mu \in (0,\infty)$. An operator $\mathbb{A}$ is *β-cocoercive* or *β-inverse strongly monotone* if $\beta > 0$ and

$$\langle u - v, x - y\rangle \ge \beta\|u - v\|^2 \qquad \forall\,(x,u),(y,v) \in \mathbb{A}.$$

We say $\mathbb{A}$ is cocoercive if it is β-cocoercive for some unspecified $\beta \in (0,\infty)$. Cocoercivity is the dual property of strong monotonicity; $\mathbb{A}$ is β-cocoercive if and only if $\mathbb{A}^{-1}$ is β-strongly monotone. Clearly, strongly monotone and cocoercive operators are monotone.

We can more concisely express μ-strong monotonicity as

$$\langle \mathbb{A}x - \mathbb{A}y, x - y\rangle \geq \mu\|x - y\|^2 \qquad \forall x,y \in \mathbb{R}^n,$$

and, when $\mathbb{A}$ is a priori known or assumed to be single-valued, express β-cocoercivity as

$$\langle \mathbb{A}x - \mathbb{A}y, x - y\rangle \geq \beta\|\mathbb{A}x - \mathbb{A}y\|^2 \qquad \forall x,y \in \mathbb{R}^n.$$

When $\mathbb{A}$ is β-cocoercive, the Cauchy–Schwartz inequality tells us

$$(1/\beta)\|x - y\| \geq \|\mathbb{A}x - \mathbb{A}y\| \qquad \forall x,y \in \mathbb{R}^n.$$

that is, $\mathbb{A}$ is $(1/\beta)$-Lipschitz. Therefore, cocoercive operators are single-valued. The converse is not true. The single-valued operator $\mathbb{A}\colon \mathbb{R}^2 \to \mathbb{R}^2$ defined as

$$\mathbb{A}(x_1,x_2) = \begin{bmatrix} 0 & 1 \\ -1 & 0 \end{bmatrix} \begin{bmatrix} x_1 \\ x_2 \end{bmatrix} = \begin{bmatrix} x_2 \\ -x_1 \end{bmatrix}$$

is an example of an operator that is maximal monotone and Lipschitz, but not cocoercive since $\langle \mathbb{A}x - \mathbb{A}y, x - y\rangle = 0, \forall x,y \in \mathbb{R}^n$.

We say $\mathbb{A}$ is maximal μ-strongly monotone if there is no other μ-strongly monotone operator $\mathbb{B}$ such that $\operatorname{Gra}\mathbb{A} \subset \operatorname{Gra}\mathbb{B}$ properly. We say $\mathbb{A}$ is maximal β-cocoercive if there is no other β-cocoercive operator $\mathbb{B}$ such that $\operatorname{Gra}\mathbb{A} \subset \operatorname{Gra}\mathbb{B}$ properly. Maximal cocoercivity is the dual property of maximal strong monotonicity; $\mathbb{A}$ is maximal β-cocoercive if and only if $\mathbb{A}^{-1}$ is maximal β-strongly monotone. A β-cocoercive operator $\mathbb{A}$ is maximal if and only if $\operatorname{dom}\mathbb{A} = \mathbb{R}^n$. (We show this fact in §10.3 as Theorem 15.) Since a β-cocoercive operator is single-valued, the statement "$\mathbb{A}\colon \mathbb{R}^n \to \mathbb{R}^n$ is β-cocoercive" is equivalent to "$\mathbb{A}\colon \mathbb{R}^n \rightrightarrows \mathbb{R}^n$ is maximal β-cocoercive" since the notation $\mathbb{A}\colon \mathbb{R}^n \to \mathbb{R}^n$ implicitly assumes $\operatorname{dom}\mathbb{A} = \mathbb{R}^n$. For further discussion, see §10 and Exercises 10.11 and 10.12.

Assume f is CCP. Then f is μ-strongly convex if and only if ∂f is μ-strongly monotone, and f is L-smooth if and only if ∂f is $(1/L)$-cocoercive. Since ∂f is μ-strongly monotone if and only if $(\partial f)^{-1} = \partial f^*$ is μ-cocoercive, f is μ-strongly convex if and only if f^* is $(1/\mu)$-smooth.

The notion of Lipschitz continuity and cocoercivity coincide for subdifferential operators of convex functions: ∂f is L-Lipschitz if and only if ∂f is $(1/L)$-cocoercive. This result is known as the *Baillon–Haddad theorem*.

Example 2.3 An operator on $\mathbb{R}$ is monotone if its graph is a nondecreasing curve in $\mathbb{R}^2$. If it has vertical portions, the operator is multi-valued. If it is continuous with no end points, then it is maximal. If its slope is at least μ everywhere, then it is μ-strongly monotone. If its slope is never more than L, then it is L-Lipschitz. The notion of Lipschitz continuity and cocoercivity coincide for operators on $\mathbb{R}$.

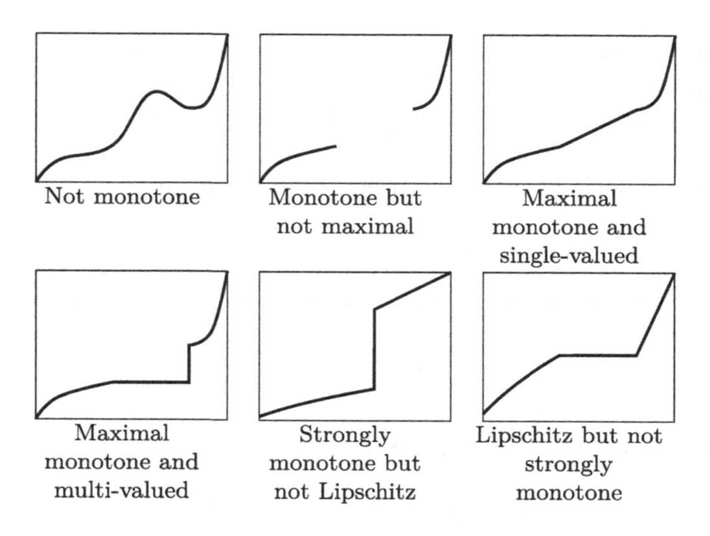

2.2.2 Operations Preserving (Maximal) Monotonicity

If $\mathbb{T}$ is (maximal) monotone, then $\mathbb{S}(x) = y + \alpha\mathbb{T}(x + z)$ for any $\alpha > 0$ and $y, z \in \mathbb{R}^n$ is (maximal) monotone. If $\mathbb{T}$ is (maximal) monotone, then $\mathbb{T}^{-1}$ is (maximal) monotone. If $\mathbb{T}$ and $\mathbb{S}$ are monotone, then $\mathbb{T} + \mathbb{S}$ is monotone. If $\mathbb{T}$ and $\mathbb{S}$ are maximal monotone and if $\operatorname{dom}\mathbb{T} \cap \operatorname{int}\operatorname{dom}\mathbb{S} \neq \emptyset$, then $\mathbb{T} + \mathbb{S}$ is maximal monotone. If $\mathbb{T}\colon \mathbb{R}^n \rightrightarrows \mathbb{R}^n$ is monotone and $M \in \mathbb{R}^{n\times m}$, then $M^\intercal\mathbb{T}M$ is a monotone operator on $\mathbb{R}^m$. If $\mathbb{T}$ is maximal and $\mathcal{R}(M) \cap \operatorname{int}\operatorname{dom}\mathbb{T} \neq \emptyset$, then $M^\intercal\mathbb{T}M$ is maximal. See §10 for proofs of maximality.

If $\mathbb{R}\colon \mathbb{R}^n \rightrightarrows \mathbb{R}^n$ and $\mathbb{S}\colon \mathbb{R}^m \rightrightarrows \mathbb{R}^m$, then the operator $\mathbb{T}\colon \mathbb{R}^{n+m} \rightrightarrows \mathbb{R}^{n+m}$ defined by

$$\mathbb{T}(x, y) = \{(u, v) \mid u \in \mathbb{R}x,\ v \in \mathbb{S}y\}$$

is (maximal) monotone if $\mathbb{R}$ and $\mathbb{S}$ are. We call $\mathbb{T}$ the *concatenation* of $\mathbb{R}$ and $\mathbb{S}$ and use the notation

$$\mathbb{T} = \begin{bmatrix}\mathbb{R}\\ \mathbb{S}\end{bmatrix}, \qquad \mathbb{T}(x, y) = \begin{bmatrix}\mathbb{R}x\\ \mathbb{S}y\end{bmatrix}.$$

If $\mathbb{T}$ is μ-strongly monotone and $\mathbb{S}$ is monotone, then $\mathbb{T} + \mathbb{S}$ is μ-strongly monotone and $\alpha\mathbb{T}$ is $(\alpha\mu)$-strongly monotone for $\alpha > 0$. If $\mathbb{T}\colon \mathbb{R}^n \rightrightarrows \mathbb{R}^n$ is μ-strongly monotone and $M \in \mathbb{R}^{n\times m}$ has rank m (so $n \geq m$), then $M^\intercal\mathbb{T}M$ is $(\mu\sigma_{\min}^2(M))$-strongly monotone. If $\mathbb{T}\colon \mathbb{R}^n \to \mathbb{R}^n$ is L-Lipschitz and $M \in \mathbb{R}^{n\times m}$, then $M^\intercal\mathbb{T}M$ is $(L\sigma_{\max}^2(M))$-Lipschitz.

2.2.3 Examples

Affine Operators

An affine operator $\mathbb{T}(x) = Ax + b$ is maximal monotone if and only if $A + A^\intercal \succeq 0$. It is a subdifferential operator of a CCP function if and only if $A = A^\intercal$ and $A \succeq 0$. It is $\lambda_{\min}(A + A^\intercal)/2$-strongly monotone and $\sigma_{\max}(A)$-Lipschitz.

Continuous Operators

We say an operator $\mathbb{T}\colon \mathbb{R}^n \rightrightarrows \mathbb{R}^n$ is continuous if $\operatorname{dom} \mathbb{T} = \mathbb{R}^n$, $\mathbb{T}$ is single-valued, and $\mathbb{T}$ is continuous as a function. A continuous monotone operator $\mathbb{T}\colon \mathbb{R}^n \to \mathbb{R}^n$ is maximal. See Exercise 2.4 for a proof. Therefore maximality comes into question only with discontinuous or set-valued operators.

Differentiable Operators

We say anoperator is differentiable if it is single-valued, continuous, and differentiable. A differentiable operator $\mathbb{T}\colon \mathbb{R}^n \to \mathbb{R}^n$ is monotone if and only if $D\mathbb{T}(x) + D\mathbb{T}(x)^\intercal \succeq 0$ for all $x \in \mathbb{R}^n$, where $D\mathbb{T}(x)$ is the $n \times n$ Jacobian matrix evaluated at x. It is μ-strongly monotone if and only if $D\mathbb{T}(x) + D\mathbb{T}(x)^\intercal \succeq 2\mu I$ for all x, and it is L-Lipschitz if and only if $\sigma_{\max}(D\mathbb{T}(x)) \leq L$ for all x. See Exercises 2.7 and 2.8 for proofs.

If a monotone operator $\mathbb{T}$ is differentiable with continuous $D\mathbb{T}$, then $\mathbb{T}$ is a subdifferential operator of a CCP function if and only if $D\mathbb{T}(x)$ is symmetric for all $x \in \mathbb{R}^n$. When $n = 3$, this condition is equivalent to the so-called curl-less (or irrotational) condition discussed in the context of electromagnetic potentials.

Saddle Subdifferential

Let $\mathbf{L}\colon \mathbb{R}^n \times \mathbb{R}^m \to \mathbb{R} \cup \{\pm\infty\}$ be a convex-concave saddle function, that is, $\mathbf{L}(x,u)$ is convex in x for fixed u and concave in u for fixed x. The *saddle subdifferential operator* $\partial\mathbf{L}\colon \mathbb{R}^n \times \mathbb{R}^m \rightrightarrows \mathbb{R}^n \times \mathbb{R}^m$ is defined as

$$\partial\mathbf{L}(x,u) = \begin{bmatrix} \partial_x \mathbf{L}(x,u) \\ \partial_u(-\mathbf{L}(x,u)) \end{bmatrix}. \tag{2.3}$$

To clarify, ∂_x and ∂_u respectively denote the subgradients with respect to x and u. To clarify, $\partial\mathbf{L}(x,u)$ is nonempty if both $\partial_x\mathbf{L}(x,u)$ and $\partial_u(-\mathbf{L}(x,u))$ are nonempty. $\operatorname{Zer}\partial\mathbf{L}$ is the set of saddle points of $\mathbf{L}$, that is, $0 \in \partial\mathbf{L}(x^\star,u^\star)$ if and only if $(x^\star,u^\star)$ is a saddle point of $\mathbf{L}$.

For most well-behaved convex-concave saddle functions, their saddle subdifferentials are maximal monotone. Specifically, "closed proper" convex-concave saddle functions have maximal monotone saddle subdifferentials. (See the bibliographical notes section.) In this book, we avoid this notion, as it is usually straightforward to verify the maximality of saddle subdifferentials on a case-by-case basis.

As a technical note, we adopt the convention $+\infty - \infty = -\infty + \infty = -\infty$ in saddle functions. We do encounter $+\infty - \infty$ in certain cases such as the Lagrangians for DRS (2.17), PDHG (1.8), and Condat–Vũ (3.12). The specific value that we ascribe to $+\infty - \infty$ does not matter, but we define it for concreteness.

KKT Operator

Consider the problem

$$\begin{array}{ll} \underset{x}{\text{minimize}} & f_0(x) \\ \text{subject to} & f_i(x) \leq 0, \quad i = 1,\ldots,m \\ & h_i(x) = 0, \quad i = 1,\ldots,p, \end{array}$$

where $f_0, \ldots, f_m$ are CCP and $h_1, \ldots, h_p$ are affine. The associated Lagrangian

$$\mathbf{L}(x, \lambda, \nu) = f_0(x) + \sum_{i=1}^{m} \lambda_i f_i(x) + \sum_{i=1}^{p} \nu_i h_i(x) - \delta_{\mathbb{R}_+^m}(\lambda),$$

where $\mathbb{R}_+^m$ denotes the nonnegative orthant, is a convex-concave saddle function, and we define the Karush–Kuhn–Tucker (KKT) operator as

$$\mathbb{T}(x, \lambda, \nu) = \begin{bmatrix} \partial_x \mathbf{L}(x, \lambda, \nu) \\ -\mathbb{F}(x) + \mathbb{N}_{\mathbb{R}_+^m}(\lambda) \\ -\mathbb{H}(x) \end{bmatrix} = \begin{bmatrix} \partial_x \mathbf{L}(x, \lambda, \nu) \\ \partial_\lambda(-\mathbf{L}(x, \lambda, \nu)) \\ \partial_\nu(-\mathbf{L}(x, \lambda, \nu)) \end{bmatrix},$$

where

$$\mathbb{F}(x) = \begin{bmatrix} f_1(x) \\ \vdots \\ f_m(x) \end{bmatrix}, \qquad \mathbb{H}(x) = \begin{bmatrix} h_1(x) \\ \vdots \\ h_p(x) \end{bmatrix}.$$

$\mathbb{T}$ is monotone, since it is a special case of the saddle subdifferential. Arguments based on total duality tell us that $0 \in \mathbb{T}(x^\star, \lambda^\star, \nu^\star)$ if and only if $x^\star$ solves the primal problem, $(\lambda^\star, \nu^\star)$ solves the dual problem, and strong duality holds.

2.2.4 Monotone Inclusion Problem

Monotone inclusion problems are problems of the form

$$\underset{x \in \mathbb{R}^n}{\text{find}} \quad 0 \in \mathbb{A}x,$$

where $\mathbb{A}$ is monotone. Many interesting problems can be formulated as monotone inclusion problems. For example, minimizing a convex function f is equivalent to finding a zero of ∂f.

2.3 NONEXPANSIVE AND AVERAGED OPERATORS

Nonexpansive and Contractive Operators

We say an operator $\mathbb{T}$ is nonexpansive if

$$\|\mathbb{T}x - \mathbb{T}y\| \leq \|x - y\| \qquad \forall x, y \in \operatorname{dom} \mathbb{T}.$$

In other words, $\mathbb{T}$ is nonexpansive if it is 1-Lipschitz. We say $\mathbb{T}$ is a contraction if it is L-Lipschitz with $L < 1$. Mapping a pair of points by a contraction reduces the distance between them; mapping them by a nonexpansive operator does not increase the distance between them.

If $\mathbb{T}$ and $\mathbb{S}$ are nonexpansive, then $\mathbb{TS}$ is nonexpansive. If $\mathbb{T}$ or $\mathbb{S}$ is furthermore contractive, then $\mathbb{TS}$ is contractive. If $\mathbb{T}$ and $\mathbb{S}$ are nonexpansive, then $\theta\mathbb{T} + (1-\theta)\mathbb{S}$ with $\theta \in [0, 1]$, a convex combination, is nonexpansive. If $\mathbb{T}$ is furthermore contractive and $\theta \in (0, 1]$, then $\theta\mathbb{T} + (1-\theta)\mathbb{S}$ is contractive.

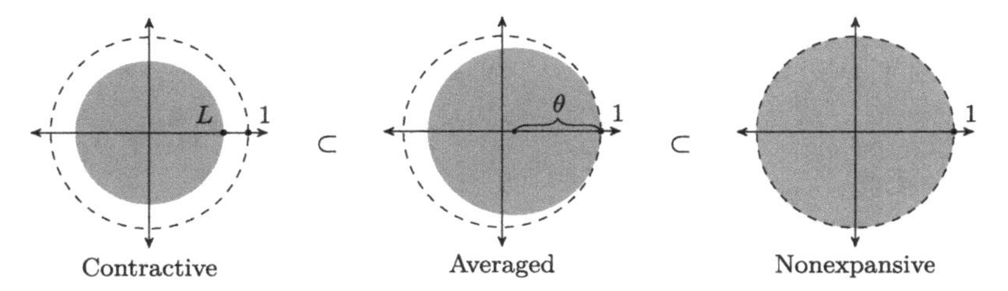

Figure 2.1 Illustration of classes of contractive, averaged, and nonexpansive operators. The figure illustrates the relationship: contractive ⊂ averaged ⊂ nonexpansive. The precise meaning of these figures will be defined in §13.

Averaged Operators

For $\theta \in (0, 1)$, we say an operator $\mathbb{T}$ is *θ-averaged* if $\mathbb{T} = (1-\theta)\mathbb{I} + \theta\mathbb{S}$ for some nonexpansive operator $\mathbb{S}$. We say an operator is averaged if it is θ-averaged for some unspecified $\theta \in (0, 1)$. In other words, taking a weighted average (convex combination) of $\mathbb{I}$ and a nonexpansive operator gives an averaged operator. We say an operator is *firmly nonexpansive* if it is (1/2)-averaged. See Figure 2.1. When operators $\mathbb{T}$ and $\mathbb{S}$ are averaged, the composition $\mathbb{T}\mathbb{S}$ is averaged. We prove this as Theorem 27 later in §13.

Averagedness is the central notion in establishing convergence for many splitting methods. In fact, Theorems 1, 2, and 3, the main convergence theorems of Part I, are based on the notion of averagedness.

2.4 FIXED-POINT ITERATION

We now discuss the first meta-algorithm of this book: the fixed-point iteration. Using the fixed-point iteration involves two steps. First, find a suitable operator whose fixed points are solutions to a monotone inclusion problem of interest. Second, show that the iteration converges to a fixed point.

2.4.1 Fixed Points

We say x is a *fixed point* of $\mathbb{T}$ if $x = \mathbb{T}x$, and write

$$\operatorname{Fix}\mathbb{T} = \{x \mid x = \mathbb{T}x\} = (\mathbb{I} - \mathbb{T})^{-1}(0)$$

for the set of fixed points of $\mathbb{T}$. If $\mathbb{T}$ is nonexpansive and $\operatorname{dom}\mathbb{T} = \mathbb{R}^n$, then its set of fixed points is closed and convex. Certainly, $\operatorname{Fix}\mathbb{T}$ can be empty (for example, $\mathbb{T}x = x + 1$ on $\mathbb{R}$) or contain many points (for example, $\mathbb{T}x = |x|$ on $\mathbb{R}$).

Let us show $\operatorname{Fix}\mathbb{T}$ is closed and convex when $\mathbb{T}\colon \mathbb{R}^n \to \mathbb{R}^n$ is nonexpansive. That $\operatorname{Fix}\mathbb{T}$ is closed follows from the fact that $\mathbb{T} - \mathbb{I}$ is a continuous function. Now suppose $x, y \in \operatorname{Fix}\mathbb{T}$ and $\theta \in [0, 1]$. We will show that $z = \theta x + (1-\theta)y \in \operatorname{Fix}\mathbb{T}$. Since $\mathbb{T}$ is nonexpansive, we have

$$\|\mathbb{T}z - x\| \le \|z - x\| = (1-\theta)\|y - x\|,$$

and similarly, we have

$$\|\mathbb{T}z - y\| \leq \theta\|y - x\|.$$

So, the triangle inequality

$$\|x - y\| \leq \|\mathbb{T}z - x\| + \|\mathbb{T}z - y\|$$

holds with equality, which means the previous inequalities hold with equality and $\mathbb{T}z$ is on the line segment between x and y. From $\|\mathbb{T}z - y\| = \theta\|y - x\|$, we conclude that $\mathbb{T}z = \theta x + (1-\theta)y = z$. Thus $z \in \operatorname{Fix}\mathbb{T}$.

2.4.2 Fixed-Point Iteration

The algorithm *fixed-point iteration* (FPI), also called the *Picard iteration*, is

$$x^{k+1} = \mathbb{T}x^k$$

for $k = 0, 1, \ldots$, where $x^0 \in \mathbb{R}^n$ is some starting point and $\mathbb{T}\colon \mathbb{R}^n \to \mathbb{R}^n$ is single-valued. The FPI is used to find a fixed point of $\mathbb{T}$. Clearly, the algorithm stays at a fixed point if it starts at a fixed point. For the sake of brevity, we will usually omit stating that $x^0 \in \mathbb{R}^n$ is some starting point and that $k = 0, 1, \ldots$ when we write an FPI.

In general, an FPI need not converge, even if we assume $\mathbb{T}$ is nonexpansive. For example, this is the case when $\mathbb{T}$ is a rotation about some line or a reflection through a plane. We provide two conditions that guarantee convergence, although these are not the only possible approaches.

Contractive Operators

Suppose that $\mathbb{T}\colon \mathbb{R}^n \to \mathbb{R}^n$ is a contraction with Lipschitz constant $L < 1$. In this setting, FPI is also called the *contraction mapping algorithm*. For $x^\star \in \operatorname{Fix}\mathbb{T}$, we have

$$\|x^k - x^\star\| \leq L\|x^{k-1} - x^\star\| \leq \cdots \leq L^k\|x^0 - x^\star\|.$$

This is the basis of the classic Banach fixed-point theorem; see Exercise 2.14.

So, when $\mathbb{T}$ is a contraction, the convergence analysis is very simple. In many optimization setups, however, a contraction is too much to ask for, and we need an approach to establish convergence under weaker assumptions.

Averaged Operators

Suppose $\mathbb{T}\colon \mathbb{R}^n \to \mathbb{R}^n$ is averaged. In this setting, FPI is also called the *averaged* or *Krasnosel'skiĭ–Mann* iteration, and it converges to a solution if one exists.

Theorem 1 Assume $\mathbb{T}\colon \mathbb{R}^n \to \mathbb{R}^n$ is θ-averaged with $\theta \in (0,1)$ and $\operatorname{Fix}\mathbb{T} \neq \emptyset$. Then $x^{k+1} = \mathbb{T}x^k$ with any starting point $x^0 \in \mathbb{R}^n$ converges to one fixed point, that is,

$$x^k \to x^\star$$

for some $x^\star \in \operatorname{Fix}\mathbb{T}$. The quantities $\operatorname{dist}(x^k, \operatorname{Fix}\mathbb{T})$, $\|x^{k+1} - x^k\|$, and $\|x^k - x^\star\|$ for any $x^\star \in \operatorname{Fix}\mathbb{T}$ are monotonically nonincreasing with k. Finally, we have

$$\operatorname{dist}(x^k, \operatorname{Fix}\mathbb{T}) \to 0$$

and

$$\|x^{k+1} - x^k\|^2 \leq \frac{\theta}{(k+1)(1-\theta)}\operatorname{dist}^2(x^0, \operatorname{Fix}\mathbb{T}).$$

To find a fixed point of a nonexpansive operator $\mathbb{T}$ that is not necessarily averaged, we can perform FPI on the averaged operator $(1-\theta)\mathbb{I}+\theta\mathbb{T}$ with $\theta \in (0,1)$. $\mathbb{T}$ and $(1-\theta)\mathbb{I}+\theta\mathbb{T}$ share the same set of fixed points, that is, $\operatorname{Fix}\mathbb{T} = \operatorname{Fix}((1-\theta)\mathbb{I} + \theta\mathbb{T})$. This ensures the iteration converges, with essentially no additional computational cost.

Example 2.4 Consider $\mathbb{T}\colon \mathbb{R}^2 \to \mathbb{R}^2$ defined as

$$\mathbb{T}x = \begin{bmatrix} -0.5 & 0 \\ 0 & 1 \end{bmatrix} x = \left(\frac{3}{4}\begin{bmatrix} -1 & 0 \\ 0 & 1 \end{bmatrix} + \frac{1}{4}\begin{bmatrix} 1 & 0 \\ 0 & 1 \end{bmatrix}\right) x.$$

This is a $(3/4)$-averaged operator with $\operatorname{Fix}\mathbb{T} = \{(0,z) \mid z \in \mathbb{R}\}$.

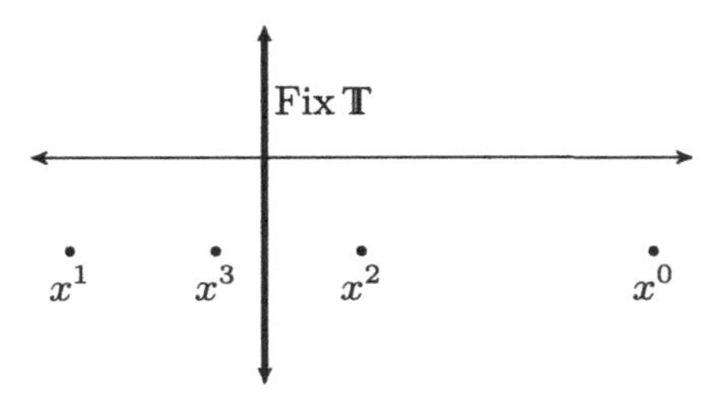

We can see that FPI with respect to $\mathbb{T}$ converges to one of the fixed points and that the limit depends on the starting point x^0.

Proof of Theorem 1. Before we begin the proof in earnest, we summarize the core idea of the proof. Assume we have *nonnegative* scalar sequences $V^0, V^1, \ldots$ and $S^0, S^1, \ldots$. (To clarify, the superscripts denote iteration count, not exponents.) Say we establish the inequality

$$V^{k+1} \leq V^k - S^k$$

for $k = 0,1,2,\ldots$. Such an inequality has two useful consequences. The first is that V^k is monotonically nonincreasing, although there is no guarantee that V^k decreases to 0.

The second is that $S^k \to 0$. To see why, sum both sides from 0 to k to get

$$\sum_{i=0}^{k} S^i \leq V^0 - V^{k+1} \leq V^0.$$

Taking $k \to \infty$ gives us

$$\sum_{i=0}^{\infty} S^i \leq V^0 < \infty,$$

and we say the sequence $S^0, S^1, \ldots$ is *summable*. Nonnegative summable sequences converge to 0, so $S^k \to 0$. If, furthermore, we can show that $S^0, S^1, \ldots$ is nonincreasing, then

$$(k+1)S^k \leq \sum_{i=0}^{k} S^i \leq V^0,$$

and hence $S^k \leq \frac{1}{k+1}V^0$. As an aside, we call V^k the *Lyapunov function* and S^k the *summable term*.

The proof technique of showing that a Lyapunov function produces a summable term, which converges to zero, is called the *summability argument*.

Stage 1 Note

$$\|(1-\theta)x + \theta y\|^2 = (1-\theta)\|x\|^2 + \theta\|y\|^2 - \theta(1-\theta)\|x-y\|^2,$$

for all $\theta \in \mathbb{R}$, $x, y \in \mathbb{R}^n$. Verifying the identity is a matter of expanding both sides.

Write $\mathbb{T} = (1-\theta)\mathbb{I} + \theta\mathbb{S}$, where $\mathbb{S}$ is nonexpansive. Write the FPI as

$$x^{k+1} = \mathbb{T}x^k = (1-\theta)x^k + \theta\mathbb{S}x^k.$$

For any $x^\star \in \operatorname{Fix}\mathbb{T}$, we use the previous identity to get

$$\begin{aligned}\|x^{k+1} - x^\star\|^2 &= (1-\theta)\|x^k - x^\star\|^2 + \theta\|\mathbb{S}(x^k) - x^\star\|^2 - \theta(1-\theta)\|\mathbb{S}(x^k) - x^k\|^2 \\ &\leq (1-\theta)\|x^k - x^\star\|^2 + \theta\|x^k - x^\star\|^2 - \theta(1-\theta)\|\mathbb{S}(x^k) - x^k\|^2 \\ &= \underbrace{\|x^k - x^\star\|^2}_{=V^k} - \underbrace{\theta(1-\theta)\|\mathbb{S}(x^k) - x^k\|^2}_{=S^k},\end{aligned} \tag{2.4}$$

where we used nonexpansiveness of $\mathbb{S}$ in the second line.

We now establish the monotonic decreases. The core inequality (2.4) tells us

$$\|x^{k+1} - x^\star\| \leq \|x^k - x^\star\|$$

for any $x^\star \in \operatorname{Fix}\mathbb{T}$, that is, the distance of the iterates to any fixed point is monotonically nonincreasing. Minimizing both sides with respect to $x^\star \in \operatorname{Fix}\mathbb{T}$ gives us

$$\operatorname{dist}(x^{k+1}, \operatorname{Fix}\mathbb{T}) \leq \operatorname{dist}(x^k, \operatorname{Fix}\mathbb{T}),$$

that is, the distance of the iterates to the set of fixed points is monotonically nonincreasing. As another aside, an algorithm is said to be *Fejér monotone* if the distance of the iterates to the solution set is monotonically nonincreasing.

We call $\mathbb{T}(x^k)-x^k = x^{k+1}-x^k$ the *fixed-point residual*. If $\mathbb{T}(x^k)-x^k = 0$, the FPI is at a fixed point, and the iteration stops, so one can use $\|\mathbb{T}(x^k)-x^k\|$ as a measure of progress of the FPI. Since $\mathbb{T}$ is nonexpansive, we have

$$\|x^{k+1} - x^k\| = \|\mathbb{T}x^k - \mathbb{T}x^{k-1}\| \leq \|x^k - x^{k-1}\|,$$

that is, the magnitude of the fixed-point residual is monotonically nonincreasing.

Using the monotonic decrease of $\|x^{k+1} - x^k\|$, we obtain a rate of convergence for $\|x^{k+1} - x^k\| \to 0$. Summing the inequality (2.4) from 0 to k gives us

$$\|x^{k+1} - x^\star\|^2 \leq \|x^0 - x^\star\|^2 - \frac{1-\theta}{\theta}\sum_{j=0}^{k}\|\mathbb{T}x^j - x^j\|^2.$$

Reorganizing, we get

$$\sum_{j=0}^{k}\|\mathbb{T}x^j - x^j\|^2 \leq \frac{\theta}{1-\theta}\|x^0 - x^\star\|^2 - \frac{\theta}{1-\theta}\|x^{k+1} - x^\star\|^2.$$

With the monotonic decrease of $\|x^{k+1} - x^k\|$ we get

$$(k+1)\|x^{k+1} - x^k\|^2 \leq \sum_{j=0}^{k}\|x^{j+1} - x^j\|^2 \leq \frac{\theta}{1-\theta}\|x^0 - x^\star\|^2,$$

and we conclude that

$$\|x^{k+1} - x^k\|^2 \leq \frac{\theta}{(k+1)(1-\theta)}\|x^0 - x^\star\|^2.$$

Minimizing the right-hand side with respect to $x^\star \in \operatorname{Fix}\mathbb{T}$, we get

$$\|x^{k+1} - x^k\|^2 \leq \frac{\theta}{(k+1)(1-\theta)}\operatorname{dist}^2(x^0, \operatorname{Fix}\mathbb{T}).$$

Stage 2 We now show $x^k \to x^\star$ for some $x^\star \in \operatorname{Fix}\mathbb{T}$. Consider any $\tilde{x}^\star \in \operatorname{Fix}\mathbb{T}$. Then (2.4) tells us that $x^0, x^1, \ldots$ lie within the compact set $\{x \mid \|x - \tilde{x}^\star\| \leq \|x^0 - \tilde{x}^\star\|\}$, and $x^0, x^1, \ldots$ has an accumulation point $x^\star$. Let x^{k_j} be a subsequence such that $x^{k_j} \to x^\star$. Then $(\mathbb{T} - \mathbb{I})(x^k) \to 0$ implies $(\mathbb{T} - \mathbb{I})(x^{k_j}) \to 0$. Since $\mathbb{T} - \mathbb{I}$ is continuous, $x^{k_j} \to x^\star$ and $(\mathbb{T} - \mathbb{I})(x^{k_j}) \to 0$ implies $(\mathbb{T} - \mathbb{I})(x^\star) = 0$. In other words, $x^\star \in \operatorname{Fix}\mathbb{T}$. Finally, applying (2.4) to this accumulation point $x^\star \in \operatorname{Fix}\mathbb{T}$, we conclude that $\|x^k - x^\star\|$ monotonically decreases to 0, that is, the entire sequence converges to $x^\star$. □

Termination Criterion

Although we avoid the discussion of termination criterion throughout this book for the sake of simplicity, detecting when an iterate is a sufficiently accurate approximation of the solution is essential for a practical iterative method. We simply point out that $\|x^{k+1} - x^k\| < \varepsilon$ for some small $\varepsilon > 0$ can generally be used as a termination criterion. Specific setups may have other termination criteria that better capture the particular goals of the setup.

2.4.3 Methods

Gradient Descent
Consider the problem

$$\underset{x\in\mathbb{R}^n}{\text{minimize}} \quad f(x).$$

Assume f is CCP and differentiable. Then x is a solution if and only if

$$0 = \nabla f(x) \quad \Leftrightarrow \quad x = (\mathbb{I} - \alpha\nabla f)(x)$$

for any nonzero $\alpha \in \mathbb{R}$. In other words, x is a solution if and only if it is a fixed point of the operator $\mathbb{I} - \alpha\nabla f$.

The FPI for this setup is

$$x^{k+1} = x^k - \alpha\nabla f(x^k).$$

This algorithm is called the *gradient method* or *gradient descent*, and α is called the *stepsize*.

Now assume f is L-smooth. By the cocoercivity inequality,

$$\begin{aligned}
&\|(\mathbb{I} - (2/L)\nabla f)x - (\mathbb{I} - (2/L)\nabla f)y\|^2 \\
&= \|x-y\|^2 - \frac{4}{L}\left(\langle x-y, \nabla f(x) - \nabla f(y)\rangle - \frac{1}{L}\|\nabla f(x) - \nabla f(y)\|^2\right) \\
&\leq \|x-y\|^2.
\end{aligned}$$

Therefore, $\mathbb{I} - \alpha\nabla f$ is averaged for $\alpha \in (0, 2/L)$ since

$$\mathbb{I} - \alpha\nabla f = (1-\theta)\mathbb{I} + \theta(\mathbb{I} - (2/L)\nabla f),$$

where $\theta = \alpha L/2 < 1$. Consequently, $x^k \to x^\star$ for some solution $x^\star$, if one exists, with rate

$$\|\nabla f(x^k)\|^2 = O(1/k),$$

for any

$$\alpha \in (0, 2/L). \tag{2.5}$$

If we furthermore assume f is strongly convex, we can show the iteration is a contraction.

Forward Step Method
Consider the problem

$$\underset{x\in\mathbb{R}^n}{\text{find}} \quad 0 = \mathbb{F}(x),$$

where $\mathbb{F}\colon \mathbb{R}^n \to \mathbb{R}^n$.

By the same argument as for gradient descent, x is a solution if and only if it is a fixed point of $\mathbb{I} - \alpha\mathbb{F}$ for any nonzero $\alpha \in \mathbb{R}$. The FPI for this setup is

$$x^{k+1} = x^k - \alpha\mathbb{F}x^k,$$

which we call the *forward step method*.

The forward step method converges if $\mathbb{F}$ is β-cocoercive and $\alpha \in (0, 2\beta)$. The forward step iteration is a contraction for small enough $\alpha > 0$ if $\mathbb{F}$ is strongly monotone and Lipschitz.

However, the method does not necessarily converge if $\mathbb{F}$ is merely monotone and Lipschitz. The operator

$$\mathbb{F}(x, y) = \begin{bmatrix} 0 & 1 \\ -1 & 0 \end{bmatrix} \begin{bmatrix} x \\ y \end{bmatrix}$$

is such an example, since the 2×2 matrix representing $\mathbb{I} - \alpha\mathbb{F}$ has singular values strictly greater than 1 for any $\alpha \neq 0$. (This operator arises as, say, the KKT operator of the problem of minimizing x subject to $x = 0$.) The scaled relative graphs of §13 will provide the geometric intuition of this counterexample.

Dual Ascent

Consider the primal-dual problem pair (1.6) and (1.7),

$$\begin{array}{ll} \underset{x \in \mathbb{R}^n}{\text{minimize}} & f(x) \\ \text{subject to} & Ax = b, \end{array} \qquad \underset{u \in \mathbb{R}^m}{\text{maximize}} \quad -f^*(-A^\intercal u) - b^\intercal u,$$

generated by the Lagrangian (1.5)

$$\mathbf{L}(x, u) = f(x) + \langle u, Ax - b \rangle.$$

Define $g(u) = f^*(-A^\intercal u) + b^\intercal u$. By the discussion of §1.3.8, if f is μ-strongly convex, then f^* is differentiable and ∇f^* is $(1/\mu)$-Lipschitz. By the discussion of §2.2.2,

$$\nabla g(u) = -A\nabla f^*(-A^\intercal u) + b$$

is Lipschitz with parameters $\sigma^2_{\max}(A)/\mu$.

Using (2.2), write the gradient method applied to g, the FPI on $\mathbb{I} - \alpha\nabla g$, as

$$\begin{aligned} x^{k+1} &= \underset{x}{\text{argmin}}\, \mathbf{L}(x, u^k) \\ u^{k+1} &= u^k + \alpha(Ax^{k+1} - b). \end{aligned}$$

The first step is minimizing the Lagrangian, and the second is a multiplier update. This method is called the *Uzawa method* or *dual ascent*. If f is μ-strongly convex, total duality holds, and $0 < \alpha < 2\mu/\sigma^2_{\max}(A)$, then $x^k \to x^\star$ and $u^k \to u^\star$. See Exercise 2.17.

2.5 RESOLVENTS

The *resolvent* of an operator $\mathbb{A}$ is defined as

$$\mathbb{J}_{\mathbb{A}} = (\mathbb{I} + \mathbb{A})^{-1}.$$

The *reflected resolvent*, also called the *Cayley operator* or the *reflection operator*, of $\mathbb{A}$ is defined as

$$\mathbb{R}_{\mathbb{A}} = 2\mathbb{J}_{\mathbb{A}} - \mathbb{I}.$$

Often, we will use $\mathbb{J}_{\alpha\mathbb{A}}$ and $\mathbb{R}_{\alpha\mathbb{A}}$ with $\alpha > 0$. If $\mathbb{A}$ is maximal monotone, $\mathbb{R}_{\mathbb{A}}$ is a nonexpansive (single-valued) with $\operatorname{dom} \mathbb{R}_{\mathbb{A}} = \mathbb{R}^n$, and $\mathbb{J}_{\mathbb{A}}$ is a $(1/2)$-averaged with $\operatorname{dom} \mathbb{J}_{\mathbb{A}} = \mathbb{R}^n$.

Let us prove nonexpansiveness. Assume we have $(x, u), (y, v) \in \mathbb{J}_{\mathbb{A}}$. By definition of resolvents, we have

$$x \in u + \mathbb{A}u, \qquad y \in v + \mathbb{A}v.$$

By monotonicity of $\mathbb{A}$,

$$\langle (x - u) - (y - v), u - v \rangle \geq 0$$

and

$$\begin{aligned} \|(2u - x) - (2v - y)\|^2 &= \|x - y\|^2 - 4\langle (x - u) - (y - v), u - v \rangle \\ &\leq \|x - y\|^2. \end{aligned}$$

This proves $\mathbb{R}_{\mathbb{A}}$ is nonexpansive and therefore single-valued, and $\mathbb{J}_{\mathbb{A}} = (1/2)\mathbb{I} + (1/2)\mathbb{R}_{\mathbb{A}}$ is $(1/2)$-averaged.

The *Minty surjectivity theorem* states that $\operatorname{dom} \mathbb{J}_{\mathbb{A}} = \mathbb{R}^n$ when $\mathbb{A}$ is maximal monotone. This result is easy to intuitively see in 1D but is nontrivial in higher dimensions. We prove this in §10.

Zero Set of a Maximal Monotone Operator

Using resolvents, we can quickly show $\operatorname{Zer} \mathbb{A}$ is a closed convex set when $\mathbb{A}$ is maximal monotone. Since

$$0 \in \mathbb{A}x \quad \Leftrightarrow \quad x \in x + \mathbb{A}x \quad \Leftrightarrow \quad \mathbb{J}_{\mathbb{A}}x = x,$$

we have $\operatorname{Zer} \mathbb{A} = \operatorname{Fix} \mathbb{J}_{\mathbb{A}}$. Since $\mathbb{J}_{\mathbb{A}}$ is nonexpansive, $\operatorname{Fix} \mathbb{J}_{\mathbb{A}}$ is a closed convex set. Note that this proof relies on maximality through the condition $\operatorname{dom} \mathbb{J}_{\mathbb{A}} = \mathbb{R}^n$.

Example 2.5 When $\mathbb{A}$ is a monotone linear operator represented by a symmetric matrix, it is easier to see why $\mathbb{J}_{\mathbb{A}}$ and $\mathbb{R}_{\mathbb{A}}$ are nonexpansive. In this case, $\mathbb{A}$ has eigenvalues in $[0, \infty)$ and $\mathbb{J}_{\mathbb{A}} = (\mathbb{I} + \mathbb{A})^{-1}$ has eigenvalues in $(0, 1]$. The reflected resolvent,

$$\mathbb{R}_{\mathbb{A}} = 2\mathbb{J}_{\mathbb{A}} - \mathbb{I} = (\mathbb{I} - \mathbb{A})(\mathbb{I} + \mathbb{A})^{-1} = (\mathbb{I} + \mathbb{A})^{-1}(\mathbb{I} - \mathbb{A}),$$

also called the Cayley transform of $\mathbb{A}$, has eigenvalues in $(-1, 1]$.

Example 2.6 Let $z \in \mathbb{C}$ be a complex number. We can identify z with a linear operator from $\mathbb{C}$ to $\mathbb{C}$ defined by multiplication, that is, we can view z as the operator that maps $x \mapsto zx$ for any $x \in \mathbb{C}$. We equip the set of complex numbers with the inner product $\langle x, y \rangle = \operatorname{Re} x\overline{y}$ for any $x, y \in \mathbb{C}$, where $\overline{y}$ is the complex conjugate of y. Then $z \in \mathbb{C}$ is a monotone operator if and only if $\operatorname{Re} z \geq 0$.

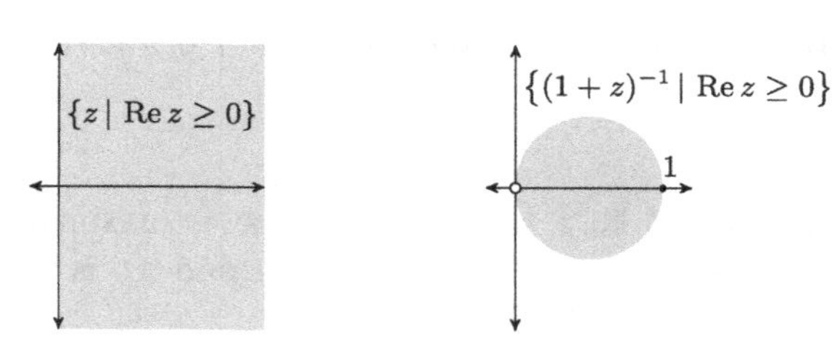

So a monotone z is a complex number on the right half-plane, and its resolvent $(1+z)^{-1}$ is a complex number within the disk with center $1/2$ and radius $1/2$ except for the origin.

2.5.1 Examples

Subdifferential

When f is CCP and $\alpha > 0$, we have

$$\mathbb{J}_{\alpha\partial f} = \mathrm{Prox}_{\alpha f}.$$

This follows from

$$\begin{aligned} z = (I+\alpha\partial f)^{-1}(x) \quad &\Leftrightarrow \quad z + \alpha\partial f(z) \ni x \\ &\Leftrightarrow \quad 0 \in \partial_z\left(\alpha f(z) + \frac{1}{2}\|z-x\|^2\right) \\ &\Leftrightarrow \quad z = \operatorname*{argmin}_z \left\{\alpha f(z) + \frac{1}{2}\|z-x\|^2\right\} \\ &\Leftrightarrow \quad z = \mathrm{Prox}_{\alpha f}(x). \end{aligned}$$

Subdifferential of Conjugate

Let $g(u) = f^*(A^\intercal u)$, and assume f is CCP and $\operatorname{ri}\operatorname{dom} f^* \cap \mathcal{R}(A^\intercal) \neq \emptyset$. Then

$$v = \mathrm{Prox}_{\alpha g}(u) \quad \Leftrightarrow \quad \begin{array}{l} x \in \operatorname{argmin}_x \left\{f(x) - \langle u, Ax\rangle + \frac{\alpha}{2}\|Ax\|^2\right\} \\ v = u - \alpha Ax. \end{array} \tag{2.6}$$

This follows from

$$\begin{aligned} v = (I+\alpha\partial g)^{-1}(u) \quad &\Leftrightarrow \quad v + \alpha A\partial f^*(A^\intercal v) \ni u \\ &\Leftrightarrow \quad v + \alpha Ax = u,\ x \in \partial f^*(A^\intercal v) \\ &\Leftrightarrow \quad v = u - \alpha Ax,\ \partial f(x) \ni A^\intercal v \\ &\Leftrightarrow \quad v = u - \alpha Ax,\ \partial f(x) \ni A^\intercal(u - \alpha Ax) \\ &\Leftrightarrow \quad v = u - \alpha Ax,\ x \in \operatorname*{argmin}_x \left\{f(x) - \langle u, Ax\rangle + \frac{\alpha}{2}\|Ax\|^2\right\}. \end{aligned}$$

Projection

Let $C \subset \mathbb{R}^n$ be a nonempty closed convex set. Remember from §1 that δ_C is the indicator function of C, $\mathbb{N}_C$ is the normal cone operator of C, and Π_C is the projection onto C. These satisfy the following properties: $\delta_C = \alpha\delta_C$ and $\mathbb{N}_C = \alpha\mathbb{N}_C$ for any $\alpha > 0$; $\partial\delta_C = \mathbb{N}_C$; and $\mathbb{J}_{\mathbb{N}_C} = \mathrm{Prox}_{\delta_C} = \Pi_C$.

KKT Operator for Linearly Constrained Problems

Consider the Lagrangian

$$\mathbf{L}(x,u) = f(x) + \langle u, Ax - b\rangle,$$

which generates the primal problem

$$\begin{array}{ll} \underset{x\in\mathbb{R}^n}{\text{minimize}} & f(x) \\ \text{subject to} & Ax = b. \end{array}$$

We can compute its resolvent with

$$\mathbb{J}_{\alpha\partial\mathbf{L}}(x,u) = (y,v) \quad \Leftrightarrow \quad \begin{array}{l} y = \text{argmin}_z \left\{\mathbf{L}_\alpha(z,u) + \frac{1}{2\alpha}\|z-x\|^2\right\} \\ v = u + \alpha(Ay-b), \end{array} \tag{2.7}$$

where $\mathbf{L}_\alpha = f(x) + \langle u, Ax-b\rangle + \frac{\alpha}{2}\|Ax-b\|^2$ is the augmented Lagrangian of (1.11).

Let us show this. For any $\alpha > 0$, we have

$$\begin{aligned} \mathbb{J}_{\alpha\partial\mathbf{L}}(x,u) = (y,v) \quad &\Leftrightarrow \quad \begin{bmatrix} x \\ u \end{bmatrix} \in \begin{bmatrix} y \\ v \end{bmatrix} + \alpha \begin{bmatrix} \partial f(y) + A^\intercal v \\ b - Ay \end{bmatrix} \\ &\Leftrightarrow \quad \begin{bmatrix} x \\ u \end{bmatrix} \in \alpha \begin{bmatrix} \partial f(y) \\ b \end{bmatrix} + \begin{bmatrix} I & \alpha A^\intercal \\ -\alpha A & I \end{bmatrix} \begin{bmatrix} y \\ v \end{bmatrix}. \end{aligned}$$

We left-multiply the invertible matrix

$$\begin{bmatrix} I & -\alpha A^\intercal \\ 0 & I \end{bmatrix}$$

to get

$$\Leftrightarrow \quad \begin{bmatrix} x - \alpha A^\intercal u \\ u \end{bmatrix} \in \alpha \begin{bmatrix} \partial f(y) - \alpha A^\intercal b \\ b \end{bmatrix} + \begin{bmatrix} I + \alpha^2 A^\intercal A & 0 \\ -\alpha A & I \end{bmatrix} \begin{bmatrix} y \\ v \end{bmatrix}.$$

We call this the *Gaussian elimination technique* and discuss it in more detail in §3.4. Now that the first line of the inclusion is independent of v, we can compute y first and then compute v. Reorganizing, we get

$$\begin{aligned} 0 \in \partial f(y) + A^\intercal u + \alpha A^\intercal(Ay-b) + (1/\alpha)(y-x) \\ v = u + \alpha(Ay-b), \end{aligned}$$

and we have the formula

$$\begin{aligned} &y = \underset{z}{\text{argmin}} \left\{ f(z) + \langle u, Az-b\rangle + \frac{\alpha}{2}\|Az-b\|^2 + \frac{1}{2\alpha}\|z-x\|^2 \right\} \\ &v = u + \alpha(Ay-b). \end{aligned}$$

2.5.2 Basic Identities

Resolvent Identities

If $\mathbb{A}$ is maximal monotone, $\alpha > 0$, and $\mathbb{B}(x) = \mathbb{A}(x) + t$, then

$$\mathbb{J}_{\alpha\mathbb{B}}(u) = \mathbb{J}_{\alpha\mathbb{A}}(u - \alpha t). \tag{2.8}$$

This follows from

$$\begin{aligned} \mathbb{J}_{\alpha\mathbb{B}}u = v \quad &\Leftrightarrow \quad u \in v + \alpha\mathbb{B}v \\ &\Leftrightarrow \quad u - \alpha t \in v + \alpha\mathbb{A}v \\ &\Leftrightarrow \quad v = \mathbb{J}_{\alpha\mathbb{A}}(u - \alpha t). \end{aligned}$$

With similar calculations, one can show that if $\mathbb{A}$ is maximal monotone, $\alpha > 0$, and $\mathbb{B}(x) = \mathbb{A}(x - t)$, then

$$\mathbb{J}_{\alpha\mathbb{B}}(u) = \mathbb{J}_{\alpha\mathbb{A}}(u - t) + t, \tag{2.9}$$

and if $\mathbb{A}$ is maximal monotone, $\alpha > 0$, and $\mathbb{B}(x) = -\mathbb{A}(t - x)$, then

$$\mathbb{J}_{\alpha\mathbb{B}}(u) = t - \mathbb{J}_{\alpha\mathbb{A}}(t - u). \tag{2.10}$$

The *inverse resolvent identity* states

$$\mathbb{J}_{\alpha^{-1}\mathbb{A}}(x) + \alpha^{-1}\mathbb{J}_{\alpha\mathbb{A}^{-1}}(\alpha x) = x, \tag{2.11}$$

for maximal monotone $\mathbb{A}$ and $\alpha > 0$. This follows from

$$\begin{aligned} x - \mathbb{J}_{\alpha^{-1}\mathbb{A}}x = y \quad &\Leftrightarrow \quad x \in x - y + \alpha^{-1}\mathbb{A}(x - y) \\ &\Leftrightarrow \quad \alpha y \in \mathbb{A}(x - y) \\ &\Leftrightarrow \quad \mathbb{A}^{-1}(\alpha y) \ni x - y \\ &\Leftrightarrow \quad (\mathbb{I} + \alpha\mathbb{A}^{-1})(\alpha y) \ni \alpha x \\ &\Leftrightarrow \quad y = (1/\alpha)\mathbb{J}_{\alpha\mathbb{A}^{-1}}(\alpha x). \end{aligned}$$

When $\alpha = 1$, we get the further elegant formula

$$\mathbb{J}_{\mathbb{A}} + \mathbb{J}_{\mathbb{A}^{-1}} = \mathbb{I}.$$

The *Moreau identity*, a special case, states that for any CCP f,

$$\mathrm{Prox}_f + \mathrm{Prox}_{f^*} = \mathbb{I},$$

or more generally,

$$\mathrm{Prox}_{\alpha^{-1}f}(x) + \alpha^{-1}\mathrm{Prox}_{\alpha f^*}(\alpha x) = x. \tag{2.12}$$

An important practical consequence of the Moreau identity is that $\mathrm{Prox}_{\alpha f}$ and $\mathrm{Prox}_{\alpha f^*}$ require essentially the same computational cost. In other words, if you can compute $\mathrm{Prox}_{\alpha f}$, then you can compute $\mathrm{Prox}_{\alpha f^*}$, and vice versa.

Reflected Resolvent Identities

If $\mathbb{A}$ is maximal monotone and single-valued and $\alpha > 0$, we have

$$\mathbb{R}_{\alpha\mathbb{A}} = (\mathbb{I} - \alpha\mathbb{A})(\mathbb{I} + \alpha\mathbb{A})^{-1}.$$

This follows from

$$\begin{aligned} \mathbb{R}_{\alpha\mathbb{A}} &= 2(\mathbb{I} + \alpha\mathbb{A})^{-1} - \mathbb{I} \\ &= 2(\mathbb{I} + \alpha\mathbb{A})^{-1} - (\mathbb{I} + \alpha\mathbb{A})(\mathbb{I} + \alpha\mathbb{A})^{-1} \\ &= (\mathbb{I} - \alpha\mathbb{A})(\mathbb{I} + \alpha\mathbb{A})^{-1}, \end{aligned}$$

where we used the result of Exercise 2.1 in the second equality.

If $\mathbb{A}$ is maximal monotone (but not necessarily single-valued) and $\alpha > 0$, we have

$$\mathbb{R}_{\alpha\mathbb{A}}(\mathbb{I} + \alpha\mathbb{A}) = \mathbb{I} - \alpha\mathbb{A}. \tag{2.13}$$

Let us prove this. Since $(\mathbb{I} + \alpha\mathbb{A})^{-1}$ is single-valued, for any $x \in \mathrm{dom}\,\mathbb{A}$ we have

$$\mathbb{R}_{\alpha\mathbb{A}}(\mathbb{I} + \alpha\mathbb{A})(x) = 2(\mathbb{I} + \alpha\mathbb{A})^{-1}(\mathbb{I} + \alpha\mathbb{A})(x) - (\mathbb{I} + \alpha\mathbb{A})(x)$$

$$= 2\mathbb{I}(x) - (\mathbb{I} + \alpha\mathbb{A})(x)$$
$$= (\mathbb{I} - \alpha\mathbb{A})(x),$$

where we used the result of Exercise 2.1 in the second equality. For any $x \notin \operatorname{dom}\mathbb{A}$, both sides are empty sets.

2.6 PROXIMAL POINT METHOD

Consider the problem

$$\underset{x\in\mathbb{R}^n}{\text{find}} \quad 0 \in \mathbb{A}x,$$

where $\mathbb{A}$ is maximal monotone. This problem is equivalent to finding a fixed point of $\mathbb{J}_{\alpha\mathbb{A}}$, since $\operatorname{Zer}\mathbb{A} = \operatorname{Fix}\mathbb{J}_{\alpha\mathbb{A}}$ for any $\alpha > 0$. The FPI

$$x^{k+1} = \mathbb{J}_{\alpha\mathbb{A}}(x^k),$$

called the *proximal point method* (PPM) or *proximal minimization*, converges to a solution if one exists, since $\mathbb{J}_{\alpha\mathbb{A}}$ is averaged.

2.6.1 Methods of Multipliers

Consider the primal-dual problem pair,

$$\begin{array}{ll} \underset{x\in\mathbb{R}^n}{\text{minimize}} & f(x) \\ \text{subject to} & Ax = b, \end{array} \qquad \underset{u\in\mathbb{R}^m}{\text{maximize}} \quad -f^*(-A^\intercal u) - b^\intercal u,$$

of (1.6) and (1.7) generated by the Lagrangian $\mathbf{L}(x,u) = f(x) + \langle u, Ax-b\rangle$. The associated augmented Lagrangian discussed in Example 1.11 is

$$\mathbf{L}_\alpha(x,u) = f(x) + \langle u, Ax-b\rangle + \frac{\alpha}{2}\|Ax-b\|^2.$$

Method of Multipliers

Assume $\mathcal{R}(A^\intercal)\cap \operatorname{ri}\operatorname{dom} f^* \neq \emptyset$. Write $g(u) = f^*(-A^\intercal u) + b^\intercal u$ for the dual function. Using (2.6) and (2.8), we can write the FPI $u^{k+1} = \mathbb{J}_{\alpha\partial g}(u^k)$ with $\alpha > 0$ as

$$\begin{aligned} x^{k+1} &\in \operatorname*{argmin}_x \mathbf{L}_\alpha(x,u^k) \\ u^{k+1} &= u^k + \alpha(Ax^{k+1} - b), \end{aligned}$$

which is called the *method of multipliers*, also known as the *augmented Lagrangian method* or *ALM*. The first step is minimizing the augmented Lagrangian, and the second is a multiplier update.

If a dual solution exists and $\alpha > 0$, then $u^k \to u^\star$. If we further assume f is strictly convex, we can show $x^k \to x^\star$. See Exercises 2.18 and 10.4.

Proximal Method of Multipliers

Using (2.7), we can write the FPI $(x^{k+1}, u^{k+1}) = \mathbb{J}_{\alpha\partial\mathbf{L}}(x^k, u^k)$ with $\alpha > 0$ as

$$\begin{aligned} x^{k+1} &= \operatorname*{argmin}_x \left\{ \mathbf{L}_\alpha(x,u^k) + \frac{1}{2\alpha}\|x - x^k\|^2 \right\} \\ u^{k+1} &= u^k + \alpha(Ax^{k+1} - b), \end{aligned}$$

which is called the *proximal method of multipliers*, also *the proximal augmented Lagrangian method*. The first step is minimizing the augmented Lagrangian with an additional proximal term, and the second is a multiplier update. If total duality holds and $\alpha > 0$, then $x^k \to x^\star$ and $u^k \to u^\star$.

The proximal method of multipliers becomes useful when it is combined with the linearization technique. We discuss this in §3.5.

2.7 OPERATOR SPLITTING

Consider the monotone inclusion problems of finding an $x \in \operatorname{Zer}(\mathbb{A}+\mathbb{B})$ or $x \in \operatorname{Zer}(\mathbb{A}+\mathbb{B}+\mathbb{C})$, where $\mathbb{A}$, $\mathbb{B}$, and $\mathbb{C}$ are maximal monotone. In this section, we present a few *base splitting schemes*, which transform these monotone inclusion problems into fixed-point equations with averaged operators constructed from $\mathbb{A}$, $\mathbb{B}$, $\mathbb{C}$, and their resolvents.

The key technique is to formulate a given optimization problem as a monotone inclusion problem, apply one of the base splitting schemes, and use the fixed-point iteration discussed in §2.4.2, or the randomized coordinate or asynchronous variants of §5 and §6. The main message of Part I of this book is that a wide range of methods can be derived and analyzed through this unified approach.

2.7.1 Base Splitting Schemes

Forward-Backward and Backward-Forward Splitting

Consider the problem

$$\underset{x \in \mathbb{R}^n}{\text{find}} \quad 0 \in (\mathbb{A} + \mathbb{B})x,$$

where $\mathbb{A}$ and $\mathbb{B}$ are maximal monotone and $\mathbb{A}$ is single-valued. Then for any $\alpha > 0$, we have

$$\begin{aligned} 0 \in (\mathbb{A} + \mathbb{B})x \quad &\Leftrightarrow \quad 0 \in (\mathbb{I} + \alpha\mathbb{B})x - (\mathbb{I} - \alpha\mathbb{A})x \\ &\Leftrightarrow \quad (\mathbb{I} + \alpha\mathbb{B})x \ni (\mathbb{I} - \alpha\mathbb{A})x \\ &\Leftrightarrow \quad x = \mathbb{J}_{\alpha\mathbb{B}}(\mathbb{I} - \alpha\mathbb{A})x. \end{aligned}$$

So, x is a solution if and only if it is a fixed point of $\mathbb{J}_{\alpha\mathbb{B}}(\mathbb{I} - \alpha\mathbb{A})$. This splitting is called *forward-backward splitting* (FBS).

Assume $\mathbb{A}$ is β-cocoercive and $\alpha \in (0, 2\beta)$. Then the *forward step* $\mathbb{I} - \alpha\mathbb{A}$ and the *backward step* $(\mathbb{I} + \alpha\mathbb{B})^{-1}$ are averaged. So, the composition $\mathbb{J}_{\alpha\mathbb{B}}(\mathbb{I} - \alpha\mathbb{A})$ is an averaged operator.

The FPI with FBS

$$x^{k+1} = \mathbb{J}_{\alpha\mathbb{B}}(x^k - \alpha\mathbb{A}x^k)$$

converges if $\alpha \in (0, 2\beta)$ and $\operatorname{Zer}(\mathbb{A} + \mathbb{B}) \neq \emptyset$.

We can also consider a similar splitting with a permuted order:

$$\begin{aligned} 0 \in (\mathbb{A} + \mathbb{B})x \quad &\Leftrightarrow \quad (\mathbb{I} + \alpha\mathbb{B})x \ni (\mathbb{I} - \alpha\mathbb{A})x \\ &\Leftrightarrow \quad z = (\mathbb{I} - \alpha\mathbb{A})x,\ z \in (\mathbb{I} + \alpha\mathbb{B})x \\ &\Leftrightarrow \quad z = (\mathbb{I} - \alpha\mathbb{A})x,\ \mathbb{J}_{\alpha\mathbb{B}}z = x \\ &\Leftrightarrow \quad z = (\mathbb{I} - \alpha\mathbb{A})\mathbb{J}_{\alpha\mathbb{B}}z,\ \mathbb{J}_{\alpha\mathbb{B}}z = x. \end{aligned}$$

So, x is a solution if and only if there is a $z \in \operatorname{Fix}(\mathbb{I}-\alpha\mathbb{A})\mathbb{J}_{\alpha\mathbb{B}}$ and $x = \mathbb{J}_{\alpha\mathbb{B}}z$. This splitting is called *backward-forward splitting* (BFS).

The FPI with BFS

$$\begin{aligned} x^{k+1} &= \mathbb{J}_{\alpha\mathbb{B}}z^k \\ z^{k+1} &= x^{k+1} - \alpha\mathbb{A}x^{k+1} \end{aligned}$$

converges if $\alpha \in (0, 2\beta)$ and $\operatorname{Zer}(\mathbb{A}+\mathbb{B}) \neq \emptyset$.

Since BFS is FBS with the order permuted, BFS may seem like an unnecessary complication. In fact, the FPIs with FBS and BFS have the same iterates if the starting points x^0 for FBS and z^0 for BFS are matched in the sense that $x^0 = \mathbb{J}_{\alpha\mathbb{B}}z^0$. However, we will later see that BFS can be more natural to work with when using the randomized or asynchronous coordinate fixed-point iterations of §5 and §6.

Peaceman–Rachford and Douglas–Rachford Splitting

Consider the problem

$$\underset{x \in \mathbb{R}^n}{\text{find}} \quad 0 \in (\mathbb{A}+\mathbb{B})x,$$

where $\mathbb{A}$ and $\mathbb{B}$ are maximal monotone.

For any $\alpha > 0$, we have

$$\begin{aligned} 0 \in (\mathbb{A}+\mathbb{B})x \quad &\Leftrightarrow \quad 0 \in (\mathbb{I}+\alpha\mathbb{A})x - (\mathbb{I}-\alpha\mathbb{B})x \\ &\Leftrightarrow \quad 0 \in (\mathbb{I}+\alpha\mathbb{A})x - \mathbb{R}_{\alpha\mathbb{B}}(\mathbb{I}+\alpha\mathbb{B})x \\ &\Leftrightarrow \quad 0 \in (\mathbb{I}+\alpha\mathbb{A})x - \mathbb{R}_{\alpha\mathbb{B}}z,\ z \in (\mathbb{I}+\alpha\mathbb{B})x \\ &\Leftrightarrow \quad \mathbb{R}_{\alpha\mathbb{B}}z \in (\mathbb{I}+\alpha\mathbb{A})\mathbb{J}_{\alpha\mathbb{B}}z,\ x = \mathbb{J}_{\alpha\mathbb{B}}z \\ &\Leftrightarrow \quad \mathbb{J}_{\alpha\mathbb{A}}\mathbb{R}_{\alpha\mathbb{B}}z = \mathbb{J}_{\alpha\mathbb{B}}z,\ x = \mathbb{J}_{\alpha\mathbb{B}}z \\ &\Leftrightarrow \quad \mathbb{R}_{\alpha\mathbb{A}}\mathbb{R}_{\alpha\mathbb{B}}z = z,\ x = \mathbb{J}_{\alpha\mathbb{B}}z, \end{aligned}$$

where we have used (2.13). So x is a solution if and only if there is a $z \in \operatorname{Fix}\mathbb{R}_{\alpha\mathbb{A}}\mathbb{R}_{\alpha\mathbb{B}}$ and $x = \mathbb{J}_{\alpha\mathbb{B}}z$. This splitting is called *Peaceman–Rachford splitting* (PRS).

Since the operator $\mathbb{R}_{\alpha\mathbb{A}}\mathbb{R}_{\alpha\mathbb{B}}$ is merely nonexpansive, the FPI with PRS

$$z^{k+1} = \mathbb{R}_{\alpha\mathbb{A}}\mathbb{R}_{\alpha\mathbb{B}}(z^k) \tag{2.14}$$

is not guaranteed to converge. See Exercise 2.27.

To ensure convergence, we average. For any $\alpha > 0$, we have

$$0 \in (\mathbb{A}+\mathbb{B})x \quad \Leftrightarrow \quad \left(\frac{1}{2}\mathbb{I} + \frac{1}{2}\mathbb{R}_{\alpha\mathbb{A}}\mathbb{R}_{\alpha\mathbb{B}}\right)(z) = z,\ x = \mathbb{J}_{\alpha\mathbb{B}}(z).$$

This splitting is called *Douglas–Rachford splitting* (DRS).

The FPI with DRS

$$\begin{aligned} x^{k+1/2} &= \mathbb{J}_{\alpha\mathbb{B}}(z^k) \\ x^{k+1} &= \mathbb{J}_{\alpha\mathbb{A}}(2x^{k+1/2} - z^k) \\ z^{k+1} &= z^k + x^{k+1} - x^{k+1/2} \end{aligned}$$

converges for any $\alpha > 0$ if $\operatorname{Zer}(\mathbb{A}+\mathbb{B}) \neq \emptyset$. See Exercise 2.26.

We can think of the $x^{k+1/2}$- and x^{k+1}-iterates as estimates of a solution with different properties. For example, if $\mathbb{J}_{\alpha\mathbb{B}}$ is a projection onto a constraint set, $x^{k+1/2}$-iterates satisfy these constraints exactly.

Davis–Yin Splitting

Consider the problem

$$\underset{x\in\mathbb{R}^n}{\text{find}} \quad 0 \in (\mathbb{A}+\mathbb{B}+\mathbb{C})x,$$

where $\mathbb{A}$, $\mathbb{B}$, and $\mathbb{C}$ are maximal monotone, and $\mathbb{C}$ is single-valued.

Then for any $\alpha > 0$, we have

$$\begin{aligned}
0 \in (\mathbb{A}+\mathbb{B}+\mathbb{C})x \quad &\Leftrightarrow \quad 0 \in (\mathbb{I}+\alpha\mathbb{A})x - (\mathbb{I}-\alpha\mathbb{B})x + \alpha\mathbb{C}x \\
&\Leftrightarrow \quad 0 \in (\mathbb{I}+\alpha\mathbb{A})x - \mathbb{R}_{\alpha\mathbb{B}}(\mathbb{I}+\alpha\mathbb{B})x + \alpha\mathbb{C}x \\
&\Leftrightarrow \quad 0 \in (\mathbb{I}+\alpha\mathbb{A})x - \mathbb{R}_{\alpha\mathbb{B}}z + \alpha\mathbb{C}x,\ z \in (\mathbb{I}+\alpha\mathbb{B})x \\
&\Leftrightarrow \quad (\mathbb{R}_{\alpha\mathbb{B}} - \alpha\mathbb{C}\mathbb{J}_{\alpha\mathbb{B}})z \in (\mathbb{I}+\alpha\mathbb{A})\mathbb{J}_{\alpha\mathbb{B}}z, \\
&\qquad\ \ x = \mathbb{J}_{\alpha\mathbb{B}}z \\
&\Leftrightarrow \quad \mathbb{J}_{\alpha\mathbb{A}}(\mathbb{R}_{\alpha\mathbb{B}} - \alpha\mathbb{C}\mathbb{J}_{\alpha\mathbb{B}})z = \mathbb{J}_{\alpha\mathbb{B}}z,\ x = \mathbb{J}_{\alpha\mathbb{B}}z \\
&\Leftrightarrow \quad (\mathbb{R}_{\alpha\mathbb{A}}(\mathbb{R}_{\alpha\mathbb{B}} - \alpha\mathbb{C}\mathbb{J}_{\alpha\mathbb{B}}) - \alpha\mathbb{C}\mathbb{J}_{\alpha\mathbb{B}})z = z, \\
&\qquad\ \ x = \mathbb{J}_{\alpha\mathbb{B}}z \\
&\Leftrightarrow \quad ((1/2)\mathbb{I} + (1/2)\mathbb{T})z = z,\ x = \mathbb{J}_{\alpha\mathbb{B}}z, \\
&\qquad\ \ \mathbb{T} = \mathbb{R}_{\alpha\mathbb{A}}(\mathbb{R}_{\alpha\mathbb{B}} - \alpha\mathbb{C}\mathbb{J}_{\alpha\mathbb{B}}) - \alpha\mathbb{C}\mathbb{J}_{\alpha\mathbb{B}}.
\end{aligned}$$

So, x is a solution if and only if there is a $z \in \operatorname{Fix}((1/2)\mathbb{I} + (1/2)\mathbb{T})$ and $x = \mathbb{J}_{\alpha\mathbb{B}}z$. This splitting is called *Davis–Yin splitting* (DYS). We can also write

$$(1/2)\mathbb{I} + (1/2)\mathbb{T} = \mathbb{I} - \mathbb{J}_{\alpha\mathbb{B}} + \mathbb{J}_{\alpha\mathbb{A}}(\mathbb{R}_{\alpha\mathbb{B}} - \alpha\mathbb{C}\mathbb{J}_{\alpha\mathbb{B}}).$$

Assume $\mathbb{C}$ is β-cocoercive and $\alpha \in (0, 2\beta)$, then $(1/2)\mathbb{I} + (1/2)\mathbb{T}$ is averaged. We prove this in §13 as Theorem 28. $\mathbb{T}$ itself may not be nonexpansive. The FPI with DYS

$$\begin{aligned}
x^{k+1/2} &= \mathbb{J}_{\alpha\mathbb{B}}(z^k) \\
x^{k+1} &= \mathbb{J}_{\alpha\mathbb{A}}(2x^{k+1/2} - z^k - \alpha\mathbb{C}x^{k+1/2}) \\
z^{k+1} &= z^k + x^{k+1} - x^{k+1/2}
\end{aligned}$$

converges for $\alpha \in (0, 2\beta)$ if $\operatorname{Zer}(\mathbb{A}+\mathbb{B}+\mathbb{C}) \neq \emptyset$. Note that DYS reduces to BFS when $\mathbb{A} = 0$, to FBS when $\mathbb{B} = 0$, and to DRS when $\mathbb{C} = 0$.

2.7.2 Splitting for Convex Optimization and Total Duality

In §3, we combine the base splittings with various techniques to derive a wide range of methods. In this section, we directly apply the base splittings to convex optimization problems as is.

Proximal Gradient Method

Consider the problem

$$\underset{x\in\mathbb{R}^n}{\text{minimize}} \quad f(x) + g(x),$$

where f and g are CCP functions on $\mathbb{R}^n$ and f is differentiable. Then x is a solution if and only if $x \in \operatorname{Zer}(\nabla f + \partial g)$.

The FPI with FBS is

$$x^{k+1} = \operatorname{Prox}_{\alpha g}(x^k - \alpha \nabla f(x^k)),$$

which is also called the *proximal gradient method*. Assume a primal solution exists, f is L-smooth, and $\alpha \in (0, 2/L)$. Then $x^k \to x^\star$.

We can write the proximal gradient method equivalently as

$$x^{k+1} = \operatorname*{argmin}_x \left\{ f(x^k) + \langle \nabla f(x^k), x - x^k \rangle + g(x) + \frac{1}{2\alpha} \|x - x^k\|_2^2 \right\}.$$

So, the proximal gradient method uses a first-order approximation of f about x^k.

When $g = \delta_C$ for some nonempty convex set C, the proximal gradient method reduces to the *projected gradient method*:

$$x^{k+1} = \Pi_C(x^k - \alpha \nabla f(x^k)).$$

DRS for Convex Optimization and Total Duality

Consider the primal-dual problem pair

$$\underset{x \in \mathbb{R}^n}{\text{minimize}} \quad f(x) + g(x) \tag{2.15}$$

and

$$\underset{u \in \mathbb{R}^n}{\text{maximize}} \quad -f^*(-u) - g^*(u) \tag{2.16}$$

generated by the Lagrangian

$$\mathbf{L}(x, u) = f(x) + \langle x, u \rangle - g^*(u), \tag{2.17}$$

where f and g are CCP functions on $\mathbb{R}^n$.

As we soon prove, the primal problem is equivalent to

$$\underset{x \in \mathbb{R}^n}{\text{find}} \quad 0 \in (\partial f + \partial g)x$$

when total duality holds. The FPI with DRS is

$$\begin{aligned} x^{k+1/2} &= \operatorname{Prox}_{\alpha g}(z^k) \\ x^{k+1} &= \operatorname{Prox}_{\alpha f}(2x^{k+1/2} - z^k) \\ z^{k+1} &= z^k + x^{k+1} - x^{k+1/2}. \end{aligned} \tag{2.18}$$

Assume total duality holds and $\alpha > 0$. Then $x^k \to x^\star$ and $x^{k+1/2} \to x^\star$. In §9, we furthermore show that fixed points are of the form $z^\star = x^\star + \alpha u^\star$. So, $z^k \to x^\star + \alpha u^\star$.

The FPI with DRS requires f and g to be CCP, and the method converges for all $\alpha > 0$. In contrast, the proximal gradient method furthermore requires f to be L-smooth, and the parameter α must lie within a specific range. DRS is useful when evaluating $\operatorname{Prox}_{\alpha f}$ and $\operatorname{Prox}_{\alpha g}$ is easy but evaluating $\operatorname{Prox}_{\alpha(f+g)}$ is not. The proximal gradient method is useful when evaluating ∇f and $\operatorname{Prox}_{\alpha g}$ is easy. The proximal point method is useful when evaluating $\operatorname{Prox}_{\alpha(f+g)}$ is easy.

Note that although the primal problem (2.15) is symmetric in f and g, the dual problem (2.16) is not. Swapping the roles of f and g changes the sign of the dual variable. The algorithm (2.18) is also not symmetric in f and g, and swapping the roles of f and g changes the sign of the dual variable in $z^k \to x^\star + \alpha u^\star$.

DYS for Convex Optimization and Total Duality

Consider the primal-dual problem pair

$$\underset{x\in\mathbb{R}^n}{\text{minimize}} \quad f(x) + g(x) + h(x)$$

and

$$\underset{u\in\mathbb{R}^n}{\text{maximize}} \quad -(f+h)^*(-u) - g^*(u)$$

generated by the Lagrangian

$$\mathbf{L}(x,u) = f(x) + h(x) + \langle x,u\rangle - g^*(u).$$

The FPI with DYS is

$$\begin{aligned} x^{k+1/2} &= \mathrm{Prox}_{\alpha g}(z^k) \\ x^{k+1} &= \mathrm{Prox}_{\alpha f}(2x^{k+1/2} - z^k - \alpha\nabla h(x^{k+1/2})) \\ z^{k+1} &= z^k + x^{k+1} - x^{k+1/2}. \end{aligned}$$

Assume total duality holds, h is L-smooth, and $\alpha \in (0, 2/L)$. Then $x^k \to x^\star$ and $x^{k+1/2} \to x^\star$. In §9, we furthermore show that fixed points are of the form $z^\star = x^\star + \alpha u^\star$. So, $z^k \to x^\star + \alpha u^\star$.

Necessity and Sufficiency of Total Duality

The following equivalence summarizes the role of total duality in splitting methods:

$$\mathrm{argmin}(f+g) = \mathrm{Zer}\,(\partial f + \partial g) \neq \emptyset \quad \Leftrightarrow \quad \text{total duality holds between (2.15) and (2.16).}$$

Therefore, we can write

$$\underset{x\in\mathbb{R}^n}{\text{minimize}} \quad f(x) + g(x) \quad \Leftrightarrow \quad \underset{x\in\mathbb{R}^n}{\text{find}} \quad 0 \in (\partial f + \partial g)(x)$$

when total duality holds. This fact explains why total duality is required for the convergence of so many operator splitting methods.

Let us see why. First, assume that total duality holds. Then $x^\star \in \mathrm{argmin}(f+g)$ if and only if $(x^\star, u^\star)$ is a saddle point of

$$\mathbf{L}(x,u) = f(x) + \langle x,u\rangle - g^*(u)$$

for some $u^\star \in \mathbb{R}^n$, and

$$\begin{aligned} (x^\star, u^\star) \text{ is a saddle point of } \mathbf{L} \quad &\Leftrightarrow \quad 0 \in \partial\mathbf{L}(x^\star, u^\star) \\ &\Leftrightarrow \quad 0 \in \partial_x\mathbf{L}(x^\star, u^\star),\ 0 \in \partial_u(-\mathbf{L})(x^\star, u^\star) \\ &\Leftrightarrow \quad -u^\star \in \partial f(x^\star),\ u^\star \in \partial g(x^\star) \\ &\Leftrightarrow \quad 0 \in (\partial f + \partial g)(x^\star). \end{aligned}$$

We conclude that $\mathrm{argmin}(f+g) = \mathrm{Zer}\,(\partial f + \partial g) \neq \emptyset$.

Next, assume $\operatorname{argmin}(f+g) = \operatorname{Zer}(\partial f + \partial g) \neq \emptyset$. Then any $x^\star \in \operatorname{argmin}(f+g)$ satisfies $0 \in (\partial f + \partial g)(x^\star)$. By a similar chain of arguments, $(x^\star, u^\star)$ is a saddle point of $\mathbf{L}$ for some $u^\star \in \mathbb{R}^n$, and we conclude that total duality holds.

2.7.3 Discussion

Fixed-Point Encoding

A *fixed-point encoding* establishes a correspondence between solutions of a monotone inclusion problem and fixed points of a related operator. The splittings we discussed are fixed-point encodings.

Upon reading §2.7.1, one may ask why there is no "forward-forward" splitting. A "forward-forward splitting" of the form $\mathbb{I} - \alpha(\mathbb{A} + \mathbb{B})$ is an instance of the forward-step method. A "forward-forward splitting" of the form $(\mathbb{I} - \alpha\mathbb{A})(\mathbb{I} - \beta\mathbb{B})$ would not be a valid fixed-point encoding; that is, we cannot recover a zero of $A + B$ from a fixed point of $(\mathbb{I} - \alpha\mathbb{A})(\mathbb{I} - \beta\mathbb{B})$. Likewise, a "backward-backward splitting" of the form $\mathbb{J}_{\alpha\mathbb{A}}\mathbb{J}_{\alpha\mathbb{B}}$ is not a valid fixed-point encoding. See Exercise 2.28.

Why Use the Resolvent?

The splittings we discuss use resolvents or direct evaluations of single-valued operators. Why do we not use other operators such as $(\mathbb{I} - \alpha\mathbb{A})^{-1}$? One reason is computational convenience. The resolvent is often easy to evaluate for many interesting operators, while evaluating something like $(\mathbb{I} - \alpha\partial f)^{-1}$ is often difficult.

Another reason is that only single-valued operators are, in a sense, algorithmically actionable. On a computer, we can compute and store a vector in $\mathbb{R}^n$, but we cannot store a subset of $\mathbb{R}^n$ in most cases. While multi-valued operators are a useful mathematical concept, single-valued operators, such as resolvents, are more algorithmically useful.

The Role of Maximality

An FPI $x^{k+1} = \mathbb{T}x^k$ becomes undefined if its iterates ever escape the domain of $\mathbb{T}$. In §2.4.2, we implicitly assumed $\operatorname{dom}\mathbb{T} = \mathbb{R}^n$ through stating $\mathbb{T}\colon \mathbb{R}^n \to \mathbb{R}^n$. When the operators are maximal monotone, FPIs defined with resolvents do not run into this issue.

So, we assume maximality out of theoretical necessity, but in practice the non-maximal monotone operators, such as the gradient operator of a nonconvex function, are usually ones we cannot efficiently compute the resolvent for anyway. In other words, there is little need to consider resolvents of non-maximal monotone operators, theoretically or practically.

Computational Efficiency

These base splitting methods are useful when the operators used in the splitting are efficient to compute. For example, although the convergence of DRS iteration

$$z^{k+1} = \left(\frac{1}{2}\mathbb{I} + \frac{1}{2}\mathbb{R}_{\alpha\mathbb{A}}\mathbb{R}_{\alpha\mathbb{B}}\right) z^k$$

does not depend on the value of α, it is most useful when $\mathbb{R}_{\alpha\mathbb{A}}$ and $\mathbb{R}_{\alpha\mathbb{B}}$ can be computed efficiently.

For a given optimization problem, there is often more than one applicable method. The trick is to find a method using computationally efficient split components.

2.7.4 Methods

LASSO and ISTA

Consider the problem

$$\underset{x\in\mathbb{R}^n}{\text{minimize}} \quad \frac{1}{2}\|Ax-b\|^2+\lambda\|x\|_1,$$

for $A\in\mathbb{R}^{m\times n}$, $b\in\mathbb{R}^m$, and $\lambda>0$. This particular optimization problem is called LASSO. Let $S(x;\kappa)$ be the soft-thresholding operator of Example 1.12.

The FPI with DRS

$$\begin{aligned} x^{k+1/2} &= (I+\alpha A^\intercal A)^{-1}(z^k+\alpha A^\intercal b)\\ x^{k+1} &= S(2x^{k+1/2}-z^k;\alpha\lambda)\\ z^{k+1} &= z^k+x^{k+1}-x^{k+1/2}\end{aligned}$$

converges for any $\alpha>0$.

The FPI with FBS

$$x^{k+1}=S(x^k-\alpha A^\intercal(Ax^k-b);\alpha\lambda)$$

converges for $0<\alpha<2/\lambda_{\max}(A^\intercal A)$. This particular instance of the proximal gradient method is called the Iterative Shrinkage-Thresholding Algorithm (ISTA).

Note that DRS uses the matrix inverse $(I+\alpha A^\intercal A)^{-1}$, while FBS does not. When m and n are large, computing the matrix inverse can be prohibitively expensive. Therefore, FBS is the more computationally effective splitting for large-scale LASSO problems.

Consensus Technique

Consider the problem

$$\underset{x\in\mathbb{R}^n}{\text{minimize}} \quad \sum_{i=1}^m g_i(x),$$

where $g_1,\dots,g_m$ are CCP functions on $\mathbb{R}^n$. This problem is equivalent to

$$\begin{array}{ll}\underset{\mathbf{x}\in\mathbb{R}^{nm}}{\text{minimize}} & \sum_{i=1}^m g_i(x_i)\\ \text{subject to} & \mathbf{x}\in C,\end{array}$$

where $\mathbf{x}=(x_1,\dots,x_m)$ and

$$C=\{(x_1,\dots,x_m)\in\mathbb{R}^{nm}\,|\,x_1=\cdots=x_m\} \tag{2.19}$$

is the *consensus set*. In turn, this problem is equivalent to

$$\underset{\mathbf{x}\in\mathbb{R}^{nm}}{\text{find}} \quad 0\in\begin{bmatrix}\partial g_1(x_1)\\ \vdots\\ \partial g_m(x_m)\end{bmatrix}+\mathbf{N}_C(\mathbf{x}),$$

assuming $\bigcap_{i=1}^m \operatorname{int}\operatorname{dom} g_i\neq\emptyset$.

The projection onto the consensus set is simple averaging:

$$\Pi_C \mathbf{x} = \overline{\mathbf{x}} = (\overline{x}, \overline{x}, \ldots, \overline{x}), \qquad \overline{x} = \frac{1}{m}\sum_{i=1}^{m} x_i.$$

Define $\overline{\mathbf{z}}^k = \Pi_C \mathbf{z}^k$. The FPI with DRS for this setup

$$\begin{aligned} x_i^{k+1} &= \mathrm{Prox}_{\alpha g_i}(2\overline{z}^k - z_i^k - \alpha \nabla f_i(\overline{z}^k)) \qquad \text{for } i = 1, \ldots, m, \\ \mathbf{z}^{k+1} &= \mathbf{z}^k + \mathbf{x}^{k+1} - \overline{\mathbf{z}}^k, \end{aligned}$$

converges for any $\alpha > 0$, if $\bigcap_{i=1}^m \operatorname{int}\operatorname{dom} g_i \neq \emptyset$ and a solution exists. Since $\mathrm{Prox}_{\alpha g_i}$ for $i = 1, \ldots, m$ can be evaluated independently, this method is well-suited for parallel and distributed computing, which we discuss in §4.2.1 and §11.1.

The use of the consensus set (2.19) is called the *consensus technique* and it can more generally solve

$$\underset{x \in \mathbb{R}^n}{\text{find}} \quad 0 \in \sum_{i=1}^{m} \mathbb{A}_i x,$$

where $\mathbb{A}_1, \ldots, \mathbb{A}_m$ are maximal monotone. See Exercise 2.36.

Forward-Douglas–Rachford

Consider the problem

$$\underset{x \in \mathbb{R}^n}{\text{minimize}} \quad \sum_{i=1}^{m} (f_i(x) + g_i(x)),$$

where $g_1, \ldots, g_m$ are CCP and $f_1, \ldots, f_m$ are L-smooth. With the consensus technique, we can recast the problem into

$$\begin{array}{ll} \underset{\mathbf{x} \in \mathbb{R}^{nm}}{\text{minimize}} & \sum_{i=1}^{m} f_i(x_i) + \sum_{i=1}^{m} g_i(x_i) \\ \text{subject to} & \mathbf{x} \in C, \end{array}$$

where we use the same notation as we did for consensus optimization.

The FPI with DYS for this setup

$$\begin{aligned} x_i^{k+1} &= \mathrm{Prox}_{\alpha g_i}(2\overline{z}^k - z_i^k) \qquad \text{for } i = 1, \ldots, m, \\ \mathbf{z}^{k+1} &= \mathbf{z}^k + \mathbf{x}^{k+1} - \overline{\mathbf{z}}^k \end{aligned}$$

is called *generalized forward-backward* or *forward-Douglas–Rachford*. This method converges if total duality holds, $\bigcap_{i=1}^m \operatorname{int}\operatorname{dom} g_i \neq \emptyset$, and $\alpha \in (0, 2/L)$.

2.8 VARIABLE METRIC METHODS

In the theory we have developed so far, the Euclidean norm plays a special role. In the definition of the proximal operator

$$\mathrm{Prox}_f(x) = \underset{z}{\operatorname{argmin}} \left\{ f(z) + \frac{1}{2}\|z - x\|^2 \right\},$$

the $(1/2)\|z - x\|^2$ term, called the *proximal term*, is defined with the Euclidean norm. Theorem 1 is stated in terms of the Euclidean norm. *Variable metric* methods generalize many of the notions we have discussed so far with the M-norm.

One reason to consider this generalization is preconditioning. A good choice of the norm $\|\cdot\|_M$ can reduce the number of iterations needed for convergence. Variable metric methods are also useful when an operator $\mathbb{A}$ has structure and a well-chosen M cancels certain terms to make $(M+\mathbb{A})^{-1}$ easy to evaluate. We explore this technique thoroughly in §3.3.

Despite the name variable *metric* methods, the generalization works only with M-norms since they are the norms induced by the inner product $\langle x, y\rangle_M = x^\intercal My$. The analysis of this section does not extend to other metrics, such as the ℓ^1-norm.

Variable Metric Proximal Point Method

Let $\mathbb{A}$ be maximal monotone and $M \succ 0$. Then $M^{-1/2}\mathbb{A}M^{-1/2}$ is maximal monotone and the proximal point method

$$y^{k+1} = (\mathbb{I} + M^{-1/2}\mathbb{A}M^{-1/2})^{-1}y^k$$

converges.

With the change of variables $x^k = M^{-1/2}y^k$, we get

$$\begin{aligned}(\mathbb{I} + M^{-1/2}\mathbb{A}M^{-1/2})y^{k+1} &\ni y^k\\ (\mathbb{I} + M^{-1}\mathbb{A})x^{k+1} &\ni x^k.\end{aligned}$$

This gives us

$$\begin{aligned}x^{k+1} &= \mathbb{J}_{M^{-1}\mathbb{A}}x^k\\ &= (M + \mathbb{A})^{-1}Mx^k.\end{aligned}$$

We call this the *variable metric PPM*. The iterates x^k inherit the convergence properties of y^k. For example, the fact that $\|y^k - y^\star\|$ is monotonically nonincreasing translates to the fact that $\|x^k - x^\star\|_M$ is monotonically nonincreasing. Likewise, $\|x^{k+1} - x^k\|_M \to 0$ monotonically at rate $O(1/k)$.

When $\mathbb{A} = \partial f$, then

$$\mathbb{J}_{M^{-1}\partial f}(x) = \underset{z\in\mathbb{R}^d}{\operatorname{argmin}}\left\{f(z) + \frac{1}{2}\|z - x\|_M^2\right\}.$$

We can interpret the variable metric PPM as PPM performed with the norm $\|\cdot\|_M$ instead of the Euclidean norm.

Variable Metric Forward-Backward Splitting

Let $\mathbb{A}$ and $\mathbb{B}$ be maximal monotone and let $\mathbb{A}$ be single-valued. Then with the same reasoning, we can use a change of variables to write the FBS FPI with respect to $M^{-1/2}\mathbb{A}M^{-1/2}$ and $M^{-1/2}\mathbb{B}M^{-1/2}$ as

$$\begin{aligned}x^{k+1} &= (M + \mathbb{B})^{-1}(M - \mathbb{A})x^k\\ &= \mathbb{J}_{M^{-1}\mathbb{B}}(\mathbb{I} - M^{-1}\mathbb{A})x^k.\end{aligned}$$

We call this splitting *variable metric FBS*. This method converges if $\mathbb{I} - M^{-1/2}\mathbb{A}M^{-1/2}$ is averaged.

When $\mathbb{A} = \nabla f$ and $\mathbb{B} = \partial g$, then

$$\mathbb{J}_{M^{-1}\partial g}(\mathbb{I} - M^{-1}\nabla f)x = \underset{z\in\mathbb{R}^d}{\operatorname{argmin}} \left\{ g(z) + \langle \nabla f(x), z\rangle + \frac{1}{2}\|z - x\|_M^2 \right\}.$$

We can interpret the variable metric FBS as the proximal gradient method performed with the norm $\|\cdot\|_M$ instead of the Euclidean norm.

If $\mathbb{A}$ is β-cocoercive, then $M^{-1/2}\mathbb{A}M^{-1/2}$ is $(\beta/\|M^{-1}\|)$-cocoercive. See Exercise 2.9. Therefore, the FPI with variable metric FBS converges if $\|M^{-1}\| < 2\beta$.

Averagedness with Respect to $\|\cdot\|_M$

Assume $M \succ 0$. We say $\mathbb{T}$ is nonexpansive in $\|\cdot\|_M$ if

$$\|\mathbb{T}x - \mathbb{T}y\|_M \le \|x - y\|_M \qquad \forall x, y \in \operatorname{dom}\mathbb{T}.$$

For $\theta \in (0,1)$, we say $\mathbb{T}$ is θ-averaged in $\|\cdot\|_M$ if $\mathbb{T} = (1-\theta)\mathbb{I} + \theta\mathbb{S}$ for some $\mathbb{S}$ that is nonexpansive in $\|\cdot\|_M$. We say $\mathbb{T}$ is averaged in $\|\cdot\|_M$ if it is θ-averaged in $\|\cdot\|_M$ for some unspecified $\theta \in (0,1)$.

The operator $M^{-1/2}\mathbb{T}M^{-1/2}$ is nonexpansive (in $\|\cdot\|$) if and only if $M^{-1}\mathbb{T}$ is nonexpansive in $\|\cdot\|_M$. This is easy to verify since

$$\|M^{-1/2}\mathbb{T}M^{-1/2}x - M^{-1/2}\mathbb{T}M^{-1/2}y\|^2 \le \|x - y\|^2$$

is equivalent to

$$\|M^{-1}\mathbb{T}\tilde{x} - M^{-1}\mathbb{T}\tilde{y}\|_M^2 \le \|\tilde{x} - \tilde{y}\|_M^2$$

with the change of variables $M^{-1/2}x = \tilde{x}$ and $M^{-1/2}y = \tilde{y}$.

2.9 COMMONLY USED FORMULAS

For later convenience, we list a few commonly used formulas derived in this section.

- If $g(y) = f^*(A^\intercal y)$, where f is CCP and $\mathcal{R}(A^\intercal) \cap \operatorname{ri}\operatorname{dom} f^* \neq \emptyset$, then

$$u \in \partial g(y) \quad \Leftrightarrow \quad \begin{aligned} &x \in \operatorname{argmin}_z \{f(z) - \langle y, Az\rangle\} \\ &u = Ax. \end{aligned} \tag{2.2}$$

- If $g(y) = f^*(A^\intercal y)$, where f is CCP and $\mathcal{R}(A^\intercal) \cap \operatorname{ri}\operatorname{dom} f^* \neq \emptyset$, then

$$v = \operatorname{Prox}_{\alpha g}(u) \quad \Leftrightarrow \quad \begin{aligned} &x \in \operatorname{argmin}_x \{f(x) - \langle u, Ax\rangle + \tfrac{\alpha}{2}\|Ax\|^2\} \\ &v = u - \alpha Ax. \end{aligned} \tag{2.6}$$

- Let $\mathbf{L}(x,u) = f(x) + \langle u, Ax - b\rangle$ and let $\mathbf{L}_\alpha$ be the augmented Lagrangian of (1.11). Then

$$\mathbb{J}_{\alpha\partial\mathbf{L}}(x,u) = (y,v) \quad \Leftrightarrow \quad \begin{aligned} &y = \operatorname{argmin}_z \{\mathbf{L}_\alpha(z,u) + \tfrac{1}{2\alpha}\|z - x\|^2\} \\ &v = u + \alpha(Ay - b). \end{aligned} \tag{2.7}$$

- If $\mathbb{B}(x) = \mathbb{A}(x) + t$, where $\mathbb{A}$ is maximal monotone and $\alpha > 0$, then

$$\mathbb{J}_{\alpha\mathbb{B}}(u) = \mathbb{J}_{\alpha\mathbb{A}}(u - \alpha t). \tag{2.8}$$

- If $\mathbb{B}(x) = \mathbb{A}(x - t)$, where $\mathbb{A}$ is maximal monotone and $\alpha > 0$, then

$$\mathbb{J}_{\alpha\mathbb{B}}(u) = \mathbb{J}_{\alpha\mathbb{A}}(u - t) + t. \tag{2.9}$$

- If $\mathbb{B}(x) = -\mathbb{A}(t - x)$, where $\mathbb{A}$ is maximal monotone and $\alpha > 0$, then

$$\mathbb{J}_{\alpha\mathbb{B}}(u) = t - \mathbb{J}_{\alpha\mathbb{A}}(t - u). \tag{2.10}$$

- Inverse resolvent identity: If $\mathbb{A}$ is maximal monotone and $\alpha > 0$, then

$$\mathbb{J}_{\alpha^{-1}\mathbb{A}}(x) + \alpha^{-1}\mathbb{J}_{\alpha\mathbb{A}^{-1}}(\alpha x) = x. \tag{2.11}$$

- Moreau identity: If f is CCP and $\alpha > 0$, then

$$\mathrm{Prox}_{\alpha^{-1}f}(x) + \alpha^{-1}\mathrm{Prox}_{\alpha f^*}(\alpha x) = x. \tag{2.12}$$

BIBLIOGRAPHICAL NOTES

There are many classical and recent review papers based on the core insight that monotone operators serve as an elegant and unifying abstraction in the analysis of optimization algorithms: Lemaire and Penot in 1989 [LP89], Iusem in 1999 [Ius99], Combettes in 2004 [Com04], Combettes and Wajs in 2005 [CW05], Combettes and Pesquet in 2011[CP11b], Combettes, Condat, Pesquet, and Vũ in 2014 [CCPV14], Komodakis and Pesquet in 2015 [KP15], Clason and Valkonen in 2020 [CV20], and Condat, Kitahara, Contreras, and Hirabayashi in 2020 [CKCH22]. This book is largely influenced by these prior treatments.

Early Development: Basic Notions The notion of monotonicity was first formalized by Zarantonello in 1960 [Zar60]. The fact that derivatives of convex functions on $\mathbb{R}$ are nondecreasing was established by Jensen in 1906 [Jen06], and this monotonicity property was extended to gradients of convex functions on higher-dimensional spaces by Kačurovskiĭ in 1960 [Kac60] and Minty in 1962 [Min62]. The notion of *maximal* monotonicity was first established by Minty in 1962 [Min62]. Maximal monotonicity of subdifferentials of CCP functions on Hilbert spaces (and thus on $\mathbb{R}^n$) was established by Minty in 1964 [Min64] and Moreau in 1965 [Mor65]. This maximality result was generalized to convex functions on Banach spaces by Rockafellar [Roc66, Roc70b].

Fenchel's identity (2.1) was first presented by Fenchel in 1951 in his lectures [Fen53, Section 5]. The proximal operator was first introduced by Moreau in 1962 [Mor62, Mor65], and the Moreau identity was introduced in 1965 [Mor65]. The proof of $\mathrm{dom}\,\mathbb{J}_{\mathbb{A}} = \mathbb{R}^n$ when $\mathbb{A}$ is maximal monotone, the Minty surjectivity theorem, was established by Minty in 1962 [Min62]. The $(1/2)$-averagedness of resolvents was first discussed by Browder and Petryshyn in 1967 [BP67].

The study of convex-concave saddle functions and their saddle subdifferentials was pioneered by Rockafellar. His work started in the 1960s [Roc64, Roc68], and the maximal monotonicity of "closed proper" saddle subdifferentials was established in 1970 [Roc70a].

The augmented Lagrangian was used in [Hes69, Pow69] and later studied by Rockafellar in the late 1970s [Roc76b, Roc78].

Early Development: Methods Gradient descent dates back to Cauchy in 1847 [Cau47]. Fixed-point iterations date back to Picard, Lindelöf, and Banach in the late 1800s and early 1900s [Pic90, Lin94, Ban22]. The proximal point method was first studied in the 1970s [Mar70, Mar72b, Roc76b, BL78], and its convergence rate in terms of function values was later studied by Güler in 1991 [Gül91]. The method of multipliers was first presented in 1969 by Hestenes and Powell [Hes69, Pow69] and was interpreted as an instance of PPM by Rockafellar in 1973 [Roc73]. Dual ascent was first presented by Uzawa in 1972 [AHU58] and was later further studied by Tseng, Bertsekas, and Tsitsiklis [TB87, Tse90a]. The projected gradient method was first presented in the 1960s by Goldstein, Levitin, and Polyak [Gol64, LP66]. The forward step method is due to Bruck in 1977 [Bru77] and forward-backward splitting in its operator theoretic form was first presented in the 70s by Bruck and Passty [Bru77, Pas79]. In modern literature, FBS applied to the sum of two convex functions has been referred to as the proximal-gradient method [CW05].

Peaceman–Rachford and Douglas–Rachford splitting methods were first presented as splitting methods to solve the heat equation in 1955 and 1956 [PR55, DR56]. In 1979, Lions and Mercier generalized the technique to a sum of two maximal monotone operators [LM79]. The effort of combining Douglas–Rachford and Forward–Backward splitting schemes was initiated by Raguet, Fadili, and Peyré [RFP13, Rag19], extended by Briceño-Arias [Bri15], and completed by Davis and Yin [DY17b] as they proved averagedness in the general case with two maximal monotone operators and one cocoercive operator. This splitting method, which we refer to as Davis–Yin splitting, is also called the Forward-Douglas–Rachford splitting.

As we explore further in §3, many of the splitting methods are intimately connected. Since the DRS operator is firmly nonexpansive, it is a resolvent of a maximal monotone operator, and this was first pointed out by Lawrence and Spingarn in 1987 [LS87] and later by Eckstein and Bertsekas in 1992 [EB92]. That the gradient update can be viewed as the proximal operator of the function's first-order approximation, as discussed in §2.7.2, was first identified by Polyak in 1987 [Pol87].

Fixed-Point Iteration The FPI analyzed in Theorem 1 is also called the Krasnosel'skiĭ–Mann iteration. In 1953, Mann showed that the FPI converges when $n = 1$, $C \subset \mathbb{R}$ is a compact interval, and $T\colon C \to C$ is 1/2-averaged. In 1955, Krasnosel'skiĭ established convergence when $C \subset \mathbb{R}^n$ is compact and $T\colon C \to C$ is 1/2-averaged [Kra55]. In 1957, Schaefer extended Krasnosel'skiĭ's result to θ-averaged operators with $\theta \in (0, 1)$ [Sch57]. The general convergence result of Theorem 1 (without any compactness assumption) is due to Martinet's 1972 work [Mar72a, Théorème 5.5.2]. A key component of our (and Martinet's) proof is the subsequence convergence argument of Stage 2, which is due to Opial's 1967 work [Opi67]. In fact, Theorem 1 of [Opi67] captures this subsequence argument and is known as Opial's lemma. The notion of averaged operators was first formally defined in 1978 by Baillon, Bruck, and Reich [BBR78].

Infinite-Dimensional Analysis Although we focus on finite-dimensional spaces in this book, much of the monotone operator theory is developed in the infinite-dimensional setup, where a new set of interesting challenges arise. For example, the convergence $x^k \to x^\star$ of Theorem 1 becomes weak when the underlying space is an infinite-dimensional Hilbert space instead of $\mathbb{R}^n$. Bauschke and Combette's textbook [BC17a] provides a thorough treatment for operators on Hilbert spaces. Works on other setups include Reich and Shoikhet's [Rei79, RS98] work studying averaged operators in Banach spaces and Goebel and Reich's work [GR84, Rei85] studying averaged operators on the Hilbert ball with the hyperbolic metric.

Forward and Backward Nomenclature and Gradient-Flow The operators $\mathbb{I} - \alpha\mathbb{A}$ and $(\mathbb{I} + \alpha\mathbb{A})^{-1}$ are respectively called forward and backward steps in analogy to the forward and backward Euler discretizations of $\dot{x}(t) = -\mathbb{A}x(t)$, a continuous-time differential equation defined for single-valued $\mathbb{A}$. This interpretation is due to Lamaire and Penot [LP89, Lem92] and Eckstein [Eck89, §3.2.2] in 1989. However, the gradient flow $\dot{x}(t) = -\nabla f(x(t))$ for functions f itself was studied earlier by Bruck in 1975 [Bru75a] and Botsaris and Jacobson in 1976 [BJ76].

Consensus Technique The first use of the consensus technique, also called the *product space trick*, seems to be due to Pierra in 1984 [Pie84] and Spingarn in 1983 through the "method of partial inverses" [Spi83, Spi85]. The use of the technique for distributed optimization and machine learning was popularized through the works of Boyd, Parikh, Chu, Peleato, and Eckstein [BPC+11, PB14b, PB14a].

Variable Metric Methods The variable metric proximal point method can be thought of as a special case of the Bregman proximal point method, which was first presented by Censor and Zenios for minimizing convex functions [CZ92] and Burachik and Iusem for monotone inclusions [BI98]. Other early work includes that of Chen and Teboulle [CT93], Bonnans, Gilbert, Lemaréchal, and Sagastizábal [BGLS95], Parente, Lotito, and Solodov [PLS08], and He and Yuan [HY12b]. Variable metric forward-backward splitting was first formalized by Combettes and Vũ [CV14]. A block coordinate extension was given by Chouzenoux, Pesquet, and Repetti [CPR16]. Liu and Yin [LY19] used variable metrics to analyze the Davis–Yin splitting for smooth nonconvex problems. Vũ [Vũ13b] proposed variable metric extensions of Tseng's forward-backward-forward splitting. Briceño-Arias and Davis [BD18] proposed variable metric extensions of their forward-backward-half forward splitting. A different approach to apply variable metrics was introduced by Burke and Qian [BQ99].

LASSO Application LASSO (least absolute shrinkage and selection operator) first introduced in geophysics literature in 1986 [SS86]. It was later independently rediscovered, popularized, and named LASSO by the statistician Tibshirani in 1996 [Tib96]. LASSO is one of the main models of compressed sensing [Don06, CT05, CT06] when the sensing is corrupted by noise or the signal to sense is approximately sparse.

Early work regarding the computation of LASSO includes [EHJT04, FNW07, HYZ08, YOGD08]. The Nesterov acceleration to the iterative soft thresholding algorithm was introduced in [BT09].

EXERCISES

2.1 *When $\mathbb{T}^{-1}$ is a left-inverse of $\mathbb{T}$.* Show that if $x \in \operatorname{dom} \mathbb{T}$ and $\mathbb{T}^{-1}$ is single-valued, then $\mathbb{T}^{-1}\mathbb{T}x = x$.

2.2 *Non-maximal subdifferential.* Consider the function f on $\mathbb{R}$ defined as

$$f(x) = \begin{cases} \infty & \text{for } x < 0 \\ 1 & \text{for } x = 0 \\ 0 & \text{for } x > 0. \end{cases}$$

Show that f is convex and proper but not closed. Show that ∂f is not maximal.

2.3 *Monotonicity of saddle subdifferential.* Assume $\mathbf{L}\colon \mathbb{R}^n \times \mathbb{R}^m \to \mathbb{R}$ and $\mathbf{L}(x,u)$ is convex-concave. Recall $\partial\mathbf{L}$ is defined in (2.3). Show that $\partial\mathbf{L}$ is monotone.
Hint. Add the four subgradient inequalities that lower bound
- $\mathbf{L}(x_2,u_1)$ with a subgradient of $\mathbf{L}(\cdot,u_1)$ at x_1
- $-\mathbf{L}(x_1,u_2)$ with a subgradient of $-\mathbf{L}(x_1,\cdot)$ at u_1
- $\mathbf{L}(x_1,u_2)$ with a subgradient of $\mathbf{L}(\cdot,u_2)$ at x_2
- $-\mathbf{L}(x_2,u_1)$ with a subgradient of $-\mathbf{L}(x_2,\cdot)$ at u_2

to show

$$\langle \partial_x\mathbf{L}(x_1,u_1) - \partial_x\mathbf{L}(x_2,u_2), x_1 - x_2\rangle + \langle \partial_u(-\mathbf{L}(x_1,u_1)) - \partial_u(-\mathbf{L}(x_2,u_2)), u_1 - u_2\rangle \geq 0.$$

2.4 *Maximality of continuous monotone operators.* Show that if $\mathbb{T}\colon \mathbb{R}^n \to \mathbb{R}^n$ is continuous and monotone, then $\mathbb{T}$ is maximal.
Hint. Assume for contradiction that there is a pair $(y,v) \notin \mathbb{T}$ such that

$$0 \leq \langle v - \mathbb{T}x, y - x\rangle$$

for all $x \in \mathbb{R}^n$. Plug in $x = y - \delta$ and use continuity of $\mathbb{T}$ to argue

$$0 \leq \langle v - \mathbb{T}(y-\delta), \delta\rangle = \langle v - \mathbb{T}y, \delta\rangle + o(\|\delta\|)$$

as $\delta \to 0$. Argue that $v = \mathbb{T}y$ and draw a contradiction.

2.5 Show that if f is a strictly convex CCP function, then (i) ∂f^* is single-valued and (ii) f^* is differentiable on $\operatorname{int}\operatorname{dom} f^*$.
Remark. Since f^* is CCP, f^* is subdifferentiable on $\operatorname{int}\operatorname{dom} f^*$ and $\partial f^*(u)$ is a singleton if and only if f^* is differentiable at u.

2.6 *Recovering a primal solution from a dual solution.* Let f be a strictly convex CCP function on $\mathbb{R}^n$, g a CCP function on $\mathbb{R}^m$, and $A \in \mathbb{R}^{m\times n}$. Consider the primal problem

$$\underset{x\in\mathbb{R}^n}{\text{minimize}} \quad f(x) + g(Ax)$$

and dual problem

$$\underset{u\in\mathbb{R}^m}{\text{maximize}} \quad -f^*(-A^\intercal u) - g^*(u)$$

generated by the Lagrangian

$$\mathbf{L}(x,u) = f(x) + \langle Ax, u\rangle - g^*(u).$$

Assume total duality holds. Show that $\nabla f^*(-A^\intercal u^\star)$ is a primal solution.
Hint. Use Exercise 2.5.
Remark. Without the strict convexity, this statement is not true. The setting $n = 1$, $m = 1$, $f(x) = 0$, $A = 1$, $g(x) = \delta_{\{0\}}(x)$, and $\mathbf{L}(x,u) = xu$ is a counterexample: $x^\star = 0$ and $u^\star = 0$ are the unique primal and dual solutions, but $\partial f^*(-u^\star) = \mathbb{R}$.

2.7 *Differentiable monotone operators.* Show that a differentiable operator $\mathbb{T}\colon \mathbb{R}^n \to \mathbb{R}^n$ is monotone if and only if $D\mathbb{T}(x) + D\mathbb{T}(x)^\intercal \succeq 0$ for all x.
Hint. Assume $\mathbb{T}$ is monotone, and use

$$D\mathbb{T}(x)v = \lim_{h\to 0}\frac{1}{h}(\mathbb{T}(x+hv) - \mathbb{T}(x))$$

to show $v^\intercal D\mathbb{T}(x)v \geq 0$ for all $v \in \mathbb{R}^n$. For the other direction, assume $D\mathbb{T}(x)+D\mathbb{T}(x)^\intercal \succeq 0$ for all x, define $g(t) = \langle x-y, \mathbb{T}(tx+(1-t)y)\rangle$, and use the mean value theorem to show

$$\langle x-y, \mathbb{T}x - \mathbb{T}y\rangle = g(1) - g(0) = g'(\xi)$$

for some $\xi \in [0,1]$.

2.8 *Differentiable Lipschitz operators.* Show that a differentiable operator $\mathbb{T}\colon \mathbb{R}^n \to \mathbb{R}^n$ is L-Lipschitz if and only if $\sigma_{\max}(D\mathbb{T}(x)) \leq L$ for all x.
Hint. Assume $\sigma_{\max}(D\mathbb{T}(x)) \leq L$, define $g(t) = \mathbb{T}(tx+(1-t)y)$, and use the mean value theorem and the Cauchy–Schwartz inequality to get

$$\|\mathbb{T}x - \mathbb{T}y\|^2 = \langle \mathbb{T}x - \mathbb{T}y, g(1) - g(0)\rangle = \langle \mathbb{T}x - \mathbb{T}y, g'(\xi)\rangle \leq \|\mathbb{T}x - \mathbb{T}y\|\|g'(\xi)\|.$$

For the other direction, assume $\mathbb{T}$ has Lipschitz parameter L and use

$$\|D\mathbb{T}(x)v\| = \lim_{h\to 0}\frac{1}{h}\|\mathbb{T}(x+hv) - \mathbb{T}(x)\|.$$

2.9 Show that if $\mathbb{T}\colon \mathbb{R}^n \to \mathbb{R}^n$ is β-cocoercive and $M \in \mathbb{R}^{n\times n}$ is symmetric positive definite, then $M^{-1/2}\mathbb{T}M^{-1/2}$ is $(\beta/\|M^{-1}\|)$-cocoercive.

2.10 *Moreau envelope.* Let f be a CCP function on $\mathbb{R}^n$. For $\beta > 0$, define the *Moreau envelope* of f of parameter β as

$$^{\beta}f(x) = \inf_{z\in\mathbb{R}^n}\left\{f(z) + \frac{1}{2\beta}\|z-x\|^2\right\}.$$

Show that
(a) $^{\beta}f(x)$ is convex and proper,
(b) $\nabla {}^{\beta}f = \beta^{-1}(\mathbb{I} - \mathrm{Prox}_{\beta f})$,
(c) $^{\beta}f(x)$ is closed, and
(d) $^{\beta}f$ is $1/\beta$-smooth.
Hint. For (a), establish closedness with $^{\beta}f(x) = f(\mathrm{Prox}_{\beta f}(x)) + \frac{1}{2\beta}\|\mathrm{Prox}_{\beta f}(x) - x\|^2$ and the fact that $\mathrm{Prox}_{\beta f}(x)$ is well defined. For (b), note that

$$^{\beta}f(x) = \frac{1}{2\beta}\|x\|^2 - \frac{1}{\beta}\sup_{z\in\mathbb{R}^n}\left\{\langle x,z\rangle - \beta f(z) - \frac{1}{2}\|z\|^2\right\},$$

and the supremum can be written as a conjugate. Take the gradient of both sides. For (c), use the fact that $^{\beta}f$ is differentiable and therefore continuous. For (d), use the Moreau identity to show $\beta\nabla {}^{\beta}f$ is a proximal operator of a convex function.

2.11 *Moreau envelope as a smooth approximation.* Let f be a CCP function on $\mathbb{R}^n$ and $\beta > 0$. Show that $\lim_{\beta\to 0} {}^{\beta}\!f(x) \to f(x)$ for all $x \in \mathbb{R}^n$.
Hint. First show that $u \in \partial f(x)$ if and only if $f(x) + f^*(u) = \langle u, x\rangle$. Then argue that for any $x \in \mathbb{R}^n$ (possibly $x \notin \operatorname{dom} f$)

$$f(x) = \sup_u \{-f^*(u) + \langle x, u\rangle\} = \sup_{(y,u)\in\partial f} \{f(y) + \langle u, x - y\rangle\}.$$

So there exists a sequence $(y^0, u^0), (y^1, u^1), \ldots$ in ∂f such that

$$f(y^k) + \langle u^k, x - y^k\rangle \to f(x).$$

Remark. This result, along with the smoothness property of Exercise 2.10 allows us to view ${}^{\beta}\!f$ as a smooth approximation of f. The interpretation of the Moreau envelope as a smooth, regularized function is due to Attouch [Att77, Lemme 1], [Att84, Theorem 2.64]. However, the analogous notion for monotone operators, known as the Moreau–Yosida approximation, was used earlier by Brezis [Bre71, Lemma 3], and the Moreau envelope itself was presented earlier yet by Moreau [Mor65]. The result of this problem was first presented by Friedlander, Goodwin, and Hoheisel [FGH21, Proposition 4].

2.12 *PPM is GD.* Show that $\operatorname{argmin} f = \operatorname{argmin} {}^{\beta}\!f$ for any $\beta > 0$. Also show that the PPM with f is equivalent to gradient descent with respect to ${}^{\beta}\!f$ for some $\beta > 0$.
Hint. Use Exercise 2.10.
Remark. This problem illustrates that the Moreau envelope is also useful as a conceptual tool for drawing connections.

2.13 *Projection onto convex sets.* Consider the *convex feasibility problem*

$$\underset{x\in\mathbb{R}^n}{\text{find}} \quad x \in C \cap D,$$

where C and D are nonempty closed convex sets. Assume $C \cap D \neq \emptyset$.

(a) The convex feasibility problem is equivalent to the optimization problem

$$\begin{array}{ll} \underset{x\in\mathbb{R}^n}{\text{minimize}} & \frac{1}{2}\operatorname{dist}^2(x, D) \\ \text{subject to} & x \in C. \end{array}$$

Show that the proximal gradient method with stepsize 1 applied to this problem is

$$x^{k+1} = \Pi_C \Pi_D x^k,$$

which is called the *alternating projections method.*

(b) The convex feasibility problem is also equivalent to the optimization problem

$$\underset{x\in\mathbb{R}^n}{\text{minimize}} \quad \tfrac{\theta}{2}\operatorname{dist}^2(x, C) + \tfrac{1-\theta}{2}\operatorname{dist}^2(x, D),$$

where $\theta \in (0, 1)$. Show that the gradient method with stepsize 1 applied to this problem is

$$x^{k+1} = \theta \Pi_C x^k + (1 - \theta)\Pi_D x^k,$$

which is called the *parallel projections method.*

(c) Show that $x^k \to x^\star \in C \cap D$ for both methods.

Hint. Note that $\frac{1}{2}\operatorname{dist}^2(x, C)$ is a Moreau envelope of δ_C.
Remark. See [BB96, BBL97, ER11] for an overview on convex feasibility problems.

2.14 *Banach fixed-point theorem.* Let $\mathbb{T}\colon \mathbb{R}^n \to \mathbb{R}^n$ be contractive. Show that $\mathbb{T}$ has a unique fixed point, that is, show that a fixed point of $\mathbb{T}$ exists and is unique.
Hint. Consider an FPI and show that $x^0, x^1, \ldots$ is a Cauchy sequence.
Remark. This result is called the *Banach fixed-point theorem* [Ban22].

2.15 *Strong monotonicity and unique zero.* Show that if $\mathbb{T}\colon \mathbb{R}^n \rightrightarrows \mathbb{R}^n$ is maximal μ-strongly monotone for some $\mu > 0$, then $\mathbb{T}$ has exactly one zero.
Hint. Use the Banach fixed-point theorem.

2.16 *Contraction factor of gradient descent.* Assume f is CCP, μ-strongly convex, L-smooth, and twice continuously differentiable. Show that $I - \alpha\nabla f$ is $\max\{|1-\alpha\mu|, |1-\alpha L|\}$-contractive for $0 < \alpha < 2/L$.
Hint. The fundamental theorem of calculus tells us

$$(I - \alpha\nabla f)(x) - (I - \alpha\nabla f)(y) = \int_0^1 (I - \alpha\nabla^2 f(tx + (1-t)))(x - y)\, dt.$$

Use the instance of Jensen's inequality

$$\left\| \int_0^1 v(t)\, dt \right\| \leq \int_0^1 \|v(t)\|\, dt,$$

where $v(t) \in \mathbb{R}^n$ for $t \in [0, 1]$.
Remark. The result still holds when f is not continuously differentiable. See §13.

2.17 *Convergence of dual ascent.* Show that dual ascent converges in the sense of $x^k \to x^\star$ and $u^k \to u^\star$, where $x^\star$ and $u^\star$ are primal and dual solutions, under the stated conditions.
Hint. Use Theorem 1 to establish $u^k \to u^\star$ and write x^k as a continuous function of u^k.
Remark. The stated conditions are f is CCP and μ-strongly convex, total duality holds, and $0 < \alpha < 2\mu/\sigma^2_{\max}(A)$.

2.18 *Method of multipliers primal solution convergence.* Show that the method of multipliers converges in the sense of $x^k \to x^\star$ under the stated conditions and strict convexity. Use the following fact: if h is a CCP function that is differentiable on $D \subseteq \mathbb{R}^n$, then $\nabla h\colon D \to \mathbb{R}^n$ is a continuous function, that is, differentiability and continuous differentiability coincide.
Remark. The stated conditions are f is CCP, $\mathcal{R}(A^\intercal) \cap \operatorname{ri}\operatorname{dom} f^* \neq \emptyset$, a dual solution exists, $\alpha > 0$, and $\mathbf{L}_\alpha(x, u) = f(x) + \langle u, Ax - b\rangle + \frac{\alpha}{2}\|Ax - b\|^2$.
Hint. Consider the primal problem

$$\begin{array}{ll} \underset{u\in\mathbb{R}^m,\, v\in\mathbb{R}^n}{\text{minimize}} & f^*(v) + b^\intercal u \\ \text{subject to} & -v - A^\intercal u = 0 \end{array}$$

generated by the Lagrangian $\tilde{\mathbf{L}}(v, u, x) = f^*(v) + b^\intercal u - \langle x, v + A^\intercal u\rangle$, and use Slater's constraint qualification to show that $\mathcal{R}(A^\intercal) \cap \operatorname{ri}\operatorname{dom} f^* \neq \emptyset$ implies strong duality and the existence of a primal solution for the primal-dual problem pair generated by $\mathbf{L}$. Use Exercise 2.5 to write $x^k = \sigma(u^k)$, where $\sigma\colon \mathbb{R}^m \to \mathbb{R}^n$ is a continuous function.
Remark. The derivation of (2.6) or Exercise 1.5 establishes $\operatorname{argmin}_x \mathbf{L}_\alpha(x, u^k) \neq \emptyset$, that is, $x^{k+1} \in \operatorname{argmin}_x \mathbf{L}_\alpha(x, u^k)$ is well defined for any $u^k \in \mathbb{R}^m$.

2.19 *Contraction factor of dual ascent.* Consider dual ascent. Assume f is μ-strongly convex, L-smooth, CCP, and $0 < \alpha < 2\mu/\sigma^2_{\max}(A)$. Using Exercise 2.16, show that dual ascent converges with contraction factor

$$\max\{|1 - \alpha\sigma^2_{\max}(A)/\mu|, |1 - \alpha\sigma^2_{\min}(A)/L|\}.$$

2.20 *Lyapunov analysis without summability.* Let $\mathbb{T}\colon \mathbb{R}^n \to \mathbb{R}^n$ be θ-averaged, and consider the fixed-point iteration $x^{k+1} = \mathbb{T}x^k$. Consider the Lyapunov function

$$V^k = k\frac{1-\theta}{\theta}\|x^k - x^{k-1}\|^2 + \|x^k - x^\star\|^2.$$

Show that

$$V^{k+1} \leq V^k$$

for $k = 0, 1, \ldots$. Use this inequality, instead of the summability argument, to prove Theorem 1.

2.21 *When there is no fixed point.* Assume $\mathbb{T}\colon \mathbb{R}^n \to \mathbb{R}^n$ is averaged and $\operatorname{Fix}\mathbb{T} = \emptyset$. Prove that sequence $x^{k+1} = \mathbb{T}x^k$ satisfies $\|x^k\| \to \infty$.

Hint. Assume for contradiction that $\|x^k\| \not\to \infty$, which implies, by the Bolzano–Weierstrass theorem, that there is a subsequence $k_j \to \infty$ such that $x^{k_j} \to \bar{x}$ for some limit $\bar{x}$. Next, show $\|x^{k+1} - x^k\| \to c$ for some $c \geq 0$. Consider the cases $c = 0$ and $c > 0$ separately. In the $c > 0$ case, show $\mathbb{T}^{k+1}\bar{x} - \mathbb{T}^k\bar{x} = \mathbb{T}^k\bar{x} - \mathbb{T}^{k-1}\bar{x}$ and argue that $\|\mathbb{T}^k\bar{x}\| \to \infty$, where

$$\mathbb{T}^k = \underbrace{\mathbb{T} \circ \cdots \circ \mathbb{T}}_{k \text{ times}}.$$

Remark. Interestingly, this result, first proved by Roehrig and Sine [RS81], does depend on the finite-dimensionality of $\mathbb{R}^n$. If $\mathbb{T}\colon \mathcal{H} \to \mathcal{H}$ is an averaged operator on an infinite-dimensional Hilbert space $\mathcal{H}$, Browder and Petryshyn showed that $\limsup_{k\to\infty} \|x^k\| = \infty$ [BP66], but Edelstein provided a counterexample for which $\liminf_{k\to\infty} \|x^k\| = 0$ [Ede64, BGMS20].

2.22 *FPI with quasi-nonexpansive operators.* We say $\mathbb{S}$ is *quasi-nonexpansive* if

$$\|\mathbb{S}x - x^\star\|^2 \leq \|x - x^\star\|^2$$

for all $x^\star \in \operatorname{Fix}\mathbb{S}$. We say $\mathbb{T}$ is *θ-quasi-averaged* if $\mathbb{T} = (1-\theta)\mathbb{I} + \theta\mathbb{S}$ for some quasi-nonexpansive operator $\mathbb{S}$. Assume $\mathbb{T}\colon \mathbb{R}^n \to \mathbb{R}^n$ is continuous and θ-quasi-averaged with $\theta \in (0,1)$. Assume $\operatorname{Fix}\mathbb{T} \neq \emptyset$. Show that $x^{k+1} = \mathbb{T}x^k$ with any starting point $x^0 \in \mathbb{R}^n$ converges to one fixed point, that is,

$$x^k \to x^\star$$

for some $x^\star \in \operatorname{Fix}\mathbb{T}$.

2.23 *Gradient descent with varying stepsize.* Consider the problem of minimizing

$$\underset{x\in\mathbb{R}^n}{\text{minimize}} \quad f(x),$$

where f is an L-smooth CCP function. Then

$$x^{k+1} = x^k - \alpha_k \nabla f(x^k),$$

where $\alpha_0, \alpha_1, \ldots \in \mathbb{R}$, is called *gradient descent with varying stepsize*. Assume $\operatorname{argmin} f \neq \emptyset$ and

$$0 < \inf_{k=0,1,\ldots} \alpha_k \leq \sup_{k=0,1,\ldots} \alpha_k < 2/L.$$

Show

$$x^k \to x^\star \in \operatorname{argmin} f.$$

Hint. Adapt the proof of Theorem 1 to fit the current setup.

2.24 Show (2.9) and (2.10).

2.25 *Conic programs with DRS.* Consider the problem of

$$\begin{array}{ll} \underset{x\in\mathbb{R}^n}{\text{minimize}} & c^\intercal x \\ \text{subject to} & Ax = b \\ & x \in K, \end{array}$$

where $K \subset \mathbb{R}^n$ is a nonempty closed convex set. When K is a nonempty closed convex cone, the problem is said to be a *conic program*. Assume $A \in \mathbb{R}^{m\times n}$, where A has rank m and $b \in \mathbb{R}^m$. Show that the FPI with DRS is

$$\begin{aligned} x^{k+1/2} &= \Pi_K(z^k) \\ x^{k+1} &= D(2x^{k+1/2} - z^k) + v \\ z^{k+1} &= z^k + x^{k+1} - x^{k+1/2}, \end{aligned}$$

where $D = I - A^\intercal(AA^\intercal)^{-1}A$ and $v = A^\intercal(AA^\intercal)^{-1}b - \alpha Dc$.

2.26 *Convergence of DRS.* Consider the FPI with DRS. Theorem 1 implies $z^k \to z^\star$ for any $\alpha > 0$, provided that a fixed point exists. Show that this implies $x^k \to x^\star$, and $x^{k+1/2} \to x^\star$. Is $\|x^{k+3/2} - x^{k+1/2}\| \to 0$ and $\|x^{k+1} - x^k\| \to 0$ true?

2.27 *When PRS does not converge.* Consider the operators $\mathbb{A} = \mathbb{N}_{\{0\}}$ and $\mathbb{B} = 0$. Show that although a fixed point of PRS does correspond to a solution, the FPI with PRS does not converge. This example also demonstrates that the FPI with the reflected resolvent need not converge.

2.28 *Backward-backward is alternating minimization.* Consider the monotone inclusion problem

$$\underset{x\in\mathbb{R}^n}{\text{find}} \quad 0 \in (\mathbb{A} + \mathbb{B})x.$$

The *backward-backward method* is

$$x^{k+1} = \mathbb{J}_{\alpha\mathbb{A}}\mathbb{J}_{\alpha\mathbb{B}}x^k,$$

where $\alpha > 0$. Show that when $\mathbb{A} = \partial f$ and $\mathbb{B} = \partial g$, where f and g are CCP functions, we have

$$\begin{aligned} y^{k+1} &= \underset{y\in\mathbb{R}^n}{\operatorname{argmin}} \left\{ f(x^k) + g(y) + \frac{1}{2\alpha}\|x^k - y\|^2 \right\} \\ x^{k+1} &= \underset{x\in\mathbb{R}^n}{\operatorname{argmin}} \left\{ f(x) + g(y^{k+1}) + \frac{1}{2\alpha}\|x - y^{k+1}\|^2 \right\}. \end{aligned}$$

and that fixed points correspond to minimizers of

$$\underset{x\in\mathbb{R}^n}{\text{minimize}} \quad f(x) + g(y) + \tfrac{1}{2\alpha}\|x - y\|^2. \tag{2.20}$$

Finally, show that the backward-backward method converges.

Remark. This result was first published by Bauschke, Combettes, and Reich [BCR05].

2.29 *Consensus + proximable is proximable.* Let r be a CCP function on $\mathbb{R}^n$, C be the consensus set as defined in (2.19), and

$$g(x_1, \ldots, x_m) = \delta_C(x_1, \ldots, x_m) + \sum_{i=1}^m r(x_i).$$

Show that we can evaluate $\mathrm{Prox}_{\alpha g}$ with

$$\mathrm{Prox}_{\alpha g}(y_1,\dots,y_m) = (x,\dots,x), \quad x = \mathrm{Prox}_{\alpha r}\left(\frac{1}{m}\sum_{i=1}^{m} y_i\right).$$

Also, what is the proximal operator of $h(x_1,\dots,x_m) = \delta_C(x_1,\dots,x_m) + r(x_1)$?

2.30 Let $\eta \in (0,1)$ and consider the monotone inclusion problem

$$\underset{x\in\mathbb{R}^n}{\text{find}} \quad 0 \in (2(1-\eta)\mathbb{I} + \mathbb{A} + \mathbb{B})x,$$

where $\mathbb{A}$ and $\mathbb{B}$ are maximal monotone, and assume $\mathbb{A}+\mathbb{B}$ is maximal. Show that the solution can be found through the FPI $z^{k+1} = \mathbb{T}z^k$ with

$$\mathbb{T} = \frac{1}{2}\mathbb{I} + \frac{1}{2}(2\eta\mathbb{J}_{\mathbb{A}} - \mathbb{I})(2\eta\mathbb{J}_{\mathbb{B}} - \mathbb{I}).$$

Hint. Show $\mathrm{Zer}\,(2(1-\eta)\mathbb{I} + \mathbb{A} + \mathbb{B}) = \frac{1}{\eta}\mathrm{Zer}\,(\mathbb{A}^{(\eta)} + \mathbb{B}^{(\eta)})$, where $\mathbb{A}^{(\eta)} = \mathbb{A}\circ\frac{1}{\eta}\mathbb{I} + \frac{1-\eta}{\eta}\mathbb{I}$ and $\mathbb{B}^{(\eta)}$ is defined likewise.

Remark. Since $\mathrm{Zer}\,(2(1-\eta)\mathbb{I} + \mathbb{A} + \mathbb{B}) = \mathbb{J}_{\frac{1}{2(1-\eta)}(\mathbb{A}+\mathbb{B})}(0)$, a unique solution exists. This method is called the averaged alternating modified reflections (AAMR) [AAC18, AAC19].

2.31 *Further properties of the proximal operator.* Let f be a CCP function on $\mathbb{R}^n$. Show:

(a) $f(\mathrm{Prox}_{\alpha f}(x))$ is a nonincreasing function of $\alpha \in (0,\infty)$ (for a fixed $x \in \mathbb{R}^n$).
(b) $\lim_{\alpha\to\infty} f(\mathrm{Prox}_{\alpha f}(x)) = \inf_x f(x)$ (including the case $\inf_x f(x) = -\infty$).
(c) $f(\mathrm{Prox}_{\alpha f}(x)) \le f(x)$ for any $\alpha > 0$.
(d) $\lim_{\alpha\to 0^+} f(\mathrm{Prox}_{\alpha f}(x)) = f(x)$ for all $x \in \mathrm{dom}\, f$.

Hint. For (a), argue that

$$\alpha f(\mathrm{Prox}_{\alpha f}(x)) + \frac{1}{2}\|\mathrm{Prox}_{\alpha f}(x) - x\|^2 \le \alpha f(\mathrm{Prox}_{\beta f}(x)) + \frac{1}{2}\|\mathrm{Prox}_{\beta f}(x) - x\|^2$$
$$\beta f(\mathrm{Prox}_{\beta f}(x)) + \frac{1}{2}\|\mathrm{Prox}_{\beta f}(x) - x\|^2 \le \beta f(\mathrm{Prox}_{\alpha f}(x)) + \frac{1}{2}\|\mathrm{Prox}_{\alpha f}(x) - x\|^2$$

for $\alpha,\beta \in \mathbb{R}$. For (b), let $\varepsilon > 0$ and $M > \inf_x f(x)$. Let $x_{M,\varepsilon}$ be a point such that $f(x_{M,\varepsilon}) < M + \varepsilon/2$. Then

$$f(x_{M,\varepsilon}) + \frac{1}{2\alpha}\|x_{M,\varepsilon} - x\|^2 < M + \varepsilon$$

for large enough α. For (d), show

$$\alpha f(x) \ge \alpha f(\mathrm{Prox}_{\alpha f}(x)) + \frac{1}{2}\|\mathrm{Prox}_{\alpha f}(x) - x\|^2$$

and let $\alpha \to 0$.

Remark. The result of (d) is not necessarily true when $x \notin \mathrm{dom}\, f$. For example, consider $f = \delta_{\{0\}}$ and $x = 1$.

Remark. In general, one can show $\lim_{\alpha\to 0^+}\mathrm{Prox}_{\alpha f}(x) = \Pi_{\overline{\mathrm{dom}\, f}}(x)$ [FGH21, Proposition 5].

2.32 *Proximable inequality constraints.* Let f be a CCP function on $\mathbb{R}^n$ and informally assume f is proximable. Through the following steps, show that $\delta_{\{x\in\mathbb{R}^n \mid f(x)\le 0\}}$ is proximable. Show:

(a) For maximal monotone $\mathbb{A}$, $\alpha, \beta \in (0, \infty)$,

$$\mathbb{J}_{\alpha\mathbb{A}}x = \mathbb{J}_{\beta\mathbb{A}}\left(\frac{\beta}{\alpha}x + \left(1 - \frac{\beta}{\alpha}\right)\mathbb{J}_{\alpha\mathbb{A}}x\right) \qquad \forall x \in \mathbb{R}^n,$$

and

$$\|\mathbb{J}_{\alpha\mathbb{A}}x - \mathbb{J}_{\beta\mathbb{A}}x\| \leq \left|1 - \frac{\beta}{\alpha}\right| \|\mathbb{J}_{\alpha\mathbb{A}}x - x\| \qquad \forall x \in \mathbb{R}^n.$$

(b) For a fixed $x \in \mathbb{R}^n$, $f(\mathrm{Prox}_{\alpha f}(x))$ is a nonincreasing continuous function of $\alpha \in (0, \infty)$.

(c) Assume that $\mathrm{dom} f = \mathbb{R}^n$ and that there exists an $x \in \mathbb{R}^n$ such that $f(x) < 0$. Let $\alpha^\star = \inf\{\alpha > 0 \mid f(\mathrm{Prox}_{\alpha f}(x)) \leq 0\}$. Then

$$\Pi_{\{x \in \mathbb{R}^n \mid f(x) \leq 0\}}(x) = \begin{cases} x & \text{if } f(x) \leq 0 \\ \mathrm{Prox}_{\alpha^\star f}(x) & \text{otherwise.} \end{cases}$$

(d) Assume $\mathrm{dom} f = \mathbb{R}^n$ and $f(x) > 0$. Also assume $l^0, u^0 \in \mathbb{R}$ satisfy $f(\mathrm{Prox}_{l^0 f}(x)) > 0 \geq f(\mathrm{Prox}_{u^0 f}(x))$. The iteration

$$(l^{k+1}, u^{k+1}) = \begin{cases} (l^k, \frac{l^k + u^k}{2}) & \text{if } f\left(\mathrm{Prox}_{\frac{l^k+u^k}{2} f}(x)\right) \leq 0 \\ (\frac{l^k+u^k}{2}, u^k) & \text{otherwise} \end{cases}$$

converges in the sense that $l^k \to \alpha^\star$ and $u^k \to \alpha^\star$.

Hint. Show that $(\mathrm{Prox}_{\alpha^\star f}(x), \alpha^\star)$ is a saddle point of

$$\mathbf{L}(z, \lambda) = \frac{1}{2}\|z - x\|^2 + \lambda f(z) - \delta_{\mathbb{R}_+}(\lambda),$$

which implies that $\mathrm{Prox}_{\alpha^\star f}(x)$ is a solution to the primal problem generated by $\mathbf{L}$.

Hint. Use Exercise 2.31.

Remark. The result of this problem was first presented by Friedlander, Goodwin, and Hoheisel [FGH21, Corollary 13].

2.33 Consider the problem

$$\begin{array}{ll} \underset{x \in \mathbb{R}^n}{\text{minimize}} & f_0(x) \\ \text{subject to} & f_i(x) \leq 0, \quad i = 1, \ldots, m, \end{array}$$

where $f_0, \ldots, f_m$ are CCP. Assume all forms of total duality. Show that

$$\begin{aligned} x^{k+1/2} &= \mathrm{Prox}_{\alpha f_0}\left(\frac{1}{m}\sum_{i=1}^m z_i^k\right) \\ x_i^{k+1} &= \Pi_{\{x \in \mathbb{R}^n \mid f_i(x) \leq 0\}}(2x^{k+1/2} - z_i^k) \\ z_i^{k+1} &= z_i^k + x_i^{k+1} - x^{k+1/2} \qquad \text{for } i = 1, \ldots, m \end{aligned}$$

converges in the sense that $x^{k+1/2} \to x^\star$ and $x_i^{k+1} \to x^\star$ for $i = 1, \ldots, m$.

Hint. Use Exercise 2.29.

2.34 *Indicator function of a subspace.* Let $V \subseteq \mathbb{R}^n$ be a subspace and

$$V^\perp = \{u \in \mathbb{R}^n \mid \langle u, v\rangle = 0 \;\forall\, v \in V\}$$

be its orthogonal complement. Show:

(a) $(\delta_V)^* = \delta_{V^\perp}$,

(b) $\mathbb{N}_V(v) = V^\perp$ for all $v \in V$, and
(c) $\Pi_V + \Pi_{V^\perp} = \mathbb{I}$.

2.35 *Indicator function of a convex cone.* Let $K \subseteq \mathbb{R}^n$ be a nonempty closed *convex cone*, that is, K is a nonempty closed set satisfying

$$x_1, x_2 \in K \quad \Rightarrow \quad \theta_1 x_1 + \theta_2 x_2 \in K$$

for all $\theta_1, \theta_2 \geq 0$. Let

$$K^* = \{u \in \mathbb{R}^n \mid \langle u, x\rangle \geq 0 \; \forall x \in K\}$$

be the *dual cone* of K. Show:
(a) $(\delta_K)^* = \delta_{-K^*}$,
(b) $\mathbb{N}_K(x) = \{u \in -K^* \mid \langle u, x\rangle = 0\}$ for all $x \in K$, and
(c) $\Pi_K + \Pi_{-K^*} = \mathbb{I}$.
Remark. This problem subsumes Exercise 2.34.

2.36 *Consensus technique for operators.* Show that the problem

$$\underset{x \in \mathbb{R}^n}{\text{find}} \quad 0 \in \sum_{i=1}^{m} \mathbb{A}_i x,$$

where $\mathbb{A}_1, \ldots, \mathbb{A}_m$ are (multi-valued) operators, is equivalent to

$$\underset{x_1,\ldots,x_m \in \mathbb{R}^n}{\text{find}} \quad 0 \in \begin{bmatrix} \mathbb{A}_1(x_1) \\ \vdots \\ \mathbb{A}_m(x_m) \end{bmatrix} + \mathbb{N}_C(x_1, \ldots, x_m),$$

where $C = \{(x_1, \ldots, x_m) \in \mathbb{R}^{nm} \mid x_1 = \cdots = x_m\}$ is the consensus set.
Hint. Show $C^\perp = \{(u_1, \ldots, u_m) \in \mathbb{R}^{nm} \mid u_1 + \cdots + u_m = 0\}$ and use Exercise 2.34.

2.37 *Variable metric DRS.* Consider the problem

$$\underset{x \in \mathbb{R}^n}{\text{find}} \quad 0 \in (\mathbb{A} + \mathbb{B})x,$$

where $\mathbb{A}$ and $\mathbb{B}$ are maximal monotone. Assume $\text{Zer}(\mathbb{A} + \mathbb{B}) \neq \emptyset$. Let $M \in \mathbb{R}^{n \times n}$ be a symmetric positive definite matrix. Show that the FPI with *variable metric DRS*

$$\begin{aligned} x^{k+1/2} &= \mathbb{J}_{M^{-1}\mathbb{B}}(z^k) \\ x^{k+1} &= \mathbb{J}_{M^{-1}\mathbb{A}}(2x^{k+1/2} - z^k) \\ z^{k+1} &= z^k + x^{k+1} - x^{k+1/2} \end{aligned}$$

converges.

2.38 *PPXA.* Consider the problem

$$\underset{x \in \mathbb{R}^n}{\text{minimize}} \quad \sum_{i=1}^{m} g_i(x),$$

where $g_1, \ldots, g_m$ are CCP functions on $\mathbb{R}^n$. Let $\theta_1, \ldots, \theta_m \in \mathbb{R}$ be such that $\theta_i > 0$ for $i = 1, \ldots, m$ and $\theta_1 + \cdots + \theta_m = 1$. Define the weighted average

$$\overline{z}_\theta^k = \theta_1 z_1^k + \cdots + \theta_m z_m^k$$

and denote

$$\overline{\mathbf{z}}_\theta^k = (\overline{z}_\theta^k, \ldots, \overline{z}_\theta^k) \in \mathbb{R}^{mn}.$$

The algorithm *parallel proximal algorithm* (PPXA) is

$$x_i^{k+1} = \mathrm{Prox}_{(1/\theta_i)g_i}(2\overline{z}_\theta^k - z_i^k) \qquad \text{for } i = 1,\dots,m,$$
$$\mathbf{z}^{k+1} = \mathbf{z}^k + \mathbf{x}^{k+1} - \overline{\mathbf{z}}_\theta^k.$$

Assume $\bigcap_{i=1}^m \operatorname{int} \operatorname{dom} g_i \neq \emptyset$ and that a solution exists. Show that PPXA converges in the sense that there exists a solution $x^\star$ such that

$$(x_1^k,\dots,x_m^k) \to (x^\star,\dots,x^\star).$$

Hint. Consider the variable metric DRS with

$$M = \begin{bmatrix} \theta_1 I & & \\ & \ddots & \\ & & \theta_m I \end{bmatrix} \in \mathbb{R}^{mn\times mn},$$

where $I \in \mathbb{R}^{n\times n}$ is the identity matrix, and use

$$\mathbb{J}_{M^{-1}\partial f}(x) = \underset{z\in\mathbb{R}^d}{\operatorname{argmin}} \left\{ f(z) + \frac{1}{2}\|z - x\|_M^2 \right\}.$$

Remark. PPXA was presented by Combettes and Pesquet [CP08, CP11b].

3 Primal-Dual Splitting Methods

This chapter presents techniques for deriving a collection of *primal-dual methods*, methods that explicitly maintain and update both primal and dual variables. The splitting methods of §2.7.2 are limited to optimization problems of the form of minimizing $f(x)+g(x)$ or $f(x)+g(x)+h(x)$. The primal-dual methods of this chapter can solve a wider range of problems and can exploit problem structures with a high level of freedom.

With the techniques we present, we reduce a wide range of classical and modern methods into instances of other methods for which we have established convergence. Many of these connections are not at all obvious and were, in fact, discovered years after the original publications of the methods. However, they are straightforward to verify, and once a reduction is done, convergence analysis comes down to mere bookkeeping.

For many methods, we present multiple derivations. For example, we derive PDHG as a variable metric PPM, with the BCV technique, and as an instance of linearized ADMM. The different derivations provide related but distinct interpretations, and they show the intimate connection between the various primal-dual methods.

3.1 INFIMAL POSTCOMPOSITION TECHNIQUE

The *infimal postcomposition technique* uses infimal postcomposition $A \triangleright f$, which we define soon, to recast linearly constrained problems of the form

$$\begin{array}{ll} \underset{x\in\mathbb{R}^p}{\text{minimize}} & f(x) + \cdots \\ \text{subject to} & Ax + \cdots \end{array}$$

into an equivalent form without constraints

$$\underset{z\in\mathbb{R}^n}{\text{minimize}} \quad (A \triangleright f)(z) + \cdots$$

and then applies a base splitting of §2.7.2.

Infimal Postcomposition

Given a function f on $\mathbb{R}^n$ and matrix $A \in \mathbb{R}^{m\times n}$, define the function $A \triangleright f$ on $\mathbb{R}^m$ with

$$(A \triangleright f)(z) = \inf_{x\in\{x\,|\,Ax=z\}} f(x).$$

This is called the *infimal postcomposition of f by A* or the *image of f under A*. If f is CCP and $\mathcal{R}(A^\intercal) \cap \operatorname{ri}\operatorname{dom} f^* \neq \emptyset$, then $A \triangleright f$ is CCP.

The infimal postcomposition arises due to the formula

$$(A \triangleright f)^*(u) = f^*(A^\intercal u), \tag{3.1}$$

which follows from

$$\begin{aligned}
(A \triangleright f)^*(u) &= \sup_{z\in\mathbb{R}^m} \left\{ \langle u, z\rangle - \inf_{x\in\mathbb{R}^n} \left\{ f(x) + \delta_{\{x \mid Ax=z\}}(x) \right\} \right\} \\
&= -\inf_{z\in\mathbb{R}^m} \left\{ -\langle u, z\rangle + \inf_{x\in\mathbb{R}^n} \left\{ f(x) + \delta_{\{x \mid Ax=z\}}(x) \right\} \right\} \\
&= -\inf_{x\in\mathbb{R}^n, z\in\mathbb{R}^m} \left\{ f(x) + \delta_{\{x \mid Ax=z\}}(x) - \langle u, z\rangle \right\} \\
&= -\inf_{x\in\mathbb{R}^n} \left\{ f(x) - \langle u, Ax\rangle \right\} \\
&= f^*(A^\intercal u).
\end{aligned}$$

If $\mathcal{R}(A^\intercal) \cap \operatorname{ri}\operatorname{dom} f^* \neq \emptyset$, then

$$\begin{aligned}
&x \in \operatorname*{argmin}_x \left\{ f(x) + (1/2)\|Ax - y\|^2 \right\} \\
&z = Ax
\end{aligned} \quad \Leftrightarrow \quad z = \operatorname{Prox}_{A \triangleright f}(y) \tag{3.2}$$

and the argmin of the left-hand side exists. (The argmin_x may not be unique, but $z = Ax$ is unique.) See Exercise 3.1 for a proof.

Alternating Direction Method of Multipliers (ADMM)

Let f and g be CCP, $A \in \mathbb{R}^{n\times p}$, $B \in \mathbb{R}^{n\times q}$, and $c \in \mathbb{R}^n$. Consider the primal

$$\begin{array}{ll}
\underset{x\in\mathbb{R}^p,\, y\in\mathbb{R}^q}{\text{minimize}} & f(x) + g(y) \\
\text{subject to} & Ax + By = c
\end{array} \tag{3.3}$$

and the dual problem

$$\underset{u\in\mathbb{R}^n}{\text{maximize}} \quad -f^*(-A^\intercal u) - g^*(-B^\intercal u) - c^\intercal u \tag{3.4}$$

generated by the Lagrangian

$$\mathbf{L}(x, y, u) = f(x) + g(y) + \langle u, Ax + By - c\rangle. \tag{3.5}$$

Assume the regularity conditions

$$\begin{aligned}
&\mathcal{R}(A^\intercal) \cap \operatorname{ri}\operatorname{dom} f^* \neq \emptyset \\
&\mathcal{R}(B^\intercal) \cap \operatorname{ri}\operatorname{dom} g^* \neq \emptyset.
\end{aligned} \tag{3.6}$$

We will use the augmented Lagrangian:

$$\mathbf{L}_\rho(x, y, u) = f(x) + g(y) + \langle u, Ax + By - c\rangle + \frac{\rho}{2}\|Ax + By - c\|^2. \tag{3.7}$$

The primal problem (3.3) is equivalent to

$$\underset{z\in\mathbb{R}^n}{\text{minimize}} \quad \underbrace{(A \triangleright f)(z)}_{=\tilde{f}(z)} + \underbrace{(B \triangleright g)(c - z)}_{=\tilde{g}(z)},$$

which is in the required form. We apply DRS to the equivalent primal problem. The FPI with respect to $\frac{1}{2}\mathbb{I} + \frac{1}{2}\mathbb{R}_{\alpha^{-1}\partial\tilde{f}}\mathbb{R}_{\alpha^{-1}\partial\tilde{g}}$ is

$$\begin{aligned} z^{k+1/2} &= \mathrm{Prox}_{\alpha^{-1}\tilde{g}}(\zeta^k) \\ z^{k+1} &= \mathrm{Prox}_{\alpha^{-1}\tilde{f}}(2z^{k+1/2} - \zeta^k) \\ \zeta^{k+1} &= \zeta^k + z^{k+1} - z^{k+1/2}. \end{aligned}$$

We introduce and substitute the variables x^k, y^k, and u^k defined implicitly by $z^{k+1/2} = c - By^{k+1}$, $z^{k+1} = Ax^{k+2}$, and $\zeta^k = \alpha^{-1}u^k + Ax^{k+1}$ and use (3.2) to get

$$\begin{aligned} y^{k+1} &\in \underset{y}{\mathrm{argmin}} \left\{ g(y) + \langle u^k, Ax^{k+1} + By - c\rangle + \frac{\alpha}{2}\|Ax^{k+1} + By - c\|^2 \right\} \\ x^{k+2} &\in \underset{x}{\mathrm{argmin}} \left\{ f(x) + \langle u^{k+1}, Ax + By^{k+1} - c\rangle + \frac{\alpha}{2}\|Ax + By^{k+1} - c\|^2 \right\} \\ u^{k+1} &= u^k + \alpha(Ax^{k+1} + By^{k+1} - c). \end{aligned}$$

Reordering the updates to get the dependency right, we get

$$\begin{aligned} x^{k+1} &\in \underset{x}{\mathrm{argmin}} \left\{ f(x) + \langle u^k, Ax + By^k - c\rangle + \frac{\alpha}{2}\|Ax + By^k - c\|^2 \right\} \\ y^{k+1} &\in \underset{y}{\mathrm{argmin}} \left\{ g(y) + \langle u^k, Ax^{k+1} + By - c\rangle + \frac{\alpha}{2}\|Ax^{k+1} + By - c\|^2 \right\} \\ u^{k+1} &= u^k + \alpha(Ax^{k+1} + By^{k+1} - c). \end{aligned}$$

Using the augmented Lagrangian (see Example 1.11), we can write the updates more concisely as

$$x^{k+1} \in \underset{x}{\mathrm{argmin}}\, \mathbf{L}_\alpha(x, y^k, u^k) \tag{3.8a}$$

$$y^{k+1} \in \underset{y}{\mathrm{argmin}}\, \mathbf{L}_\alpha(x^{k+1}, y, u^k) \tag{3.8b}$$

$$u^{k+1} = u^k + \alpha(Ax^{k+1} + By^{k+1} - c). \tag{3.8c}$$

This method is called the *alternating direction methods of multipliers (ADMM)*.

At this point, we have completed the core of the convergence analysis; we have reduced ADMM to an instance of DRS applied to an equivalent transformation of (3.3). What remains is the bookkeeping, where we check whether the necessary conditions are met and translate the convergence of DRS into the convergence of ADMM.

Convergence Analysis When applying DRS to convex optimization, we require total duality defined for a specific Lagrangian for convergence. In the current setup, the Lagrangian for which we need total duality is not **L**.

DRS applied to the equivalent primal problem requires total duality between

$$\underset{z\in\mathbb{R}^n}{\text{minimize}} \quad \underbrace{(A \rhd f)(z)}_{=\tilde{f}(z)} + \underbrace{(B \rhd g)(c - z)}_{=\tilde{g}(z)}$$

and

$$\underset{u\in\mathbb{R}^n}{\text{maximize}} \quad -f^*(-A^\intercal u) - g^*(-B^\intercal u) - c^\intercal u$$

generated by the Lagrangian

$$\tilde{\mathbf{L}}(z,u) = (A \rhd f)(z) + \langle z,u \rangle - g^*(-B^\intercal u) - c^\intercal u.$$

If the original primal and dual problems have solutions $(x^\star, y^\star)$ and $u^\star$ for which strong duality holds, the equivalent primal and dual problems generated by $\tilde{\mathbf{L}}(z,u)$ have solutions $Ax^\star$ and $u^\star$ for which strong duality holds. In other words, total duality of the original problems implies total duality of the equivalent problems. Total duality implies that the FPI with DRS converges, and this translates to the following convergence results.

If total duality between (3.3) and (3.4) holds, the regularity condition (3.6) holds, and $\alpha > 0$, then ADMM is well defined, $Ax^k \to Ax^\star$, and $By^k \to By^\star$.

Regularity Condition The assumed regularity condition (3.6) serves two purposes: It ensures that $A \rhd f$ and $B \rhd g$ are CCP functions, and that the minimizers defining the iterations exist. (DRS applied to non-CCP (but convex) functions can run into pathologies.) While (3.6) is a sufficient condition that ensures our analysis is valid, it is not necessary. See the bibliographical notes section, Exercise 3.1, and §8 for further discussion.

3.2 DUALIZATION TECHNIQUE

The *dualization technique* is to apply base splittings to the dual problems. Certain primal problems with linear equality constraints have dual problems already of the form of minimizing $\tilde{f}(u) + \tilde{g}(u)$. We have seen this technique in the derivation of the method of multipliers.

Alternating Direction Method of Multipliers (ADMM)

With the dualization technique, we provide an alternate derivation and analysis of ADMM. Again consider the problems (3.3) and (3.4) generated by the Lagrangian (3.5). Apply DRS to the dual. Write $\tilde{f}(u) = f^*(-A^\intercal u)$ and $\tilde{g}(u) = g^*(-B^\intercal u) + c^\intercal u$, and the FPI with $\frac{1}{2}\mathbb{I} + \frac{1}{2}\mathbb{R}_{\alpha\partial\tilde{f}}\mathbb{R}_{\alpha\partial\tilde{g}}$ is

$$\begin{aligned}
\mu^{k+1/2} &= \mathbb{J}_{\alpha\partial\tilde{g}}(\psi^k)\\
\mu^{k+1} &= \mathbb{J}_{\alpha\partial\tilde{f}}(2\mu^{k+1/2} - \psi^k)\\
\psi^{k+1} &= \psi^k + \mu^{k+1} - \mu^{k+1/2}.
\end{aligned}$$

Using (2.6) and (2.8), we write out the resolvent evaluations more explicitly as

$$\begin{aligned}
\tilde{y}^{k+1} &\in \operatorname*{argmin}_y \left\{ g(y) + \langle \psi^k - \alpha c, By \rangle + \frac{\alpha}{2}\|By\|_2^2 \right\}\\
\mu^{k+1/2} &= \psi^k + \alpha(B\tilde{y}^{k+1} - c)\\
\tilde{x}^{k+1} &\in \operatorname*{argmin}_x \left\{ f(x) + \langle \psi^k + 2\alpha(B\tilde{y}^{k+1} - c), Ax \rangle + \frac{\alpha}{2}\|Ax\|_2^2 \right\}\\
\mu^{k+1} &= \psi^k + \alpha A\tilde{x}^{k+1} + 2\alpha(B\tilde{y}^{k+1} - c)\\
\psi^{k+1} &= \psi^k + \alpha(A\tilde{x}^{k+1} + B\tilde{y}^{k+1} - c).
\end{aligned}$$

Remove $\mu^{k+1/2}$ and μ^{k+1}, as they no longer have any explicit dependence. Reorganizing, we get

$$\begin{aligned}
\tilde{y}^{k+1} &\in \underset{y}{\operatorname{argmin}} \left\{ g(y) + \langle \psi^k - \alpha A\tilde{x}^k, By\rangle + \frac{\alpha}{2}\|A\tilde{x}^k + By - c\|_2^2 \right\} \\
\tilde{x}^{k+1} &\in \underset{x}{\operatorname{argmin}} \left\{ f(x) + \langle \psi^k + \alpha(B\tilde{y}^{k+1} - c), Ax\rangle + \frac{\alpha}{2}\|Ax + B\tilde{y}^{k+1} - c\|_2^2 \right\} \\
\psi^{k+1} &= \psi^k + \alpha(A\tilde{x}^{k+1} + B\tilde{y}^{k+1} - c).
\end{aligned}$$

Next, substitute $u^k = \psi^k - \alpha A\tilde{x}^k$:

$$\begin{aligned}
\tilde{y}^{k+1} &\in \underset{y}{\operatorname{argmin}} \left\{ g(y) + \langle u^k, By\rangle + \frac{\alpha}{2}\|A\tilde{x}^k + By - c\|_2^2 \right\} \\
\tilde{x}^{k+1} &\in \underset{x}{\operatorname{argmin}} \left\{ f(x) + \langle u^{k+1}, Ax\rangle + \frac{\alpha}{2}\|Ax + B\tilde{y}^{k+1} - c\|_2^2 \right\} \\
u^{k+1} &= u^k + \alpha(A\tilde{x}^k + B\tilde{y}^{k+1} - c).
\end{aligned}$$

Finally, we swap the order of the u^{k+1} and $\tilde{x}^{k+1}$ update to get the correct dependency and substitute $x^{k+1} = \tilde{x}^k$ and $y^k = \tilde{y}^k$ to recover ADMM:

$$\begin{aligned}
x^{k+1} &\in \underset{x}{\operatorname{argmin}}\, \mathbf{L}_\alpha(x, y^k, u^k) \\
y^{k+1} &\in \underset{y}{\operatorname{argmin}}\, \mathbf{L}_\alpha(x^{k+1}, y, u^k) \\
u^{k+1} &= u^k + \alpha(Ax^{k+1} + By^{k+1} - c).
\end{aligned}$$

If total duality, (3.6), and $\alpha > 0$ hold, then $u^k \to u^\star$, $Ax^k \to Ax^\star$, and $By^k \to By^\star$.

Convergence Analysis The previous analysis of §3.1 established that $Ax^k \to Ax^\star$ and $By^k \to By^\star$. Since $\mu^{k+1/2} \to u^\star$, this implies $\psi^k \to u^\star + \alpha Ax^\star$. Therefore, we conclude $u^k \to u^\star$.

Alternating Minimization Algorithm (AMA)

Again consider the problems (3.3) and (3.4) generated by the Lagrangian (3.5). Again, the dual problem (3.4) is

$$\underset{u\in\mathbb{R}^n}{\text{maximize}} \quad \underbrace{-f^*(-A^\intercal u)}_{=\tilde{f}(u)} - \underbrace{(g^*(-B^\intercal u) + c^\intercal u)}_{=\tilde{g}(u)}.$$

We furthermore assume f is μ-strongly convex. This implies $f^*(-A^\intercal u)$ is $(\lambda_{\max}(A^\intercal A)/\mu)$-smooth. Assume the regularity condition (3.6).

We apply FBS to the dual problem. The FPI with $(\mathbb{I} + \alpha\partial\tilde{g})^{-1}(\mathbb{I} - \alpha\nabla\tilde{f})$ is

$$\begin{aligned}
u^{k+1/2} &= u^k - \alpha\nabla\tilde{f}(u^k) \\
u^{k+1} &= (I + \alpha\partial\tilde{g})^{-1}(u^{k+1/2}).
\end{aligned}$$

Using (2.2), (2.6), and (2.8) and assuming $\mathcal{R}(B) \cap \operatorname{ri}\operatorname{dom} g^* \neq \emptyset$, we write the iteration as

$$\begin{aligned} x^{k+1} &= \underset{x}{\operatorname{argmin}} \left\{ f(x) + \langle u^k, Ax \rangle \right\} \\ u^{k+1/2} &= u^k + \alpha A x^{k+1} \\ y^{k+1} &\in \underset{y}{\operatorname{argmin}} \left\{ g(y) + \langle u^{k+1/2} - \alpha c, By \rangle + \frac{\alpha}{2} \|By\|^2 \right\} \\ u^{k+1} &= u^{k+1/2} + \alpha B y^{k+1} - \alpha c. \end{aligned}$$

Eliminate $u^{k+1/2}$ and use the Lagrangian (3.5) and augmented Lagrangian (3.7) to write the iteration as

$$\begin{aligned} x^{k+1} &= \underset{x}{\operatorname{argmin}} \, \mathbf{L}(x, y^k, u^k) \\ y^{k+1} &\in \underset{y}{\operatorname{argmin}} \, \mathbf{L}_\alpha(x^{k+1}, y, u^k) \\ u^{k+1} &= u^k + \alpha(Ax^{k+1} + By^{k+1} - c). \end{aligned}$$

This method is called the *alternating minimization algorithm* (AMA) or *dual proximal gradient*. If total duality, regularity conditions of (3.6), μ-strong convexity of f, and $\alpha \in (0, 2\mu/\lambda_{\max}(A^\intercal A))$ hold, then $u^k \to u^\star$, $x^k \to x^\star$, and $By^k \to By^\star$.

Convergence Analysis Under the stated assumptions, the convergence of FBS tells us $u^k \to u^\star$. Since $(x^\star, y^\star, u^\star)$ is a saddle point, we have $x^\star = \operatorname{argmin}_x \mathbf{L}(x, y^\star, u^\star)$, which implies $0 \in \partial f(x^\star) + A^\intercal u^\star$, which in turn implies $x^\star = \nabla f^*(-A^\intercal u^\star)$. Since $x^{k+1} = \nabla f^*(-A^\intercal u^k)$, and since ∇f^* is a continuous operator, $u^k \to u^\star$ implies $x^k \to x^\star$. Finally, $u^k \to u^\star$ implies $u^{k+1} - u^k \to 0$, which in turn implies $Ax^{k+1} + By^{k+1} - c \to 0$. Combining this with $x^k \to x^\star$ implies $By^k \to By^\star$.

3.3 VARIABLE METRIC TECHNIQUE

In §3.1 and §3.2, we transformed a given optimization problem into another equivalent optimization problem and applied the splittings. In the following two sections, we apply splittings to the saddle subdifferentials.

In this section, we present the *variable metric technique*. Its key insight is to use variable metric PPM or variable metric FBS with a metric M carefully chosen to cancel out certain terms and thereby simplify the update.

PDHG

Let f and g be CCP functions and $A \in \mathbb{R}^{m \times n}$. Consider the problem pair (1.9) and (1.10)

$$\underset{x \in \mathbb{R}^n}{\text{minimize}} \quad f(x) + g(Ax), \qquad \underset{u \in \mathbb{R}^m}{\text{maximize}} \quad -f^*(-A^\intercal u) - g^*(u)$$

generated by the Lagrangian (1.8),

$$\mathbf{L}(x, u) = f(x) + \langle u, Ax \rangle - g^*(u).$$

We apply the variable metric PPM to the saddle subdifferential

$$\partial \mathbf{L}(x, u) = \begin{bmatrix} 0 & A^\intercal \\ -A & 0 \end{bmatrix} \begin{bmatrix} x \\ u \end{bmatrix} + \begin{bmatrix} \partial f(x) \\ \partial g^*(u) \end{bmatrix}.$$

The matrix

$$M = \begin{bmatrix} (1/\alpha)I & -A^\intercal \\ -A & (1/\beta)I \end{bmatrix} \tag{3.9}$$

satisfies $M \succ 0$ if $\alpha, \beta > 0$ and $\alpha\beta\lambda_{\max}(A^\intercal A) < 1$.

The FPI with $(M + \partial\mathbf{L})^{-1}M$ is

$$\begin{bmatrix} x^{k+1} \\ u^{k+1} \end{bmatrix} = \left(\begin{bmatrix} (1/\alpha)I & 0 \\ -2A & (1/\beta)I \end{bmatrix} + \begin{bmatrix} \partial f \\ \partial g^* \end{bmatrix} \right)^{-1} \begin{bmatrix} (1/\alpha)x^k - A^\intercal u^k \\ -Ax^k + (1/\beta)u^k \end{bmatrix},$$

which is equivalent to

$$\begin{bmatrix} (1/\alpha)I & 0 \\ -2A & (1/\beta)I \end{bmatrix} \begin{bmatrix} x^{k+1} \\ u^{k+1} \end{bmatrix} + \begin{bmatrix} \partial f(x^{k+1}) \\ \partial g^*(u^{k+1}) \end{bmatrix} \ni \begin{bmatrix} (1/\alpha)x^k - A^\intercal u^k \\ -Ax^k + (1/\beta)u^k \end{bmatrix}.$$

Because the linear system of the resolvent is lower triangular, we can compute x^{k+1} from the upper inclusion and then, by substituting x^{k+1} into the lower inclusion, compute u^{k+1}:

$$\begin{aligned} x^{k+1} &= \mathrm{Prox}_{\alpha f}(x^k - \alpha A^\intercal u^k) \\ u^{k+1} &= \mathrm{Prox}_{\beta g^*}(u^k + \beta A(2x^{k+1} - x^k)). \end{aligned}$$

This method is called the *primal-dual hybrid gradient* (PDHG) or *Chambolle–Pock*. If total duality holds, $\alpha > 0$, $\beta > 0$, and $\alpha\beta\lambda_{\max}(A^\intercal A) < 1$, then $x^k \to x^\star$ and $u^k \to u^\star$.

There is another version PDHG that uses a similar but different M and obtains u^{k+1} before x^{k+1}. See Exercise 3.5.

Choice of Metric Although PDHG is derived from PPM, which is technically not an operator splitting, PDHG is a splitting since it deals with f and g separately. Using the variable metric M to obtain a lower triangular system is crucial. For example, although the FPI $(x^{k+1}, u^{k+1}) = (\mathbb{I} + \partial\mathbf{L})^{-1}(x^k, u^k)$ would converge in theory, it is not computationally useful; the off-diagonal terms $A^\intercal$ and $-A$ of $\partial\mathbf{L}$ couple the x^{k+1} and u^{k+1}-updates, so they must be computed simultaneously. With no splitting, a single iteration is no easier than solving the whole problem itself.

Condat–Vũ

Consider the primal problem

$$\underset{x \in \mathbb{R}^n}{\text{minimize}} \quad f(x) + h(x) + g(Ax) \tag{3.10}$$

and its dual problem

$$\underset{u \in \mathbb{R}^m}{\text{maximize}} \quad -(f + h)^*(-A^\intercal u) - g^*(u), \tag{3.11}$$

where f, g, and h are CCP, h is differentiable, and $A \in \mathbb{R}^{m \times n}$. The Lagrangian that generates these problems is

$$\mathbf{L}(x, u) = f(x) + h(x) + \langle u, Ax \rangle - g^*(u). \tag{3.12}$$

This generalizes the PDHG setup, as it allows the additional differentiable function h.

We apply the variable metric FBS to the saddle subdifferential $\partial\mathbf{L}$ with the metric M defined in (3.9). We split the saddle subdifferential into

$$\partial\mathbf{L}(x,u) = \underbrace{\begin{bmatrix}\nabla h(x)\\ 0\end{bmatrix}}_{=\mathbb{H}(x,u)} + \underbrace{\begin{bmatrix}0 & A^\intercal\\ -A & 0\end{bmatrix}\begin{bmatrix}x\\ u\end{bmatrix} + \begin{bmatrix}\partial f(x)\\ \partial g^*(u)\end{bmatrix}}_{=\mathbb{F}(x,u)}.$$

The FPI with $(x^{k+1},u^{k+1}) = (M+\mathbb{F})^{-1}(M-\mathbb{H})(x^k,u^k)$ is

$$\begin{bmatrix}x^{k+1}\\ u^{k+1}\end{bmatrix} = \left(\begin{bmatrix}(1/\alpha)I & 0\\ -2A & (1/\beta)I\end{bmatrix} + \begin{bmatrix}\partial f\\ \partial g^*\end{bmatrix}\right)^{-1}\begin{bmatrix}(1/\alpha)x^k - A^\intercal u^k - \nabla h(x^k)\\ -Ax^k + (1/\beta)u^k\end{bmatrix},$$

which we write as

$$\begin{aligned}
x^{k+1} &= \mathrm{Prox}_{\alpha f}(x^k - \alpha A^\intercal u^k - \alpha\nabla h(x^k))\\
u^{k+1} &= \mathrm{Prox}_{\beta g^*}(u^k + \beta A(2x^{k+1} - x^k)).
\end{aligned}$$

This method is called *Condat–Vũ*. (See Exercise 3.5 for the other version of *Condat–Vũ*.) If total duality holds, h is L-smooth, $\alpha > 0$, $\beta > 0$, and

$$\alpha L/2 + \alpha\beta\lambda_{\max}(A^\intercal A) < 1, \tag{3.13}$$

then $x^k \to x^\star$ and $u^k \to u^\star$.

Convergence Analysis First, note that $M \succ 0$ under the assumption $\alpha, \beta > 0$ and (3.13). With basic computation, we get

$$M^{-1} = \begin{bmatrix}\alpha(I - \alpha\beta A^\intercal A)^{-1} & \alpha\beta A^\intercal(I - \alpha\beta AA^\intercal)^{-1}\\ \alpha\beta A(I - \alpha\beta A^\intercal A)^{-1} & \beta(I - \alpha\beta AA^\intercal)^{-1}\end{bmatrix}.$$

Let

$$\theta = \frac{2}{L}\left(\frac{1}{\alpha} - \beta\lambda_{\max}(A^\intercal A)\right) > 1.$$

Note that the condition $\theta > 1$ equivalent to $\alpha L/2 + \alpha\beta\lambda_{\max}(A^\intercal A) < 1$. Then

$$\theta\left(\frac{1}{\alpha}I - \beta A^\intercal A\right)^{-1} \preceq \theta\left(\frac{1}{\alpha} - \beta\lambda_{\max}(A^\intercal A)\right)^{-1} I = \frac{2}{L}I.$$

If $\mathbb{I} - \theta M^{-1}\mathbb{H}$ is nonexpansive in $\|\cdot\|_M$, then $\mathbb{I} - M^{-1}\mathbb{H}$ is averaged in $\|\cdot\|_M$, and Condat–Vũ, a variable metric FBS, converges. (Nonexpansiveness and averagedness in $\|\cdot\|_M$ were discussed in §2.8.) Nonexpansiveness of $\mathbb{I} - \theta M^{-1}\mathbb{H}$ in $\|\cdot\|_M$ follows from

$$\begin{aligned}
&\|(\mathbb{I} - \theta M^{-1}\mathbb{H})(x,u) - (\mathbb{I} - \theta M^{-1}\mathbb{H})(y,v)\|_M^2\\
&= \|(x,u) - (y,v)\|_M^2\\
&\quad - 2\theta\langle (x,u) - (y,v), \mathbb{H}(x,u) - \mathbb{H}(y,v)\rangle + \theta^2\|\mathbb{H}(x,u) - \mathbb{H}(y,v)\|_{M^{-1}}^2\\
&= \|(x,u) - (y,v)\|_M^2\\
&\quad - 2\theta\langle x - y, \nabla h(x) - \nabla h(y)\rangle + \theta^2\|\nabla h(x) - \nabla h(y)\|_{\alpha(I-\alpha\beta A^\intercal A)^{-1}}^2\\
&\le \|(x,u) - (y,v)\|_M^2\\
&\quad - (2\theta/L)\|\nabla h(x) - \nabla h(y)\|^2 + \theta^2\|\nabla h(x) - \nabla h(y)\|_{(\alpha^{-1}I-\beta A^\intercal A)^{-1}}^2\\
&\le \|(x,u) - (y,v)\|_M^2.
\end{aligned}$$

Example 3.1 In computational tomography (CT), the medical device measures the Radon transform of a patient. The Radon transform is a linear operator $R \in \mathbb{R}^{m\times n}$ and $b \in \mathbb{R}^m$ is the measurement. It is often the case that $m < n$, that is, there are more unknowns than measurements, and $b \approx Rx^{\text{true}}$, that is, the measurement is corrupted by small noise. Given the measurement b, the image is reconstructed by solving the optimization problem

$$\underset{x\in\mathbb{R}^n}{\text{minimize}} \quad \tfrac{1}{2}\|Rx - b\|^2 + \lambda\|Dx\|_1,$$

where the optimization variable $x \in \mathbb{R}^n$ represents the 2D or 3D image to recover, D is the 2D or 3D finite difference operator, and $\lambda > 0$. Although R and D are very large matrices, the evaluation of matrix-vector products with R, D, $R^\intercal$, and $D^\intercal$ are efficient. To solve this problem, we can transform the given problem into

$$\underset{x\in\mathbb{R}^n}{\text{minimize}} \quad 0(x) + g(Ax),$$

where

$$A = \begin{bmatrix} R \\ (\beta/\alpha)D \end{bmatrix}, \qquad 0(x) = 0, \qquad g(y,z) = \frac{1}{2}\|y - b\|^2 + (\lambda\alpha/\beta)\|z\|_1,$$

for any $\alpha, \beta > 0$, and apply PDHG to get

$$\begin{aligned}
x^{k+1} &= x^k - (1/\alpha)(\alpha R^\intercal u^k + \beta D^\intercal v^k) \\
u^{k+1} &= \frac{1}{1+\alpha}(u^k + \alpha R(2x^{k+1} - x^k) - \alpha b) \\
v^{k+1} &= \Pi_{[-\lambda\alpha/\beta,\lambda\alpha/\beta]}\left(v^k + \beta D(2x^{k+1} - x^k)\right),
\end{aligned}$$

where $\Pi_{[-\lambda\alpha/\beta,\lambda\alpha/\beta]}$ is applied elementwise. The computational bottleneck of this algorithm is computing $R^\intercal u^k$ and $R(2x^{k+1} - x^k)$. (Computing $D^\intercal v^k$ and $D(2x^{k+1} - x^k)$ costs much less.) In particular, this algorithm does not utilize any matrix inverses.
To further clarify, $x \in \mathbb{R}^n$ is a 2D or 3D image reshaped into a length n vector. Explicitly forming the matrices R and D is infeasible as they are too large, but there are efficient algorithms for computing matrix-vector products with R, D, $R^\intercal$, and $D^\intercal$. In particular, the application of $R^\intercal$ is called backprojection.

3.4 GAUSSIAN ELIMINATION TECHNIQUE

The *Gaussian elimination technique* reduces a system of inclusions into an upper or lower triangular form through multiplying both sides by an invertible matrix. The lower or upper triangular system is then solved sequentially, in a split manner.

Proximal Method of Multipliers with Function Linearization

Consider the constrained problem

$$\begin{array}{ll} \underset{x\in\mathbb{R}^n}{\text{minimize}} & f(x) + h(x) \\ \text{subject to} & Ax = b, \end{array} \tag{3.14}$$

where $A \in \mathbb{R}^{m\times n}$, $b \in \mathbb{R}^m$, f and h are CCP, and h is differentiable.
The corresponding Lagrangian is

$$\mathbf{L}(x,u) = f(x) + h(x) + \langle u, Ax - b\rangle.$$

We split the saddle subdifferential into

$$\partial\mathbf{L}(x,u) = \underbrace{\begin{bmatrix}\nabla h(x)\\ b\end{bmatrix}}_{=\mathbb{H}(x,u)} + \underbrace{\begin{bmatrix}0 & A^\intercal\\ -A & 0\end{bmatrix}\begin{bmatrix}x\\ u\end{bmatrix} + \begin{bmatrix}\partial f(x)\\ 0\end{bmatrix}}_{=\mathbb{G}(x,u)}. \tag{3.15}$$

The FPI with $(\mathbb{I}+\alpha\mathbb{G})^{-1}(\mathbb{I}-\alpha\mathbb{H})$ is described by

$$\begin{bmatrix}I & \alpha A^\intercal\\ -\alpha A & I\end{bmatrix}\begin{bmatrix}x^{k+1}\\ u^{k+1}\end{bmatrix} + \begin{bmatrix}\alpha\partial f(x^{k+1})\\ 0\end{bmatrix} \ni \begin{bmatrix}x^k - \alpha\nabla h(x^k)\\ u^k - \alpha b\end{bmatrix}.$$

At first sight, this system may not seem useful, as the x^{k+1} and u^{k+1}-updates are seemingly coupled. However, left-multiply the system with the invertible matrix

$$\begin{bmatrix}I & -\alpha A^\intercal\\ 0 & I\end{bmatrix},$$

which corresponds to Gaussian elimination, and get

$$\begin{bmatrix}I+\alpha^2 A^\intercal A & 0\\ -\alpha A & I\end{bmatrix}\begin{bmatrix}x^{k+1}\\ u^{k+1}\end{bmatrix} + \begin{bmatrix}\alpha\partial f(x^{k+1})\\ 0\end{bmatrix}$$
$$\ni \begin{bmatrix}x^k - \alpha\nabla h(x^k) - \alpha A^\intercal(u^k - \alpha b)\\ u^k - \alpha b\end{bmatrix},$$

a lower-triangular system. Now we compute x^{k+1} first and then compute u^{k+1}:

$$x^{k+1} = \underset{x}{\operatorname{argmin}}\left\{f(x) + \langle\nabla h(x^k), x\rangle + \langle u^k, Ax - b\rangle + \frac{\alpha}{2}\|Ax-b\|^2 + \frac{1}{2\alpha}\|x - x^k\|^2\right\} \tag{3.16a}$$

$$u^{k+1} = u^k + \alpha(Ax^{k+1} - b). \tag{3.16b}$$

This is called the proximal method of multipliers with function linearization. If total duality holds, h is L-smooth, and $\alpha \in (0, 2/L)$, then $x^k \to x^\star$ and $u^k \to u^\star$.

Using Gaussian Elimination with Inclusions

It is important to keep in mind that row operations of Gaussian elimination can only be performed using rows with single-valued operators. Given the system of inclusions

$$\begin{aligned}\mathbb{A}z &\ni b\\ \mathbb{B}z &= c,\end{aligned}$$

where $\mathbb{B}$ is single-valued, we can multiply M by the equation of the second row and add it to the first row to obtain

$$\begin{aligned}\mathbb{A}z + M\mathbb{B}z &\ni b + Mc\\ \mathbb{B}z &= c.\end{aligned}$$

This is equivalent to the original system of inclusions, as we can multiply $-M$ by the second row and add it to the first row to recover the original system.

Given the system of inclusions

$$\begin{aligned}\mathbb{A}z &\ni b\\ \mathbb{B}z &\ni c,\end{aligned}$$

where $\mathbb{B}$ is not necessarily single-valued, we can multiply M by the inclusion of the second row and add it to the first row to obtain

$$\begin{aligned} \mathbb{A}z + M\mathbb{B}z &\ni b + Mc \\ \mathbb{B}z &\ni c. \end{aligned}$$

However, while this inclusion is a consequence of the original inclusion, it is not equivalent; if we multiply $-M$ by the second row and add it to the first row, we get

$$\begin{aligned} \mathbb{A}z + M\mathbb{B}z - M\mathbb{B}z &\ni b \\ \mathbb{B}z &\ni c, \end{aligned}$$

and this is not equivalent to the original system.

PAPC/PDFP^{2}O

Consider the Lagrangian (3.12) in the special case of $f = 0$. This gives us the problems

$$\underset{x\in\mathbb{R}^n}{\text{minimize}} \quad h(x) + g(Ax) \qquad\qquad \underset{u\in\mathbb{R}^m}{\text{maximize}} \quad -h^*(-A^\intercal u) - g^*(u), \tag{3.17}$$

where h is differentiable, and the Lagrangian

$$\mathbf{L}(x,u) = h(x) + \langle u, Ax\rangle - g^*(u).$$

We apply the variable metric FBS to the saddle subdifferential $\partial\mathbf{L}$ and use the Gaussian elimination technique to evaluate the resolvent. (So we combine the two techniques.) We split the saddle subdifferential into

$$\partial\mathbf{L}(x,u) = \underbrace{\begin{bmatrix} \nabla h(x) \\ 0 \end{bmatrix}}_{=\mathbb{H}(x,u)} + \underbrace{\begin{bmatrix} 0 & A^\intercal \\ -A & 0 \end{bmatrix}\begin{bmatrix} x \\ u \end{bmatrix} + \begin{bmatrix} 0 \\ \partial g^*(u) \end{bmatrix}}_{=\mathbb{G}(x,u)}.$$

The matrix

$$M = \begin{bmatrix} (1/\alpha)I & 0 \\ 0 & (1/\beta)I - \alpha AA^\intercal \end{bmatrix}$$

satisfies $M \succ 0$ if $\alpha\beta\lambda_{\max}(A^\intercal A) < 1$.

The FPI with $(M + \mathbb{G})^{-1}(M - \mathbb{H})$ is described by

$$\begin{bmatrix} (1/\alpha)I & A^\intercal \\ -A & (1/\beta)I - \alpha AA^\intercal \end{bmatrix}\begin{bmatrix} x^{k+1} \\ u^{k+1} \end{bmatrix} + \begin{bmatrix} 0 \\ \partial g^*(u^{k+1}) \end{bmatrix} \ni \begin{bmatrix} (1/\alpha)x^k - \nabla h(x^k) \\ (1/\beta)u^k - \alpha AA^\intercal u^k \end{bmatrix}.$$

Left-multiply the system by the invertible matrix

$$\begin{bmatrix} I & 0 \\ \alpha A & I \end{bmatrix},$$

which corresponds to Gaussian elimination, and get

$$\begin{aligned} &\begin{bmatrix} (1/\alpha)I & A^\intercal \\ 0 & (1/\beta)I \end{bmatrix}\begin{bmatrix} x^{k+1} \\ u^{k+1} \end{bmatrix} + \begin{bmatrix} 0 \\ \partial g^*(u^{k+1}) \end{bmatrix} \\ &\qquad\qquad \ni \begin{bmatrix} (1/\alpha)x^k - \nabla h(x^k) \\ Ax^k - \alpha A\nabla h(x^k) + (1/\beta)u^k - \alpha AA^\intercal u^k \end{bmatrix}. \end{aligned}$$

Now that the linear system of the resolvent is upper triangular, we can compute u^{k+1} first and then compute x^{k+1}:

$$\begin{aligned} u^{k+1} &= \mathrm{Prox}_{\beta g^*}\left(u^k + \beta A(x^k - \alpha A^\top u^k - \alpha \nabla h(x^k))\right) \\ x^{k+1} &= x^k - \alpha A^\top u^{k+1} - \alpha \nabla h(x^k). \end{aligned}$$

This method is called *proximal alternating predictor corrector* (PAPC) or *primal-dual fixed point algorithm based on proximity operator* (PDFP^{2}O). If total duality holds, h is L-smooth, $\alpha > 0$, $\beta > 0$, $\alpha\beta\lambda_{\max}(A^\top A) < 1$, and $\alpha < 2/L$, then $x^k \to x^\star$ and $u^k \to u^\star$.

Example 3.2 In isotonic regression, entries of the regressor are constrained to be nondecreasing. Isotonic regression with the Huber loss solves

$$\begin{array}{ll} \underset{x \in \mathbb{R}^n}{\text{minimize}} & \ell(Ax - b) \\ \text{subject to} & x_{i+1} - x_i \geq 0 \quad \text{for } i = 1, \ldots, n-1, \end{array}$$

where $A \in \mathbb{R}^{m \times n}$, $b \in \mathbb{R}^m$, and

$$\ell(y) = \sum_{i=1}^{m} h(y_i), \qquad h(r) = \begin{cases} r^2 & \text{for } |r| \leq 1 \\ 2|r| - 1 & \text{for } |r| > 1. \end{cases}$$

For the sake of simplicity, assume n is even.
One solution method is to transform the problem into

$$\underset{x \in \mathbb{R}^n}{\text{minimize}} \quad \overbrace{\sum_{i=1,3,\ldots,n-1} \delta_{\mathbb{R}_+}(x_{i+1} - x_i)}^{\text{proximable}} + \overbrace{\sum_{i=2,4,\ldots,n-2} \delta_{\mathbb{R}_+}(x_{i+1} - x_i)}^{\text{proximable}} + \overbrace{\ell(Ax - b)}^{\text{differentiable}}$$

and use the FPI with DYS:

$$\begin{aligned} x^{k+1/2} &= \Pi_{\text{odd}}(z^k) \\ x^{k+1} &= \Pi_{\text{even}}(2x^{k+1/2} - z^k - \alpha A^\top \nabla \ell(Ax^{k+1/2} - b)) \\ z^{k+1} &= z^k + x^{k+1} - x^{k+1/2}, \end{aligned}$$

where

$$\Pi_{\text{odd}} = \mathrm{Prox}_{\sum_{i=1,3,\ldots,n-1} \delta_{\mathbb{R}_+}}, \qquad \Pi_{\text{even}} = \mathrm{Prox}_{\sum_{i=2,4,\ldots,n-2} \delta_{\mathbb{R}_+}}$$

can be evaluated efficiently by Exercise 1.7.
Another solution method is to transform the problem into

$$\underset{x \in \mathbb{R}^n}{\text{minimize}} \quad \ell(Ax - b) + \delta_{\mathbb{R}_+^{(n-1)}}(Dx),$$

where $\mathbb{R}_+^{(n-1)} = \{(u_1, \ldots, u_{n-1}) \in \mathbb{R}^{(n-1)} \mid u_i \geq 0, i = 1, \ldots, n-1\}$ is the nonnegative orthant and

$$D = \begin{bmatrix} -1 & 1 & 0 & \cdots & 0 & 0 \\ 0 & -1 & 1 & \cdots & 0 & 0 \\ \vdots & & & \ddots & & \vdots \\ 0 & 0 & 0 & \cdots & -1 & 1 \end{bmatrix} \in \mathbb{R}^{(n-1) \times n},$$

and use PAPC:

$$u^k = \Pi_{(-\infty,0]}(u^k + \beta D(x^k - \alpha D^\intercal u^k - \alpha A^\intercal \nabla \ell(Ax^k - b)))$$
$$x^k = x^k - \alpha D^\intercal u^k + \alpha A^\intercal \nabla \ell(Ax^k - b),$$

where $\Pi_{(-\infty,0]}$ is applied elementwise. Note, $(\delta_{\mathbb{R}_+^{(n-1)}})^* = \delta_{-\mathbb{R}_+^{(n-1)}}$ by Exercise 2.35.

3.5 LINEARIZATION TECHNIQUE

The *linearization technique* involves using a proximal term to cancel out a computationally inconvenient quadratic term. More specifically, consider the setup where the method's update is defined through

$$x^{k+1} = \underset{x \in \mathbb{R}^n}{\operatorname{argmin}} \left\{ f(x) + \frac{\alpha}{2}\|Ax - b\|^2 + \frac{1}{2}\|x - x^k\|_M^2 \right\}.$$

If f is proximable and we have the freedom to choose $M \succ 0$, we can choose $M = (1/\beta)I - \alpha A^\intercal A$ with $1/\beta > \alpha\lambda_{\max}(A^\intercal A)$ to get

$$\begin{aligned}
f(x)+&\frac{\alpha}{2}\|Ax - b\|^2 + \frac{1}{2}\|x - x^k\|_M^2 \\
&= f(x) - \alpha\langle Ax, b\rangle - x^\intercal M x^k + \frac{\alpha}{2}x^\intercal A^\intercal A x + \frac{1}{2}x^\intercal M x + \text{constant} \\
&= f(x) + \alpha\langle Ax^k - b, Ax\rangle - \frac{1}{\beta}\langle x^k, x\rangle + \frac{1}{2\beta}\|x\|^2 + \text{constant} \\
&= f(x) + \alpha\langle Ax^k - b, Ax\rangle + \frac{1}{2\beta}\|x - x^k\|^2 + \text{constant} \\
&= f(x) + \frac{1}{2\beta}\left\|x - \left(x^k - \alpha\beta A^\intercal(Ax^k - b)\right)\right\|^2 + \text{constant},
\end{aligned}$$

and we have

$$x^{k+1} = \operatorname{Prox}_{\beta f}\left(x^k - \alpha\beta A^\intercal(Ax^k - b)\right).$$

We call the $\|x - x^k\|_M^2$ term the "proximal term" and we choose M carefully to cancel out the quadratic term $x^\intercal A^\intercal A x$ originating from $\|Ax - b\|^2$. The linearization technique is named so because the result is as if we linearized the quadratic term

$$\frac{\alpha}{2}\|Ax - b\|^2 \approx \alpha\langle Ax, Ax^k - b\rangle + \text{constant}$$

and added $(2\beta)^{-1}\|x - x^k\|^2$ to ensure convergence.

Linearized Method of Multipliers

Consider the primal problem (1.6):

$$\begin{array}{ll}
\underset{x \in \mathbb{R}^n}{\text{minimize}} & f(x) \\
\text{subject to} & Ax = b.
\end{array}$$

Let $M \succ 0$ and $K = \alpha^{-1/2}M^{-1/2}$. Re-parameterize the problem with $x = Ky$:

$$\begin{array}{ll}
\underset{y \in \mathbb{R}^n}{\text{minimize}} & f(Ky) \\
\text{subject to} & AKy = b.
\end{array}$$

The proximal method of multipliers of §2.6.1 applied to the re-parameterized problem is

$$\begin{aligned} y^{k+1} &= \underset{y}{\operatorname{argmin}} \left\{ f(Ky) + \langle u^k, AKy \rangle + \frac{\alpha}{2}\|AKy - b\|^2 + \frac{1}{2\alpha}\|y - y^k\|^2 \right\} \\ u^{k+1} &= u^k + \alpha(AKy^{k+1} - b). \end{aligned}$$

Now we substitute back $x = Ky$ and get

$$\begin{aligned} x^{k+1} &= \underset{x}{\operatorname{argmin}} \left\{ f(x) + \langle u^k, Ax \rangle + \frac{\alpha}{2}\|Ax - b\|^2 + \frac{1}{2}\|x - x^k\|_M^2 \right\} \\ u^{k+1} &= u^k + \alpha(Ax^{k+1} - b). \end{aligned}$$

Let $M = (1/\beta)I - \alpha A^\intercal A$, where $\alpha\beta\lambda_{\max}(A^\intercal A) < 1$ so that $M \succ 0$. Then, we get

$$\begin{aligned} x^{k+1} &= \underset{x}{\operatorname{argmin}} \left\{ f(x) + \langle u^k + \alpha(Ax^k - b), Ax \rangle + \frac{1}{2\beta}\|x - x^k\|^2 \right\} \\ u^{k+1} &= u^k + \alpha(Ax^{k+1} - b) \end{aligned}$$

and we can write

$$\begin{aligned} x^{k+1} &= \operatorname{Prox}_{\beta f}\left(x^k - \beta A^\intercal(u^k + \alpha(Ax^k - b))\right) \\ u^{k+1} &= u^k + \alpha(Ax^{k+1} - b). \end{aligned}$$

This method is called *linearized method of multipliers*. If total duality holds, $\alpha > 0$, $\beta > 0$, and $\alpha\beta\lambda_{\max}(A^\intercal A) < 1$, then $x^k \to x^\star$ and $u^k \to u^\star$.

When $\operatorname{Prox}_{\beta f}$ is computationally easy to evaluate, but $\operatorname{argmin}_x\{f(x) + \frac{1}{2}\|Ax - b\|^2\}$ is not, the linearized method of multipliers can be much more effective than the (original) method of multipliers.

3.5.1 BCV Technique

When using the linearization technique, the proximal term $(1/2)\|x - x^k\|_M^2$ must come from somewhere. Sometimes we can use methods that already have a proximal term, such as the proximal method of multipliers or the proximal ADMM of Exercise 3.2. Alternatively, we can create proximal terms with the *BCV technique*, named after Bertsekas, O'Connor, and Vandenberghe.

PDHG

Consider problem (1.9):

$$\underset{x \in \mathbb{R}^n}{\text{minimize}} \quad f(x) + g(Ax).$$

This problem is equivalent to

$$\underset{x \in \mathbb{R}^n,\, \tilde{x} \in \mathbb{R}^m}{\text{minimize}} \quad \underbrace{f(x) + \delta_{\{0\}}(\tilde{x})}_{=\tilde{f}(x,\tilde{x})} + \underbrace{g(Ax + M^{1/2}\tilde{x})}_{=\tilde{g}(x,\tilde{x})},$$

for any $M \succeq 0$. This transformation is the *BCV technique*, and it will provide us with a proximal term that we can use for the linearization.

Consider the FPI with DRS:

$$(z^{k+1}, \tilde{z}^{k+1}) = \left(\frac{1}{2}\mathbb{I} + \frac{1}{2}\mathbb{R}_{\alpha\partial\tilde{g}}\mathbb{R}_{\alpha\partial\tilde{f}}\right)(z^k, \tilde{z}^k).$$

Using (2.6), we have

$$\begin{aligned}
&\mathrm{Prox}_{\alpha\tilde{g}}(x, \tilde{x}) = (y, \tilde{y})\\
&\quad\Leftrightarrow\quad u \in \operatorname*{argmin}_u \left\{ g^*(u) - \left\langle \begin{bmatrix} x \\ \tilde{x} \end{bmatrix}, \begin{bmatrix} A^\intercal \\ M^{1/2} \end{bmatrix} u \right\rangle + \frac{\alpha}{2} \left\| \begin{bmatrix} A^\intercal \\ M^{1/2} \end{bmatrix} u \right\|^2 \right\}\\
&\qquad\quad y = x - \alpha A^\intercal u\\
&\qquad\quad \tilde{y} = \tilde{x} - \alpha M^{-1/2} u
\end{aligned}$$

under the regularity condition $\operatorname{ri}\operatorname{dom} g \cap \mathcal{R}([A\ M^{1/2}]) \neq \emptyset$, and we write

$$\begin{aligned}
x^{k+1/2} &= \operatorname*{argmin}_x \left\{ f(x) + \frac{1}{2\alpha}\|x - z^k\|^2 \right\}\\
\tilde{x}^{k+1/2} &= 0\\
u^{k+1} &= \operatorname*{argmin}_u \left\{ g^*(u) - \langle A(2x^{k+1/2} - z^k) - M^{1/2}\tilde{z}^k, u\rangle + \frac{\alpha}{2}\left(\|A^\intercal u\|^2 + \|M^{1/2}u\|^2 \right) \right\}\\
x^{k+1} &= 2x^{k+1/2} - z^k - \alpha A^\intercal u^{k+1}\\
\tilde{x}^{k+1} &= -\tilde{z}^k - \alpha M^{1/2} u^{k+1}\\
z^{k+1} &= x^{k+1/2} - \alpha A^\intercal u^{k+1}\\
\tilde{z}^{k+1} &= -\alpha M^{1/2} u^{k+1}.
\end{aligned}$$

We simplify this further to get

$$\begin{aligned}
x^{k+1/2} &= \operatorname*{argmin}_x \left\{ f(x) + \frac{1}{2\alpha}\|x - (x^{k-1/2} - \alpha A^\intercal u^k)\|^2 \right\}\\
u^{k+1} &= \operatorname*{argmin}_u \left\{ g^*(u) - \langle A(2x^{k+1/2} - x^{k-1/2}), u\rangle + \frac{\alpha}{2}\|u - u^k\|^2_{(AA^\intercal + M)} \right\}.
\end{aligned}$$

We now perform linearization by setting $M = (\beta\alpha)^{-1}I - AA^\intercal$, where $\alpha\beta\lambda_{\max}(A^\intercal A) \leq 1$ so that $M \succeq 0$, and we get

$$\begin{aligned}
x^{k+1/2} &= \mathrm{Prox}_{\alpha f}(x^{k-1/2} - \alpha A^\intercal u^k)\\
u^{k+1} &= \mathrm{Prox}_{\beta g^*}(u^k + \beta A(2x^{k+1/2} - x^{k-1/2})).
\end{aligned}$$

If total duality between (1.9) and (1.10), regularity condition $\operatorname{ri}\operatorname{dom} g \cap \mathcal{R}([A\ M^{1/2}]) \neq \emptyset$, $\alpha > 0$, $\beta > 0$, and $\alpha\beta\lambda_{\max}(A^\intercal A) \leq 1$ hold, then $x^{k+1/2} \to x^\star$.

Convergence Analysis Note that the DRS in this derivation applies $\mathbb{R}_{\alpha\partial\tilde{f}}$ before $\mathbb{R}_{\alpha\partial\tilde{g}}$, which is inconsistent with the usual ordering of §2.7.2, which applies $\mathrm{Prox}_{\alpha g}$ before $\mathrm{Prox}_{\alpha f}$. Keeping this reversed order in mind, note that the Lagrangian

$$\tilde{\mathbf{L}}(x, \tilde{x}, \mu, \tilde{\mu}) = g(Ax + M^{-1/2}\tilde{x}) + \langle x, \mu\rangle + \langle \tilde{x}, \tilde{\mu}\rangle - f^*(\mu),$$

the analog of (2.17), generates the stated equivalent primal problem and the dual problem

$$\underset{\mu\in\mathbb{R}^n,\,\tilde{\mu}\in\mathbb{R}^m}{\text{maximize}} \quad -\left(\begin{bmatrix} A^\intercal \\ M^{1/2}\end{bmatrix} \triangleright g^*\right)(-\mu,-\tilde{\mu}) - f^*(\mu).$$

If (1.9) and (1.10) have solutions $x^\star$ and $u^\star$ for which strong duality holds, then the equivalent primal-dual problem pair have solutions $(x^\star, 0)$ and $(-A^\intercal u^\star, -M^{1/2}u^\star)$ for which strong duality holds. In other words, total duality of the original problems imply total duality of the equivalent problems. So the FPI with DRS converges under the stated assumptions and we conclude that $x^{k+1/2} \to x^\star$.

PD3O

Consider the primal problem (3.10)

$$\underset{x\in\mathbb{R}^n}{\text{minimize}} \quad f(x) + h(x) + g(Ax),$$

which was considered in Condat–Vũ. In particular, assume h is L-smooth. This problem is equivalent to

$$\underset{x\in\mathbb{R}^n,\,\tilde{x}\in\mathbb{R}^m}{\text{minimize}} \quad \underbrace{f(x) + \delta_{\{0\}}(\tilde{x})}_{=\tilde{f}(x,\tilde{x})} + \underbrace{g(Ax + M^{1/2}\tilde{x})}_{=\tilde{g}(x,\tilde{x})} + \underbrace{h(x)}_{=\tilde{h}(x,\tilde{x})}.$$

The DYS FPI is

$$(z^{k+1}, \tilde{z}^{k+1}) = (\mathbb{I} - \mathbb{J}_{\alpha\partial\tilde{f}} + \mathbb{J}_{\alpha\partial\tilde{g}}(\mathbb{R}_{\alpha\partial\tilde{f}} - \alpha\nabla\tilde{h}\mathbb{J}_{\alpha\partial\tilde{f}}))(z^k, \tilde{z}^k).$$

We let $M = (\beta\alpha)^{-1}I - AA^\intercal$ and get

$$\begin{aligned} x^{k+1} &= \mathrm{Prox}_{\alpha f}\left(x^k - \alpha A^\intercal u^k - \alpha\nabla h(x^k)\right) \\ u^{k+1} &= \mathrm{Prox}_{\beta g^*}\left(u^k + \beta A\left(2x^{k+1} - x^k + \alpha\nabla h(x^k) - \alpha\nabla h(x^{k+1})\right)\right). \end{aligned}$$

This method is called *primal-dual three-operator splitting (PD3O)*. If total duality holds, $\alpha > 0$, $\beta > 0$, $\alpha\beta\lambda_{\max}(A^\intercal A) \le 1$, and $\alpha < 2/L$, then $x^{k+1/2} \to x^\star$. See Exercise 3.12.

Comparison with Condat–Vũ Condat–Vũ and PD3O solve the same problem. Condat–Vũ generalizes PDHG. PD3O generalizes PAPC and PDHG. The two methods are very similar when compared side by side and have essentially identical computational costs per iteration.

The convergence criteria of the two methods slightly differ. Condat–Vũ requires the stricter condition $\alpha\beta\lambda_{\max}(A^\intercal A) + \alpha L/2 < 1$, while PD3O requires $\alpha\beta\lambda_{\max}(A^\intercal A) \le 1$ and $\alpha L/2 < 1$. This difference allows PD3O to use stepsizes that are, roughly speaking, twice as large. In some cases, this leads to PD3O converging twice as fast compared to Condat–Vũ.

Proximal ADMM

Consider the primal problem (3.3):

$$\begin{array}{ll} \underset{x\in\mathbb{R}^p,\,y\in\mathbb{R}^q}{\text{minimize}} & f(x) + g(y) \\ \text{subject to} & Ax + By = c. \end{array}$$

Let $M \succeq 0$, $N \succeq 0$, $P = \alpha^{-1/2}M^{1/2}$, and $Q = \alpha^{-1/2}N^{1/2}$. This problem is equivalent to

$$\begin{array}{ll} \underset{\substack{x\in\mathbb{R}^p,\, y\in\mathbb{R}^q \\ \tilde{x}\in\mathbb{R}^q,\, \tilde{y}\in\mathbb{R}^p}}{\text{minimize}} & f(x) + g(y) \\ \text{subject to} & \begin{bmatrix} A & 0 \\ P & 0 \\ 0 & I \end{bmatrix} \begin{bmatrix} x \\ \tilde{x} \end{bmatrix} + \begin{bmatrix} B & 0 \\ 0 & I \\ Q & 0 \end{bmatrix} \begin{bmatrix} y \\ \tilde{y} \end{bmatrix} = \begin{bmatrix} c \\ 0 \\ 0 \end{bmatrix}. \end{array}$$

Applying ADMM to this problem gives us

$$\begin{aligned} x^{k+1} &\in \underset{x\in\mathbb{R}^p}{\operatorname{argmin}} \left\{ \mathbf{L}_\alpha(x, y^k, u^k) + \langle \tilde{u}_1^k, Px\rangle + \frac{\alpha}{2}\|Px + \tilde{y}^k\|^2 \right\} \\ \tilde{x}^{k+1} &= \underset{\tilde{x}\in\mathbb{R}^q}{\operatorname{argmin}} \left\{ \langle \tilde{u}_2^k, \tilde{x}\rangle + \frac{\alpha}{2}\|\tilde{x} + Qy^k\|^2 \right\} \\ &= -Qy^k - (1/\alpha)\tilde{u}_2^k \\ y^{k+1} &\in \underset{y\in\mathbb{R}^q}{\operatorname{argmin}} \left\{ \mathbf{L}_\alpha(x^{k+1}, y, u^k) + \langle \tilde{u}_2^k, Qy\rangle + \frac{\alpha}{2}\|\tilde{x}^{k+1} + Qy\|^2 \right\} \\ \tilde{y}^{k+1} &= \underset{\tilde{y}\in\mathbb{R}^p}{\operatorname{argmin}} \left\{ \langle \tilde{u}_1^k, \tilde{y}\rangle + \frac{\alpha}{2}\|Px^{k+1} + \tilde{y}\|^2 \right\} \\ &= -Px^{k+1} - (1/\alpha)\tilde{u}_1^k \\ u^{k+1} &= u^k + \alpha(Ax^{k+1} + By^{k+1} - c) \\ \tilde{u}_1^{k+1} &= \tilde{u}_1^k + \alpha(Px^{k+1} + \tilde{y}^{k+1}) = 0 \\ \tilde{u}_2^{k+1} &= \tilde{u}_2^k + \alpha(\tilde{x}^{k+1} + Qy^{k+1}) = \alpha Q(y^{k+1} - y^k). \end{aligned}$$

We simplify this to

$$\begin{aligned} x^{k+1} &\in \underset{x}{\operatorname{argmin}} \left\{ \mathbf{L}_\alpha(x, y^k, u^k) + \frac{1}{2}\|x - x^k\|_M^2 \right\} \\ y^{k+1} &\in \underset{y}{\operatorname{argmin}} \left\{ \mathbf{L}_\alpha(x^{k+1}, y, u^k) + \frac{1}{2}\|y - y^k\|_N^2 \right\} \\ u^{k+1} &= u^k + \alpha(Ax^{k+1} + By^{k+1} - c). \end{aligned}$$

This method is called the *proximal alternating direction method of multipliers* or *proximal ADMM*. If total duality, $M \succeq 0$, $N \succeq 0$, $(\mathcal{R}(A^\intercal) + \mathcal{R}(M)) \cap \operatorname{ri}\operatorname{dom} f^* \neq \emptyset$, $(\mathcal{R}(B^\intercal) + \mathcal{R}(N)) \cap \operatorname{ri}\operatorname{dom} g^* \neq \emptyset$, and $\alpha > 0$ hold, then $u^k \to u^\star$, $Ax^k \to Ax^\star$, $Mx^k \to Mx^\star$, $By^k \to By^\star$, and $Ny^k \to Ny^\star$.

Convergence Analysis The Lagrangian

$$\mathbf{L}(x, y, u, \tilde{u}_1, \tilde{u}_2) = f(x) + g(y) + \langle u, Ax + By - c\rangle + \langle \tilde{u}_1, Px + \tilde{y}\rangle + \langle \tilde{u}_2, \tilde{x} + Qy\rangle$$

generates the equivalent primal problem. $\mathbf{L}$ generates the dual problem

$$\underset{u\in\mathbb{R}^n}{\text{maximize}} \ -f^*(-A^\intercal u - P\tilde{u}_1) - \delta_{\{0\}}(-\tilde{u}_2) - g^*(-B^\intercal u - Q\tilde{u}_2) - \delta_{\{0\}}(-\tilde{u}_1) - c^\intercal u$$

If the original problems (3.3) and (3.4) have solutions $(x^\star, y^\star)$ and $u^\star$ for which strong duality holds, then the equivalent problems have solutions $(x^\star, y^\star)$ and $(u^\star, 0)$ for which strong duality holds. In other words, total duality of the original problems implies total duality of the equivalent problems. So under the stated assumptions, ADMM applied to the equivalent problem converges, and we get the stated convergence results.

Finally, note that the equivalent dual problem resembles what we had when we applied the BCV technique to PDHG. What we did is the BCV technique applied to the dual.

Linearized ADMM

Consider the primal problem (3.3),

$$\begin{array}{ll} \underset{x \in \mathbb{R}^p,\, y \in \mathbb{R}^q}{\text{minimize}} & f(x) + g(y) \\ \text{subject to} & Ax + By = c. \end{array}$$

Let $M = (1/\beta)I - \alpha A^\intercal A$ and $N = (1/\gamma)I - \alpha B^\intercal B$. Proximal ADMM applied to this setup is

$$\begin{aligned} x^{k+1} &= \underset{x}{\operatorname{argmin}} \left\{ f(x) + \langle u^k, Ax \rangle + \alpha \langle Ax, Ax^k + By^k - c \rangle + \frac{1}{2\beta} \|x - x^k\|^2 \right\} \\ y^{k+1} &= \underset{y}{\operatorname{argmin}} \left\{ g(y) + \langle u^k, By \rangle + \alpha \langle By, Ax^{k+1} + By^k - c \rangle + \frac{1}{2\gamma} \|y - y^k\|^2 \right\} \\ u^{k+1} &= u^k + \alpha(Ax^{k+1} + By^{k+1} - c), \end{aligned}$$

which we can also write as

$$\begin{aligned} x^{k+1} &= \operatorname{Prox}_{\beta f}\left(x^k - \beta A^\intercal(u^k + \alpha(Ax^k + By^k - c))\right) \\ y^{k+1} &= \operatorname{Prox}_{\gamma g}\left(y^k - \gamma B^\intercal(u^k + \alpha(Ax^{k+1} + By^k - c))\right) \\ u^{k+1} &= u^k + \alpha(Ax^{k+1} + By^{k+1} - c). \end{aligned}$$

This method is called *linearized ADMM*. If total duality holds, $\alpha > 0$, $\beta > 0$, $\gamma > 0$, $\alpha\beta\lambda_{\max}(A^\intercal A) \le 1$, and $\alpha\gamma\lambda_{\max}(B^\intercal B) \le 1$, then $x^k \to x^\star$, $y^k \to y^\star$, and $u^k \to u^\star$.

Convergence Analysis Under the stated assumptions, the convergence results for proximal ADMM tell us $u^k \to u^\star$. We furthermore have $Ax^k \to Ax^\star$, which implies $\alpha AA^\intercal x^k \to \alpha AA^\intercal x^\star$, and $x \to Mx^\star$. Since $\alpha A^\intercal A + M = \beta^{-1}I$, we add the two convergence results to get $x^k \to x^\star$. We can show $y^k \to y^\star$ with a similar argument.

PDHG

Consider the problem

$$\begin{array}{ll} \underset{y \in \mathbb{R}^m,\, x \in \mathbb{R}^n}{\text{minimize}} & g(y) + f(x) \\ \text{subject to} & -Iy + Ax = 0, \end{array}$$

which is equivalent to (1.9), the primal problem for PDHG.

Linearized ADMM applied to this problem is

$$\begin{aligned} y^{k+1} &= \operatorname{Prox}_{\beta g}\left(y^k + \beta(u^k - \alpha(y^k - Ax^k))\right) \\ x^{k+1} &= \operatorname{Prox}_{\gamma f}\left(x^k - \gamma A^\intercal(u^k - \alpha(y^{k+1} - Ax^k))\right) \\ u^{k+1} &= u^k - \alpha(y^{k+1} - Ax^{k+1}). \end{aligned}$$

Let $\beta = 1/\alpha$ and use the Moreau identity to get

$$\begin{aligned} y^{k+1} &= (1/\alpha)u^k + Ax^k - (1/\alpha)\underbrace{\mathrm{Prox}_{\alpha g^*}\left(u^k + \alpha A x^k\right)}_{=\mu^{k+1}} \\ x^{k+1} &= \mathrm{Prox}_{\gamma f}\left(x^k - \gamma A^\top \mu^{k+1}\right) \\ u^{k+1} &= \mu^{k+1} + \alpha A(x^{k+1} - x^k), \end{aligned}$$

and we recover PDHG:

$$\begin{aligned} \mu^{k+1} &= \mathrm{Prox}_{\alpha g^*}\left(\mu^k + \alpha A(2x^k - x^{k-1})\right) \\ x^{k+1} &= \mathrm{Prox}_{\gamma f}\left(x^k - \gamma A^\top \mu^{k+1}\right). \end{aligned}$$

If total duality, $\alpha > 0$, $\gamma > 0$, and $\alpha\gamma\lambda_{\max}(A^\top A) \leq 1$ hold, then $\mu^k \to u^\star$ and $x^k \to x^\star$.

3.6 DISCUSSION

In this section, we derived and established convergence of a wide range of splitting methods through reducing them to another method for which we have already established convergence. At a detailed level, the many techniques are not obvious, and the execution often spans many lines of calculations. At a high level, however, the approach is conceptually simple, as the theoretical basis of the convergence all reduce to Theorem 1.

At this point, it is natural to ask how one should choose the appropriate optimization method among the numerous ones that have been discussed. In practice, a given problem usually has at most a few methods that apply conveniently. Among the possible options, a good rule of thumb is to first consider methods with a low per-iteration cost.

BIBLIOGRAPHICAL NOTES

One key message of this chapter is that many operator splitting methods are closely interconnected. This interconnectivity has been studied in prior works by Combettes, Condat, Pesquet, and Vũ in 2014 [CCPV14], Yan and Yin in 2016 [YY16], Moursi and Zinchenko in 2018 [MZ19], O'Connor and Vandenberghe in 2020 [OV20], and Condat, Kitahara, Contreras, and Hirabayashi in 2020 [CKCH22].

Methods In 2011, Chambolle and Pock published PDHG (the form presented in this chapter) [CP11a] and popularized the method. The method was not named in this work, so many referred to it as the Chambolle–Pock method. However, similar methods were proposed earlier by Pock, Cremers, Bischof, and Chambolle [PCBC09, Equation (21)] in 2009 and by Esser, Zhang, and Chan [EZC10, Equation (2.18) $\alpha_k \equiv 1$] in 2010. More precisely, the algorithm we (along with Chambolle and Pock) consider corresponds to the "PDHGMu" and "PDHGMp" of [EZC10]. The name "primal-dual hybrid gradient (PDHG)" was first used by Zhu and Chan in their 2008 work [ZC08] to describe a similar but different method. Nowadays, the method is more commonly referred to as PDHG rather than Chambolle–Pock, even by Chambolle and Pock themselves [CP16a,

Section 5.1]. PDHG and its variants were initially not presented as instances of the variable metric PPM; this interpretation is due to He and Yuan in 2012 [HY12b]. Boţ, Csetnek, Heinrich, and Hendrich's 2015 work [BCHH15] and Chambolle and Pock's 2016 work provide further refined analyses of PDHG [CP16b].

PAPC/PDFP^{2}O was independently proposed three different times: by Loris and Verhoeven in 2011 for the case where h is quadratic [LV11], Chen, Huang, and Zhang in 2013 under the name PDFP^{2}O [LV11], and by Drori, Sabach, and Teboulle in 2015 under the name PAPC [DST15]. Combettes, Condat, Pesquet, and Vũ reinterpreted the method as an instance of variable-metric FBS in 2014 [CCPV14]. Li and Yan [LY17] improved the convergence analysis of PDFP^{2}O/PAPC and relaxed the stepsize requirement to $\alpha\beta\lambda_{\max}(A^\intercal A) \le 4/3$.

The Condat–Vũ method was presented independently by Condat and Vũ in 2013 [Con13, Vũ13a]. PD3O was presented by Yan in 2018 [Yan18a], and it was reinterpreted as an instance of DYS by O'Connor and Vandenberghe in 2020 [OV20].

For the historical discussion of ADMM, see the bibliographical notes section of §8.

Regularity Conditions for ADMM For the ADMM iterations to be well defined, one must either assume certain regularity conditions or directly assume the subproblems are solvable. The influential review paper of Boyd et al. [BPC+11], which introduced ADMM to the broad machine learning community, mistakenly claimed that the ADMM iterations are well defined when f and g are CCP. This error was pointed out by Chen, Sun, and Toh [CST17b].

Technique To the best of our knowledge, the first published instance of the infimal postcomposition technique is due to Yan and Yin in 2016 [YY16], but the technique was likely known earlier. In particular, the insight appears in a homework problem written by Boyd in 2015 or earlier [BD15, Problem 7.1]. A thorough treatment of the infimal postcomposition can be found in [Roc70d, Section 39] or [BC17a, §12.5]. The notion is also referred to as the "image of a convex function." The earliest instances of the dualization technique are Rockafellar's 1976 work showing that the augmented Lagrangian method for a linearly constrained convex problem is PPM applied to its dual problem [Roc76b], Gabay's 1983 work showing ADMM is DRS applied to the dual [Gab83], and Tseng's 1990 work deriving AMA as FBS applied to the dual [Tse90b, Tse91]. For the historical discussion of the variable metric technique, see the bibliographical notes section of §2. The origin of the Gaussian elimination technique is unclear; to the best of our knowledge, this chapter is the first instance where the Gaussian elimination technique is articulated, and the name is due to us. However, the idea was likely known prior to the writing of this book. For the historical discussion of the linearization technique, see the bibliographical notes section of §8. The BCV technique was independently presented in the 2016 edition of Bertsekas's book [Ber16, Chapter 7.4.2], where the technique is used to obtain a version of the proximal ADMM from the regular ADMM, and by O'Connor and Vandenberghe's 2020 paper [OV20], where it is used to obtain PDHG from DRS. Although the two derivations seem different at first sight, they are, loosely speaking, equivalent under duality.

EXERCISES

3.1 *Prox of infimal postcomposition.* Let f be CCP. Show that if $\mathcal{R}(A^\intercal) \cap \operatorname{ri}\operatorname{dom} f^* \neq \emptyset$, then

$$\begin{aligned} x &\in \underset{x}{\operatorname{argmin}} \{f(x) + (1/2)\|Ax - y\|^2\} \\ z &= Ax \end{aligned} \quad \Leftrightarrow \quad z = \operatorname{Prox}_{A \triangleright f}(y),$$

and the argmin of the left-hand side exists.
Hint. Use Exercise 1.5 and show

$$\underset{z}{\operatorname{argmin}} \left\{ \inf_{x \in \{x \mid Ax = z\}} f(x) + \frac{1}{2}\|Ax - y\|^2 \right\} = \operatorname{Prox}_{A \triangleright f}(y).$$

3.2 *Proximal ADMM from KKT operator.* Consider the primal-dual problem pair (3.3) and (3.4) generated by the Lagrangian $\mathbf{L}$ of (3.5). Split the Lagrangian into

$$\mathbf{L}(x, y, u) = \underbrace{f(x) + \langle u, Ax \rangle}_{=\mathbf{L}_1(x,y,u)} + \underbrace{g(y) + \langle u, By - c \rangle}_{=\mathbf{L}_2(x,y,u)}.$$

Show that the FPI with DRS

$$(\xi^{k+1}, \zeta^{k+1}, \omega^{k+1}) = \left(\frac{1}{2}I + \frac{1}{2}R_{\alpha\partial\mathbf{L}_1} R_{\alpha\partial\mathbf{L}_2}\right)(\xi^k, \zeta^k, \omega^k)$$

simplifies to

$$\begin{aligned} x^{k+1} &= \underset{x}{\operatorname{argmin}} \left\{ \mathbf{L}_\alpha(x, y^k, u^k) + \frac{1}{2\alpha}\|x - x^k\|_2^2 \right\} \\ y^{k+1} &= \underset{y}{\operatorname{argmin}} \left\{ \mathbf{L}_\alpha(x^{k+1}, y, u^k) + \frac{1}{2\alpha}\|y - y^k\|_2^2 \right\} \\ u^{k+1} &= u^k + \alpha(Ax^{k+1} + By^{k+1} - c), \end{aligned}$$

where $\mathbf{L}_\alpha$ is the augmented Lagrangian of (3.7). Show that if total duality holds and $\alpha > 0$, then $x^k \to x^\star$, $y^k \to y^\star$, and $u^k \to u^\star$.

3.3 *ADMM primal convergence.* In the setup of ADMM, show that if g and f are strictly convex in addition to the stated convergence conditions, then $y^k \to y^\star$ and $x^k \to x^\star$, where $(x^\star, y^\star)$ is the primal solution. Use the following fact: if h is a CCP function that is differentiable on $D \subseteq \mathbb{R}^n$, then $\nabla h \colon D \to \mathbb{R}^n$ is a continuous function, that is, differentiability and continuous differentiability coincide.
Remark. The stated conditions are f and g are CCP, $\mathcal{R}(A^\intercal) \cap \operatorname{ri}\operatorname{dom} f^* \neq \emptyset$, $\mathcal{R}(B^\intercal) \cap \operatorname{ri}\operatorname{dom} g^* \neq \emptyset$, $\mathbf{L}(x, y, u) = f(x) + g(y) + \langle u, Ax + By - c \rangle$ has a saddle point, and $\alpha > 0$.

3.4 *3-block extension of ADMM with DYS.* Consider the problem

$$\begin{array}{ll} \underset{x,y,z}{\text{minimize}} & f(x) + g(y) + h(z) \\ \text{subject to} & Ax + By + Cz = d, \end{array}$$

where $x \in \mathbb{R}^p$, $y \in \mathbb{R}^q$, $z \in \mathbb{R}^r$ are the optimization variables and $A \in \mathbb{R}^{n \times p}$, $B \in \mathbb{R}^{n \times q}$, $C \in \mathbb{R}^{n \times r}$, and $d \in \mathbb{R}^n$. This is the primal problem generated by the Lagrangian

$$\mathbf{L}(x, y, z, u) = f(x) + g(y) + h(z) + \langle u, Ax + By + Cz - d \rangle.$$

Assume f, g, and h are CCP, and furthermore assume h is μ-strongly convex. Show that the dualization technique and DYS leads to the method

$$\begin{aligned}
z^{k+1} &= \operatorname*{argmin}_z \left\{ \mathbf{L}(x^k, y^k, z, u^k) \right\} \\
y^{k+1} &\in \operatorname*{argmin}_y \left\{ \mathbf{L}(x^k, y, z^{k+1}, u^k) + \frac{\alpha}{2}\|Ax^k + By + Cz^{k+1} - d\|^2 \right\} \\
x^{k+1} &\in \operatorname*{argmin}_x \left\{ \mathbf{L}(x, y^{k+1}, z^{k+1}, u^k) + \frac{\alpha}{2}\|Ax + By^{k+1} + Cz^{k+1} - d\|^2 \right\} \\
u^{k+1} &= u^k + \alpha(Ax^{k+1} + By^{k+1} + Cz^{k+1} - d).
\end{aligned}$$

Under what conditions does this method converge?

3.5 *Condat–Vũ, the other version.* In the derivation of Condat–Vũ, show that if we instead use

$$M = \begin{bmatrix} (1/\alpha)I & A^\intercal \\ A & (1/\beta)I \end{bmatrix},$$

we get the method

$$\begin{aligned}
u^{k+1} &= \mathrm{Prox}_{\beta g^*}(u^k + \beta A x^k) \\
x^{k+1} &= \mathrm{Prox}_{\alpha f}(x^k - \alpha A^\intercal(2u^{k+1} - u^k) - \alpha \nabla h(x^k)).
\end{aligned}$$

Also show that if total duality holds, h is L-smooth, $\alpha > 0$, $\beta > 0$, and (3.13) holds, then $x^k \to x^\star$ and $u^k \to u^\star$.

Remark. Doing the same with $h = 0$ gives us the other version of PDHG:

$$\begin{aligned}
u^{k+1} &= \mathrm{Prox}_{\beta g^*}(u^k + \beta A x^k) \\
x^{k+1} &= \mathrm{Prox}_{\alpha f}(x^k - \alpha A^\intercal(2u^{k+1} - u^k)).
\end{aligned}$$

3.6 *PDHG generalizes DRS.* PDHG with $A = I$ and $\beta = 1/\alpha$ is

$$\begin{aligned}
x^{k+1} &= \mathrm{Prox}_{\alpha f}(x^k - \alpha u^k) \\
u^{k+1} &= \mathrm{Prox}_{(1/\alpha)g^*}(u^k + (1/\alpha)(2x^{k+1} - x^k)).
\end{aligned}$$

DRS with $\mathrm{Prox}_{\alpha f}$ applied first is

$$\begin{aligned}
x^{k+1/2} &= \mathrm{Prox}_{\alpha f}(z^k) \\
x^{k+1} &= \mathrm{Prox}_{\alpha g}(2x^{k+1/2} - z^k) \\
z^{k+1} &= z^k + x^{k+1} - x^{k+1/2}.
\end{aligned}$$

Show that the two methods are equivalent in the sense that they generate an identical sequence of iterates after a change of variables.

Hint. For PDHG, define $\tilde{z}^k = x^k - \alpha u^k$.

Remark. The BCV technique establishes the converse, that DRS generalizes PDHG.

PD3O generalizes DYS. PD3O with $A = I$ and $\beta = 1/\alpha$ is

$$\begin{aligned}
x^{k+1} &= \mathrm{Prox}_{\alpha f}(x^k - \alpha u^k - \alpha \nabla h(x^k)) \\
u^{k+1} &= \mathrm{Prox}_{(1/\alpha)g^*}(u^k + (1/\alpha)(2x^{k+1} - x^k) + \nabla h(x^k) - \nabla h(x^{k+1})).
\end{aligned}$$

DYS with $\mathrm{Prox}_{\alpha f}$ applied first is

$$\begin{aligned}
x^{k+1/2} &= \mathrm{Prox}_{\alpha f}(z^k) \\
x^{k+1} &= \mathrm{Prox}_{\alpha g}(2x^{k+1/2} - z^k - \alpha \nabla h(x^{k+1/2})) \\
z^{k+1} &= z^k + x^{k+1} - x^{k+1/2}.
\end{aligned}$$

Show that the two methods are equivalent in the sense that they generate an identical sequence of iterates after a change of variables.
Remark. The BCV technique establishes that DYS generalizes PD3O.

3.7 *Preconditioned PDHG.* Consider the problem

$$\underset{x\in\mathbb{R}^n}{\text{minimize}} \quad f(x) + g(Ax),$$

where $A \in \mathbb{R}^{m\times n}$ and f and g are CCP, and show that

$$\begin{aligned} x^{k+1} &= (N + \partial f)^{-1}(Nx^k - A^\intercal u^k) \\ &= \underset{x}{\text{argmin}} \left\{ f(x) + \frac{1}{2}\|x - (x^k - N^{-1}A^\intercal u^k)\|_N^2 \right\} \\ u^{k+1} &= (M + \partial g^*)^{-1}(Mu^k + A(2x^{k+1} - x^k)) \\ &= \underset{u}{\text{argmin}} \left\{ g^*(u) + \frac{1}{2}\|u - (u^k + M^{-1}A(2x^{k+1} - x^k))\|_M^2 \right\}, \end{aligned} \tag{3.18}$$

where $N \in \mathbb{R}^{n\times n}$ and $M \in \mathbb{R}^{m\times m}$ are symmetric positive definite, converges when

$$\begin{bmatrix} N & -A^\intercal \\ -A & M \end{bmatrix} \succ 0.$$

Remark. When $N \neq I$ or $M \neq I$, (3.18) is called *preconditioned* PDHG. When N and M are diagonal and $N \neq I$ or $M \neq I$, (3.18) is called *diagonally preconditioned* PDHG. Preconditioning is essential for PDHG to work well in practice [PC11].

3.8 *Doubly-linearized method of multipliers.* Consider the primal problem

$$\begin{array}{ll} \underset{x\in\mathbb{R}^n}{\text{minimize}} & f(x) + h(x) \\ \text{subject to} & Ax = b, \end{array}$$

where $A \in \mathbb{R}^{m\times n}$, $b \in \mathbb{R}^m$, f and h are CCP, and h is differentiable, generated by the Lagrangian

$$\mathbf{L}(x,u) = f(x) + h(x) + \langle u, Ax - b\rangle.$$

Show that the FPI with $(M + \mathbb{G})^{-1}(M - \mathbb{H})$ with $\mathbb{G}$ and $\mathbb{H}$ defined as in (3.15) and

$$M = \begin{bmatrix} (1/\alpha)I - \beta A^\intercal A & 0 \\ 0 & (1/\beta)I \end{bmatrix}$$

gives us

$$\begin{aligned} x^{k+1} &= \text{Prox}_{\alpha f}\left(x^k - \alpha\nabla h(x^k) - \alpha A^\intercal(u^k + \beta(Ax^k - b))\right) \\ u^{k+1} &= u^k + \beta(Ax^{k+1} - b). \end{aligned}$$

Under what conditions does this method converge? Note that Condat–Vũ and PD3O can be used to solve this problem. How do the algorithms and their convergence conditions compare?
Remark. This method is presented in [LY17, Section 3]. This method is useful when f is proximable but $f + h$ is not.

3.9 *Constraint relaxation.* The constraint $Ax = b$ is equivalent to the objective function $\delta_{\{0\}}(Ax - b)$. When we do not expect $Ax = b$ to hold (due to errors or noise), we can minimize its violation $\ell(Ax - b)$ with some loss function ℓ. Consider

$$\underset{x \in \mathbb{R}^n}{\text{minimize}}\ f(x) + h(x) + \ell(Ax - b),$$

where f is CCP, h is CCP and L-smooth, and ℓ is CCP and μ-strongly convex. The primal problem is generated by the Lagrangian

$$\mathbf{L}(x, u) = f(x) + h(x) + \langle u, Ax - b \rangle - \ell^*(u).$$

Consider the decomposition

$$\partial \mathbf{L}(x, u) = \underbrace{\begin{bmatrix} \nabla h(x) \\ \nabla \ell^*(u) + b \end{bmatrix}}_{=\mathbb{H}(x,u)} + \underbrace{\begin{bmatrix} 0 & A^\intercal \\ -A & 0 \end{bmatrix} \begin{bmatrix} x \\ u \end{bmatrix} + \begin{bmatrix} \partial f(x) \\ 0 \end{bmatrix}}_{=\mathbb{G}(x,u)}.$$

Show that the FPI with $(M + \mathbb{G})^{-1}(M - \mathbb{H})$, where

$$M = \begin{bmatrix} (1/\alpha)I - \beta A^\intercal A & 0 \\ 0 & (1/\beta)I \end{bmatrix}$$

gives us

$$\begin{aligned} x^{k+1} &= \mathrm{Prox}_{\alpha f}\left(x^k - \alpha \nabla h(x^k) - \alpha A^\intercal (u^k + \beta(Ax^k - b - \nabla \ell^*(u^k)))\right) \\ u^{k+1} &= u^k + \beta\left(Ax^{k+1} - b - \nabla \ell^*(u^k)\right). \end{aligned}$$

Under what conditions does this method converge?

Remark. This method is presented in [LY17, Section 2]. When ℓ^* is not proximable, this method is applicable, while Condat–Vũ and PD3O are not. This method generalizes the method of Exercise 3.8 since $\nabla \ell^*$ vanishes when $\ell = \delta_{\{0\}}$ and $\ell^* = 0$.

3.10 *Linearized method of multipliers with BCV.* We used the linearization technique with the proximal method of multipliers to prove convergence of the linearized method of multipliers for $\alpha\beta\lambda_{\max}(A^\intercal A) < 1$. By using the BCV technique, show that in fact $u^k \to u^\star$ for $\alpha\beta\lambda_{\max}(A^\intercal A) \le 1$.

Hint. Apply ADMM to

$$\begin{array}{ll} \underset{x \in \mathbb{R}^p,\, \tilde{y} \in \mathbb{R}^p}{\text{minimize}} & f(x) \\ \text{subject to} & \begin{bmatrix} A \\ P \end{bmatrix} x + \begin{bmatrix} 0 \\ I \end{bmatrix} \tilde{y} = \begin{bmatrix} b \\ 0 \end{bmatrix}. \end{array}$$

3.11 *PD3O generalizes PAPC/PDFP^{2}O.* PD3O with $f = 0$ is

$$\begin{aligned} x^{k+1} &= x^k - \alpha A^\intercal u^k - \alpha \nabla h(x^k) \\ u^{k+1} &= \mathrm{Prox}_{\beta g^*}\left(u^k + \beta A\left(2x^{k+1} - x^k + \alpha \nabla h(x^k) - \alpha \nabla h(x^{k+1})\right)\right). \end{aligned}$$

PAPC/PDFP^{2}O is

$$\begin{aligned} u^{k+1} &= \mathrm{Prox}_{\beta g^*}\left(u^k + \beta A(x^k - \alpha A^\intercal u^k - \alpha \nabla h(x^k))\right) \\ x^{k+1} &= x^k - \alpha A^\intercal u^{k+1} - \alpha \nabla h(x^k). \end{aligned}$$

Show that the two methods are equivalent in the sense that they generate an identical sequence of iterates after a change of variables.

3.12 *PD3O.* Show the omitted derivation of PD3O. Furthermore, show that $u^k \to u^\star$, that is, show that the dual variable converges to an optimal dual solution, under the stated conditions.

Remark. The stated conditions are f and h are CCP functions on $\mathbb{R}^n$, h is L-smooth, g is a CCP function on $\mathbb{R}^m$, $A \in \mathbb{R}^{m\times n}$, total duality holds, $\alpha > 0$, $\beta > 0$, $\alpha\beta\lambda_{\max}(A^\intercal A) \le 1$, and $\alpha < 2/L$.

3.13 *Recast to LASSO.* Let h be CCP and differentiable, $A \in \mathbb{R}^{n\times p}$, and $c \in \mathbb{R}^n$. Consider the problem

$$\underset{x\in\mathbb{R}^p}{\text{minimize}} \quad \mu\|x\|_1 + h(Ax - b), \tag{3.19}$$

where $\mu > 0$ is a penalty parameter.

Let us apply the infimal postcomposition technique to obtain the equivalent problem

$$\underset{z}{\text{minimize}} \; (A \rhd \mu\|\cdot\|_1)(z) + h(z - b).$$

The FPI with FBS is

$$z^{k+1} = \mathrm{Prox}_{\alpha A \rhd \mu\|\cdot\|_1}(z^k - \alpha\nabla h(z^k - b)).$$

Show that this is equivalent to

$$\begin{aligned} c^k &= Ax^k - \alpha\nabla h(Ax^k - b) \\ x^{k+1} &\in \underset{x}{\mathrm{argmin}} \left\{ \mu\|x\|_1 + \frac{1}{2\alpha}\|Ax - c^k\|^2 \right\}. \end{aligned}$$

Under what conditions does this method converge?

Remark. The subproblem for the x^{k+1}-iterates is LASSO, which we discussed in §2.7.4. (In fact, the problem at hand is LASSO if $h = \|\cdot\|^2$.) Many sophisticated software packages can effectively solve very large LASSO problems, and the presented method can benefit from such packages.

3.14 *Linearized method of multipliers and PDHG.* Show the linearized method of multipliers equivalent to a special case of PDHG with $g = \delta_{\{0\}}$.

Hint. Start with the linearized method of multipliers and define $v^0 = u^0 + \alpha(Ax^0 - b)$ and $v^{k+1} = v^k + \alpha(A(2x^{k+1} - x^k) - b)$ and eliminate u^k.

3.15 *Chen–Teboulle is variable metric PPM.* Consider the primal problem

$$\underset{x\in\mathbb{R}^n}{\text{minimize}} \quad f(x) + g(Ax),$$

where f is a CCP function on $\mathbb{R}^n$, g is a CCP function on $\mathbb{R}^m$, and $A \in \mathbb{R}^{m\times n}$, generated by the Lagrangian (convex with respect to (x, z) and concave with respect to u)

$$\mathbf{L}(x, z, u) = f(x) + g(z) + \langle u, Ax - z\rangle.$$

Show that the *Chen–Teboulle* method

$$\begin{aligned} p^{k+1} &= u^k + \alpha(Ax^k - z^k) \\ x^{k+1} &= \mathrm{Prox}_{\alpha f}(x^k - \alpha A^\intercal p^{k+1}) \\ z^{k+1} &= \mathrm{Prox}_{\alpha g}(z^k + \alpha p^{k+1}) \\ u^{k+1} &= u^k + \alpha(Ax^{k+1} - z^{k+1}) \end{aligned}$$

is equivalent to an instance of the variable metric proximal point method on $\partial\mathbf{L}$ with

$$M = \begin{bmatrix} \alpha I & 0 & -A^\intercal \\ 0 & \alpha I & I \\ -A & I & \alpha I \end{bmatrix}.$$

Remark. The Chen–Teboulle method was published in 1994 [CT94], and this connection was pointed out by Becker in 2019 [Bec19].

3.16 *Chen–Teboulle is linearized method of multipliers.* Consider the problem

$$\begin{array}{ll} \underset{x\in\mathbb{R}^n,\, z\in\mathbb{R}^m}{\text{minimize}} & f(x) + g(z) \\ \text{subject to} & Ax - z = 0, \end{array}$$

where f is a CCP function on $\mathbb{R}^n$, g is a CCP function on $\mathbb{R}^m$, and $A \in \mathbb{R}^{m\times n}$. Show that the *Chen–Teboulle* method of Exercise 3.15 is equivalent to an instance of the linearized method of multipliers.

Remark. The Chen–Teboulle method was published in 1994 [CT94], and this connection was pointed out by Ma in 2020 [Ma20].

3.17 *Unification of PAPC/PDFP^{2}O and Condat–Vũ.* Consider the primal-dual problem pair

$$\underset{x\in\mathbb{R}^n}{\text{minimize}} \quad h(x) + g(Ax) \qquad\qquad \underset{u\in\mathbb{R}^m}{\text{maximize}} \quad -h^*(-A^\intercal u) - g^*(u),$$

where g is a CCP function on $\mathbb{R}^m$, h is a differentiable CCP function on $\mathbb{R}^n$, and $A \in \mathbb{R}^{m\times n}$, generated by the Lagrangian

$$\mathbf{L}(x,u) = h(x) + \langle u, Ax\rangle - g^*(u).$$

The PAPC/PDFP^{2}O method

$$\begin{aligned} u^{k+1} &= \mathrm{Prox}_{\beta g^*}\left(u^k + \beta A(x^k - \alpha A^\intercal u^k - \alpha\nabla h(x^k))\right) \\ x^{k+1} &= x^k - \alpha A^\intercal u^{k+1} - \alpha\nabla h(x^k) \end{aligned}$$

can be derived as the variable metric FBS with the metric matrix

$$M = \begin{bmatrix} (1/\alpha)I & 0 \\ 0 & (1/\beta)I - \alpha AA^\intercal \end{bmatrix}.$$

The second version of Condat–Vũ (cf. Exercise 3.5),

$$\begin{aligned} u^{k+1} &= \mathrm{Prox}_{\beta g^*}\left(u^k + \beta Ax^k\right) \\ x^{k+1} &= x^k - \alpha A^\intercal(2u^{k+1} - u^k) - \alpha\nabla h(x^k), \end{aligned}$$

can be derived as the variable metric FBS with the metric matrix

$$M = \begin{bmatrix} (1/\alpha)I & A^\intercal \\ A & (1/\beta)I \end{bmatrix}.$$

Let $B \in \mathbb{R}^{m\times n}$ satisfy $AB^\intercal = BA^\intercal$. In this problem, use the metric matrix

$$M = \begin{bmatrix} (1/\alpha)I & B^\intercal \\ B & (1/\beta)I + \alpha(BB^\intercal - AA^\intercal) \end{bmatrix}$$

to derive

$$\begin{aligned} u^{k+1} &= \mathrm{Prox}_{\beta g^*}\left(\beta Ax^k + (I + \alpha\beta(B-A)A^\intercal)u^k + \alpha\beta(B-A)\nabla h(x^k)\right) \\ x^{k+1} &= x^k - \alpha A^\intercal u^{k+1} - \alpha\nabla h(x^k) - \alpha B^\intercal(u^{k+1} - u^k) \end{aligned}$$

as an instance of variable metric FBS. Note that when $B = 0$, we recover PAPC, and when $B = A$, we recover Condat–Vũ. Show that the method converges if total duality holds, h is L-smooth, $\alpha > 0$, $\beta > 0$, $\alpha < \frac{2}{L}$, and

$$\alpha\beta\lambda_{\max}(B^\intercal B) < \left(\frac{2}{\alpha L} - 1\right)(1 - \alpha\beta\lambda_{\max}(A^\intercal A)).$$

You may use

$$M^{-1} = \begin{bmatrix} \alpha I + \alpha^2\beta B^\intercal(I - \alpha\beta A^\intercal A)^{-1}B & -\alpha\beta B^\intercal(I - \alpha\beta AA^\intercal)^{-1} \\ -\alpha\beta(I - \alpha\beta A^\intercal A)^{-1}B & \beta(I - \alpha\beta AA^\intercal)^{-1} \end{bmatrix}$$

without proof.

Hint. $AB^\intercal = BA^\intercal$ implies $BB^\intercal - AA^\intercal = (B - A)(A + B)^\intercal$.

Remark. This unification was presented by Ko, Yu, and Won in 2019 [KYW19].

3.18 *Variable metric DYS.* Consider the 3-operator splitting problem

$$\underset{x\in\mathbb{R}^d}{\text{find}} \quad 0 \in (\mathbb{A} + \mathbb{B} + \mathbb{C})x,$$

where $\mathbb{A}$, $\mathbb{B}$, and $\mathbb{C}$ are maximal monotone operators on $\mathbb{R}^d$ and $\operatorname{Zer}(\mathbb{A} + \mathbb{B} + \mathbb{C}) \neq \emptyset$. Show that if $M \in \mathbb{R}^{d\times d}$ is symmetric positive definite and $M^{-1/2}\mathbb{C}M^{-1/2}$ is γ-cocoercive, then

$$\begin{aligned} x^{k+1/2} &= \mathbb{J}_{\alpha M^{-1}\mathbb{B}}\left(z^k\right) \\ x^{k+1} &= \mathbb{J}_{\alpha M^{-1}\mathbb{A}}\left(2x^{k+1/2} - z^k - \alpha M^{-1}\mathbb{C}x^{k+1/2}\right) \\ z^{k+1} &= z^k + x^{k+1} - x^{k+1/2}, \end{aligned}$$

with $\alpha \in (0, 2\gamma)$ converges.

3.19 *PD3O via variable metric DYS.* Consider the problem

$$\underset{x\in\mathbb{R}^n}{\text{minimize}} \quad f(x) + h(x) + g(Ax),$$

where $h : \mathbb{R}^n \to \mathbb{R}$ is CCP and L-smooth, f is a CCP function on $\mathbb{R}^n$, g is a CCP function on $\mathbb{R}^m$, and $A \in \mathbb{R}^{m\times n}$. Assume

$$\mathbf{L}(x, u) = f(x) + h(x) + \langle u, Ax\rangle - g^*(u)$$

has a saddle point. Consider the following decomposition of $\partial\mathbf{L}$:

$$\partial\mathbf{L}(x, u) = \underbrace{\begin{bmatrix} A^\intercal u \\ -Ax + \partial g^*(u) \end{bmatrix}}_{=\mathbb{A}(x,u)} + \underbrace{\begin{bmatrix} \partial f(x) \\ 0 \end{bmatrix}}_{=\mathbb{B}(x,u)} + \underbrace{\begin{bmatrix} \nabla h(x) \\ 0 \end{bmatrix}}_{=\mathbb{C}(x,u)}.$$

Let $\alpha > 0$ and $\beta > 0$ and consider the metric matrix

$$M = \begin{bmatrix} I & 0 \\ 0 & \frac{\alpha}{\beta}I - \alpha^2 AA^\intercal \end{bmatrix}.$$

(a) Under what condition is M positive definite?

(b) Under what condition does the variable metric DYS

$$\begin{aligned} y^{k+1/2} &= \mathbb{J}_{\alpha M^{-1}\mathbb{B}}\left(z^k\right) \\ y^{k+1} &= \mathbb{J}_{\alpha M^{-1}\mathbb{A}}\left(2y^{k+1/2} - z^k - \alpha M^{-1}\mathbb{C}y^{k+1/2}\right) \\ z^{k+1} &= z^k + y^{k+1} - y^{k+1/2} \end{aligned}$$

with the given decomposition of $\partial \mathbf{L}$ converge?

(c) Show that the variable metric DYS in (b) is equivalent to PD3O:

$$x^{k+1} = \mathrm{Prox}_{\alpha f}\left(x^k - \alpha A^\intercal u^k - \alpha \nabla h(x^k)\right)$$
$$u^{k+1} = \mathrm{Prox}_{\beta g^*}\left(u^k + \beta A\left(2x^{k+1} - x^k + \alpha \nabla h(x^k) - \alpha \nabla h(x^{k+1})\right)\right).$$

Hint. For (c), use

$$z^k = \begin{bmatrix} p^k \\ q^k \end{bmatrix}, \quad y^{k+1/2} = \begin{bmatrix} x^k \\ w^k \end{bmatrix}, \quad y^{k+1} = \begin{bmatrix} r^k \\ u^k \end{bmatrix}.$$

Remark. In Exercise 3.12, we derived PD3O via the BCV technique and obtained the stepsize requirement $\alpha\beta\lambda_{\max}(AA^\intercal) \leq 1$ and $\alpha < 2/L$. Since the purpose of this problem is to obtain an alternate derivation, you may not appeal to the prior analysis in your answers for this problem.

Remark. This derivation of PD3O as variable metric DYS was first presented by Yan [Yan18b] in his presentation slides and was later formally published by Salim, Condat, Mishchenko, and Richtárik in 2020 [SCMR20].

4 Parallel Computing

In this chapter, we briefly discuss the basic notion of computational complexity and parallel computing. The notion of computational complexity we consider is, in a sense, incomplete as it accounts only for the cost of arithmetic operations, while ignoring other costs such as the cost of coordination and communication between computational agents. Nevertheless, this framework is a useful approximation for analyzing the running time of algorithms.

4.1 COMPUTATIONAL COMPLEXITY VIA FLOP COUNT

A *floating-point operation* or a *flop* is a single arithmetic operation carried out with floating-point numbers. So a single operation of addition, subtraction, multiplication, and division count as a flop. For simplicity, we also count a single evaluation of a non-elementary function such as $\exp(x)$, $\log(x)$, or $\sqrt{x}$ as a single flop.

For example, we can evaluate

$$\|x\| = \sqrt{x_1^2 + \cdots + x_n^2}$$

for $x \in \mathbb{R}^n$ with n multiplications, $n-1$ additions, and 1 square root. In total, $\|x\|$ costs $2n = \mathcal{O}(n)$ flops to compute.

The matrix-vector product Ax, where $A \in \mathbb{R}^{m\times n}$ and $x \in \mathbb{R}^n$, costs $\mathcal{O}(mn)$ flops. The matrix-matrix product AB, where $A \in \mathbb{R}^{m\times n}$ and $B \in \mathbb{R}^{n\times p}$, costs $\mathcal{O}(mnp)$ flops. When computing the ABx, where $A \in \mathbb{R}^{m\times n}$, $B \in \mathbb{R}^{n\times p}$, and $x \in \mathbb{R}^p$, it is better to use the formula $A(Bx)$, which costs $\mathcal{O}(mn+np)$ flops, instead of the formula $(AB)x$, which costs $\mathcal{O}(mnp)$ flops. Given a square matrix $A \in \mathbb{R}^{n\times n}$, the matrix inverse A^{-1} costs $\mathcal{O}(n^3)$ flops.

Modern CPUs operate at a clock speed of about 1 GHz to 5 GHz, and we can expect them to compute roughly 10^9 flops or 1 gigaflop per second. This rough estimate is quite useful in predicting the run time of an algorithm and analyzing where the computational bottleneck of an algorithm will likely be. On the other hand, it is a very rough estimate; expect a 10-fold or even a 100-fold inaccuracy.

Algorithm vs. Method

In this book, the words *algorithm* and *method* both refer to a specification of how to compute a quantity of interest. However, they are different in that a *method* is a higher-level description expressed in mathematical equations, while an *algorithm* is a more literal step-by-step procedure unambiguously describing the steps the computer takes. Although this distinction is not precise, it is useful. If an algorithm carries out the idea described by a method, we say the algorithm *implements* the method and call the algorithm an *implementation* of the method.

In a rigorous discussion, one should ascribe a flop count only to an algorithm, not to a method. As an example, consider the method

$$x^{k+1} = x^k - \alpha A^\intercal (Ax^k - b),$$

where $A \in \mathbb{R}^{m\times n}$ and $b \in \mathbb{R}^m$. The algorithm corresponding to the formula $A^\intercal(Ax^k - b)$ costs $O(mn)$ flops per iteration, while the algorithm corresponding to the formula $(A^\intercal A)x^k - A^\intercal b$ costs $O(n^2)$ flops per iteration, provided that $A^\intercal A \in \mathbb{R}^{n\times n}$ and $A^\intercal b \in \mathbb{R}^m$ have been precomputed and stored. There is often more than one way to implement a method written with mathematical equations. However, when the implementation in consideration is clear from the context, we informally ascribe the flop count to the method.

Flop-Count Operator

Define the *flop-count operator*

$$\mathcal{F}[\{x_1,\dots,x_n\} \mapsto \{y_1,\dots,y_m\} \mid \mathcal{A}]$$

as the number of flops the algorithm $\mathcal{A}$ processes to compute $\{y_1,\dots,y_m\}$, given $\{x_1,\dots,x_n\}$. Again, it is the specific algorithm $\mathcal{A}$, not a method, that determines the flop count. When the algorithm is clear from context, we suppress the dependency on $\mathcal{A}$ and write

$$\mathcal{F}[\{x_1,\dots,x_n\} \mapsto \{y_1,\dots,y_m\}].$$

When the input and/or output is a single quantity, we omit the curly braces and write

$$\mathcal{F}[x \mapsto y].$$

For example, we write

$$\mathcal{F}[x \mapsto \|x\|] = 2n = O(n)$$

and

$$\begin{aligned}\mathcal{F}[A \mapsto (I+\alpha A^\intercal A)^{-1}] &= \mathcal{F}[A \mapsto I+\alpha A^\intercal A] + \mathcal{F}[I+\alpha A^\intercal A \mapsto (I+\alpha A^\intercal A)^{-1}]\\ &= O\left(mn^2\right) + O\left(n^3\right) = O\left((m+n)n^2\right).\end{aligned}$$

As another example, consider

$$\underset{x\in\mathbb{R}^n}{\text{minimize}} \quad \frac{1}{2}\|Ax-b\|^2 + \lambda\|x\|_1,$$

where $A \in \mathbb{R}^{m\times n}$, $b \in \mathbb{R}^m$, and $\lambda > 0$. As discussed in §2.7.4, the FPI with DRS is

$$\begin{aligned} x^{k+1/2} &= (I + \alpha A^\intercal A)^{-1}(z^k + \alpha A^\intercal b) \\ x^{k+1} &= S(2x^{k+1/2} - z^k; \alpha\lambda) \\ z^{k+1} &= z^k + x^{k+1} - x^{k+1/2}, \end{aligned}$$

where S is the soft-thresholding operator. A straightfoward and naive implementation costs

$$\begin{aligned} \mathcal{F}\left[z^k \mapsto z^{k+1}\right] &= \mathcal{F}\left[A \mapsto (I + \alpha A^\intercal A)^{-1}\right] + \mathcal{F}\left[\{z^k, (I + \alpha A^\intercal A)^{-1}\} \mapsto x^{k+1/2}\right] \\ &\quad + \mathcal{F}\left[\{x^{k+1/2}, z^k\} \mapsto x^{k+1}\right] + \mathcal{F}\left[\{z^k, x^{k+1/2}, x^{k+1}\} \mapsto z^{k+1}\right] \\ &= \mathcal{O}\left((m+n)n^2\right) + \mathcal{O}\left((n+m)n\right) + \mathcal{O}\left(n\right) + \mathcal{O}\left(n\right) \\ &= \mathcal{O}\left((m+n)n^2\right) \end{aligned}$$

flops per iteration.

It is possible to reduce this cost. When $m \geq n$, precompute $(I + \alpha A^\intercal A)^{-1}$ with cost

$$\mathcal{F}\left[A \mapsto (I + \alpha A^\intercal A)^{-1}\right] = \mathcal{O}\left(mn^2\right)$$

and $\alpha A^\intercal b$ with cost

$$\mathcal{F}\left[\{\alpha, A, b\} \mapsto \alpha A^\intercal b\right] = \mathcal{O}\left(mn\right).$$

In subsequent iterations, use precomputed quantities to reduce the cost to

$$\begin{aligned} &\mathcal{F}\left[\{z^k, (I + \alpha A^\intercal A)^{-1}, \alpha A^\intercal b\} \mapsto z^{k+1}\right] \\ &= \mathcal{F}\left[\{z^k, (I + \alpha A^\intercal A)^{-1}, \alpha A^\intercal b\} \mapsto x^{k+1/2}\right] + \mathcal{F}\left[\{x^{k+1/2}, z^k\} \mapsto x^{k+1}\right] \\ &\quad + \mathcal{F}\left[\{z^k, x^{k+1/2}, x^{k+1}\} \mapsto z^{k+1}\right] \\ &= \mathcal{O}\left(n^2\right) + \mathcal{O}\left(n\right) + \mathcal{O}\left(n\right) \\ &= \mathcal{O}\left(n^2\right) \end{aligned}$$

flops per iteration.

4.2 PARALLEL COMPUTING

In parallel computing, calculations are carried out simultaneously by multiple computing units, such as multiple cores in a CPU, multiple cores in a GPU, or multiple computers connected over the Internet. An (over)simplified view of parallel computing is to think of a group of computational agents coordinating and working together to complete a single task.

Assume we have p processors. If $A, B \in \mathbb{R}^{m \times n}$ and $p \leq mn$, then $C = A + B$ can be computed with $\mathcal{O}\left(mn/p\right)$ flops for each processor. To see why, consider the following algorithm:

```
parallel for i=1,...,m, j=1,...,n {
  C[i,j] = A[i,j]+B[i,j]
}
```

The "parallel for" loop represents mn independent tasks. If p divides mn, then each of the p processors can perform exactly mn/p out of the mn tasks. Otherwise, partition the mn tasks into p groups of sizes roughly equal to mn/p and assign them to the p processors.

We say a computational task is *embarrassingly parallel* if it takes little to no effort to divide it into parallel parts. (Embarrassingly parallel is good.) For example, the computation of $v = Ax$ is embarrassingly parallel:

```
parallel for i=1,...,m {
  v[i] = 0;
  for j=1,...,n
    v[i] += A[i,j]*x[j]
}
```

Not all computational tasks benefit from parallel computing. Consider the FPI with DRS as in (2.18):

$$\begin{aligned} x^{k+1/2} &= \mathrm{Prox}_{\alpha f}(z^k) \\ x^{k+1} &= \mathrm{Prox}_{\alpha g}(2x^{k+1/2} - z^k) \\ z^{k+1} &= z^k + x^{k+1} - x^{k+1/2}. \end{aligned}$$

Since the evaluation of $\mathrm{Prox}_{\alpha g}$ depends on the evaluation of $\mathrm{Prox}_{\alpha f}$, it is in general not possible to simultaneously compute $\mathrm{Prox}_{\alpha g}$ and $\mathrm{Prox}_{\alpha f}$. When we have $p \le n$ processors, the vector sum $z^k + x^{k+1} - x^{k+1/2}$ can be split up into p independent parts, each costing $O(n/p)$ flops. However, the computational bottleneck is usually in evaluating $\mathrm{Prox}_{\alpha g}$ or $\mathrm{Prox}_{\alpha f}$. It may not be possible to use parallel computing to accelerate the evaluations of $\mathrm{Prox}_{\alpha f}$ and $\mathrm{Prox}_{\alpha g}$, and, if not, this method does not significantly benefit from parallel computing.

Parallel Flop Count Operator

Let $\mathcal{A}$ be an algorithm that utilizes p parallel computing units. More specifically, $\mathcal{A}$ can process up to p flops in parallel each step, provided that the p operations are independent. In some steps, $\mathcal{A}$ may be unable to fully utilize the p computing units and will process fewer than p flops. Define the *parallel flop-count operator*

$$\mathcal{F}_p\left[\{x_1,\ldots,x_n\} \mapsto \{y_1,\ldots,y_m\} \mid \mathcal{A}\right]$$

as the number of such steps $\mathcal{A}$ takes to compute $\{y_1,\ldots,y_m\}$, given $\{x_1,\ldots,x_n\}$. As before, we omit the dependency on $\mathcal{A}$ if the algorithm is clear from context, and we omit the curly braces when the input and/or output is a single quantity.

Parallelizable Methods and Operators

An algorithm is *parallel* if it utilizes multiple computing units and is *serial* otherwise. Loosely speaking, a method is *parallelizable* if it has a parallel implementation that provides a significant speedup using many processors ($p \gg 1$). We say a method is *serial* if it is not parallelizable. What constitutes a "significant" speedup depends on the setup. We say an operator is parellelizable if there is a parallelizable method for evaluating it.

Using the parallel flop count operator, we can express parallelizability of a method for computing $\{y_1, \ldots, y_m\}$, given $\{x_1, \ldots, x_n\}$ as

$$\mathcal{F}_p [\{x_1, \ldots, x_n\} \mapsto \{y_1, \ldots, y_m\}] \ll \mathcal{F} [\{x_1, \ldots, x_n\} \mapsto \{y_1, \ldots, y_m\}]$$

for large enough p. Again, what counts as $\ll$ depends on context, but when

$$\mathcal{F}_p [\{x_1, \ldots, x_n\} \mapsto \{y_1, \ldots, y_m\}] \sim \frac{C}{p} \mathcal{F} [\{x_1, \ldots, x_n\} \mapsto \{y_1, \ldots, y_m\}]$$

for some $C > 0$ not too large, we safely say the method is parallelizable. Likewise, an operator $\mathbb{T}$ is parallelizable if

$$\mathcal{F}_p [x \mapsto \mathbb{T}x] \ll \mathcal{F} [x \mapsto \mathbb{T}x] .$$

Parallel Reduction

Reduction combines a set of numbers into one number with an associative binary operator. A common instance of reduction is the sum

$$x_{\text{sum}} = \sum_{i=1}^{m} x_i,$$

where $x_1, \ldots, x_n \in \mathbb{R}$. With $p = 1$ processor, reduction requires $O(n)$ operations.

With $p \geq \lfloor n/2 \rfloor$ processors, reduction takes $O(\log n)$ steps. To see why, consider the example of $n = 8$ and $p = 4$. The algorithm described by the following diagram takes $\mathcal{F}_p [\{x_1, \ldots, x_8\} \mapsto x_{\text{sum}}] = 3$ steps.

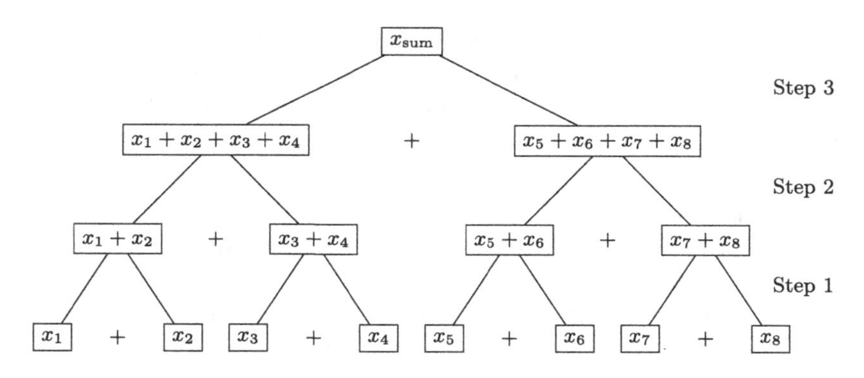

The general strategy is to have the algorithm follow a *binary tree* with $\lfloor n/2 \rfloor$ parallel operations at the bottom and with a depth of $\log_2 n$.

With $p < \lfloor n/2 \rfloor$ processors, reduction takes $O(n/p + \log p)$ steps. To see why, consider the example of $n = 40$ and $p = 4$. The algorithm described by the following diagram takes $\mathcal{F}_p [\{x_1, \ldots, x_{40}\} \mapsto x_{\text{sum}}] = 40/4 - 1 + \log_2 4 = 11$ steps.

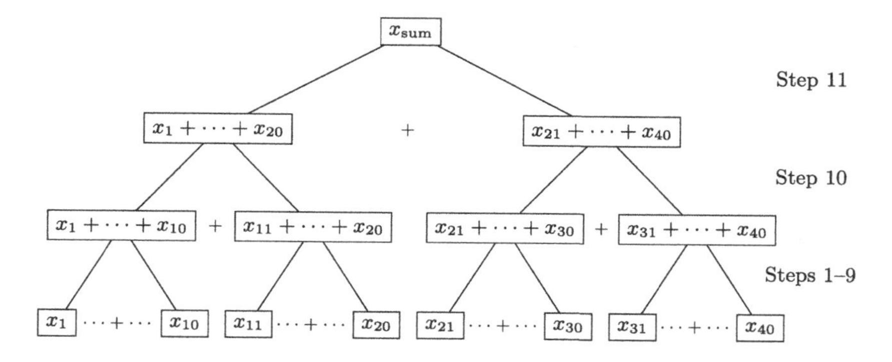

The general strategy is to partition the n numbers into p groups of sizes roughly equal to n/p, take $O(n/p)$ steps for the reduction on the p groups, and then reduce the p numbers with an additional $O(\log p)$ steps.

To summarize,

$$\mathcal{F}_p[\{x_1,\ldots,x_n\} \mapsto x_{\text{sum}}] = \begin{cases} O(n) & \text{if } p = 1 \\ O(n/p + \log p) & \text{if } 1 < p < \lfloor n/2 \rfloor \\ O(\log n) & \text{if } p \geq \lfloor n/2 \rfloor. \end{cases}$$

With a similar strategy, we can compute

- minimum and maximum of $x_1,\ldots,x_n \in \mathbb{R}$,
- arithmetic mean, geometric mean, and product of $x_1,\ldots,x_n \in \mathbb{R}$,
- $\langle x,y \rangle$ for $x,y \in \mathbb{R}^n$, and
- $\|x\|_1$ and $\|x\|_\infty$ for $x \in \mathbb{R}^n$.

Parallel Matrix-Vector Multiplication

Let $A \in \mathbb{R}^{m \times n}$ and $x \in \mathbb{R}^n$ and consider the task of computing the matrix-vector product $b = Ax$. Then

$$\mathcal{F}_p[\{A,x\} \mapsto b] = \begin{cases} O(mn) & \text{if } p = 1 \\ O(mn/p) & \text{if } p \leq m \\ O(mn/p + \log(p/m)) & m < p < mn/2 \\ O(\log n) & \text{if } mn/2 \leq p. \end{cases}$$

To see why, when $p \leq m$, we assign each processor with roughly m/p of the m independent subtasks $b_i = \sum_{j=1}^{n} A_{i,j}x_j$ for $i = 1,\ldots,m$, and when $p > m$, we assign (p/m) processors to compute $b_i = \sum_{j=1}^{n} A_{i,j}x_j$ in $O(n/(p/m) + \log(p/m))$ steps for $i = 1,\ldots,m$, with the strategy used for computing parallel reduction.

The parallel flop count of reduction on m vectors in $\mathbb{R}^n$ follows from the same reasoning: when $x_1,\ldots,x_m \in \mathbb{R}^n$,

$$\mathcal{F}_p[\{x_1,\ldots,x_n\} \mapsto x_1 + \cdots + x_n] = \begin{cases} O(mn) & \text{if } p = 1 \\ O(mn/p) & \text{if } p \leq m \\ O(mn/p + \log(p/m)) & m < p < mn/2 \\ O(\log n) & \text{if } mn/2 \leq p. \end{cases}$$

Other Costs: Coordination and Communication

For parallel computing on a multicore CPU, counting floating-point operations to analyze the computational cost of an algorithm is a useful approximation. However, this may be inadequate for other parallel computing environments.

Parallel computing on a graphics processing unit (GPU) relies on thousands of processors that are individually slower than a CPU's processor but in aggregate provide much more computing power. On GPUs, the cost of coordination and synchronization may be significant and thus should be taken into account.

In distributed and decentralized computing, which we discuss further in §11, many computers operate in parallel and communicate over slow communication channels such as the Internet. In this setup, the cost of communication may be significant and thus should be taken into account.

4.2.1 Examples: Finite-Sum Optimization

When a method relies on linear algebraic operations such as matrix-vector multiplication, it is possible to parallelize the linear algebra. In some cases, however, a method itself is parallelizable at a higher level. We discuss several methods for finite-sum minimization problems and to what extent they can be parallelized.

Sum of Smooth Functions

Consider the optimization problem

$$\underset{x\in\mathbb{R}^n}{\text{minimize}} \quad f(x) + \frac{1}{m}\sum_{i=1}^{m} h_i(x),$$

where f is CCP and $h_1,\ldots,h_m$ are differentiable and CCP. The FPI with FBS, the proximal gradient method, applied to this setup is

$$\begin{aligned} v^k &= -\frac{\alpha}{m}\sum_{i=1}^{m} \nabla h_i(x^k) \\ x^{k+1} &= \mathrm{Prox}_{\alpha f}\left(x^k + v^k\right). \end{aligned}$$

Assume computing $\mathrm{Prox}_{\alpha f}$ costs C_f flops and ∇h_i costs C_h flops (or fewer) for $i = 1,\ldots,m$. Then, for $p \le \min\{m,n\}$,

$$\begin{aligned} &\mathcal{F}_p\left[x^k \mapsto x^{k+1}\right] \\ &= \mathcal{F}_p\left[x^k \mapsto \{\nabla h_i(x^k)\}_{i=1}^m\right] + \mathcal{F}_p\left[\{\nabla h_i(x^k)\}_{i=1}^m \mapsto v^k\right] + \mathcal{F}_p\left[\{x^k, v^k\} \mapsto x^{k+1}\right] \\ &= \mathcal{O}\left(mC_h/p\right) + \mathcal{O}\left(mn/p\right) + \mathcal{O}\left(n/p + C_f\right) \\ &= \mathcal{O}\left((C_h + n)m/p + C_f\right). \end{aligned}$$

Therefore, the method is parallelizable if $C_f = \mathcal{O}\left((C_h + n)m/p\right)$.

Sum of Proximable Functions

Consider the problem

$$\underset{x\in\mathbb{R}^n}{\text{minimize}} \quad f(x) + \frac{1}{m}\sum_{i=1}^{m} g_i(x).$$

where $f, g_1, \ldots, g_m$ are CCP and proximable. Using the consensus technique, which we have seen in §2.7.4, we reformulate this problem into

$$\underset{x_1,\ldots,x_m \in \mathbb{R}^n}{\text{minimize}} \quad f(x_1) + \delta_C(x_1,\ldots,x_m) + \frac{1}{m}\sum_{i=1}^m g_i(x_i),$$

where $C = \{(x_1,\ldots,x_m) \,|\, x_1 = \cdots = x_m\}$. The FPI with DRS is

$$\begin{aligned} x^{k+1/2} &= \mathrm{Prox}_{\alpha f}\left(\frac{1}{m}\sum_{i=1}^m z_i^k\right) \\ x_i^{k+1} &= \mathrm{Prox}_{\alpha g_i}(2x^{k+1/2} - z_i^k) \\ z_i^{k+1} &= z_i^k + x_i^{k+1} - x^{k+1/2} \qquad \text{for } i = 1,\ldots,m. \end{aligned}$$

(See Exercise 2.29.) Assume computing $\mathrm{Prox}_{\alpha f}$ costs C_f flops and $\mathrm{Prox}_{\alpha g_i}$ costs C_g flops (or fewer) for $i = 1,\ldots,m$. Write $\mathbf{z}^k = (z_1^k,\ldots,z_m^k)$. Then, for $p \le m$,

$$\begin{aligned} \mathcal{F}_p\left[\mathbf{z}^k \mapsto \mathbf{z}^{k+1}\right] &= \mathcal{F}_p\left[\mathbf{z}^k \mapsto x^{k+1/2}\right] + \mathcal{F}_p\left[\{\mathbf{z}^k, x^{k+1/2}\} \mapsto \mathbf{z}^{k+1}\right] \\ &= O\left(mn/p + C_f + C_g m/p\right). \end{aligned}$$

Sum of Proximable Functions and a Strongly Convex Function

Consider the primal problem

$$\underset{x \in \mathbb{R}^n}{\text{minimize}} \quad f(x) + \sum_{i=1}^m g_i(a_i^\intercal x - b_i)$$

and the dual problem

$$\underset{u_1,\ldots,u_m \in \mathbb{R}}{\text{maximize}} \quad -f^*\left(-\sum_{i=1}^m u_i a_i\right) - \sum_{i=1}^m (g_i^*(u_i) + b_i u_i)$$

generated by the Lagrangian

$$\mathbf{L}(x, u_1,\ldots,u_m) = f(x) + \sum_{i=1}^m \langle u_i, a_i^\intercal x - b_i\rangle - \sum_{i=1}^m g_i^*(u_i),$$

where $a_1,\ldots,a_m \in \mathbb{R}^n$, $b_1,\ldots,b_m \in \mathbb{R}$, f is a strongly convex CCP function on $\mathbb{R}^n$, and $g_1,\ldots,g_m$ are proximable CCP functions on $\mathbb{R}$. The FPI with FBS, the proximal gradient method, applied to the dual is

$$\begin{aligned} x^k &= \nabla f^*\left(-\sum_{i=1}^m u_i^k a_i\right) \\ u_i^{k+1} &= \mathrm{Prox}_{\alpha g_i^*}\left(u_i^k + \alpha(a_i^\intercal x^k - b_i)\right) \qquad \text{for } i = 1,\ldots,m. \end{aligned}$$

(Since f is strongly convex, f^* is smooth.) Assume computing ∇f^* costs C_f flops and $\mathrm{Prox}_{\alpha g_i^*}$ costs C_g flops (or fewer) for $i = 1,\ldots,m$. Then, for $p \le m$ and $p \le n$,

$$\mathcal{F}_p\left[\{u_1^k,\ldots,u_m^k\} \mapsto \{u_1^{k+1},\ldots,u_m^{k+1}\}\right] = O\left((C_g + n)m/p + C_f\right).$$

Example 4.1 In the *support-vector machine* (SVM) setup, which is widely used in machine learning for classification, we solve

$$\underset{x \in \mathbb{R}^n}{\text{minimize}} \quad \frac{\lambda}{2}\|x\|^2 + \sum_{i=1}^{m} \max\{1 - y_i a_i^\intercal x, 0\},$$

where $a_1, \ldots, a_m \in \mathbb{R}^n$, $y_1, \ldots, y_n \in \{-1, 1\}$, and $\lambda > 0$. The FPI with FBS applied to the dual is

$$\begin{aligned} x^k &= \frac{1}{2\lambda}\left(-\sum_{i=1}^{m} u_i^k y_i a_i\right) \\ u_i^{k+1} &= \Pi_{[-1,0]}\left(u_i^k - \alpha(1 - y_i a_i^\intercal x^k)\right) \qquad \text{for } i = 1, \ldots, m. \end{aligned}$$

Then, this is parallelizable since

$$\mathcal{F}_p\left[\{u_1^k, \ldots, u_m^k\} \mapsto \{u_1^{k+1}, \ldots, u_m^{k+1}\}\right] = \mathcal{O}(nm/p)$$

for $p \leq \min\{m, n\}$.

4.2.2 Amdahl's Law

Imagine that for a specific problem instance, the algorithm

$$\begin{aligned} x^{k+1/2} &= x^k - \alpha \nabla f(x^k) \\ x^{k+1} &= \mathrm{Prox}_{\alpha g}(x^{k+1/2}) \end{aligned}$$

takes 6 ms to evaluate $x^{k+1/2}$ and 3 ms to evaluate x^{k+1}. So the algorithm takes 9 ms per iteration. Imagine that we reduce the computation time of $x^{k+1/2}$ from 6 ms to 0 ms. The speedup is

$$\frac{9}{3+0} = 3.$$

This thought experiment provides an upper bound to the maximum speedup achievable by reducing the computation time of $x^{k+1/2}$; the speedup is at most 3.

Amdahl's law formalizes this idea. Consider a task with a part that takes time $\eta \in [0, 1]$, in proportion, to compute. If we speed up the computation time of this part by s (through better math and parallel computing), then the total speedup is

$$S(s) = \frac{1}{1 - \eta + \eta/s}.$$

We call this formula *Amdahl's law*.

As a corollary, we have

$$S(s) \leq \frac{1}{1 - \eta},$$

that is, $1/(1-\eta)$ upper bounds the speedup we can achieve by accelerating the part. This implies that a part of an algorithm is worth accelerating only if it occupies a significant portion of the running time. Identifying the bottleneck should be the first step of an effort to accelerate an algorithm.

BIBLIOGRAPHICAL NOTES

Measuring the computational complexity of an algorithm by counting floating-point operations is standard in applied mathematics. While our flop-count operator $\mathcal{F}$ is nonstandard, it does formalize the standard considerations of randomized and asynchronous coordinate update algorithms that we discuss in §5 and §6. Our definitions of "method" and "algorithm" are also nonstandard; the two words are often used interchangeably.

We point out that the stated complexity of $O(n^3)$ for computing AB and A^{-1} when $A \in \mathbb{R}^{n\times n}$ and $B \in \mathbb{R}^{n\times n}$ is not optimal. The Strassen algorithm [Str69] costs $O(n^{2.807})$, and the Coppersmith–Winograd algorithm [CW90] costs $O\left(n^{2.375}\right)$ to compute AB and A^{-1}. The Strassen algorithm has found some practical applications, but not in convex optimization. The Coppersmith–Winograd algorithm can only provide an advantage for matrices of inordinate size and therefore has no practical use.

Amdahl's law was formalized by Amdahl in 1967 [Amd67]. Support vector machine was presented by Cortes and Vapnik in 1995 [CV95].

EXERCISES

4.1 *Matrix inversion lemma.* The *matrix inversion lemma* or the *Sherman–Woodbury–Morrison formula* states that

$$(E + BC)^{-1} = E^{-1} - E^{-1}B(I + CE^{-1}B)^{-1}CE^{-1},$$

provided that E is invertible.
Let $A \in \mathbb{R}^{m\times n}$ and $m \leq n$. Show that with a given A and a precomputation $O(m^2 n)$ flops, each iteration

$$x^{k+1} = (I + \alpha A^\intercal A)^{-1}(x^k + \alpha b)$$

can be computed with $O(mn)$ flops per iteration.

4.2 *Parallel PDHG.* Consider the problem

$$\underset{x\in\mathbb{R}^n}{\text{minimize}} \quad \sum_{i=1}^{\ell} g_i(A_i x),$$

where $A_1, \ldots, A_\ell \in \mathbb{R}^{m\times n}$ and $g_1, \ldots, g_\ell$ are CCP. Assume computing $\mathrm{Prox}_{\alpha g_i}$ costs $O(C_g)$ flops (or fewer) for $i = 1, \ldots, \ell$. Find a method that solves this problem using $O(\ell mn + \ell C_g)$ flops per iteration. Can this algorithm benefit from parallel computing?

4.3 *Parallel Condat–Vũ.* Consider the problem

$$\underset{x\in\mathbb{R}^n}{\text{minimize}} \quad f(x) + h(x) + \sum_{i=1}^{\ell} g_i(A_i x - b_i),$$

where $A_1, \ldots, A_\ell \in \mathbb{R}^{m\times n}$, $g_1, \ldots, g_\ell$ are CCP, f is CCP, and h is differentiable and CCP. Assume computing $\mathrm{Prox}_{\alpha f}$ and ∇h respectively costs C_f and C_h flops and computing $\mathrm{Prox}_{\alpha g_i}$ costs $O(C_g)$ flops (or fewer) for $i = 1, \ldots, \ell$. Find a method that solves this problem using $O(\ell mn + \ell C_g + C_f + C_h)$ flops per iteration. Can this algorithm benefit from parallel computing?

4.4 *Parallel PAPC/PDFP²O.* Consider the problem

$$\underset{x\in\mathbb{R}^n}{\text{minimize}} \quad h(x) + \frac{1}{m}\sum_{i=1}^{\ell} g_i(A_i x),$$

where $A_1,\ldots,A_\ell \in \mathbb{R}^{m\times n}$, $g_1,\ldots,g_\ell$ are CCP, and h is differentiable and CCP. Assume $\mathcal{F}[x \mapsto \nabla h] = C_h$ flops and computing $\mathcal{F}[y \mapsto \mathrm{Prox}_{\alpha g_i}(y)] \le C_g$ for $i = 1,\ldots,\ell$. Find a method that solves this problem using $O\left(\ell mn + \ell C_g + C_h\right)$ flops per iteration. Is this method parallelizable?

4.5 *Consensus technique and DYS.* Consider the problem

$$\underset{x\in\mathbb{R}^n}{\text{minimize}} \quad f(x) + \frac{1}{m}\sum_{i=1}^{m} \left(g_i(x) + h_i(x)\right),$$

where f is CCP, $g_1,\ldots,g_m$ are CCP, $h_1,\ldots,h_m$ are differentiable and CCP. The consensus technique yields the equivalent problem

$$\underset{x_1,\ldots,x_m\in\mathbb{R}^n}{\text{minimize}} \quad mf(x_1) + \delta_C(x_1,\ldots,x_m) + \sum_{i=1}^{m} \left(g_i(x_i) + h_i(x_i)\right).$$

Show that the FPI with DYS is

$$\begin{aligned} x^{k+1/2} &= \mathrm{Prox}_{\alpha f}\left(\frac{1}{m}\sum_{i=1}^{m} z_i^k\right) \\ x_i^{k+1} &= \mathrm{Prox}_{\alpha g_i}(2x^{k+1/2} - z_i^k - \alpha\nabla h_i(x^{k+1/2})) \\ z_i^{k+1} &= z_i^k + x_i^{k+1} - x^{k+1/2} \qquad \text{for } i = 1,\ldots,m. \end{aligned}$$

Assume computing $\mathrm{Prox}_{\alpha f}$ costs C_f flops, $\mathrm{Prox}_{\alpha g_i}$ costs C_g flops, and ∇h_i costs C_h flops for $i = 1,\ldots,m$. Assume the cost C_f cannot be further reduced through parallelization. What is the parallel flop count $\mathcal{F}_p\left[\{z_1^k,\ldots,z_m^k\} \mapsto \{z_1^{k+1},\ldots,z_m^{k+1}\}\right]$ for $p \le \min\{m,n\}$? For simplicity, you may assume m/p and n/p are integers.
Hint. Use Exercise 2.29.
Remark. This method was first published by Raguet under the name generalized forward-Douglas–Rachford [Rag19].

5 Randomized Coordinate Update Methods

In this chapter we present the randomized coordinate-update fixed-point iteration (RC-FPI), a randomized method that updates a randomly chosen coordinate.

5.1 RANDOMIZED COORDINATE FIXED-POINT ITERATION

Partition $x \in \mathbb{R}^n$ into m non-overlapping blocks of sizes $n_1,\dots,n_m$, so $n = n_1 + \cdots + n_m$. Write $x = (x_1,\dots,x_m)$, so $x_i \in \mathbb{R}^{n_i}$ for $i = 1,\dots,m$. Given an operator $\mathbb{T}\colon \mathbb{R}^n \to \mathbb{R}^n$, partition the output into m blocks and write

$$\mathbb{T}(x) = \begin{bmatrix} (\mathbb{T}(x))_1 \\ \vdots \\ (\mathbb{T}(x))_m \end{bmatrix},$$

so $(\mathbb{T}(x))_i \in \mathbb{R}^{n_i}$ for $i = 1,\dots,m$. For each $i = 1,\dots,m$, define $\mathbb{T}_i\colon \mathbb{R}^n \to \mathbb{R}^n$ as

$$\mathbb{T}_i(x) = \begin{bmatrix} x_1 \\ \vdots \\ x_{i-1} \\ (\mathbb{T}(x))_i \\ x_{i+1} \\ \vdots \\ x_m \end{bmatrix},$$

that is, $\mathbb{T}_i$ is $\mathbb{T}$ on the ith block and is the identity map on the other blocks. When $n_i = 1$, the ith block corresponds to a single coordinate. Some authors use the word "block" for a collection of more than one coordinate while reserving the word "coordinate" for a single coordinate. In this book, we use these two words interchangeably.

For $\mathbb{T}\colon \mathbb{R}^n \to \mathbb{R}^n$, consider the fixed-point problem

$$\underset{x\in\mathbb{R}^n}{\text{find}} \quad x = \mathbb{T}x.$$

The method *coordinate-update fixed-point iteration* (C-FPI) is

$$\begin{aligned} &\text{select } i(k) \in \{1,\dots,m\}, \\ &x^{k+1} = \mathbb{T}_{i(k)}(x^k). \end{aligned}$$

At the kth iteration, C-FPI selects an index $i(k)$ and updates only the $i(k)$th block. Specifying the selection rule for $i(k)$ fully specifies the method.

How to select $i(k)$ is not a simple question. There are many block selection rules with different advantages and disadvantages. Common selection rules include the *cyclic* rule, which selects the blocks in a cyclic order; the *essential cyclic* rule, which allows each coordinate to appear once or more in each "cycle"; the *greedy* rule, which selects the block that leads to the most progress, measured in many different ways; and the *randomized* rule, which selects blocks randomly.

In this chapter, we focus our study on the randomized rule with uniform probability, as its analysis is simplest. More specifically, we choose $i(k) \in \{1,\dots,m\}$ independently uniformly at random. Under this selection rule, C-FPI becomes *randomized coordinate-update fixed-point iteration* (RC-FPI), which we write as

$$\begin{aligned} i(k) &\sim \text{IID Uniform}\{1,\dots,m\} \\ x^{k+1} &= \mathbb{T}_{i(k)}(x^k). \end{aligned}$$

The convergence property of RC-FPI is similar to the original FPI, and the proof follows from analogous arguments. (For RC-FPI with non-uniform coordinate selection probabilities, see Exercise 5.3.)

Theorem 2 Assume $\mathbb{T}\colon \mathbb{R}^n \to \mathbb{R}^n$ is θ-averaged with $\theta \in (0,1)$ and $\operatorname{Fix}\mathbb{T} \neq \emptyset$. Assume the random indices $i(0), i(1), \dots \in \{1,\dots,m\}$ are independent and identically distributed with uniform probability. Then $x^{k+1} = \mathbb{T}_{i(k)}x^k$ with any starting point $x^0 \in \mathbb{R}^n$ converges to one fixed point with probability 1, that is,

$$x^k \to x^\star$$

with probability 1 for some $x^\star \in \operatorname{Fix}\mathbb{T}$. The quantities $\mathbb{E}\operatorname{dist}^2(x^k, \operatorname{Fix}\mathbb{T})$ and $\mathbb{E}\|x^k - x^\star\|^2$ for any $x^\star \in \operatorname{Fix}\mathbb{T}$ are monotonically nonincreasing with k. Finally, we have

$$\operatorname{dist}(x^k, \operatorname{Fix}\mathbb{T}) \to 0$$

with probability 1.

Proof. Define $\mathbb{S}$ with $\mathbb{T} = \mathbb{I} - \theta\mathbb{S}$ and $\mathbb{S}_i$ with $\mathbb{T}_i = \mathbb{I} - \theta\mathbb{S}_i$. So we have

$$\mathbb{S}_i(x) = \begin{bmatrix} 0 \\ \vdots \\ 0 \\ (\mathbb{S}(x))_i \\ 0 \\ \vdots \\ 0 \end{bmatrix}$$

for $i = 1,\dots,m$. We can alternately express the iteration $x^{k+1} = \mathbb{T}_{i(k)}x^k$ as

$$x^{k+1} = x^k - \theta\mathbb{S}_{i(k)}x^k.$$

It is straightforward to verify that $\mathbb{T}$ is θ-averaged if and only if $\mathbb{S}$ is (1/2)-cocoercive:

$$\begin{aligned}\mathbb{T} \text{ is } \theta\text{-averaged} \quad &\Leftrightarrow \quad \frac{1}{\theta}\mathbb{T} - \left(\frac{1}{\theta} - 1\right)\mathbb{I} \text{ is nonexpansive} \\ &\Leftrightarrow \quad \mathbb{I} - \mathbb{S} \text{ is nonexpansive} \\ &\Leftrightarrow \quad \|x - \mathbb{S}x - y + \mathbb{S}y\|^2 \le \|x - y\|^2 \quad \forall x, y \in \mathbb{R}^n \\ &\Leftrightarrow \quad \frac{1}{2}\|\mathbb{S}x - \mathbb{S}y\|^2 \le \langle x - y, \mathbb{S}x - \mathbb{S}y\rangle \quad \forall x, y \in \mathbb{R}^n \\ &\Leftrightarrow \quad \mathbb{S} \text{ is (1/2)-cocoercive.}\end{aligned}$$

Clearly, $x^\star = \mathbb{T}x^\star$ if and only if $0 = \mathbb{S}x^\star$. So for any $x^\star \in \operatorname{Fix}\mathbb{T} = \operatorname{Zer}\mathbb{S}$ and any $x \in \mathbb{R}^n$, we have

$$\frac{1}{2}\|\mathbb{S}x\|^2 \le \langle \mathbb{S}x, x - x^\star\rangle. \tag{5.1}$$

In Theorem 1 of §2, the sequence $x^0, x^1, \ldots$ was deterministic. In this theorem, however, each x^{k+1} is a random variable that depends on $i(k), i(k-1), \ldots, i(0)$. The initial point x^0 is not random. Write $\mathbb{E}$ for the (full) expectation with respect to all random variables random variables $i(0), i(1), \ldots$. Write $\mathbb{E}_k$ for the conditional expectation with respect to $i(k)$ conditioned on the past random variables $i(k-1), i(k-2), \ldots, i(0)$. Then, $\mathbb{E}\big[\mathbb{E}_k[X]\big] = \mathbb{E}[X]$ by the law of total expectation. Since the randomness of x^k depends only on $i(k-1), \ldots, i(0)$, not $i(k)$, we have $\mathbb{E}_k[x^k] = x^k$. Then,

$$\mathbb{E}_k[\mathbf{S}_{i(k)}x^k] = \frac{1}{m}\mathbb{S}x^k, \tag{5.2}$$

$$\mathbb{E}_k\|\mathbf{S}_{i(k)}x^k\|^2 = \frac{1}{m}\|\mathbb{S}x^k\|^2. \tag{5.3}$$

(Note that (5.3) does not follow from linearity of expectation but rather from the fact that the squared norm $\|\cdot\|^2$ is separable across the indices.)

Stage 1 For any $x^\star \in \operatorname{Fix}\mathbb{T}$, we have

$$\begin{aligned}\|x^{k+1} - x^\star\|^2 &= \|x^k - \theta\mathbf{S}_{i(k)}x^k - x^\star\|^2 \\ &= \|x^k - x^\star\|^2 - 2\theta\langle \mathbf{S}_{i(k)}x^k, x^k - x^\star\rangle + \theta^2\|\mathbf{S}_{i(k)}x^k\|^2.\end{aligned}$$

Taking conditional expectation $\mathbb{E}_k$ on both sides and using (5.2) and (5.3), we get

$$\begin{aligned}\mathbb{E}_k\|x^{k+1} - x^\star\|^2 &= \|x^k - x^\star\|^2 - 2\theta\langle \mathbb{E}_k[\mathbf{S}_{i(k)}x^k], x^k - x^\star\rangle + \theta^2\mathbb{E}_k\|\mathbf{S}_{i(k)}x^k\|^2 \\ &= \|x^k - x^\star\|^2 - \frac{2\theta}{m}\langle \mathbb{S}x^k, x^k - x^\star\rangle + \frac{\theta^2}{m}\|\mathbb{S}x^k\|^2 \\ &\le \|x^k - x^\star\|^2 - (1-\theta)\frac{\theta}{m}\|\mathbb{S}x^k\|^2,\end{aligned} \tag{5.4}$$

where the inequality follows from (5.1). Take the full expectation on both ends of (5.4) to get

$$\mathbb{E}\|x^{k+1} - x^\star\|^2 \le \mathbb{E}\|x^k - x^\star\|^2 - (1-\theta)\frac{\theta}{m}\mathbb{E}\|\mathbb{S}x^k\|^2.$$

Therefore, $\mathbb{E}\|x^k - x^\star\|^2$ is monotonically nonincreasing with k. By minimizing both sides over $x^\star \in \operatorname{Fix}\mathbb{T}$, we obtain the monotonicity of $\mathbb{E}\operatorname{dist}^2(x^k, \operatorname{Fix}\mathbb{T})$.

Stage 2 Next, we prove convergence of the iterates. Inequality (5.4) makes the sequence $(\|x^k - x^\star\|^2)_{k=0,1,\ldots}$ a nonnegative supermartingale. We apply the supermartingale convergence theorem, which we state as Theorem 29 in the appendix, to get

(i) $\sum_{k=0}^{\infty} \|\mathbb{S}x^k\|^2 < \infty$ and
(ii) $\lim_{k\to\infty} \|x^k - x^\star\|$ exists

with probability 1. Note that (i) implies $\|\mathbb{S}x^k\|^2 \to 0$ and (ii) implies x^k is bounded with probability 1. (The limit $\lim_{k\to\infty} \|x^k - x^\star\|$ is a random variable that depends on $x^\star$ and the random indices $i(0), i(1), \ldots$.) For each $x^\star$, the convergence occurs with probability 1. Next, we apply Proposition 1, which we state and prove in what follows, to conclude with probability 1 that $\lim_{k\to\infty} \|x^k - x^\star\|$ exists for all $x^\star \in \operatorname{Fix} \mathbb{T}$. The convergence $x^k \to x^\star$ with probability 1 now follows from the same argument as that of Theorem 1 with the qualifier "with probability 1" appended to each statement. □

The necessity of Proposition 1 in Theorem 2 is subtle. Since we choose $x^\star \in \operatorname{Fix} \mathbb{T}$ first and then apply the supermartingale convergence theorem, the conclusion that $\lim_{k\to\infty} \|x^k - x^\star\|$ exists with probability 1 applies to one fixed point $x^\star$ at a time. Without a formal argument, this does not immediately imply that $\lim_{k\to\infty} \|x^k - x^\star\|$ for all $x^\star \in \operatorname{Fix} \mathbb{T}$ with probability 1 in the case where $\operatorname{Fix} \mathbb{T}$ is not a singleton and therefore has uncountably many fixed points.

Proposition 1 *Let $Y \subseteq \mathbb{R}^n$ and let $x^0, x^1, \ldots$ be a random sequence. Then statement 1 implies statement 2.*

1. *For all $y \in Y$* [with probability 1, $\lim_{k\to\infty} \|x^k - y\|$ exists].
2. *With probability 1* [for all $y \in Y$, $\lim_{k\to\infty} \|x^k - y\|$ exists].

Proof of Proposition 1. This proof uses the separability of $\mathbb{R}^n$, that is, $\mathbb{R}^n$ contains a countable, dense subset.

In particular, $Y \subseteq \mathbb{R}^n$ has a countable, dense subset $\{y^1, y^2, \ldots\}$. By statement 1, given $i \in \{1, 2, \ldots\}$, there is a probability 1 event $\Omega(y^i)$ such that $\lim_{k\to\infty} \|x^k(\omega) - y^i\|$ for all $\omega \in \Omega(y^i)$. Therefore $\lim_{k\to\infty} \|x^k(\omega) - y^i\|$ exists for all $i \in \{1, 2, \ldots\}$ for $\omega \in \cap_{i=1,2,\ldots} \Omega(y^i)$, and $\cap_{i=1,2,\ldots} \Omega(y^i)$ is an event with probability 1 since it is a countable intersection of probability 1 events.

(In other words: with probability 1 [for all $i = 1, 2, \ldots$, $\lim_{k\to\infty} \|x^k - y^i\|$ exists]. The subtlety is that an *uncountable* intersection of probability 1 events may not have probability 1.)

Now pick any $y \in Y$. Statement 2 is proved if we can show $\|x^k(\omega) - y\|$ converges for $\omega \in \cap_{i=1,2,\ldots} \Omega(y^i)$. To this end, pick any $\varepsilon > 0$. Since $\{y^1, y^2, \ldots\} \subseteq Y$ is dense, there exists $y^i \in Y$ such that $\|y^i - y\| \le \varepsilon$. We get the following lower and upper bounds with the triangle inequality:

$$\|x^k(\omega) - y\| \le \|x^k(\omega) - y^i\| + \|y^i - y\| \le \|x^k(\omega) - y^i\| + \varepsilon,$$
$$\|x^k(\omega) - y\| \ge \|x^k(\omega) - y^i\| - \|y^i - y\| \ge \|x^k(\omega) - y^i\| - \varepsilon.$$

Since $\omega \in \Omega \subset \Omega(y^i)$,

$$\limsup_{k\to\infty} \|x^k(\omega) - y\| \leq \lim_{k\to\infty} \|x^k(\omega) - y^i\| + \varepsilon,$$
$$\liminf_{k} \|x^k(\omega) - y\| \geq \lim_{k\to\infty} \|x^k(\omega) - y^i\| - \varepsilon,$$

and together we have

$$0 \leq \limsup_{k} \|x^k(\omega) - y\| - \liminf_{k} \|x^k(\omega) - y\| \leq 2\varepsilon.$$

As $\varepsilon > 0$ is arbitrary, we conclude

$$\limsup_{k\to\infty} \|x^k(\omega) - y\| = \liminf_{k\to\infty} \|x^k(\omega) - y\| = \lim_{k\to\infty} \|x^k(\omega) - y\|.$$

□

In mathematical terms, the key idea of Proposition 1 is that (i) Y has a countable, dense subset, (ii) the sequence of functions $\{\|x^k - \cdot\|\}_{k\in\mathbb{N}}$ has a limit on the countable, dense subset of Y, and (iii) if an equicontinuous sequence of functions has a limit on the dense subset of a metric space, then the limit exists on the entire metric space.

5.2 COORDINATE AND EXTENDED COORDINATE-FRIENDLY OPERATORS

While Theorem 2 is true regardless of the computational structure of $\mathbb{T}$, the RC-FPI is computationally effective when $\mathbb{T}$ is coordinate-friendly or extended coordinate-friendly.

5.2.1 Coordinate-Friendly Operators

Let $z = (z_1, \ldots, z_m) \in \mathbb{R}^n$ and $z_i \in \mathbb{R}^{n_i}$ for $i = 1, \ldots, m$. If

$$\max_{i=1,\ldots,m} \mathcal{F}[x \mapsto z_i] \ll \mathcal{F}[x \mapsto z],$$

then we say the method is *coordinate-friendly*. What counts as $\ll$ depends on context, but when

$$\mathcal{F}[x \mapsto z_i] \sim \frac{C}{m}\mathcal{F}[x \mapsto \{z_1, \ldots, z_m\}] \quad \text{for } i = 1, \ldots, m,$$

for some $C > 0$ not too large, we safely say the method is coordinate-friendly. Again, some authors use the terminology "block coordinate-friendly" and reserve "coordinate-friendly" for when $n_1 = \cdots = n_m = 1$, but we do not make this distinction.

If a method for $x \mapsto z$ is coordinate-friendly,

$$\mathcal{F}_p[x \mapsto z] = \max_{i=1,\ldots,m} \mathcal{F}[x \mapsto z_i] \ll \mathcal{F}[x \mapsto z],$$

for $p \geq m$. So coordinate-friendly methods are parallelizable.

Finally, we say an operator $\mathbb{T}\colon \mathbb{R}^n \to \mathbb{R}^n$ is coordinate-friendly if there is a coordinate-friendly method for computing $x \mapsto \mathbb{T}_i x$ for $i = 1, \ldots, m$.

Example 5.1 An affine operator $\mathbb{T}x = Ax + b$, where $A \in \mathbb{R}^{n\times n}$ and $b \in \mathbb{R}^n$, is coordinate-friendly if $n_i \ll n$ for $i = 1,\dots,m$, since

$$\mathcal{F}[x \mapsto \mathbb{T}x] \sim 2n^2$$
$$\mathcal{F}[x \mapsto \mathbb{T}_i x] \sim 2nn_i.$$

Example 5.2 We say $\mathbb{T}\colon \mathbb{R}^n \to \mathbb{R}^n$ is a *separable operator* if

$$\mathbb{T}(x) = (\mathbb{U}_1(x_1),\dots,\mathbb{U}_m(x_m)),$$

where $\mathbb{U}_i\colon \mathbb{R}^{n_i} \to \mathbb{R}^{n_i}$ for $i = 1,\dots,m$. Separable operators are coordinate-friendly if $\max_{i=1,\dots,m} \mathcal{F}[x_i \mapsto \mathbb{U}_i(x_i)] \ll \mathcal{F}[x \mapsto \mathbb{T}(x)]$. Multiplication by a (block) diagonal matrix is an example.

A function $f\colon \mathbb{R}^n \to \overline{\mathbb{R}}$ is a *separable function* if it is of the form

$$f(x) = \sum_{i=1}^{m} f_i(x_i),$$

where $f_i\colon \mathbb{R}^{n_i} \to \overline{\mathbb{R}}$ for $i = 1,\dots,m$. If f is separable and differentiable, then ∇f is separable. If f is separable and CCP, then Prox_f is separable.

In optimization problems, a *separable constraint* is of the form

$$x_i \in C_i \quad \text{for } i = 1,\dots,m.$$

The projection onto a separable constraint is a separable operator. A common example is the *box constraint*, which is of the form

$$a_i \le x_i \le b_i \quad \text{for } i = 1,\dots,m,$$

where $n_1 = \cdots = n_m = 1$, $a_i \in [-\infty,\infty)$, $b_i \in (-\infty,\infty]$, and $a_i \le b_i$.

5.2.2 Extended Coordinate-Friendly

An operator $\mathbb{T}\colon \mathbb{R}^n \to \mathbb{R}^n$ is *extended coordinate-friendly* if there is an auxiliary quantity $y(x)$ such that

$$\max_{i=1,\dots,m} \mathcal{F}\left[\{x, y(x)\} \mapsto \{\mathbb{T}_i x, y(\mathbb{T}_i x)\}\right] \ll \mathcal{F}\left[x \mapsto \mathbb{T}x\right].$$

In other words, computing $\mathbb{T}_i(x)$ is efficient so long as the auxiliary quantity $y(x)$ is maintained in memory.

More Coordinate Notation

We continue to use the notation $x = (x_1,\dots,x_m)$ with $x_i \in \mathbb{R}^{n_i}$ for $i = 1,\dots,m$. Given a matrix $A \in \mathbb{R}^{r\times n}$, let

$$A_{:,i} \in \mathbb{R}^{r\times n_i}$$

be the submatrix consisting of the columns of A corresponding to the ith block for $i = 1,\dots,m$. So

$$A = \begin{bmatrix} A_{:,1} & \cdots & A_{:,m} \end{bmatrix}.$$

Under this notation, we have

$$Ax = A_{:,1}x_1 + \cdots + A_{:,m}x_m.$$

Write $A^{\mathsf{T}}_{:,i} = (A_{:,i})^{\mathsf{T}} \in \mathbb{R}^{n_i \times r}$ for $i = 1, \ldots, m$. When f is differentiable, we write

$$\nabla f(x) = \begin{bmatrix} \nabla_1 f(x) \\ \vdots \\ \nabla_m f(x) \end{bmatrix},$$

so $(\nabla f(x))_i = \nabla_i f(x)$ for $i = 1, \ldots, m$.

Example: Gradient Descent on Least Squares
Consider the least-squares problem

$$\underset{x \in \mathbb{R}^n}{\text{minimize}} \quad \frac{1}{2}\|Ax - b\|^2,$$

where $A \in \mathbb{R}^{r \times n}$ and $b \in \mathbb{R}^r$. Consider the gradient descent operator

$$\mathbb{T}(x) = x - \alpha A^{\mathsf{T}}(Ax - b).$$

When $r \ll n$, $\mathbb{T}$ is parallelizable and *not* coordinate-friendly, but extended coordinate-friendly.

Without parallelization, evaluation of $\mathbb{T}$ costs

$$\mathcal{F}[x \mapsto \mathbb{T}x] = O(rn).$$

$\mathbb{T}$ is parallelizable assuming $p \leq \min\{r, n\}$ since

$$\begin{aligned} \mathcal{F}_p[x \mapsto \mathbb{T}x] &= \mathcal{F}_p[\{A, x\} \mapsto Ax] + \mathcal{F}_p[\{A^{\mathsf{T}}, Ax\} \mapsto A^{\mathsf{T}}(Ax)] \\ &= O(rn/p). \end{aligned}$$

$\mathbb{T}$ is not coordinate-friendly since

$$\begin{aligned} \mathcal{F}[x \mapsto \mathbb{T}_i x] &= \mathcal{F}[x \mapsto Ax] + \mathcal{F}[Ax \mapsto \mathbb{T}_i x] \\ &= O(rn) + O(rn_i) \\ &= O(rn). \end{aligned}$$

However, $\mathbb{T}$ is extended coordinate-friendly when we maintain the auxiliary quantity Ax, since

$$\mathcal{F}[\{x, Ax\} \mapsto \{\mathbb{T}_i x, A(\mathbb{T}_i x)\}] = O(rn_i)$$

if we use the formula

$$A(\mathbb{T}_i x) = Ax + A_{:,i}((\mathbb{T}x)_i - x_i).$$

Therefore the C-FPI with $\mathbb{T}$

$$\begin{aligned} x_{i(k)}^{k+1} &= x_{i(k)}^k - \alpha A^{\mathsf{T}}_{:,i(k)}(y^k - b) \\ x_j^{k+1} &= x_j^k \qquad \text{for } j \neq i(k) \\ y^{k+1} &= y^k + A_{:,i(k)}(x_{i(k)}^{k+1} - x_{i(k)}^k), \end{aligned}$$

costs $O(rn_{i(k)})$ flops per iteration. Note that the "step" $x_j^{k+1} = x_j^k$ for $j \neq i(k)$ requires no flops. We initialize $x^0 = 0$ and $y = Ax^0 = 0$.

The other approach of precomputing $A^\intercal A$ and $A^\intercal b$ and using the formula $\mathbb{T}(x) = x - \alpha((A^\intercal A)x - A^\intercal b)$ is not effective when, as before, $r \ll n$. Precomputing

$$\mathcal{F}[\{A,b\} \mapsto \{A^\intercal A, A^\intercal b\}] = O\left(rn^2\right)$$

can be prohibitively expensive, and

$$\mathcal{F}[\{x^k, A^\intercal A, A^\intercal b\} \mapsto x_{i(k)}^{k+1}] = O\left(nn_{i(k)}\right)$$

is larger than $O\left(rn_{i(k)}\right)$.

5.3 METHODS

In this section, we present several instances of the RC-FPI. When writing the iterations, we only specify the updated block. It is implied that the selection rule for $i(k)$ is IID uniform and that the other blocks are not updated.

Coordinate Gradient Descent

Consider the problem

$$\underset{x \in \mathbb{R}^n}{\text{minimize}} \quad f(x),$$

where f is differentiable. Then, the RC-FPI applied to $\mathbb{I} - \alpha \nabla f$ is

$$x_{i(k)}^{k+1} = x_{i(k)}^k - \alpha \nabla_{i(k)} f(x^k),$$

which is called the *randomized/stochastic coordinate gradient descent/method*. The method converges if a minimizer exists, f is L-smooth, and $\alpha \in (0, 2/L)$.

In general, $\mathbb{I} - \alpha \nabla f$ need not be extended coordinate-friendly. However, one setup from machine learning that does lead to an extended coordinate-friendly operator is

$$f(x) = \sum_{j=1}^{r} \ell_j(a_j^\intercal x - b_j),$$

where $a_1, \ldots, a_r \in \mathbb{R}^n$, $b_1, \ldots, b_r \in \mathbb{R}$, and $\ell_1, \ldots, \ell_r$ are differentiable CCP functions on $\mathbb{R}$. When $\ell_j(x) = (1/2)x^2$ for $j = 1, \ldots, r$, the problem reduces to the familiar least-squares problem. Write

$$A = \begin{bmatrix} - & a_1^\intercal & - \\ & \vdots & \\ - & a_r^\intercal & - \end{bmatrix} \in \mathbb{R}^{r \times n}, \qquad \ell(y) = \sum_{j=1}^{r} \ell_j(y_j).$$

Then

$$\nabla \ell(x) = (\ell_1'(x_1), \ldots, \ell_r'(x_r)).$$

Then, randomized coordinate gradient descent with $y^k = Ax^k$

$$\begin{aligned} x_{i(k)}^{k+1} &= x_{i(k)}^k - \alpha A_{:,i(k)}^\intercal \nabla \ell(y^k - b) \\ y^{k+1} &= y^k + A_{:,i(k)}(x_{i(k)}^{k+1} - x_{i(k)}^k) \end{aligned}$$

has cost per iteration of $O\left(rn_{i(k)}\right)$, if $\max_{j=1,\ldots,r} \mathcal{F}[x \mapsto \ell_j'(x)] = O(1)$.

Coordinate Gradient Descent with Block-wise Stepsize

Consider the problem

$$\underset{x\in\mathbb{R}^n}{\text{minimize}} \quad f(x),$$

where f is L-smooth. For any diagonal matrix

$$D = \begin{bmatrix} \beta_1 I_{n_1} & & & \\ & \beta_2 I_{n_2} & & \\ & & \ddots & \\ & & & \beta_m I_{n_m} \end{bmatrix},$$

where $\beta_i > 0$ and $I_{n_i} \in \mathbb{R}^{n_i \times n_i}$ is the $n_i \times n_i$ identity matrix for $i = 1, \dots, m$, the stated problem is equivalent to

$$\underset{x\in\mathbb{R}^n}{\text{minimize}} \quad f(Dx).$$

Randomized coordinate gradient method applied to the equivalent problem is

$$x_{i(k)}^{k+1} = x_{i(k)}^{k} - \alpha_{i(k)} \nabla_{i(k)} f(x^k),$$

where $\alpha_{i(k)} = \alpha\beta_{i(k)}$. Using a non-uniform block-wise stepsize is usually necessary for the randomized coordinate gradient method to be faster than the (full deterministic) gradient method in practice.

Coordinate Proximal-Gradient Descent

Consider the problem

$$\underset{x\in\mathbb{R}^n}{\text{minimize}} \quad f(x) + \sum_{i=1}^{m} g_i(x_i),$$

where f is CCP and differentiable and $g_1, \dots, g_m$ are CCP. In other words, consider the problem of minimizing the sum of a differentiable function and a separable function. Write

$$g(x) = \sum_{i=1}^{m} g_i(x_i).$$

Since g is separable, so is $\mathrm{Prox}_{\alpha g}$.

The RC-FPI with the FBS operator $\mathrm{Prox}_{\alpha g}(I - \alpha\nabla f)$ is

$$x_{i(k)}^{k+1} = \mathrm{Prox}_{\alpha g_{i(k)}}\big(x_{i(k)}^{k} - \alpha \nabla_{i(k)} f(x^k)\big),$$

which is called the *coordinate proximal-gradient descent/method*. This method converges if a minimizer exists, f is L-smooth, and $\alpha \in (0, 2/L)$. With the same block-wise stepsize argument, we can get

$$x_{i(k)}^{k+1} = \mathrm{Prox}_{\alpha_{i(k)} g_{i(k)}}\big(x_{i(k)}^{k} - \alpha_{i(k)} \nabla_{i(k)} f(x^k)\big),$$

where $\alpha_1, \dots, \alpha_m > 0$. As before, it is important in practice to use non-uniform block-wise stepsizes to achieve a speedup.

In general, when g is not separable, there is no way to implement the RC-FPI with $\mathrm{Prox}_{\alpha g}(I - \alpha\nabla f)$ efficiently. The evaluation of even a single coordinate of $\mathrm{Prox}_{\alpha g}$ requires the full output of $x - \alpha\nabla f(x)$.

Stochastic Dual Coordinate Ascent

Consider the problem

$$\underset{x\in\mathbb{R}^r}{\text{minimize}} \quad g(x) + \sum_{i=1}^{n} \ell_i(a_i^\intercal x - b_i),$$

where g is a strongly convex CCP function on $\mathbb{R}^r$ (so g^* is smooth) and ℓ_i is a CCP function on $\mathbb{R}$ for $i = 1,\dots,n$. Write

$$A = \begin{bmatrix} - & a_1^\intercal & - \\ & \vdots & \\ - & a_n^\intercal & - \end{bmatrix} \in \mathbb{R}^{n\times r}, \qquad b = \begin{bmatrix} b_1 \\ \vdots \\ b_n \end{bmatrix} \in \mathbb{R}^n.$$

This primal problem is generated by the Lagrangian

$$\mathbf{L}(x,u) = g(x) + \langle u, Ax - b\rangle - \sum_{i=1}^{n} \ell_i^*(u_i).$$

The corresponding dual problem is

$$\underset{u\in\mathbb{R}^n}{\text{maximize}} \quad -g^*\left(-A^\intercal u\right) - b^\intercal u - \textstyle\sum_{i=1}^{n} \ell_i^*(u_i).$$

The randomized coordinate proximal-gradient method applied to the dual problem is

$$\begin{aligned} u_{i(k)}^{k+1} &= \mathrm{Prox}_{\alpha_{i(k)}\ell_{i(k)}^*}\left(u_{i(k)}^k + \alpha_{i(k)}\left(A_{i(k),:}\nabla g^*(y^k) - b_{i(k)}\right)\right) \\ y^{k+1} &= y^k - A_{i(k),:}^\intercal (u_{i(k)}^{k+1} - u_{i(k)}^k), \end{aligned}$$

which is a variation of *stochastic dual coordinate ascent (SDCA)*. Assume $\mathcal{F}[y \mapsto \nabla g^*(y)] = O(r)$ and $\max_{i=1,\dots,n} \mathcal{F}[u \mapsto \mathrm{Prox}_{\alpha_i \ell_i^*}(u)] = O(1)$. Then, the operator is extended coordinate-friendly when we maintain $y^k = -A^\intercal u^k$, and we have a cost of $O\left(rn_{i(k)}\right)$ per iteration. (One can recover the primal solution with $\nabla g^*(y^k)$. See Exercise 2.6.)

Note that each iteration of coordinate update to the dual accesses $A_{i(k),:}$, a block of rows, while each iteration of coordinate update to the primal accesses $A_{:,i(k)}$, a block of columns. In machine learning, a row of A is a training sample, and it may be convenient to use it without splitting it into parts. In such cases, the dual approach is preferred.

MISO/Finito

Consider the optimization problem

$$\underset{x\in\mathbb{R}^n}{\text{minimize}} \quad r(x) + \frac{1}{m}\sum_{i=1}^{m} f_i(x),$$

where $r, f_1,\dots,f_m$ are CCP and $f_1,\dots,f_m$ are differentiable.

We use the consensus technique of §4.2.1 to get the equivalent problem

$$\underset{\mathbf{x}\in\mathbb{R}^{nm}}{\text{minimize}} \quad \delta_C(\mathbf{x}) + \sum_{i=1}^{m} \left(r(x_i) + f_i(x_i)\right),$$

where $\mathbf{x} = (x_1, \ldots, x_m)$ and C is the consensus set (2.19). Write

$$f(\mathbf{x}) = \sum_{i=1}^{m} f_i(x_i)$$

$$g(\mathbf{x}) = \delta_C(\mathbf{x}) + \sum_{i=1}^{m} r(x_i).$$

Using Exercise 2.29, we can evaluate $\mathrm{Prox}_{\alpha g}$ with

$$\mathrm{Prox}_{\alpha g}(y_1, \ldots, y_m) = (x, \ldots, x), \quad x = \mathrm{Prox}_{\alpha r}\left(\frac{1}{m}\sum_{i=1}^{m} y_i\right).$$

Both FBS and BFS operators are extended coordinate-friendly with the auxiliary quantity $\bar{z}^k$ maintained. The RC-FPI with the BFS operator $(\mathbb{I} - \alpha\nabla f)\mathrm{Prox}_{\alpha g}$ is

$$\begin{aligned} x^k &= \mathrm{Prox}_{\alpha r}\left(\bar{z}^k\right) \\ z_{i(k)}^{k+1} &= x^k - \alpha\nabla f_{i(k)}(x^k) \\ \bar{z}^{k+1} &= \bar{z}^k + \frac{1}{m}\left(z_{i(k)}^{k+1} - z_{i(k)}^k\right). \end{aligned}$$

The RC-FPI with the FBS operator $\mathrm{Prox}_{\alpha g}(\mathbb{I} - \alpha\nabla f)$ is

$$\begin{aligned} x_{i(k)}^{k+1} &= \mathrm{Prox}_{\alpha r}(\bar{z}^k) \\ \bar{z}^{k+1} &= \bar{z}^k + \frac{1}{m}\left(x_{i(k)}^{k+1} - x_{i(k)}^k - \alpha(\nabla f_{i(k)}(x_{i(k)}^{k+1}) - \nabla f_{i(k)}(x_{i(k)}^k))\right), \end{aligned}$$

where $\bar{z}^k = \frac{1}{m}\sum_{i=1}^{m}(x_i^k - \alpha\nabla f_{i(k)}(x_i^k))$. These two methods are equivalent and they are both called *minimization by incremental surrogate optimization (MISO)* or Finito. They converge if a solution exists and $\alpha \in (0, 2/L)$.

Of the two, the method from BFS has a minor and subtle advantage, as one can initialize $(z_1^0, \ldots, z_m^0) = (0, \ldots, 0)$ and $\bar{z}^0 = 0$ as the starting point. For the method from FBS, the starting point $(x_1^0, \ldots, x_m^0) \in \mathbb{R}^{nm}$ can be arbitrary, but we need to compute

$$\bar{z}^0 = \frac{1}{m}\sum_{i=1}^{m}\left(x_i^0 - \alpha\nabla f_i(x_i^0)\right)$$

before starting the iterations.

Conic Programs with Many Small Cones

Consider the problem

$$\begin{aligned} \underset{x \in \mathbb{R}^n}{\text{minimize}} \quad & c^\intercal x \\ \text{subject to} \quad & Ax = b \\ & x \in Q_1 \times \cdots \times Q_m, \end{aligned}$$

where $Q_i \subseteq \mathbb{R}^{n_i}$ is a nonempty closed convex set for $i = 1, \ldots, m$, $A \in \mathbb{R}^{r \times n}$ has rank r, and $b \in \mathbb{R}^r$. (The constraint is equivalent to $x_i \in Q_i$ for $i = 1, \ldots, m$.) When $Q_1, \ldots, Q_m$ are convex cones, this problem is called a *conic program*.

Consider the equivalent problem,

$$\underset{x\in\mathbb{R}^n}{\text{minimize}} \quad \underbrace{c^\intercal x + \delta_{\{x\,|\,Ax=b\}}(x)}_{=f(x)} + \underbrace{\delta_{Q_1\times\cdots\times Q_m}(x)}_{=g(x)}.$$

A naive implementation of RC-FPI with DRS is

$$\begin{aligned} x_i^{k+1/2} &= \Pi_{Q_i}(z_i^k) \qquad \text{for } i=1,\dots,m \\ z_{i(k)}^{k+1} &= z_{i(k)}^k + D_{i(k),:}(2x^{k+1/2} - z^k) + v_{i(k)} - x_{i(k)}^{k+1/2}, \end{aligned}$$

where $D = I - A^\intercal(AA^\intercal)^{-1}A$ and $v = A^\intercal(AA^\intercal)^{-1}b - \alpha Dc$ as discussed in Exercise 2.25. Assume $\mathcal{F}[x_i \mapsto \Pi_{Q_i}x_i] = C_i$ for $i = 1,\dots,m$. This method costs $O\left(C_1 + \cdots + C_n + nn_{i(k)}\right)$ per iteration.

A better way is to utilize the extended coordinate-friendly structure with $y^k = D2x^{k+1/2} - z^k$:

$$\begin{aligned} x_{i(k)}^{k+1/2} &= \Pi_{Q_{i(k)}}(z_{i(k)}^k) \\ z_{i(k)}^{k+1} &= z_{i(k)}^k + y_{i(k)}^k + v_{i(k)} - x_{i(k)}^{k+1/2} \\ y^{k+1} &= D_{:,i(k)}\left(2\Pi_{Q_{i(k)}}(z_{i(k)}^{k+1}) - 2x_{i(k)}^{k+1/2} - z_{i(k)}^{k+1} + z_{i(k)}^k\right). \end{aligned}$$

This implementation costs $O\left(C_{i(k)} + nn_{i(k)}\right)$ flops per iteration.

5.4 DISCUSSION

In practice, RC-FPI may provide a greater speed than FPI when the operator is extended coordinate-friendly and when RC-FPI uses coordinate-wise stepsizes that are larger than the stepsize used by FPI.

When comparing RC-FPI and FPI, it is useful to compare one iteration of FPI with m iterations of RC-FPI, which we call an *epoch*. In certain coordinate-friendly setups, an epoch of RC-FPI and an iteration of FPI have similar computational costs.

Theorems 1 and 2 guarantee a similar amount of reduction in fixed-point residual with one iteration of FPI and with one epoch of RC-FPI. However, this does not necessarily mean one iteration of FPI and one epoch of RC-FPI actually make the same amount of progress. In practice, RC-FPI and FPI often converge much faster than what Theorems 1 and 2 guarantee. If so, the similarity in the guarantees does not have much bearing on the similarity or difference of the actual performances.

When comparing optimization methods, the ability to use larger stepsizes often, but not always, translates to a speedup. Loosely speaking, there are cases where the stepsize limitation is the bottleneck of the algorithm, and alleviating it leads to a speedup. In some cases, there is theoretical support for this observation. For example, an epoch of the randomized coordinate gradient method achieves a greater reduction in function value than an iteration of the (deterministic full) gradient method when different stepsizes are used for the different blocks. Such analyses directly utilize the subgradient inequality (1.2) rather than the resulting monotonicity inequality.

In general, RC-FPI may offer no speedup. There are empirical examples where epochs of RC-FPI and iterations of FPI converge at the same rate. We are not aware of any cases where epochs of RC-FPI make less progress than iterations of FPI.

Nevertheless, studying RC-FPI without any guarantee of speedup is still useful, since it serves as a precursor to the asynchronous FPI we discuss in §6. In parallel computing, asynchrony can increase the number of iterations run per unit time. Even if an epoch of asynchronous FPI makes the same amount of progress as an iteration of FPI, asynchronous FPI can make more progress per unit time.

BIBLIOGRAPHICAL NOTES

Coordinate update is a classical technique with a history almost as long as that of the field of optimization itself. The technique has enjoyed increased popularity in recent years due to the rising demand for large-scale optimization. As a complete survey of the subject is beyond the scope of this section, we refer interested readers to the following recent reviews. Lange, Chi, and Zhou's 2014 paper [LCZ14] provides a review from a statistician's perspective; Wright's 2015 paper [Wri15] gives an in-depth review of coordinate descent methods; Shi, Tu, Xu, and Yin's 2016 paper [STXY16] also provides a thorough review on coordinate descent methods; and Peng et al.'s 2016 paper [PWX⁺16] provides a thorough review of coordinate update methods.

Coordinate Descent and Coordinate Update Methods Coordinate descent methods date back to Hildreth's 1957 work using the cyclic coordinate selection rule [Hil57]. By coordinate "descent" methods, we refer to unconstrained optimization methods that reduce (descend) the function value by updating one coordinate at a time [D'E59, War63, LT92, GS00, Tse01, BT13]. Variations of coordinate descent methods include proximal coordinate descent methods [Aus92, GS00, RHL13, XY13], which update one coordinate at a time in a proximal-point setup, and prox-linear coordinate descent methods [TY09, YT11, YTT11, Nes12, BT13, XY13, SXB14, XY17, FR15, Xu15, ZXC⁺16, HWRL17], which perform a forward-backward-type update, one coordinate at a time.

Verkama's 1996 work using a randomized coordinate update for fixed-point iterations is the first instance of coordinate update methods [Ver96]. We use the term coordinate *update* for methods solving constrained optimization problems by updating one coordinate at a time [PR15, ZX15, ZX17, PWX⁺16, GXZ19, XYLC19]. As such methods are usually primal-dual, they do not monotonically "descend" in function value. A general framework of RC-FPI was set up by Combettes and Pesquet in 2015 and 2018 [CP15, CP19], and, in fact, the proof of Theorem 2 closely follows the presentation of [CP15].

The notion of coordinate-friendly operators was first articulated by Peng et al. in 2016 [PWX⁺16], although the notion had been implicitly used in many prior works. Stochastic dual coordinate ascent was presented in [SZ13]. MISO/Finito was independently presented in [Mai13, DDC14], and further convergence analysis was presented in [QSMR19].

Coordinate Selection Rules For a thorough review of the vast literature on coordinate selection rules, see the 2016 paper by Shi, Tu, Xu, and Yin [STXY16]. The various coordinate selection rules considered in the literature include the cyclic selection rule [D'E59, Zad70, LT92, GS00, TY09, Bon11, BT13, HWRL17, RHL13, ST13, XY17, SH15, SY21]; IID uniform random selection rule [ST09, ST11, Nes12, SZ13, RT14, FR15]; independent and random but non-uniform selection rule (also referred to as "arbitrary sampling") [RT14, PN15, RT16, QR16a, QR16b]; independent and random but non-uniform selection rules with probabilities inversely proportional to the coordinate-wise Lipschitz constants (also referred to as "importance sampling") [Zha04, LL10, Nes12, RT14, ZX15, ZX17], random permutation selection rules that access the coordinates in a cyclic fashion but with the order shuffled every epoch [LW19, NJN19, HS19, SY21, SLY20, WL20, RGP20]; and greedy selection rules [Tse90a, LT92, SK03, WL08, LO09, TY09, DRT11, CHLZ12, PYY13, NSL+15, LUZ15].

EXERCISES

5.1 The RC-FPI can be generalized to the setup where p agents independently and randomly select and update indices.
Define $\mathbb{S}_i = \mathbb{I} - \mathbb{T}_i$ for $i = 1,\dots,m$. The method parallel RC-FPI is

$$i(k,w) \sim \text{IID Uniform}\{1,\dots,m\} \qquad \text{for } w = 1,\dots,p$$
$$x^{k+1} = x^k - \sum_{w=1}^{p} \mathbb{S}_{i(k,w)}(x^k).$$

The computation of $\mathbb{S}_{i(k,1)}(x^k),\dots,\mathbb{S}_{i(k,p)}(x^k)$ can be parallelized by p computational agents. We do not require $i(k,1),\dots,i(k,p)$ to be distinct. Prove convergence for parallel RC-FPI.

5.2 *Projection onto second-order cone.* Consider the second-order cone

$$Q = \left\{x \in \mathbb{R}^n \mid \sqrt{x_1^2 + \cdots + x_{n-1}^2} \le x_n\right\}.$$

Write $x_{1:(n-1)} = (x_1,\dots,x_{n-1})$. Show that $\Pi_Q(x) = (\rho(x)x_{1:(n-1)}, \sigma(x)x_n)$, where the coefficients $\rho(x), \sigma(x)$ are given by

$$(\rho(x),\sigma(x)) = \begin{cases} (0,0) & \text{if } -\|x_{1:(n-1)}\| \ge x_n \\ (1,1) & \text{if } \|x_{1:(n-1)}\| \le x_n \\ \frac{1}{2}(\|x_{1:(n-1)}\| + x_n)\left(\frac{1}{\|x_{1:(n-1)}\|}, \frac{1}{x_n}\right) & \text{otherwise.} \end{cases}$$

5.3 *Non-uniform selection rules.* Assume $i(k)$ is an IID random variable with $\text{Prob}[i(k) = j] = p_j$, where $p_1,\dots,p_m > 0$ and $p_1 + \cdots + p_m = 1$. Assume T is θ-averaged, $\theta \in (0,1)$, and $\text{Fix}\, T \neq \emptyset$. Define $S_j = I - T_j$ for $j = 1,\dots,m$. Modify the proof of Theorem 2 to show that

$$x^{k+1} = x^k - \frac{\alpha}{p_{i(k)}} S_{i(k)}(x^k)$$

converges. Provide a condition on α that ensures convergence.

5.4 *Coordinate minimization can fail.* Generally speaking, the approach of updating one coordinate at a time does not always work. Consider the problem

$$\text{minimize} \quad f(x_1,\dots,x_m),$$

where f is a CCP function. The method *coordinate minimization* can be described as

$$x_{i(k)}^{k+1} \in \underset{z_{i(k)} \in \mathbb{R}^{n_{i(k)}}}{\operatorname{argmin}} f(x_1^k, \ldots, x_{i(k)-1}^k, z_{i(k)}, x_{i(k)+1}^k, \ldots, x_m^k),$$

where $i(k)$ is chosen with some selection rule. Coordinate minimization converges under very general assumptions, but such assumptions require f to be differentiable. Consider the counterexample, $g \colon \mathbb{R}^2 \to \mathbb{R}$, defined as $g(x, y) = |x + y| + 2|x - y|$. First, show that g is a CCP function with the unique minimizer $(0, 0)$. Then, show that any (β, β), where $\beta \in \mathbb{R}$, is a fixed point of the coordinate minimization method. Finally, show that

$$\partial g \neq \begin{bmatrix} \partial_x g \\ \partial_y g \end{bmatrix},$$

where ∂_x is the subdifferential with respect to x while y is fixed, and vice versa.

5.5 *Logistic regression with MISO/Finito.* Consider the problem

$$\underset{x \in \mathbb{R}^n}{\text{minimize}} \quad \sum_{j=1}^{m} \log(1 + \exp(-y_j a_j^\intercal x)),$$

where $a_1, \ldots, a_m \in \mathbb{R}^n$ and $y_1, \ldots, y_m \in \{-1, +1\}$. Describe gradient descent and MISO/Finito applied to this problem. What are their flop counts per iteration?

6 Asynchronous Coordinate Update Methods

Let $\mathbb{T}: \mathbb{R}^n \to \mathbb{R}^n$ be a θ-averaged operator and define $\mathbb{S}: \mathbb{R}^n \to \mathbb{R}^n$ with $\mathbb{T} = \mathbb{I} - \theta\mathbb{S}$. Partition $x \in \mathbb{R}^n$ into m blocks $(x_1, \ldots, x_m)$ and define $\mathbb{T}_1, \ldots, \mathbb{T}_m$ as we did in §5.1. Define $\mathbb{S}_i: \mathbb{R}^n \to \mathbb{R}^n$ with $\mathbb{T}_i = \mathbb{I} - \theta\mathbb{S}_i$ for $i = 1, \ldots, m$. We can implement

$$x^{k+1} = x^k - \eta\mathbb{S}x^k,$$

where $\eta > 0$, with multiple computational agents simultaneously running the following code:

```
// p agents run the while loop simultaneously
// x and s are vectors in shared memory
WHILE (not converged) {
  1. WHILE (not all indices processed) {
      Select index i not yet processed
      Read x
      Write s[i] = eta*S[i](x)
    }
  2. Synchronize: wait for all agents
  -----------------------------------------------------
  3. WHILE (not all indices processed) {
      Select index i not yet processed
      Write x[i] = x[i] - s[i]
    }
  4. Synchronize: wait for all agents
  -----------------------------------------------------
}
```

This algorithm (deterministically) computes the iteration $x^{k+1} = x^k - \eta\mathbb{S}x^k$. We call Steps 2 and 4 *synchronization barriers*, and we say the algorithm is *synchronous parallel*.

Step 1 reads from, but does not write to the variable x. Step 3 reads from and writes to the variable x. Step 2 prevents agents finished with Step 1 from proceeding to Step 3 and changing x while the other agents are still using x in Step 1. Similarly, Step 4 prevents agents finished with Step 3 from proceeding to Step 1 and reading x while other agents are still changing x. The synchronization barriers of Steps 2 and 4 divide the outer while-loop into two parts; at a given time all agents are in Steps 1–2 or all agents are in Steps 3–4.

Cost of Synchrony

Synchronization barriers can be a significant computational overhead. When the number of computing agents is large, it becomes difficult to ensure all parallel tasks start and end at the same time. In distributed computing, agents are often not equally powerful. In a modern multitasking system, agents may not be equally available. When there are faster and slower agents, the faster ones will wait idly for the slower ones. Furthermore, the synchronization barrier is itself an algorithm with a cost.

Communication congestion is another cost of synchrony. In our chapter-opening algorithm, multiple agents simultaneously write data in Steps 1 and 3. Writing requires accessing memory and, in a distributed system, communication over a network. Although modern systems allow multiple agents to share the bandwidths of memory and network, simultaneous communication can cause congestion and slowdown.

As the number of computing agents increases, the cost of synchrony quickly becomes a significant factor in obtaining scalable parallel methods.

Asynchronous Parallelism

We say an algorithm is *asynchronous parallel* if it avoids synchronization barriers. Let us simply remove the synchronization barriers of the previous algorithm:

```
// p agents run the while loop asynchronously
// x and s are vectors in shared memory
WHILE (not converged) {
  1. Select i from Uniform{1,2,...,m}
  2. Read x
  3. Compute s[i] = eta*S[i](x)
  4. Write x[i] = x[i] - s[i] //Incorrect!
}
```

Now the agents run in a completely uncoordinated fashion, and the computational cost and inefficiency of synchronization are eliminated. (The computations of the agents are still related through the shared variable x.) Now, the number of updates performed is determined by the aggregate computing power and bandwidths rather than the slowest agent and the worst bottleneck.

However, does this algorithm work? In general, simply removing synchronization barriers does not lead to a working asynchronous method. The asynchrony must be carefully considered and designed around.

This algorithm implements neither the FPI $x^{k+1} = (\mathbb{I} - \eta\mathbb{S})x^k$ nor the RC-FPI $x^{k+1} = (\mathbb{I} - \eta\mathbb{S}_{i(k)})x^k$. By the time an agent is ready to perform Step 4, other agents may have updated x, rendering the value of x used to compute Step 3 outdated. In this case, we say the information is *stale*.

We soon see that we can account for the impact of stale information in the convergence analysis when we enforce *exclusive access* in Step 4. We say an agent has exclusive access to a variable stored in shared memory if no other agent can read from or write to it simultaneously.

6.1 ASYNCHRONOUS FIXED-POINT ITERATION

We first present the *asynchronous coordinate-update fixed-point iteration* (AC-FPI) with an *operational definition*:

```
// p agents run the while loop asynchronously
// x and s are vectors in shared memory
WHILE (not converged) {
  1. Select i from Uniform{1,2,...,m}
  2. Read x
  3. Compute s[i] = eta*S[i](x)
  4. Exclusively read x[i] and write x[i] = x[i] - s[i]
}
```

While an agent is updating the block x[i] in Step 4, other agents cannot access x[i]. Exclusive access can be implemented with standard parallel computing techniques such as atomic operations, mutexes, or semaphores. If $(\mathbb{S}x)_i$ depends on only some components of x, Step 2 needs to read only the necessary parts of x.

AC-FPI removes explicit synchronization barriers, although it still requires exclusive access in writing to the individual blocks of x^k. When there are many more blocks than agents, that is, $p \ll m$, it is rare for an agent to wait for the release of a block's exclusive access, and most, albeit not all, idle time is eliminated.

To mathematically analyze the AC-FPI, we need a *mathematical definition*. For asynchronous algorithms, there is more than one valid approach to defining the "iterates," We present one here, and Exercise 6.5 presents another.

Define x^0 to be the state of x before the start of the algorithm. Write $x^k = (x_1^k, \ldots, x_m^k)$ for the kth iterate and define the iteration count to increment by 1 when an agent completes an update of x in global memory, that is, when an agent completes Step 4. When the iteration counter is incremented to k, if no agent is updating (i.e., writing to) x[j], then x_j^k is the state of x[j] at that time; and if an agent is updating x[j]; then x_j^k is what x[j] used to be right before the agent currently writing to the block started the update. See Figure 6.1. If two or more agents finish updating different blocks at the same time, we break the tie arbitrarily. (The exclusive access of Step 4 prevents multiple agents from updating the same block concurrently. However, different agents may concurrently update different blocks.)

Write $i(k)$ for the index of the kth update. As in §5, we consider the IID random coordinate selection rule as it leads to the simplest theoretical analysis.

As we have discussed, the value of x read in Step 2 may become stale by the time the agent performs Step 4. Write $\hat{x}^k$ for the stale value of x used for the update of x^k to

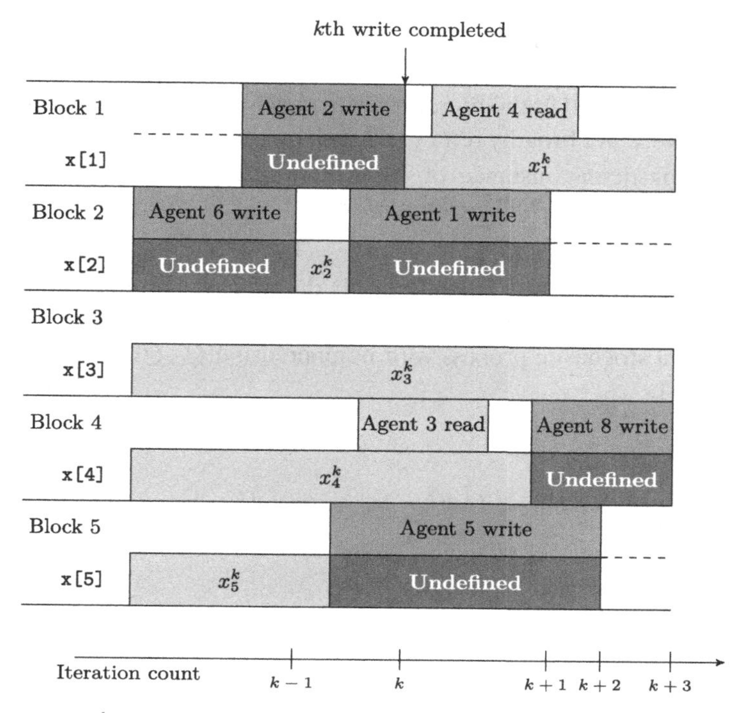

Figure 6.1 The iterate x^k is defined at the time when Agent 2 completes writing to Block 1. Since Blocks 3 and 4 are not being updated at that time, x_3^k and x_4^k are the state of `x[3]` and `x[4]` at the time. Since Blocks 2 and 5 are being updated at the time, x_2^k and x_5^k are the state of `x[2]` and `x[5]` before the writes had begun.

x^{k+1}. In other words, $x^{k+1} = x^k - \eta \mathbf{S}_{i(k)} \hat{x}^k$. It is possible that $\hat{x}^k \neq x^\ell$ for any $\ell = 0, \ldots, k$ since other agents can update blocks while $\hat{x}^k$ is being read block-by-block in Step 2. We discuss this issue further in §6.2 and illustrate it in Figure 6.2. In Step 4, each block is accessed exclusively, but the entire x is not. Therefore, we consider a coordinate-by-coordinate notion of staleness. Write

$$\hat{x}^k = \left(x_1^{k-d_1(k)}, \ldots, x_m^{k-d_m(k)}\right),$$

to denote that the ith block of $\hat{x}^k$ is outdated by $d_i(k) \geq 0$ iterations for $i = 1, \ldots, m$. We call $d_1(k), \ldots, d_m(k)$ the *block delays*. We call

$$\boldsymbol{d}(k) = (d_1(k), \ldots, d_m(k)) \in \mathbb{N}_+^m$$

the *vector delay* and write $\hat{x}^k = x^{k-\boldsymbol{d}(k)}$.

Finally, we write the mathematical definition of the AC-FPI:

$$x^{k+1} = x^k - \eta \mathbf{S}_{i(k)} x^{k-\boldsymbol{d}(k)}. \tag{6.1}$$

AC-FPI is a stochastic algorithm realized by the random variables $i(0), i(1), \ldots$ and $\boldsymbol{d}(0), \boldsymbol{d}(1), \ldots$. Randomness of the indices $i(0), i(1), \ldots$ is injected by design; they come from the random selection of Step 1. Randomness of the delays $\boldsymbol{d}(0), \boldsymbol{d}(1), \ldots$ comes from the randomness of $i(0), i(1), \ldots$ and the randomness of the agents' computation time.

ARock and Convergence of the AC-FPI

The AC-FPI is very general. The update $x^{k+1} = x^k - \eta \mathbb{S}_{i(k)} x^{k-\boldsymbol{d}(k)}$ models many asynchronous algorithms considered in the literature with consistent reads and writes, which we discuss further in §6.2. We broadly refer to all such methods as instances of the AC-FPI.

We analyze a particular instance of AC-FPI, which we call *ARock*. We call the following assumptions the *ARock assumptions*:

- $i(0), i(1), \ldots$ are independently and identically distributed with uniform probability,
- $i(k)$ and $\boldsymbol{d}(\ell)$ are mutually independent for $k = 0, 1, \ldots$ and $\ell \leq k$, and
- $\boldsymbol{d}(0), \boldsymbol{d}(1), \ldots$ is a stochastic process with nonincreasing $Q_0, Q_1, \ldots \in [0, 1]$ such that for every $k = 0, 1, \ldots,$

$$\text{Prob}\left[\max_{i=1,\ldots,m} d_i(k) \geq \ell \,\middle|\, \boldsymbol{d}(k-1), \ldots, \boldsymbol{d}(0), i(k-1), \ldots, i(0)\right] \leq Q_\ell,$$

$$\sum_{\ell=1}^{\infty} \ell (Q_\ell)^{1/2} < \infty. \tag{6.2}$$

This summability assumption is very mild. See Exercise 6.2.

Theorem 3 Assume $\mathbb{S} : \mathbb{R}^n \to \mathbb{R}^n$ is (1/2)-cocoercive or, equivalently, that $\mathbb{T} = \mathbb{I} - \theta \mathbb{S}$ is θ-averaged with $\theta \in (0, 1)$. Assume $\operatorname{Fix} \mathbb{T} \neq \emptyset$. Under the ARock assumptions, the AC-FPI $x^{k+1} = x^k - \eta \mathbb{S}_{i(k)} x^{k-\boldsymbol{d}(k)}$ with any starting point $x^0 \in \mathbb{R}^n$ and stepsize η obeying

$$0 < \eta < \left(1 + \frac{2}{\sqrt{m}} \sum_{\ell=1}^{\infty} Q_\ell^{1/2}\right)^{-1}$$

converges to one fixed point with probability 1, that is,

$$x^k \to x^\star$$

with probability 1 for some $x^\star \in \operatorname{Fix} \mathbb{T}$. Furthermore, with probability 1,

$$\operatorname{dist}(x^k, \operatorname{Fix} \mathbb{T}) \to 0.$$

Given a fixed distribution of delays, that is, fixed values of $Q_0, Q_1, \ldots$, larger m is *more favorable*. The interpretation is that the staleness becomes less harmful as the number of blocks grows. In fact, if we let $m \to \infty$ with fixed $Q_0, Q_1, \ldots$, the stepsize requirement of Theorem 3 becomes the same as that of Theorem 2.

In practice, staleness may cause AC-FPI to make less progress per iteration, although convergence is ensured by Theorem 3. Therefore, synchronous and asynchronous parallel methods represent a trade-off between better and faster iterations.

6.1.1 Discussion of Assumptions

Exclusive Access

In the operational definition of the AC-FPI, Step 4 requires exclusive access. Otherwise, we would not be able to use the notation

$$\hat{x}^k = x^{k-\boldsymbol{d}(k)} = (x_1^{k-d_1(k)}, \ldots, x_m^{k-d_m(k)}),$$

as an agent can read a block while another agent is halfway through writing to it.

Independence

We do not assume $\boldsymbol{d}(0), \boldsymbol{d}(1), \ldots$ is an independent sequence; it usually is a dependent sequence. For example, it is likely that $\hat{x}^k$ and $\hat{x}^{k+1}$ are read at close points in time, and this makes $\boldsymbol{d}(k)$ and $\boldsymbol{d}(k+1)$ highly correlated.

We do not assume $i(k)$ and $\boldsymbol{d}(\ell)$ are independent for $k < \ell$; they usually are dependent. For example, if $i(k) = j$, then $d_j(k+1) > 0$ is very likely.

We do assume that the sequence $i(0), i(1), \ldots$ is IID. When index `i` is sampled in Step 1 of the AC-FPI, we do not yet know which iteration count k the index will be associated with. If the blocks have non-uniform computational costs, the choice of index affects the iteration count the update is assigned to and the IID assumption is violated. For example, if the jth block takes longer to compute, then $i(0) = j$ will have a lower probability than, say, $i(1000) = j$. When the computational cost of each block is equal, then the IID sampling of Step 1 makes $i(0), i(1), \ldots$ an IID sequence.

We do assume $i(k)$ and $\boldsymbol{d}(k)$ are independent for $k = 0, 1, \ldots$. This is realistic if the computational costs of the blocks are uniform. On the other hand, if, for example, the jth block is much more expensive to compute than others and $i(k) = j$, then it is likely that $\boldsymbol{d}(k)$ contains large delays.

Delays

A common assumption in the literature is that the delays are bounded:

$$\max\{d_1(k), \ldots, d_m(k)\} \le D \quad \text{for } k = 0, 1, \ldots$$

for some $D < \infty$. While this bounded-delay assumption would simplify Stage 2 of our proof of Theorem 3, it is not necessary. Therefore, we do not make this assumption.

6.1.2 Proof of Theorem 3

Proof. Write $\mathbb{E}$ for the total expectation. Write $\mathbb{E}_{i(k)}$ for the expectation over $i(k)$ conditioned on $\boldsymbol{d}(k), \ldots, \boldsymbol{d}(0), i(k-1), \ldots, i(0)$. Write $\mathbb{E}_{\boldsymbol{d}(k)}$ for the expectation over $\boldsymbol{d}(k)$ conditioned on $\boldsymbol{d}(k-1), \ldots, \boldsymbol{d}(0), i(k-1), \ldots, i(0)$. Write $\mathbb{E}_{i(k),\boldsymbol{d}(k)}$ for the expectation over $i(k)$ and $\boldsymbol{d}(k)$ conditioned on $\boldsymbol{d}(k-1), \ldots, \boldsymbol{d}(0), i(k-1), \ldots, i(0)$. Note that the random variables $\boldsymbol{d}(k-1), \ldots, \boldsymbol{d}(0), i(k-1), \ldots, i(0)$ completely determine $x^k, \ldots, x^1$.

Stage 1 We define the Lyapunov function

$$V^k = \|x^k - x^\star\|^2 + \frac{1}{m} \sum_{d=1}^{\infty} c_d \|x^{k-d+1} - x^{k-d}\|^2,$$

where $x^\star \in \operatorname{Fix}\mathbb{T}$. We set $x^0 = x^{-1} = x^{-2} = \cdots$, which effectively truncates the sum to be finite. The coefficients $c_d \geq 0$ for $d = 0, 1, \ldots$ will be determined later. Clearly, $V^k \geq 0$. We have $V^k < \infty$ since the infinite sum has only finitely many nonzero terms for a fixed $k < \infty$. As an aside, if one assumes the delays are bounded, the infinite sum can be replaced with a finite sum. The first stage of the proof is to show the key inequality

$$\mathbb{E}_{i(k),\boldsymbol{d}(k)} V^{k+1} \leq V^k - \frac{\eta}{m}\left(1 - \eta\left(1 + \frac{2}{\sqrt{m}}\sum_{d=1}^{\infty} Q_d^{1/2}\right)\right)\mathbb{E}_{\boldsymbol{d}(k)}\|\mathbb{S}x^{k-\boldsymbol{d}(k)}\|^2. \tag{6.3}$$

Using the mathematical definition of AC-FPI, we have

$$\begin{aligned}\|x^{k+1} - x^\star\|^2 &= \|x^k - \eta\mathbb{S}_{i(k)}x^{k-\boldsymbol{d}(k)} - x^\star\|^2 \\ &= \|x^k - x^\star\|^2 - 2\eta\langle\mathbb{S}_{i(k)}x^{k-\boldsymbol{d}(k)}, x^k - x^\star\rangle + \eta^2\|\mathbb{S}_{i(k)}x^{k-\boldsymbol{d}(k)}\|^2.\end{aligned}$$

The independence between $i(k)$ and $\boldsymbol{d}(k)$ gives us

$$\mathbb{E}_{i(k)}\mathbb{S}_{i(k)}x^{k-\boldsymbol{d}(k)} = \frac{1}{m}\mathbb{S}x^{k-\boldsymbol{d}(k)}, \quad \mathbb{E}_{i(k)}\|\mathbb{S}_{i(k)}x^{k-\boldsymbol{d}(k)}\|^2 = \frac{1}{m}\|\mathbb{S}x^{k-\boldsymbol{d}(k)}\|^2.$$

Therefore,

$$\mathbb{E}_{i(k)}\|x^{k+1} - x^\star\|^2 = \|x^k - x^\star\|^2 - \frac{2\eta}{m}\langle\mathbb{S}x^{k-\boldsymbol{d}(k)}, x^k - x^\star\rangle + \frac{\eta^2}{m}\|\mathbb{S}x^{k-\boldsymbol{d}(k)}\|^2. \tag{6.4}$$

Using $(1/2)$-cocoercivity of $\mathbb{S}$, bound the inner-product term as

$$\begin{aligned}-2\langle\mathbb{S}x^{k-\boldsymbol{d}(k)}, x^k - x^\star\rangle &= -2\langle\mathbb{S}x^{k-\boldsymbol{d}(k)}, x^{k-\boldsymbol{d}(k)} - x^\star\rangle - 2\langle\mathbb{S}x^{k-\boldsymbol{d}(k)}, x^k - x^{k-\boldsymbol{d}(k)}\rangle \\ &\leq -\|\mathbb{S}x^{k-\boldsymbol{d}(k)}\|^2 - 2\langle\mathbb{S}x^{k-\boldsymbol{d}(k)}, x^k - x^{k-\boldsymbol{d}(k)}\rangle.\end{aligned} \tag{6.5}$$

Since the blocks have different delays, we decompose the second term of (6.5) over the blocks as

$$\begin{aligned}-2\langle\mathbb{S}x^{k-\boldsymbol{d}(k)}, x^k - x^{k-\boldsymbol{d}(k)}\rangle &= 2\sum_{i=1}^{m}\langle(-\mathbb{S}x^{k-\boldsymbol{d}(k)})_i, x_i^k - x_i^{k-d_i(k)}\rangle \\ &= \sum_{i=1}^{m}\sum_{d=1}^{d_i(k)} 2\langle(-\mathbb{S}x^{k-\boldsymbol{d}(k)})_i, x_i^{k-d+1} - x_i^{k-d}\rangle.\end{aligned}$$

For each term in the summation, we apply Young's inequality

$$-2\langle u, v\rangle \leq \frac{1}{\varepsilon}\|u\|^2 + \varepsilon\|v\|^2 \qquad \forall\,\varepsilon > 0$$

to get

$$2\langle(-\mathbb{S}x^{k-\boldsymbol{d}(k)})_i, x_i^{k-d+1} - x_i^{k-d}\rangle \leq \frac{\eta}{\varepsilon_d}\|(\mathbb{S}x^{k-\boldsymbol{d}(k)})_i\|^2 + \frac{\varepsilon_d}{\eta}\|x_i^{k-d+1} - x_i^{k-d}\|^2,$$

where we choose $\varepsilon_d > 0$ later. Define $\tau(k) = \max_{i=1,\ldots,m} d_i(k)$. Using $d_i(k) \leq \tau(k)$ and swapping the orders of sums, we get

$$\begin{aligned}&-2\langle\mathbb{S}x^{k-\boldsymbol{d}(k)}, x^k - x^{k-\boldsymbol{d}(k)}\rangle \\ &\leq \sum_{i=1}^{m}\sum_{d=1}^{\tau(k)}\left(\frac{\eta}{\varepsilon_d}\|(\mathbb{S}x^{k-\boldsymbol{d}(k)})_i\|^2 + \frac{\varepsilon_d}{\eta}\|x_i^{k-d+1} - x_i^{k-d}\|^2\right)\end{aligned}$$

$$= \eta \left(\sum_{d=1}^{\tau(k)} \varepsilon_d^{-1} \right) \|\mathbf{S}x^{k-\boldsymbol{d}(k)}\|^2 + \frac{1}{\eta} \left(\sum_{d=1}^{\tau(k)} \varepsilon_d \|x^{k-d+1} - x^{k-d}\|^2 \right). \tag{6.6}$$

Substituting (6.6) into (6.5) and substituting (6.5) into (6.4), we get

$$\mathbb{E}_{i(k)} \|x^{k+1} - x^\star\|^2 \le \|x^k - x^\star\|^2 - \frac{\eta}{m} \left(1 - \eta - \eta \sum_{d=1}^{\tau(k)} \varepsilon_d^{-1} \right) \|\mathbf{S}x^{k-\boldsymbol{d}(k)}\|^2$$
$$+ \frac{1}{m} \sum_{d=1}^{\tau(k)} \varepsilon_d \|x^{k-d+1} - x^{k-d}\|^2. \tag{6.7}$$

By the definition of V^k,

$$\mathbb{E}_{i(k)} V^{k+1} = \mathbb{E}_{i(k)} \|x^{k+1} - x^\star\|^2 + \frac{1}{m} \mathbb{E}_{i(k)} \sum_{d=1}^{\infty} c_d \|x^{k-d+2} - x^{k-d+1}\|^2$$
$$= \mathbb{E}_{i(k)} \|x^{k+1} - x^\star\|^2 + \frac{c_1}{m} \mathbb{E}_{i(k)} \|x^{k+1} - x^k\|^2 + \frac{1}{m} \sum_{d=2}^{\infty} c_d \|x^{k-d+2} - x^{k-d+1}\|^2.$$

We bound $\mathbb{E}_{i(k)} \|x^{k+1} - x^\star\|^2$ by (6.7), substitute

$$\mathbb{E}_{i(k)} \|x^{k+1} - x^k\|^2 = \mathbb{E}_{i(k)} \|\eta \mathbf{S}_{i(k)} x^{k-\boldsymbol{d}(k)}\|^2 = \frac{\eta^2}{m} \|\mathbf{S}x^{k-\boldsymbol{d}(k)}\|^2,$$

and decrement the summation index to get

$$\mathbb{E}_{i(k)} V^{k+1} \le (\text{RHS of (6.7)}) + \frac{c_1 \eta^2}{m^2} \|\mathbf{S}x^{k-\boldsymbol{d}(k)}\|^2 + \frac{1}{m} \sum_{d=1}^{\infty} c_{d+1} \|x^{k-d+1} - x^{k-d}\|^2$$
$$= \|x^k - x^\star\|^2 - \frac{\eta}{m} \left(1 - \eta - \frac{c_1 \eta}{m} - \eta \sum_{d=1}^{\tau(k)} \varepsilon_d^{-1} \right) \|\mathbf{S}x^{k-\boldsymbol{d}(k)}\|^2$$
$$+ \frac{1}{m} \left(\sum_{d=1}^{\tau(k)} \varepsilon_d \|x^{k-d+1} - x^{k-d}\|^2 + \sum_{d=1}^{\infty} c_{d+1} \|x^{k-d+1} - x^{k-d}\|^2 \right).$$

We now choose

$$\varepsilon_d = \frac{m^{1/2}}{Q_d^{1/2}} \quad \text{and} \quad c_d = \sum_{\ell=d}^{\infty} \varepsilon_\ell Q_\ell = m^{1/2} \sum_{\ell=d}^{\infty} Q_\ell^{1/2}, \quad d = 1, 2, \ldots.$$

By the assumption (6.2), $c_d < \infty$ for all d. Since

$$\mathbb{E}_{\boldsymbol{d}(k)} \sum_{d=1}^{\tau(k)} \varepsilon_d \|x^{k-d+1} - x^{k-d}\|^2 = \sum_{\ell=1}^{\infty} \text{Prob}[\tau(k) = \ell] \sum_{d=1}^{\ell} \varepsilon_d \|x^{k-d+1} - x^{k-d}\|^2$$
$$= \sum_{d=1}^{\infty} \varepsilon_d \text{Prob}[\tau(k) \ge d] \|x^{k-d+1} - x^{k-d}\|^2$$
$$\le \sum_{d=1}^{\infty} \varepsilon_d Q_d \|x^{k-d+1} - x^{k-d}\|^2$$

and since $c_d = \varepsilon_d Q_d + c_{d+1}$, we obtain

$$\mathbb{E}_{i(k),\boldsymbol{d}(k)} V^{k+1} \leq \|x^k - x^\star\|^2 - \frac{\eta}{m}\mathbb{E}_{\boldsymbol{d}(k)}\left[\left(1 - \eta - \frac{c_1\eta}{m} - \eta\sum_{d=1}^{\tau(k)}\varepsilon_d^{-1}\right)\|\mathbb{S}x^{k-\boldsymbol{d}(k)}\|^2\right]$$

$$+ \frac{1}{m}\sum_{d=1}^{\infty} c_d\|x^{k-d+1} - x^{k-d}\|^2,$$

$$\leq V^k - \frac{\eta}{m}\left(1 - \eta - \frac{c_1\eta}{m} - \eta\sum_{d=1}^{\infty}\varepsilon_d^{-1}\right)\mathbb{E}_{\boldsymbol{d}(k)}\|\mathbb{S}x^{k-\boldsymbol{d}(k)}\|^2,$$

$$= V^k - \frac{\eta}{m}\underbrace{\left(1 - \eta\left(1 + \frac{2}{\sqrt{m}}\sum_{d=1}^{\infty} Q_d^{1/2}\right)\right)}_{>0 \text{ by assumption on } \eta \text{ in Theorem 3.}}\mathbb{E}_{\boldsymbol{d}(k)}\|\mathbb{S}x^{k-\boldsymbol{d}(k)}\|^2,$$

which is (6.3). As an aside, the coefficients c_d and ε_d are carefully chosen to construct the Lyapunov function, rather than being given by the algorithm or the assumptions.

Stage 2 To make the dependence on $x^\star \in \operatorname{Fix}\mathbb{T}$ explicit, write

$$V^k(x^\star) = \|x^k - x^\star\|^2 + \frac{1}{m}\sum_{d=1}^{\infty} c_d\|x^{k-d+1} - x^{k-d}\|^2.$$

We apply Theorem 29, the supermartingale convergence theorem, to (6.3), apply the arguments of Proposition 1, and use $\|x^k - x^\star\|^2 \leq V^k$ to get

(i) $\mathbb{E}_{\boldsymbol{d}(k)}\|\mathbb{S}x^{k-\boldsymbol{d}(k)}\|^2 \to 0$,
(ii) $V^k(x^\star) \to V^\infty(x^\star)$ for all $x^\star \in \operatorname{Fix}\mathbb{T}$,
(iii) $\|x^k\| < B$ for all $k = 0, 1, \ldots$ for some $B < \infty$,

with probability 1.

Write $\mathcal{F}_k$ for the σ-algebra generated by $\boldsymbol{d}(k), \ldots, \boldsymbol{d}(0), i(k), \ldots, i(0)$. Let $L > 0$ be large enough such that $1 - Q_L > 0$, which exists by assumption (6.2). This implies

$$\operatorname{Prob}\left[\max_{i=1,\ldots,m} d_i(k) < L \,\middle|\, \mathcal{F}_{k-1}\right] \geq 1 - Q_L > 0$$

for all $k = 0, 1, \ldots$. Let $\boldsymbol{b}(k)$ be an $\mathcal{F}_{k-1}$-measurable random variable defined as

$$\boldsymbol{b}(k) = \operatorname*{argmax}_{\boldsymbol{b}<L}\left\{\operatorname{Prob}\left[\boldsymbol{d}(k) = \boldsymbol{b} \,\middle|\, \mathcal{F}_{k-1}\right]\right\},$$

where $\operatorname{argmax}_{\boldsymbol{b}<L}$ is the maximizer over all $\boldsymbol{b} = (b_1, \ldots, b_m) \in \mathbb{N}_+^m$ satisfying $\max_{i=1,\ldots,m} b_i < L$. When the argmax is not unique, we break ties in some deterministic manner, say with the lexicographical ordering on $\mathbb{N}_+^m$. Then

$$\operatorname{Prob}\left[\boldsymbol{d}(k) = \boldsymbol{b}(k) \,\middle|\, \mathcal{F}_{k-1}\right] \geq \frac{1 - Q_L}{L^m} > 0$$

since the event $\max_{i=1,\ldots,m} d_i(k) < L$ has probability at least $1 - Q_L$ with L^m possible realizations of $\boldsymbol{d}(k)$, and $\boldsymbol{b}(k)$ is defined as the most likely among the realizations. By the second Borel–Cantelli lemma, version II [Dur10, Theorem 5.3.2], for each $D \in \mathbb{N}_+$, there exists a subsequence $k_j \to \infty$ such that $\boldsymbol{d}(k_j+\ell) = \boldsymbol{b}(k_j+\ell)$ for $\ell = 0,1,\ldots,D-1$. Since x^{k_j} is bounded by (iii), there is a further subsequence $k_j' \to \infty$ such that $x^{k_j'} \to \bar{x}$.

By (i), we have

$$\underbrace{\mathbb{E}_{\boldsymbol{d}(k)}\|\mathbb{S}x^{k-\boldsymbol{d}(k)}\|^2}_{\to 0} = \mathbb{E}\left[\|\mathbb{S}x^{k-\boldsymbol{d}(k)}\|^2 \,\middle|\, \mathcal{F}_{k-1}\right]$$
$$\geq \mathbb{E}\left[1_{\{\boldsymbol{d}(k)=\boldsymbol{b}(k)\}}\|\mathbb{S}x^{k-\boldsymbol{b}(k)}\|^2 \,\middle|\, \mathcal{F}_{k-1}\right] \geq \frac{1-Q_L}{L^m}\|\mathbb{S}x^{k-\boldsymbol{b}(k)}\|^2 \to 0$$

as $k \to \infty$, so $\|\mathbb{S}x^{k-\boldsymbol{b}(k)}\|^2 \to 0$. Since

$$\|x^{k_j'+\ell+1} - x^{k_j'+\ell}\|^2 = \eta^2\|\mathbb{S}_{i(k_j'+\ell)}x^{k_j'+\ell-\boldsymbol{b}(k_j'+\ell)}\|^2 \leq \eta^2\|\mathbb{S}x^{k_j'+\ell-\boldsymbol{b}(k_j'+\ell)}\|^2 \to 0$$

for $\ell = 0,1,\ldots,D-1$, we have

$$(x^{k_j'}, x^{k_j'+1}, \ldots, x^{k_j'+D-1}) \to (\bar{x},\bar{x},\ldots,\bar{x}) \in (\mathbb{R}^n)^D.$$

If $D > L$, then $x^{k_j'+D-1-\boldsymbol{b}(k_j'+D-1)} \to \bar{x}$, and $\mathbb{S}x^{k_j'+D-1-\boldsymbol{b}(k_j'+D-1)} \to 0$ implies $\mathbb{S}\bar{x} = 0$ by continuity of $\mathbb{S}$.

Stage 3 Given $D \in \mathbb{N}_+$ such that $D > L$, consider a subsequence $k_j \to \infty$ such that

$$(x^{k_j}, x^{k_j+1}, \ldots, x^{k_j+(D-1)}) \to (\bar{x}_D, \bar{x}_D, \ldots, \bar{x}_D) \in (\mathbb{R}^n)^D.$$

We write $\bar{x}_D$ to make explicit the fact that the limit may depend on the choice of D. Since

$$V^{k_j}(\bar{x}_D) \to V^\infty(\bar{x}_D)$$

by (ii), we have

$$V^\infty(\bar{x}_D) = \lim_{k_j\to\infty}\frac{1}{m}\sum_{d=D}^\infty c_d\|x^{k_j-d+D+1} - x^{k_j-d+D}\|^2 \leq \frac{2B^2}{m}\sum_{d=D}^\infty c_d.$$

Therefore

$$\limsup_{k\to\infty}\|x^k - \bar{x}_D\|^2 \leq \lim_{k\to\infty} V^k(\bar{x}_D) \leq \frac{2B^2}{m}\sum_{d=D}^\infty c_d. \tag{6.8}$$

By (6.2), we have

$$\sum_{d=1}^\infty c_d = m^{1/2}\sum_{d=1}^\infty\sum_{\ell=d}^\infty Q_\ell^{1/2} = m^{1/2}\sum_{\ell=1}^\infty \ell Q_\ell^{1/2} < \infty.$$

Therefore,

$$\sum_{d=D}^\infty c_d \to 0 \qquad \text{as } D \to \infty.$$

For any $D \in \mathbb{N}_+$, (6.8) implies the accumulation points of x^k reside in the closed ball centered at $\bar{x}_D$ with a radius that goes to 0 as $D \to \infty$. The intersection of these balls contains a single accumulation point x^∞. (The intersection cannot be empty, as the bounded sequence x^k must have at least one accumulation point.) □

6.2 EXTENDED COORDINATE-FRIENDLY OPERATORS AND EXCLUSIVE MEMORY ACCESS

Let $\mathbb{T}: \mathbb{R}^n \to \mathbb{R}^n$ be an extended coordinate-friendly operator with the auxiliary quantity $y(x)$. Throughout this section, consider the specific case $y(x) = Ax$. We can compute $x^{k+1} = \mathbb{T}x^k$ with the following parallel synchronous algorithm:

```
// multiple agents run the while loop simultaneously
// x, y, and s in shared memory
// S=(1/theta)*(I-T)
WHILE (not converged) {
  1. WHILE (not all indices processed) {
       Select index i not yet processed
       Read x,y
       Compute s[i] = eta*S[i](x) using y
     }
  2. Synchronize: wait for all agents to finish
  ------------------------------------------------------
  3. WHILE (not all indices processed) {
       a. Select index i not yet processed
       b. y = y - A[:,i]*s[i] (Sequential, any order)
       c. x[i] = x[i] - s[i]
     }
  4. Synchronize: wait for all agents to finish
  ------------------------------------------------------
}
```

We require Step 3b to be sequential so that two or more agents do not overwrite each other's updates. Step 3b can be parallelized by concurrently updating different coordinates of y. As long as each component is sequentially updated, the algorithm is correct.

Now consider removing the synchronization barrier:

```
// p agents run the while loop asynchronously
// x, y, and s in shared memory
WHILE (not converged) {
  Select i from Uniform{1,2,...,m}
  Read x,y
  Compute s[i] = eta*S[i](x) using y
  Read y and write y = y - A[:,i]*s[i] //Incorrect!
  Exclusively read x[i] and write x[i] = x[i] - s[i]
}
```

This method is not an instance of ARock, due to race conditions. A *race condition* is a negative behavior of a parallel method whose result depends on the order in which the agents complete their tasks.

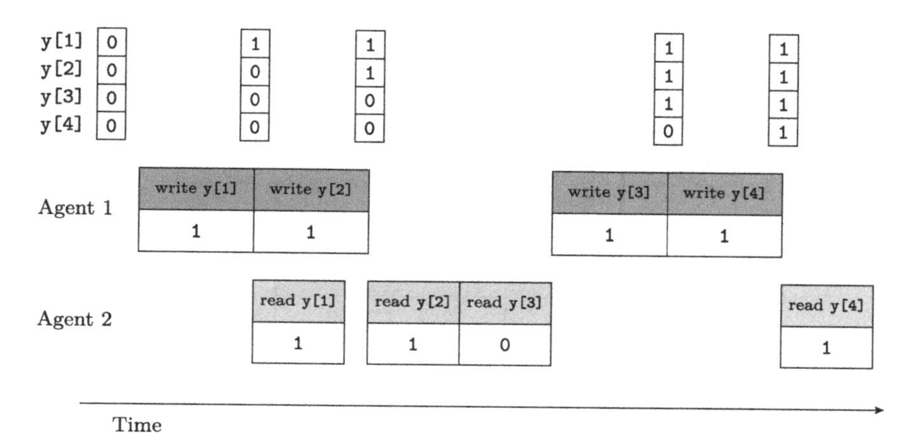

Figure 6.2 Example of an inconsistent read. Agent 2 reads y=[1,1,0,1], which was never an actual state of y in memory. AC-FPI requires consistent reads within a single block but does allow inconsistent reads on x across different blocks. (The delay within a single block must be the same, but different blocks may have different delays.)

In particular, reads and writes on y can be inconsistent. If an agent reads a block of memory while another agent writes to it, the read may retrieve partially old, partially new data. This race condition is called an *inconsistent read*. See Figure 6.2. When two agents write to the same block of memory, they may overwrite one another, resulting in data partially from one agent and partially from the other. This race condition is called an *inconsistent write*. See Figure 6.3. We can prevent inconsistent reads and writes by enforcing exclusive access of the block of memory an agent is writing to. When multiple agents read from the same block of memory but none are writing to it, there is no need for exclusive access. In this algorithm, exclusive access on y can prevent inconsistent reads and writes.

Inconsistency between x and y is another possible race condition. For this method to be an instance of ARock, the x and y that an agent reads must be related through the relationship y=A*x. This may fail to hold if x and y are updated separately. We can prevent this by enforcing exclusive access for the whole (x,y) pair:

```
// p agents run the while loop asynchronously
// x, y, and s in shared memory
WHILE (not converged) {
  1. Select i from Uniform{1,2,...,m}
  2. Read x,y
  3. Compute s[i] = eta*S[i](x) using y
  4. dy[i] = A[:,i]*s[i]
  5. Acquire exclusive access to (x,y)
  6. Read y and write y = y - dy[i]
     Read x[i] and write x[i] = x[i] - s[i]
  7. Release exclusive access to (x,y)
}
```

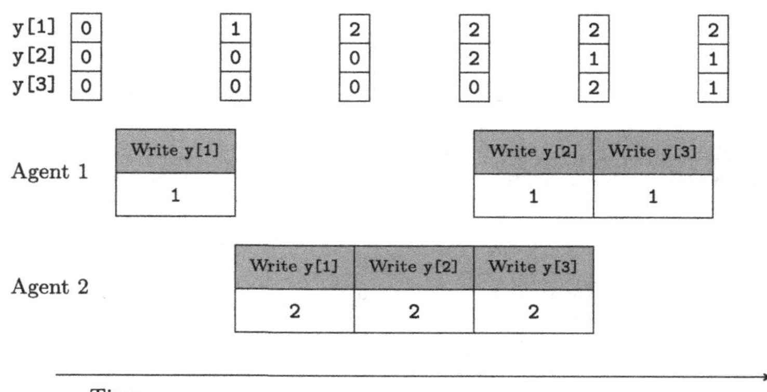

Figure 6.3 Example of an inconsistent write. The two writes of Agents 1 and 2 partially overwrite each other, and y=[2,1,1] is the resulting state. An inconsistent write can occur when multiple agents attempt to concurrently write to the same block. We enforce exclusive access to prevent inconsistent writes.

In some setups, exclusive access of Steps 5 through 7 can be a bottleneck. On a case-by-case basis, it may be possible to perform a specialized analysis to allow for inconsistency between x and y.

6.3 SERVER-WORKER FRAMEWORK

The discussion so far has been based on a *shared memory system*, where multiple agents freely access variables stored in shared memory. A single computer with a multicore CPU is modeled well by a shared memory system.

In the *server-worker framework*, a *server*, or *parameter server*, is a dedicated agent that collects, updates, and distributes variables over a network connected to *workers*, the computational agents working in parallel. The parameter server can also perform minimal computation, so long as the server can keep up with the total throughput of the workers. A cluster of multiple computers connected to a central server node over a network is modeled well by the parameter server framework.

We can use one server and m workers to compute $x^{k+1} = x^k - \eta \mathbb{S} x^k$ synchronously: the server runs

```
// Server code
WHILE (not converged) {
  Broadcast x to workers
  WHILE (not all indices processed) {
    Pick any arrived, unprocessed s_i
    x[i] = x[i] - s_i
  }
}
```

and the workers run

```
// Worker code, i = agent number
WHILE (not converged) {
  x << receive from server (wait until receive)
  s_i = eta*S[i](x)
  s_i >> server
}
```

Each iteration starts with the server broadcasting x to all the workers. Then, the server waits to receive s_i and updates x[i] upon the arrival of s_i. Once all indices are processed, the server starts a new iteration.

This synchronous parallel algorithm has several potential sources of inefficiencies. Between a broadcast and the first arrival of s_i, the server is idle. Then, workers upload the s_i's around the same time, due to synchrony, and can cause a computational and communication bottleneck. Also, a single *straggler*, a worker taking significantly longer to process its work, can slow down the entire algorithm.

An asynchronous implementation in the server-worker framework can avoid the inefficiencies of synchronization. The server broadcasts x at a certain interval, and it runs

```
// Async server code
// Queue holds s_i's. Queue is first-in-first-out
WHILE (not converged) {
  WHILE (before next broadcast schedule)
    s_i =  Queue.pop() (if empty, wait until nonempty)
    x[i] = x[i] - s_i
  Broadcast(x)
}
```

The workers run

```
// Async worker code, i = agent number
// Buffer holds only most recent x received from server
WHILE (not converged) {
  x << Buffer.read()
  s_i = eta*S[i](x)
  s_i >> server's queue
}
```

The Queue stores the s_i from workers. Between the broadcasts, the server processes the updates in the Queue on a first-in-first-out basis. Each worker has its Buffer that holds the most recent copy of x received from the server, and uses it to compute s_i.

The server must process the received `s_i` sufficiently fast, as otherwise the server cannot keep up with the workers and the `Queue` will overflow. The broadcast should be sufficiently frequent so that the workers do not process the same `x` too often. Inconsistent reads and writes do not arise in this setup, so there is no need for exclusive memory access.

We can still model this asynchronous algorithm with AC-FPI (6.1). Specifically, k increments whenever the server updates a block, and x^k is the copy of `x` in the server memory after the kth update. If the arrival times of `s_1`,...,`s_m` are m independent and identical Poisson processes, then $i(0), i(1), \ldots$ is an IID random sequence. In this case, this algorithm is an instance of ARock and we can apply Theorem 3. In general, $i(0), i(1), \ldots$ is not an IID random sequence, and Theorem 3 does not apply.

6.4 METHODS

We now present instances of AC-FPI on shared memory systems and on the parameter server framework. The ARock assumptions are approximately, but not fully satisfied by these algorithms.

6.4.1 Asynchronous Coordinate Gradient Descent

Consider the problem

$$\underset{x \in \mathbb{R}^n}{\text{minimize}} \quad f\left(\sum_{i=1}^{m} A_{:,i} x_i - b\right) + \sum_{i=1}^{m} g_i(x_i),$$

where $g_1, \ldots, g_m$ are CCP functions on $\mathbb{R}^{n_1}, \ldots, \mathbb{R}^{n_m}$, f is a CCP function on $\mathbb{R}^r$, $b \in \mathbb{R}^r$, and

$$A = \begin{bmatrix} A_{:,1} & A_{:,2} & \cdots & A_{:,m} \end{bmatrix} \in \mathbb{R}^{r \times n}.$$

Assume for simplicity that we have $p = m$ agents. (If we have more blocks than agents, we can consolidate the m blocks into p groups.) Assume the ith agent has access to x_i^k, $\mathrm{Prox}_{\alpha g_i}$, $A_{:,i}$, ∇f for $i = 1, \ldots, m$.

The RC-FPI with the FBS operator is

$$\begin{aligned} x_{i(k)}^{k+1} &= \mathrm{Prox}_{\alpha g_{i(k)}}\left(x_{i(k)}^k - \alpha A_{:,i(k)}^{\intercal} \nabla f(y^k)\right) \\ y^{k+1} &= y^k + A_{:,i(k)}(x_{i(k)}^{k+1} - x_{i(k)}^k), \end{aligned}$$

where we initialize $y^0 = Ax^0 - b$. The corresponding AC-FPI is

$$\begin{aligned} s_{i(k)}^k &= \eta\left(\hat{x}_{i(k)}^k - \mathrm{Prox}_{\alpha g_{i(k)}}\left(\hat{x}_{i(k)}^k - \alpha A_{:,i(k)}^{\intercal} \nabla f(\hat{y}^k)\right)\right) \\ x_{i(k)}^{k+1} &= x_{i(k)}^k - s_{i(k)}^k \\ y^{k+1} &= y^k - A_{:,i(k)} s_{i(k)}^k, \end{aligned}$$

where

$$\hat{x}_{i(k)}^k = x_{i(k)}^{k-d_{i(k)}(k)}, \qquad \hat{y}^k = A_{:,1}x_1^{k-d_1(k)} + \cdots + A_{:,m}x_m^{k-d_m(k)}.$$

In a shared memory system, we can implement AC-FPI with

```
// Shared memory code
// Initialize x=0, y=-b
// Pr_i = prox_{alpha*g_i}, G_f = gradient of f
WHILE (not converged) {
  //i = agent number
  Read y
  s[i] = eta*(x[i] - Pr_i(x[i] - alpha*A[:,i]'*G_f(y)))
  del[i] = -A[:,i]*s[i]
  Acquire exclusive access to y
  y = y + del[i]
  Release exclusive access to y
  x[i] = x[i] - s[i]
}
```

Although there is a momentary inconsistency between y and x[i] (after y is updated but before x[i] is updated), this inconsistency makes no difference since each x[i] is read and updated by agent i only, so to other agents, y and x[i] are effectively updated simultaneously.

In a parameter server framework, we can implement AC-FPI with the server running

```
// Server code
// Initialize y=-b
WHILE (not converged) {
  WHILE (before next broadcast schedule)
    y = y + Queue.pop() (if empty, wait until nonempty)
  Broadcast(y)
}
```

and the m agents running

```
// Worker code
// Initialize x(1)=...=x(m)=0
// Pr_i = Prox_{alpha*g_i}, G_f = gradient of f
WHILE (not converged) {
  //i = agent number
  y << last received from server
  s_i = eta*(x_i - Pr_i(x_i - alpha*A[:,i]'*G_f(y)))
  (-A[:,i]*s_i) >> server's Queue
  x_i = x_i - s_i
}
```

The value of y received from the parameter server may be inconsistent with x if the parameter server has not yet processed the worker's previous upload to the buffer. To ensure consistency, the worker can wait until it is broadcast a y with a timestamp certifying that the worker's previous upload to the buffer has been incorporated. See Exercise 6.3.

In computational models where each agent always updates the same block, the independence assumption does not hold even if all agents are equally powerful and the computational costs of all blocks are identical. See Exercise 6.1.

6.4.2 Asynchronous ADMM

Consider the optimization problem

$$\underset{x\in\mathbb{R}^n}{\text{minimize}} \quad \frac{1}{m}\sum_{i=1}^{m} f_i(x) + g(x).$$

We recast this problem into

$$\begin{array}{ll}\underset{x_1,\ldots,x_m,y\in\mathbb{R}^n}{\text{minimize}} & \frac{1}{m}\sum_{i=1}^{m} f_i(x_i) + g(y)\\ \text{subject to} & \underbrace{\begin{bmatrix} I & 0 & \ldots & 0\\ 0 & I & \ldots & 0\\ & & \ddots & \\ 0 & 0 & \ldots & I\end{bmatrix}}_{=A}\begin{bmatrix}x_1\\x_2\\ \vdots\\ x_m\end{bmatrix} - \underbrace{\begin{bmatrix}I\\I\\ \vdots\\ I\end{bmatrix}}_{=B} y = 0.\end{array} \tag{6.9}$$

Define $f(x_1,\ldots,x_m) = (1/m)(f_1(x_1) + \cdots + f_m(x_m))$. As we did in §3.2, consider the dual problem

$$\underset{\nu}{\text{minimize}} \quad \tilde{f}(\nu) + \tilde{g}(\nu),$$

where $\tilde{f}(\nu) = f^*(-A^\intercal \nu)$ and $\tilde{g}(\nu) = g^*(-B^\intercal \nu)$. The PRS operator (2.14) applied to the dual problem is

$$w^{k+1} = (2\mathrm{Prox}_{\alpha\tilde{f}} - \mathbb{I})(2\mathrm{Prox}_{\alpha\tilde{g}} - \mathbb{I})w^k.$$

The FPI with the PRS operator averaged by $\eta \in (0,1)$ is

$$\begin{aligned} y^{k+1} &= \underset{y\in\mathbb{R}^n}{\mathrm{argmin}}\left\{ g(y) - y^\intercal \sum_{j=1}^{m} w_j^k + \frac{\alpha m}{2}\|y\|^2 \right\}\\ x_i^{k+1} &= \underset{x\in\mathbb{R}^n}{\mathrm{argmin}}\left\{ \frac{1}{m} f_i(x) + x^\intercal\left(w_i^k - 2\alpha y^{k+1}\right) + \frac{\alpha}{2}\|x\|^2 \right\} \quad \text{for } i = 1,\ldots,m\\ w_i^{k+1} &= w_i^k + 2\eta\alpha(x_i^{k+1} - y^{k+1}) \quad \text{for } i = 1,\ldots,m. \end{aligned}$$

With the change of variables $w^k = \alpha u^k$ and $\rho = 1/(\alpha m)$, we get

$$\begin{aligned} y^{k+1} &= \mathrm{Prox}_{\rho g}\left(\frac{1}{m}\sum_{j=1}^{m} u_j^k\right)\\ x_i^{k+1} &= \mathrm{Prox}_{\rho f_i}\left(2y^{k+1} - u_i^k\right) \quad \text{for } i = 1,\ldots,m \end{aligned}$$

$$u_i^{k+1} = u_i^k + 2\eta(x_i^{k+1} - y^{k+1}) \quad \text{for } i = 1,\ldots,m.$$

The corresponding AC-FPI is

$$\begin{aligned} y^{k+1} &= \operatorname{Prox}_{\rho g}\left((1/m)\hat{u}^k_{\text{sum}}\right) \\ x^{k+1}_{i(k)} &= \operatorname{Prox}_{\rho f_{i(k)}}\left(2y^{k+1} - \hat{u}^k_{i(k)}\right) \\ u^{k+1}_{i(k)} &= u^k_{i(k)} + 2\eta(x^{k+1}_{i(k)} - y^{k+1}) \end{aligned}$$

where

$$\hat{u}^k_{i(k)} = u^{k-d_{i(k)}(k)}_{i(k)}, \qquad \hat{u}^k_{\text{sum}} = u_1^{k-d_1(k)} + \cdots + u_m^{k-d_m(k)}.$$

Consider a parameter server framework. Assume for simplicity that we have $p = m$ workers. Assume the parameter server has access to $\operatorname{Prox}_{g/(\alpha m)}$. Assume the ith agent has access to u_i^k and $\operatorname{Prox}_{f_i/\alpha}$, for $i = 1,\ldots,m$. We can implement ARock with the server running

```
// Parameter server code
// Initialize u_sum=0
WHILE (not converged) {
  WHILE (before next broadcast schedule)
    s = Queue.pop() (if empty, wait until nonempty)
    u_sum = u_sum + s
  y = Prox_{rho*g}(u_sum/m)
  Broadcast(y)
}
```

and the m agents running

```
// Worker code
// Initialize u[1]=...=u[m]=0
WHILE (not converged) {
  //i = agent number
  y << last received from server
  x_i = Prox_{rho*f_i}(2*y-u_i)
  2*eta*(x_i-y) >> server's Queue
  u_i = u_i + 2*eta*(x_i - y)
}
```

6.5 EXCLUSIVE MEMORY ACCESS

We now discuss how to implement exclusive memory access using atomic operations and mutual exclusion locks. We limit the discussion to a superficial level, just enough to provide clarity on the behavior and the implementation of AC-FPI. For a more thorough discussion on the lower-level considerations of concurrent programming, we refer to readers to standard resources such as [Ray13].

6.5.1 Atomic Operations

An operation of a computational agent is *atomic* if the whole operation is guaranteed to complete without interruption (or never start) in the presence of contention from other agents; if an atomic operation consists of multiple steps, other agents will not observe intermediate results. In most modern systems, reading and writing a single number (represented as a 32- or 64-bit floating-point number) is an atomic operation.

Consider the case where all blocks are single coordinates, that is, $m = n$ and $n_1 = \cdots = n_m = 1$. Then we can implement the AC-FPI as follows:

```
// p agents run the while loop asynchronously
WHILE (not converged) {
  1. Select i from Uniform{1,2,...,m}
  2. for j = 1,...,m
       read x[j]
  3. Compute s[i] = -eta*S[i](x)
  4. x[i] += s[i] (atomic with compare-and-swap)
}
```

In Step 2, each coordinate `x[j]` is read consistently, although different coordinates may have different delays.

Step 4 uses the *increment* operator `+=`, defined by `a+=b` being equivalent to `a=a+b`. The increment operator reads from `a` and `b`, computes the sum, and writes to `a`. Despite several erroneous claims in the asynchronous optimization literature, `+=` is often *not* atomic in many CPUs and GPUs; when multiple agents simultaneously increment the same variable, one agent can overwrite another's increment.

The atomic increment can be implemented via compare-and-swap:

```
// atomic a += b
do {
  old <- a
} while ( !compare_and_swap(a, old, old+b) )
```

Most modern CPUs and GPUs support the *compare-and-swap* instruction as an atomic operation. It corresponds to the following:

```
// atomic execution
// input num is passed by reference and is modifiable
function compare_and_swap(num, old, new) {
```

```
  if num != old
    return false
  num <- new
  return true
}
```

In other words, if `num` is equal to `old`, then write `new` to `num` and return `true`, but otherwise do nothing and return `false`.

In the general case where the blocks represent more than one coordinate, this approach is no longer valid. The reads of Step 2 are no longer guaranteed to be consistent, as it is possible for `x[j]` to be updated by another agent while it is being read. Moreover, the atomic increment via compare-and-swap is no longer possible, as compare-and-swap is usually supported only for data types of size 64-bits or smaller.

6.5.2 Mutual Exclusion Lock

A *mutual exclusion lock* or *mutex* is a synchronization object for concurrent programming with a *lock* and an *unlock* method. A mutex is *acquired* by at most one agent at any given time. An agent acquires a mutex with the `lock()` method. If the mutex is available, `lock()` returns immediately and acquires the mutex. Otherwise (if another agent has locked the mutex and has not yet unlocked it), `lock()` waits until the mutex becomes available and then acquires the mutex. An agent *releases* a mutex with the `unlock()` method. The `unlock()` method returns immediately. If other agents are waiting to acquire the mutex, then one of the waiting agents acquires the mutex upon the unlock. Mutexes can be implemented using the compare-and-swap operation, although it is usually better to rely on implementations provided by standard libraries.

An inefficient way to implement exclusive access in the AC-FPI is as follows:

```
// AC-FPI with one mutex. Inefficient!
WHILE (not converged) {
  1. Select i from Uniform{1,2,...,m}
  2. mutex.lock()
     Read x
     mutex.unlock()
  3. Compute s[i] = eta*S[i](x)
  4. mutex.lock()
     x[i] = x[i] - s[i]
     mutex.unlock()
}
```

With this locking mechanism, access to the shared variable `x` can be a significant bottleneck, as at most one agent is allowed to access `x` at a time.

Rather, it is more efficient to use separate mutexes for all blocks:

```
// AC-FPI with mutex for each block
WHILE (not converged) {
  1. Select i from Uniform{1,2,...,m}
  2. for j = 1,...,m
       mutex[j].lock()
       read x[j]
       mutex[j].unlock()
  3. Compute s[i] = eta*S[i](x)
  4. mutex[i].lock()
     x[i] = x[i] - s[i]
     mutex[i].unlock()
}
```

This mechanism ensures the reads and writes of all blocks are consistent, while allowing multiple agents to concurrently operate on separate blocks. However, Step 2 is still inefficient, as it prevents multiple agents from concurrently reading the same block.

6.5.3 Readers-Writers Lock

A *readers-writers lock* allows concurrent access for reads while enforcing exclusive access for writes. The following class implements a readers-writers lock.

```
class rw_lock {
 private:
  int b
  mutex MUTEX_b, MUTEX_W
 public:
  //constructor
  rw_lock() {
    b = 0
  }
  //member functions
  function read_lock() {
    MUTEX_b.lock()
    b++
    If b==1, MUTEX_W.lock()
    MUTEX_b.unlock()
  }
  function read_unlock() {
    MUTEX_b.lock()
    b--
    If b==0, MUTEX_W.unlock()
    MUTEX_b.unlock()
  }
  function write_lock() {
```

```
    MUTEX_W.lock()
  }
  function write_unlock() {
    MUTEX_W.unlock()
  }
}
```

Now we can implement the AC-FPI with readers-writers locks:

```
// AC-FPI with readers-writers locks
WHILE (not converged) {
  1. Select i from Uniform{1,2,...,m}
  2. for j = 1,...,m
       rw_lock[j].read_lock()
       read x[j]
       rw_lock[j].read_unlock()
  3. Compute s[i] = eta*S[i](x)
  4. rw_lock[i].write_lock()
     x[i] = x[i] - s[i]
     rw_lock[i].write_unlock()
}
```

This mechanism ensures that the reads and writes of all blocks are consistent, allows multiple agents to concurrently operate on separate blocks, and allows multiple agents to concurrently read from the same block.

This implementation of the readers-writers lock prioritizes readers: if there are many readers, a writer must wait until there are no more readers, while if there are many writers, a reader can acquire the lock while other writers are waiting. To reduce the staleness as much as possible, one could use a readers-writers lock prioritizing writers. However, such locks are more complex and allow for less concurrency.

BIBLIOGRAPHICAL NOTES

Stage 1 of the main proof relies on a construction of a Lyapunov function. The main insight of this construction was first presented in the 2016 ARock paper by Peng, Xu, Yan, and Yin in 2016 [PXYY16]. The specific construction follows the 2016 work of Hannah and Yin [HY18]. The arguments of Stages 2 and 3 for establishing almost sure convergence is new.

Classical Asynchronous Methods Asynchronous methods for solving linear systems of equations were first proposed by Rosenfeld in 1969 [Ros69] and Chazan and Miranker in 1969 in [CM69]. In 1983, Bertsekas analyzed FPIs that are (almost surely) nonexpansive for all choices of coordinates and delays [Ber83], and this line of analysis was generalized to FPIs that are "pseudo-nonexpansive" by Tseng, Bertsekas, and Tsitsiklis in 1990 [TBT90]. However, this setup is comparatively much more restrictive, since the AC-FPI under the ARock assumption is nonexpansive *in expectation*. Frommer and Szyld provides a comprehensive review for classical asynchronous methods prior to 2000 [FS00].

Modern Asynchronous Optimization Methods Arguably, the most influential modern work on asynchronous optimization is Recht, Re, Wright, and Niu's 2011 paper with the unusual title "Hogwild" [RRWN11]. The paper used asynchronous SGD to obtain a practical speedup, analyzed the convergence theoretically, and popularized asynchronous methods in machine learning. The theoretical convergence properties of asynchronous SGD was further analyzed by Chaturapruek, Duchi, and Ré in 2015 [CDR15]. The study of many other asynchronous optimization methods followed Hogwild: asynchronous ADMM was studied by Wei and Ozdaglar in 2013 [WO13] and Zhang and Kwok in 2014 [ZK14a]; asynchronous coordinate descent was studied by Liu et al. in 2015 [LWR+14, LWR+15], Liu and Wright in 2015 [LW15], Lian, Huang, Li, and Liu in 2015 [LHLL15], and Hsieh, Yu, and Dhillon in 2015 [HYD15]; asynchronous SAGA was studied by Leblond, Pedregosa, Lacoste-Julien in 2017 [LPL17, PLL17]; non-convex asynchronous methods were studied by Cannelli, Scutari, Facchinei, and Kungurtsev in 2016 and 2017 [CSFK16, CFKS17a]; asynchronous proximal alternating linearized minimization was studied by Davis in 2016 [Dav16]; and finally, asynchronous fixed-point iterations and monotone operator methods were studied in the "ARock paper" by Peng, Xu, Yan, and Yin in 2016 [PXYY16]. These works all rely on the bounded delay assumption.

ARock with unbounded delays was studied by Hannah and Yin in 2018 [HY18]. The unbounded delay assumption is partially justified by Peng, Xu, Yan, and Yin's 2019 work reporting that asynchronous delays empirically follow a Poisson distribution in some setups [PXYY19]. Hannah and Yin showed in 2017 that, under the ARock assumptions and a certain computational model, the AC-FPI theoretically provides a speedup by executing more iterations individually providing improvements not much worse than that of synchronous iterations [HY17]. Asynchronous optimization with unbounded delays was also studied in the machine learning context by Zhou et al. in 2018 [ZMB+18].

Our analysis crucially relies on the independence assumption between $i(k)$ and $\boldsymbol{d}(k)$ for $k = 0, 1, \ldots$; but, as discussed in §6.1.1, this assumption is not always realistic. Several past works relax this assumption, but they are only able to prove weaker results [SHY17, CFKS17b, LPL17].

System-Level Discussions As discussed in §6.3 and §6.5, the analysis of the asynchronous methods is inextricably linked to lower-level considerations of how the algorithms are implemented and executed. The parameter server framework was introduced in the machine learning community, and Smola and Narayanamurthy's 2010 paper [SN10] and Li et al.'s 2014 paper [LAP+14] are early references. We refer readers interested in a thorough treatment of classical concurrent (asynchronous) programming to Raynal's book [Ray13]. In fact, the pseudocode of §6.5.3 was taken from [Ray13, p. 76].

EXERCISES

6.1 *Violation of independence.* Consider the asynchronous coordinate gradient descent of §6.4.1 implemented on a parameter server framework. In this setup, m agents or workers are assigned to the m blocks, and there is no random selection of the blocks. Why are $i(k)$ and $\boldsymbol{d}(k)$ not independent? Why is $i(0), i(1), \ldots$ not an independent sequence?

6.2 *Delay summability condition.* Define $\tau(k) = \max_{i=1,\ldots,m} d_i(k)$. Let $\varepsilon > 0$. Assume

$$\mathbb{E}\left[\tau(k)^{5+\varepsilon} \,\middle|\, \boldsymbol{d}(k-1), \ldots, \boldsymbol{d}(0), i(k-1), \ldots, i(0)\right] < C < \infty \qquad \forall\, k = 0, 1, \ldots,$$

where C is a constant independent of k. Show there exists a nonincreasing sequence $Q_\ell \in [0,1]$ with $\ell = 0, 1, \ldots$ satisfying (6.2).

6.3 *Timestamp for consistency.* Consider the asynchronous coordinate gradient descent of §6.4.1 implemented on a parameter server framework. When can inconsistency between y and x arise?

Assume the parameter server broadcasts, along with y, a timestamp that specifies the time at which the server received the update that was last incorporated into y. How can workers use this information to ensure consistency between y and x?

6.4 *Non-uniform selection probabilities.* Assume the ARock assumptions with the following modification: $i(0), i(1), \ldots$ are independently and identically distributed with probability

$$\text{Prob}\,[i(k) = j] = p_j$$

for $j = 1, \ldots, m$, where $p_1, \ldots, p_m \geq 0$ and $\sum_{i=1}^m p_i = 1$. Show convergence of the AC-FPI

$$x^{k+1} = x^k - \frac{\eta}{p_{i(k)}} \mathbf{S}_{i(k)} x^{k-\boldsymbol{d}(k)}.$$

6.5 *After-read labeling of iterates.* We had defined the iteration count to increment when an agent completes writing to the x variable. We call this definition of iterates the *after-write labeling*. If the computational costs of the blocks are unequal, then $i(k)$ and $\boldsymbol{d}(k)$ can be dependent under the after-write labeling.

Consider the AC-FPI

```
// p agents run the while loop asynchronously
// x and s are vectors in shared memory
WHILE (not converged) {
   1. Select i from Uniform{1,2,...,m}
   2. Read x
   3. Compute s[i] = eta*S[i](x)
   4. Exclusively read x[i] and write x[i] = x[i] - s[i]
}
```

with the *after-read labeling*: starting from $k = 0$, the iteration count increments from k to $k+1$ when an agent completes Step 1, and let $\hat{x}^k$ be the copy x read just before this increment. Starting from x^0, sequentially define $x^1, x^2, \ldots$ via

$$x^{k+1} = x^k - \eta \mathbf{S}_{i(k)} \hat{x}^k.$$

(a) The "iterates" $x^0, x^1, \ldots$ do not necessarily represent the state of x in shared memory at some points in time, even if the agents do not concurrently perform Step 4. Explain why.

(b) We can no longer define a delay vector $\boldsymbol{d}(k)$ with $\hat{x}^k = x^{k-\boldsymbol{d}(k)}$. Why?
Hint. Think of

$$x^k = x^0 + \sum \text{first } k \text{ updates}, \qquad \hat{x}^k = x^0 + \sum \text{updates completed before the } k\text{th read}.$$

(c) By default, Step 2 reads the entire vector x. If S[i](x) does not depend on all of x and we instead read only the components of x that are necessary for computing S[i](x), then $i(k)$ and $\hat{x}^k$ may no longer be conditionally independent. Explain why.

Remark. This "after-read" labeling technique was introduced by Leblond, Pedregosa, and Lacoste-Julien [LPL17].

6.6 *Convergence analysis of after-read labeling.* Consider the after-read labeling of Exercise 6.5. To analyze convergence of the AC-FPI mathematically described by the after-read labeling under the ARock assumptions, how should we modify Stage 1 of the proof of Theorem 3?

6.7 *Extended coordinate-friendly operators with after-read labeling.* Let $\mathbf{T}: \mathbb{R}^n \to \mathbb{R}^n$ be an extended coordinate-friendly operator with the auxiliary quantity $y(x) = Ax$. Consider the after-write and after-read labeling of Exercise 6.5. Consider the following code, which is almost the same as the code of §6.2 but different in how we enforce exclusive access.

```
// p agents run the while loop asynchronously
// x, y, and s in shared memory
WHILE (not converged) {
  1. Select i from Uniform{1,2,...,m}
  2. Read x,y
  3. Compute s[i] = eta*S[i](x) using y
  4. dy[i] = A[:,i]*s[i]
  5. Acquire exclusive access to y
     Read y and write y = y - dy[i]
     Release exclusive access to y
  6. Acquire exclusive access to x[i]
     Read x[i] and write x[i] = x[i] - s[i]
     Release exclusive access to x[i]
}
```

The "acquire" and "release" can be implemented by creating mutexes, one each for y, x[1],..., x[m], and by locking and unlocking them.

(a) With after-write labeling, why are x and y inconsistent?

(b) With after-read labeling, why is the inconsistency resolved?

PART TWO Additional Topics

7 Stochastic Optimization

Consider the finite sum minimization problem

$$\underset{x \in \mathbb{R}^n}{\text{minimize}} \quad \frac{1}{N} \sum_{i=1}^{N} f_i(x),$$

where f_i is a CCP function on $\mathbb{R}^n$ with $\operatorname{dom} f_i = \mathbb{R}^n$ for $i = 1, \ldots, N$. When $f_1, \ldots, f_N$ are differentiable, we can apply the celebrated *stochastic gradient descent* (SGD),

$$x^{k+1} = x^k - \alpha_k \nabla f_{i(k)}(x^k).$$

When $f_1, \ldots, f_N$ are not differentiable, we can apply the *stochastic subgradient method*:

$$x^{k+1} \in x^k - \alpha_k \partial f_{i(k)}(x^k).$$

For both methods, $\alpha_0, \alpha_1, \ldots \in \mathbb{R}$ are the stepsizes and $i(k) \in \{1, \ldots, N\}$ is chosen independently uniformly at random.

For the more general problem

$$\underset{x \in \mathbb{R}^n}{\text{minimize}} \quad \frac{1}{N} \sum_{i=1}^{N} f_i(x) + g(x),$$

where f_i is a CCP function on $\mathbb{R}^n$ with $\operatorname{dom} f_i = \mathbb{R}^n$ for $i = 1, \ldots, N$ and g is a CCP function on $\mathbb{R}^n$, we can apply the *stochastic proximal subgradient method*:

$$x^{k+1} \in \operatorname{Prox}_{\alpha_k g}(x^k - \alpha_k \partial f_{i(k)}(x^k)).$$

In particular, when $g = \delta_C$ for a nonempty closed convex set C, that is, g represents the constraint $x \in C$, the method reduces to the *stochastic projected subgradient method*:

$$x^{k+1} \in \Pi_C(x^k - \alpha_k \partial f_{i(k)}(x^k)).$$

These stochastic (sub)gradient methods are used widely for many large-scale optimization problems, especially those arising in machine learning. In this section, we study generalizations of these methods to the setup of monotone inclusions and establish convergence.

7.1 STOCHASTIC FORWARD-BACKWARD METHOD

Consider the problem

$$\underset{x \in \mathbb{R}^n}{\text{find}} \quad 0 \in \left(\frac{1}{N} \sum_{i=1}^{N} \mathbb{A}_i + \mathbb{B} \right) x,$$

where $\mathbb{A}_i \colon \mathbb{R}^n \rightrightarrows \mathbb{R}^n$ is maximal monotone with $\operatorname{dom} \mathbb{A}_i = \mathbb{R}^n$ for $i = 1, \ldots, N$ and $\mathbb{B} \colon \mathbb{R}^n \rightrightarrows \mathbb{R}^n$ is maximal monotone. Consider the *stochastic forward-backward method (SFB)*,

$$x^{k+1} \in \mathbb{J}_{\alpha_k \mathbb{B}}(\mathbb{I} - \alpha_k \mathbb{A}_{i(k)}) x^k,$$

where $\alpha_k > 0$ and $i(k) \in \{1, \ldots, N\}$ is chosen independently uniformly at random. Since $\mathbb{A}_{i(k)}$ may be multi-valued, the method is defined with an inclusion. One can equivalently write

$$\begin{aligned} a^k &\in \mathbb{A}_{i(k)}(x^k) \\ x^{k+1} &= \mathbb{J}_{\alpha_k \mathbb{B}}(x^k - \alpha_k a^k). \end{aligned}$$

We do not make any assumptions on the selection $a^k \in \mathbb{A}_{i(k)}(x^k)$; we can choose a^k deterministically or randomly to be any element within the set $\mathbb{A}_{i(k)}(x^k)$.

In this section, we analyze the convergence of SFB. In general, SFB may not converge, so we establish convergence using demipositivity or averaging.

Basic Assumptions

For notational simplicity, define

$$\mathbb{A} = \frac{1}{N} \sum_{i=1}^{N} \mathbb{A}_i.$$

Assume $\mathbb{A}_i$ is maximal monotone and $\operatorname{dom} \mathbb{A}_i = \mathbb{R}^n$ for $i = 1, \ldots, N$. Assume $\mathbb{B} \colon \mathbb{R}^n \rightrightarrows \mathbb{R}^n$ is maximal monotone. Assume $\operatorname{Zer}(\mathbb{A} + \mathbb{B}) \neq \emptyset$. Assume the random indices $i(0), i(1), \ldots \in \{1, \ldots, N\}$ are independent and identically distributed with uniform probability. Assume there are nonnegative constants $C_1, C_2 < \infty$ such that

$$\frac{1}{N} \sum_{i=1}^{N} \|\mathbb{A}_i x\|^2 \le \frac{C_1}{2} \|x\|^2 + C_2 \qquad \forall x \in \operatorname{dom} \mathbb{B}. \tag{7.1}$$

Assume the positive sequence $\alpha_0, \alpha_1, \ldots$ satisfies

$$\cdots \le \alpha_1 \le \alpha_0, \qquad \sum_{k=0}^{\infty} \alpha_k = \infty, \qquad \sum_{k=0}^{\infty} \alpha_k^2 < \infty. \tag{7.2}$$

Examples of α_k that satisfy the conditions in (7.2) include

$$\alpha_k = \frac{C}{(k+1)^p}, \quad k = 0, 1, \ldots,$$

for $\frac{1}{2} < p \le 1$.

Convergence with Demipositivity

We say a maximal monotone operator $\mathbb{A}\colon \mathbb{R}^n \rightrightarrows \mathbb{R}^n$ is *demipositive* if there is an $x^\star \in \operatorname{Zer} \mathbb{A}$ such that

$$\langle \mathbb{A}x, x - x^\star \rangle > 0 \qquad \forall x \notin \operatorname{Zer} \mathbb{A}.$$

Strongly monotone, cocoercive, and subdifferential operators of CCP functions are demipositive; for any $x \notin \operatorname{Zer} \mathbb{A}$ and $x^\star \in \operatorname{Zer} \mathbb{A}$, if $\mathbb{A}$ is μ-strongly monotone, then

$$\langle \mathbb{A}x, x - x^\star \rangle \geq \mu \|x - x^\star\|^2 > 0;$$

if $\mathbb{A}$ is β-cocoercive, then

$$\langle \mathbb{A}x, x - x^\star \rangle \geq \beta \|\mathbb{A}x\|^2 > 0;$$

and if $\mathbb{A} = \partial f$ for some CCP function f, then

$$\langle \mathbb{A}x, x - x^\star \rangle \geq f(x) - f(x^\star) > 0.$$

Also, if $\mathbb{A}$ is maximal monotone and $\operatorname{int}(\operatorname{Zer} \mathbb{A}) \neq \emptyset$, that is, if $\operatorname{Zer} \mathbb{A}$ has an interior, then $\mathbb{A}$ is demipositive. See Exercise 7.5.

However, not all monotone operators are demipositive. The single-valued operator $\mathbb{A}\colon \mathbb{R}^2 \to \mathbb{R}^2$ defined as

$$\mathbb{A}(x, y) = \begin{bmatrix} 0 & 1 \\ -1 & 0 \end{bmatrix} \begin{bmatrix} x \\ y \end{bmatrix}$$

is monotone but not demipositive. A quick computational experiment shows SFB with this $\mathbb{A}$ and $\mathbb{B} = \mathbb{O}$ does not converge; the iterates cycle around the zero 0 without converging to it.

Theorem 4 Assume the outlined "basic assumptions." Assume $\mathbb{A} + \mathbb{B}$ is demipositive. Then $x^{k+1} \in \mathbb{J}_{\alpha_k \mathbb{B}}(\mathbb{I} - \alpha_k \mathbb{A}_{i(k)})(x^k)$ with any starting point x_0 converges to a zero, that is, $x^k \to x^\star \in \operatorname{Zer} \mathbb{A}$, with probability 1.

Convergence with Averaging

Consider the same SFB iteration

$$x^{k+1} \in \mathbb{J}_{\alpha_k \mathbb{B}}(\mathbb{I} - \alpha_k \mathbb{A}_{i(k)}) x^k,$$

but we compute the *averaged iterates*:

$$\bar{x}^k = \frac{\sum_{j=0}^{k} \alpha_j x^j}{\sum_{j=0}^{k} \alpha_j}. \tag{7.3}$$

This technique is also called *Polyak–Ruppert averaging*. The averaged iterates converge, that is, $\bar{x}^k \to x^\star \in \operatorname{Zer} \mathbb{A}$, without demipositivity.

Theorem 5 Assume the outlined "basic assumptions." Then the averaged iterates $\bar{x}^k$ of $x^{k+1} \in \mathbb{J}_{\alpha_k \mathbb{B}}(\mathbb{I} - \alpha_k \mathbb{A}_{i(k)})(x^k)$, defined by (7.3), with any starting point x_0 converge to a zero, that is, $\bar{x}^k \to x^\star \in \operatorname{Zer} \mathbb{A}$, with probability 1.

In addition to being convergent without demipositivity, averaging has several advantages. One is that averaging can provide a good *rate* of convergence in specific setups. See Exercise 7.1. Another is that averaging makes the convergence rate more robust to the choice of stepsizes. See Exercise 7.3. However, a drawback of averaging is that it may slow down the algorithm when the (non-averaged) iterates are converging to the solution quickly.

7.1.1 Convergence Proofs

We first present the proof of Theorem 4, which relies on a summability argument similar to what we have seen before.

Proof of Theorem 4. Define $\widetilde{\mathbb{A}}_{i(k)}x^k \in \mathbb{R}^n$ with $x^{k+1} = \mathbb{J}_{\alpha_k\mathbb{B}}(x^k - \alpha_k\widetilde{\mathbb{A}}_{i(k)}x^k)$ and $\widetilde{\mathbb{B}}x^{k+1} = (1/\alpha_k)(x^k - x^{k+1}) - \widetilde{\mathbb{A}}_{i(k)}x^k$. This implies $\widetilde{\mathbb{A}}_{i(k)}x^k \in \mathbb{A}_{i(k)}x^k$ and $\widetilde{\mathbb{B}}x^{k+1} \in \mathbb{B}x^{k+1}$. Let $\widetilde{\mathbb{A}}x^k = \mathbb{E}_{i(k)}[\widetilde{\mathbb{A}}_{i(k)}x^k] \in \mathbb{A}x^k$ for $k = 0,1,\ldots$. Let $x^\star \in \operatorname{Zer}(\mathbb{A}+\mathbb{B})$ for which the demipositivity property holds. Let $\widetilde{\mathbb{A}}x^\star \in \mathbb{R}^n$ be an element such that $\widetilde{\mathbb{A}}x^\star \in \mathbb{A}x^\star$ and $-\widetilde{\mathbb{A}}x^\star \in \mathbb{B}x^\star$. Let $\widetilde{\mathbb{B}}x^\star = -\widetilde{\mathbb{A}}x^\star$. Note that $x^{k+1} = x^k - \alpha_k\left(\widetilde{\mathbb{A}}_{i(k)}x^k + \widetilde{\mathbb{B}}x^{k+1}\right)$.

Then we have

$$\begin{aligned}\|x^{k+1}-x^\star\|^2 &= \|x^k - x^\star\|^2 + \alpha_k^2\|\widetilde{\mathbb{A}}_{i(k)}x^k - \widetilde{\mathbb{A}}x^\star\|^2 - \alpha_k^2\|\widetilde{\mathbb{B}}x^{k+1} - \widetilde{\mathbb{B}}x^\star\|^2\\ &\quad - 2\alpha_k\langle\widetilde{\mathbb{A}}_{i(k)}x^k - \widetilde{\mathbb{A}}x^\star, x^k - x^\star\rangle - 2\alpha_k\langle\widetilde{\mathbb{B}}x^{k+1} - \widetilde{\mathbb{B}}x^\star, x^{k+1}-x^\star\rangle.\end{aligned}$$

With assumption (7.1), we have

$$\begin{aligned}\|\widetilde{\mathbb{A}}_{i(k)}x^k - \widetilde{\mathbb{A}}x^\star\|^2 &\le 2\|\widetilde{\mathbb{A}}_{i(k)}x^k\|^2 + 2\|\widetilde{\mathbb{A}}x^\star\|^2\\ &\le C_1(\|x^k\|^2 + \|x^\star\|^2) + 4C_2\\ &\le C_1(2\|x^k - x^\star\|^2 + 3\|x^\star\|^2) + 4C_2,\end{aligned}$$

and we have

$$\begin{aligned}\|x^{k+1}-x^\star\|^2 &\le (1+2C_1\alpha_k^2)\|x^k - x^\star\|^2 + 3C_1\alpha_k^2\|x^\star\|^2 + 4C_2\alpha_k^2\\ &\quad - 2\alpha_k\langle\widetilde{\mathbb{A}}_{i(k)}x^k - \widetilde{\mathbb{A}}x^\star, x^k - x^\star\rangle - 2\alpha_k\langle\widetilde{\mathbb{B}}x^{k+1} - \widetilde{\mathbb{B}}x^\star, x^{k+1}-x^\star\rangle.\end{aligned}$$

Define

$$V^k = \|x^k - x^\star\|^2 + 2\alpha_k\langle\widetilde{\mathbb{B}}x^k - \widetilde{\mathbb{B}}x^\star, x^k - x^\star\rangle$$

for $k = 0,1,\ldots$. Then,

$$\begin{aligned}V^{k+1} &= \|x^{k+1}-x^\star\|^2 + 2\alpha_{k+1}\langle\widetilde{\mathbb{B}}x^{k+1} - \widetilde{\mathbb{B}}x^\star, x^{k+1}-x^\star\rangle\\ &\le \|x^{k+1}-x^\star\|^2 + 2\alpha_k\langle\widetilde{\mathbb{B}}x^{k+1} - \widetilde{\mathbb{B}}x^\star, x^{k+1}-x^\star\rangle\\ &\le (1+2C_1\alpha_k^2)V^k + 3C_1\alpha_k^2\|x^\star\|^2 + 4C_2\alpha_k^2 - 2\alpha_k\langle\widetilde{\mathbb{A}}_{i(k)}x^k + \widetilde{\mathbb{B}}x^k, x^k - x^\star\rangle.\end{aligned}$$

Write $\mathbb{E}_k$ for the conditional expectation with respect to $i(k)$ conditioned on $i(k-1)$, $i(k-2),\ldots,i(0)$. Then,

$$\mathbb{E}_kV^{k+1} \le (1+2C_1\alpha_k^2)V^k + 3C_1\alpha_k^2\|x^\star\|^2 + 4C_2\alpha_k^2 - 2\alpha_k\langle\widetilde{\mathbb{A}}x^k + \widetilde{\mathbb{B}}x^k, x^k - x^\star\rangle.$$

Now apply Theorem 30, the Robbins–Siegmund quasimartingale convergence theorem, to get that for any $x^\star \in \operatorname{Zer}(\mathbb{A}+\mathbb{B})$,

(i) $\sum_{k=0}^{\infty} \alpha_k \langle \widetilde{\mathbb{A}} x^k + \widetilde{\mathbb{B}} x^k, x^k - x^\star \rangle < \infty$,
(ii) $\lim_{k\to\infty} V^k$ exists

with probability 1. Since $\sum_k \alpha_k = \infty$ and $\langle \widetilde{\mathbb{A}} x^k + \widetilde{\mathbb{B}} x^k, x^k - x^\star \rangle \geq 0$, we conclude from (i) that

$$\liminf_k \langle \widetilde{\mathbb{A}} x^k + \widetilde{\mathbb{B}} x^k, x^k - x^\star \rangle = 0.$$

Since $\|x^k - x^\star\|^2 \leq V^k$ and V^k is bounded, we also conclude that x^k is bounded with probability 1. By (7.1), boundedness of x^k implies $\widetilde{\mathbb{A}} x^k \in \mathbb{A} x^k$ is bounded. By (7.1), boundedness of x^k implies $\widetilde{\mathbb{A}}_{i(k)} x^k$ is bounded, which implies $\widetilde{\mathbb{B}} x^{k+1}$ is bounded. Therefore, there exists a subsequence $k_j \to \infty$ such that

$$x^{k_j} \to x^\infty, \qquad \widetilde{\mathbb{A}} x^{k_j} \to a^\infty, \qquad \widetilde{\mathbb{B}} x^{k_j} \to b^\infty, \qquad \langle \widetilde{\mathbb{A}} x^{k_j} + \widetilde{\mathbb{B}} x^{k_j}, x^{k_j} - x^\star \rangle \to 0.$$

Since $\mathbb{A} + \mathbb{B}$ is maximal monotone, its graph is closed (see Exercise 10.3), and $a^\infty + b^\infty \in (\mathbb{A} + \mathbb{B}) x^\infty$. Therefore

$$0 = \langle a^\infty + b^\infty, x^\infty - x^\star \rangle$$

and

$$0 \in \langle (\mathbb{A} + \mathbb{B}) x^\infty, x^\infty - x^\star \rangle.$$

By demipositivity, we conclude $x^\infty \in \operatorname{Zer}(\mathbb{A} + \mathbb{B})$.

Note that demipositivity was invoked only in the final step. By repeating the application of Theorem 30 for all $x^\star \in \operatorname{Zer}(\mathbb{A} + \mathbb{B})$, we can show (i) and (ii) for all $x^\star \in \operatorname{Zer}(\mathbb{A} + \mathbb{B})$, not just the $x^\star$ for which the demipositivity property holds. By (i) and monotonicity of $\mathbb{A}$ and $\mathbb{B}$, we have

$$\alpha_k \langle \widetilde{\mathbb{B}} x^k - \widetilde{\mathbb{B}} x^\star, x^k - x^\star \rangle \leq \alpha_k \langle \widetilde{\mathbb{A}} x^k + \widetilde{\mathbb{B}} x^k, x^k - x^\star \rangle \to 0.$$

Therefore, (ii) implies $\lim_{k\to\infty} \|x^k - x^\star\|^2$ exists. We apply Proposition 1 to conclude $\lim_{k\to\infty} \|x^k - y\|^2$ exists for all $y \in \operatorname{Zer}(\mathbb{A} + \mathbb{B})$, including $y = x^\infty$. Therefore, $\|x^k - x^\infty\| \to 0$, that is, the entire sequence converges to x^∞. □

Next, we present the proof of Theorem 5, which is more elaborate and relies on most of the steps performed for the proof of Theorem 4. We first present two intermediate lemmas.

Lemma 1 *Let* $\mathbb{A} \colon \mathbb{R}^n \rightrightarrows \mathbb{R}^n$ *be maximal monotone. If*

$$\inf_{(x,u)\in\mathbb{A}} \langle u, x - \bar{x} \rangle \geq 0,$$

then $\bar{x} \in \operatorname{Zer} \mathbb{A}$.

Proof. The infimum can be rephrased as

$$\langle \mathbb{A} x - 0, x - \bar{x} \rangle \geq 0 \qquad \forall x \in \mathbb{R}^n.$$

Maximality of $\mathbb{A}$ implies $0 \in \mathbb{A} \bar{x}$. □

Lemma 2 *Let $\alpha_0, \alpha_1, \ldots$ be a non-summable positive sequence and $x^0, x^1, \ldots$ be vectors in $\mathbb{R}^n$. Define $\bar{x}^k$ as in (7.3). Assume there is a nonempty closed convex set Q such that (i) any convergent subsequence of $\bar{x}^k$ converges to a limit in Q; (ii) $\lim_{k\to\infty} \|x^k - q\|$ exists for all $q \in Q$. Then $\bar{x}^k$ converges to a single limit in Q.*

Proof. For any $q_1, q_2 \in Q$,

$$\left\|x^k - \frac{q_1 + q_2}{2}\right\|^2 = \|x^k - q_1\|^2 + \left\|\frac{q_1 - q_2}{2}\right\|^2 + \langle x^k - q_1, q_1 - q_2\rangle.$$

Then, $\langle x^k - q_1, q_1 - q_2\rangle$ has a limit and

$$\lim_{k\to\infty} \langle x^k - q_1, q_1 - q_2\rangle = \lim_{k\to\infty} \langle \bar{x}^k - q_1, q_1 - q_2\rangle,$$

since $\alpha_0, \alpha_1, \ldots$ is a non-summable positive sequence. Next, let σ_1 and σ_2 be two accumulation points of $\bar{x}^k$. Then,

$$\langle \sigma_1 - q_1, q_1 - q_2\rangle = \langle \sigma_2 - q_1, q_1 - q_2\rangle,$$

and by letting $q_1 = \sigma_1$ and $q_2 = \sigma_2$, we get $\|\sigma_1 - \sigma_2\|^2 = 0$. Therefore, $\bar{x}^k$ has only one limit. □

Continuous-time illustration of the proof of Theorem 5. Before we prove Theorem 5, we present a deterministic continuous-time analysis that motivates the stochastic discrete-time analysis for Theorem 5. Analyzing convergence in continuous-time (which is often easier) and then translating the analysis to the discrete-time setup is a common technique for understanding and finding proofs.

Assume $\mathbb{A}$ is single-valued and $\mathbb{B} = \mathbb{O}$. Consider the differential equation

$$\dot{x}(t) = -\mathbb{A}x(t), \qquad x(0) = x^0.$$

Then, for any $y \in \mathbb{R}^n$, we have

$$\begin{aligned}\frac{d}{dt}\frac{1}{2}\|x(t) - y\|^2 &= \langle \dot{x}(t), x(t) - y\rangle = -\langle \mathbb{A}x(t), x(t) - y\rangle \\ &\leq -\langle \mathbb{A}y, x(t) - y\rangle.\end{aligned}$$

When $y = x^\star \in \operatorname{Zer} \mathbb{A}$, we have $\frac{d}{dt}\frac{1}{2}\|x(t) - x^\star\|^2 \leq 0$. So $\lim_{t\to\infty} \|x(t) - x^\star\|^2$ exists, since $\|x(t) - x^\star\|^2$ is nonincreasing and nonnegative, for all $x^\star \in \operatorname{Zer} \mathbb{A}$ and $x(t)$ is bounded.

Integrate $\frac{d}{dt}\frac{1}{2}\|x(t) - y\|^2 \leq -\langle \mathbb{A}y, x(t) - y\rangle$ from $t = 0$ to $t = T$ and divide by T to get

$$\frac{1}{T}\left(\frac{1}{2}\|x(T) - y\|^2 - \frac{1}{2}\|x(0) - y\|^2\right) \leq -\langle \mathbb{A}y, \bar{x}(T) - y\rangle,$$

where $\bar{x}(T) = \frac{1}{T}\int_0^T x(t)\, dt$. So

$$\langle \mathbb{A}y, \bar{x}(T) - y\rangle \leq -\frac{1}{T}\left(\frac{1}{2}\|x(T) - y\|^2 - \frac{1}{2}\|x(0) - y\|^2\right)$$

and

$$\limsup_{T\to\infty} \langle \mathbb{A}y, \bar{x}(T) - y\rangle \leq 0.$$

By Lemma 1, all accumulation points of $\bar{x}(t)$ are in $\operatorname{Zer} \mathbb{A}$. With a variation of Lemma 2, we conclude that $\bar{x}(t)$ converges to a single limit.

Proof of Theorem 5. We use the same notation as in the proof of Theorem 4. By arguments of the proof of Theorem 4, with probability 1, $\lim_{k\to\infty}\|x^k - x^\star\|$ exists for all $x^\star \in \mathrm{Zer}(\mathbb{A}+\mathbb{B})$. In particular, we know that x^k, $\widetilde{\mathbb{A}}x^k$, $\widetilde{\mathbb{B}}x^k$ are bounded sequences with probability 1.

Define

$$\xi^k = \widetilde{\mathbb{A}}x^k - \widetilde{\mathbb{A}}_{i(k)}x^k,$$

and consider the martingale $\sum_{j=0}^{k}\alpha_j\xi^j$. The martingale differences are summable since

$$\begin{aligned}\sum_{k=0}^{\infty}\alpha_k^2\mathbb{E}_k\left[\|\xi^k\|^2\right] &= \sum_{k=0}^{\infty}\alpha_k^2\mathbb{E}_k\left[\|\widetilde{\mathbb{A}}x^k - \widetilde{\mathbb{A}}_{i(k)}x^k\|^2\right]\\ &\le \sum_{k=0}^{\infty}2\alpha_k^2\mathbb{E}_k\left[\|\widetilde{\mathbb{A}}x^k\|^2 + \|\widetilde{\mathbb{A}}_{i(k)}x^k\|^2\right]\\ &\le \sum_{k=0}^{\infty}2\alpha_k^2\left(C_1\|x^k\|^2 + 2C_2\right) < \infty,\end{aligned}$$

where we used (7.1) and the fact that x^k is bounded. By Theorem 31, a martingale convergence theorem,

$$\sum_{k=1}^{\infty}\alpha_k\xi^k$$

exists with probability 1. Let $K_1 \in \mathbb{N}_+$. Define $y^{K_1} = x^{K_1}$ and

$$y^{k+1} = y^k - \alpha_k\widetilde{\mathbb{A}}x^k - \alpha_k\widetilde{\mathbb{B}}x^{k+1}$$

for $k = K_1, K_1+1, \ldots$. Define

$$\varepsilon_{K_1} = \sup_{k\ge K_1}\|y^k - x^k\| = \sup_{k\ge K_1}\left\|\sum_{j=K_1}^{k}\alpha_j\xi^j\right\|.$$

Since $\sum_{k=1}^{\infty}\alpha_k\xi^k$ exists, $\varepsilon_{K_1} \to 0$ as $K_1 \to \infty$. (We use the y^k-sequence because the differences $y^{k+1}-y^k$ are defined with $\widetilde{\mathbb{A}}x^k$ rather than $\widetilde{\mathbb{A}}_{i(k)}x^k$, and because the difference between x^k and y^k is small, bounded by ε_{K_1}.)

Let $K_2 > K_1$. Define

$$\bar{x}^{(K_1,K_2)} = \frac{\sum_{k=K_1}^{K_2}\alpha_k x^k}{\sum_{k=K_1}^{K_2}\alpha_k},$$

that is, $\bar{x}^{(K_1,K_2)}$ is the averaged iterate with the averaging starting from K_1 and ending at K_2. Note that $\bar{x}^{K_2} = \bar{x}^{(0,K_2)}$, and that $\bar{x}^{(K_1+1,K_2)}$ and $\bar{x}^{K_2}$ share the same accumulation points as $K_2 \to \infty$.

Let $x \in \mathbb{R}^n$, $\widetilde{\mathbb{A}}x \in \mathbb{A}x$, and $\widetilde{\mathbb{B}}x \in \mathbb{B}x$ be arbitrary. For $k \ge K_1$, we have

$$\begin{aligned}\|y^{k+1} - x\|^2 &= \|y^k - x\|^2 + \alpha_k^2\|\widetilde{\mathbb{A}}x^k\|^2 - \alpha_k^2\|\widetilde{\mathbb{B}}x^{k+1}\|^2\\ &\quad - 2\alpha_k\langle\widetilde{\mathbb{A}}x^k, y^k - x\rangle - 2\alpha_k\langle\widetilde{\mathbb{B}}x^{k+1}, y^{k+1} - x\rangle\\ &= \|y^k - x\|^2 + \alpha_k^2\|\widetilde{\mathbb{A}}x^k\|^2 - \alpha_k^2\|\widetilde{\mathbb{B}}x^{k+1}\|^2\\ &\quad - 2\alpha_k\langle\widetilde{\mathbb{A}}x^k, x^k - x\rangle - 2\alpha_k\langle\widetilde{\mathbb{B}}x^{k+1}, x^{k+1} - x\rangle\\ &\quad - 2\alpha_k\langle\widetilde{\mathbb{A}}x^k, y^k - x^k\rangle - 2\alpha_k\langle\widetilde{\mathbb{B}}x^{k+1}, y^{k+1} - x^{k+1}\rangle\end{aligned}$$

$$\begin{aligned}&\leq \|y^k - x\|^2 + \alpha_k^2\|\widetilde{\mathbb{A}}x^k\|^2 - \alpha_k^2\|\widetilde{\mathbb{B}}x^{k+1}\|^2\\&\quad - 2\alpha_k\langle\widetilde{\mathbb{A}}x^k, x^k - x\rangle - 2\alpha_k\langle\widetilde{\mathbb{B}}x^{k+1}, x^{k+1} - x\rangle\\&\quad + 2\alpha_k\varepsilon_{K_1}\|\widetilde{\mathbb{A}}x^k\| + 2\alpha_k\varepsilon_{K_1}\|\widetilde{\mathbb{B}}x^{k+1}\|,\end{aligned}$$

where we used the Cauchy–Schwartz inequality. Let $M < \infty$ be a bound for

$$\max\left\{\|x\|, \|x^k\|, \|\widetilde{\mathbb{A}}x^k\|, \|\widetilde{\mathbb{B}}x^k\|, \sum_{k=0}^{\infty}\alpha_k^2\right\} \leq M \qquad \forall k = 0, 1, \ldots.$$

Then, we have

$$\begin{aligned}\|y^{k+1} - x\|^2 &\leq \|y^k - x\|^2 + \alpha_k^2 M^2 + 4\alpha_k\varepsilon_{K_1}M\\&\quad - 2\alpha_k\langle\widetilde{\mathbb{A}}x^k, x^k - x\rangle - 2\alpha_k\langle\widetilde{\mathbb{B}}x^{k+1}, x^{k+1} - x\rangle\\&\leq \|y^k - x\|^2 + \alpha_k^2 M^2 + 4\alpha_k\varepsilon_{K_1}M\\&\quad - 2\alpha_k\langle\widetilde{\mathbb{A}}x, x^k - x\rangle - 2\alpha_k\langle\widetilde{\mathbb{B}}x, x^{k+1} - x\rangle,\end{aligned}$$

where we used monotonicity of $\mathbb{A}$ and $\mathbb{B}$. Sum over $k = K_1, \ldots, K_2$ to get

$$\begin{aligned}&2\sum_{k=K_1}^{K_2}\alpha_k\langle\widetilde{\mathbb{A}}x, x^k - x\rangle + 2\sum_{k=K_1+1}^{K_2+1}\alpha_k\langle\widetilde{\mathbb{B}}x, x^k - x\rangle\\&\qquad\leq \|x^{K_1} - x\|^2 + M^3 + 4\varepsilon_{K_1}M\sum_{k=K_1}^{K_2}\alpha_k\\&\qquad\quad - 2\sum_{k=K_1}^{K_2}(\alpha_k - \alpha_{k+1})\langle\widetilde{\mathbb{B}}x, x^{k+1} - x\rangle.\end{aligned}$$

Dividing by $2\sum_{k=K_1+1}^{K_2}\alpha_k$, we get

$$\begin{aligned}\langle\widetilde{\mathbb{A}}x + \widetilde{\mathbb{B}}x, \bar{x}^{(K_1+1,K_2)} - x\rangle &\leq O\left(1/\sum_{k=K_1+1}^{K_2}\alpha_k\right) + 2\varepsilon_{K_1}M\left(\frac{\alpha_{K_1}}{\sum_{k=K_1+1}^{K_2}\alpha_k} + 1\right)\\&\quad + \frac{M^2\alpha_{K_1}}{\sum_{k=K_1+1}^{K_2}\alpha_k},\end{aligned}$$

where we used $-\langle\widetilde{\mathbb{B}}x, x^{k+1} - x\rangle \leq M^2$ and the fact that $\sum_{k=K_1}^{K_2}(\alpha_k - \alpha_{k+1})$ forms a telescoping sum. Therefore,

$$\limsup_{K_2\to\infty}\langle\widetilde{\mathbb{A}}x + \widetilde{\mathbb{B}}x, \bar{x}^{(K_1+1,K_2)} - x\rangle \leq 4\varepsilon_{K_1}M.$$

The right-hand side goes to 0 as $K_1 \to \infty$. Since the sequences $\bar{x}^{(K_1+1,K_2)}$ and $\bar{x}^{K_2}$ share the same accumulation points as $K_2 \to \infty$, we conclude

$$\limsup_{k\to\infty}\langle\widetilde{\mathbb{A}}x + \widetilde{\mathbb{B}}x, \bar{x}^k - x\rangle \leq 0.$$

Since $x \in \mathbb{R}^n$, $\widetilde{\mathbb{A}}x \in \mathbb{A}x$, and $\widetilde{\mathbb{B}}x \in \mathbb{B}x$ are arbitrary, Lemma 1 implies all accumulation points of $\bar{x}^k$ are in $\operatorname{Zer}(\mathbb{A} + \mathbb{B})$. Finally, Lemma 2 shows that $\bar{x}^k$ converges to a single limit in $\operatorname{Zer}(\mathbb{A} + \mathbb{B})$. □

Subgradient Methods
The problem

$$\begin{array}{ll}\underset{x\in\mathbb{R}^n}{\text{minimize}} & f(x)\\ \text{subject to} & x\in C,\end{array}$$

where f is a CCP function on $\mathbb{R}^n$ and $C \subseteq \mathbb{R}^n$ is a nonempty closed convex set, can be solved with the *(projected) subgradient method*:

$$x^{k+1} \in \Pi_C(x^k - \alpha_k \partial f(x^k)).$$

Likewise, the problem

$$\underset{x\in\mathbb{R}^n}{\text{minimize}} \quad f(x) + g(x),$$

where f and g are CCP can be solved with the *proximal subgradient method*:

$$x^{k+1} \in \mathrm{Prox}_{\alpha_k g}(x^k - \alpha_k \partial f(x^k)).$$

Convergence of these subgradient methods are ensured by Theorem 5; if a minimizer exists,

$$\|\partial f(x)\|^2 \leq \frac{C_1}{2}\|x\|^2 + C_2 \qquad \forall x\in\mathbb{R}^n$$

holds, and (7.2) holds, then $x^k \to x^\star$. Note that these methods have no stochasticity.

Stochastic Proximal Subgradient Method
For the stochastic proximal subgradient method

$$x^{k+1} \in \mathrm{Prox}_{\alpha_k g}(x^k - \alpha_k \partial f_{i(k)}(x^k)),$$

we apply Theorem 4. If a minimizer exists,

$$\frac{1}{N}\sum_{i=1}^{N}\|\partial f_i(x)\|^2 \leq \frac{C_1}{2}\|x\|^2 + C_2 \qquad \forall x\in \operatorname{dom} g$$

holds, and (7.2) holds, then $x^k \to x^\star$ with probability 1.

Stochastic Proximal Simultaneous Gradient Method
Again, consider the problem of finding a saddle point of

$$\mathbf{L}(x,u) = f(x) - g(u) + \frac{1}{N}\sum_{i=1}^{N}\mathbf{L}_i(x,u),$$

where f is a CCP function on $\mathbb{R}^n$, g is a CCP function on $\mathbb{R}^m$, and $\mathbf{L}_i\colon \mathbb{R}^{n+m} \to \mathbb{R}$ is convex-concave for $i = 1,\dots,N$. The SFB with $\partial\mathbf{L}$,

$$\begin{aligned} x^{k+1} &\in \mathrm{Prox}_{\alpha_k f}(x^k - \alpha_k \partial_x \mathbf{L}_{i(k)}(x^k,u^k))\\ u^{k+1} &\in \mathrm{Prox}_{\alpha_k g}(u^k - \alpha_k \partial_u(-\mathbf{L}_{i(k)})(x^k,u^k)),\end{aligned}$$

is called that *stochastic proximal simultaneous subgradient method.* (The method "simultaneously" updates the x and the u.)

Assume a saddle point exists, (7.1) holds, and (7.2) holds. The averaged iterates defined as in (7.3) converge, that is, $\bar{x}^k \to x^\star$ and $\bar{u}^k \to u^\star$, with probability 1. If $\mathbf{L}$ is furthermore strictly convex-concave, that is, $\mathbf{L}(x,u)$ is strictly convex in x for fixed u and strictly concave in u for fixed x, then $x^k \to x^\star$ and $u^k \to u^\star$ with probability 1.

Stochastic Condat–Vũ

Consider the problem

$$\underset{x \in \mathbb{R}^n}{\text{minimize}} \quad f(x) + g(Ax) + \tfrac{1}{N}\textstyle\sum_{i=1}^N h_i(x),$$

where f, g, and $h_1,\dots,h_N$ are CCP, $h_1,\dots,h_N$ are differentiable, and $A \in \mathbb{R}^{m\times n}$. The Lagrangian generating this primal problem is

$$\mathbf{L}(x,u) = f(x) + \frac{1}{N}\sum_{i=1}^N h_i(x) + \langle u, Ax\rangle - g^*(u).$$

We split the $\partial\mathbf{L}$ into

$$\partial\mathbf{L}(x,u) = \frac{1}{N}\sum_{i=1}^N \underbrace{\begin{bmatrix} \nabla h_{i(k)}(x) \\ 0 \end{bmatrix}}_{=\mathbb{H}_{i(k)}(x,u)} + \underbrace{\begin{bmatrix} 0 & A^\intercal \\ -A & 0 \end{bmatrix}\begin{bmatrix} x \\ u \end{bmatrix} + \begin{bmatrix} \partial f(x) \\ \partial g^*(u) \end{bmatrix}}_{=\mathbb{F}(x,u)}.$$

The variable metric version of SFB with

$$M = \begin{bmatrix} (1/\alpha)I & -A^\intercal \\ -A & (1/\beta)I \end{bmatrix}$$

and $\mathbb{A}_i = \mathbb{I} - (M - \mathbb{H}_i)(M + \mathbb{F})^{-1}$ for $i = 1,\dots,N$ and $\mathbb{B} = \mathbb{0}$ is

$$\begin{aligned}
x^{k+1} &= \mathrm{Prox}_{\beta f}(z^k) \\
u^{k+1} &= \mathrm{Prox}_{\gamma g^*}(w^k + 2\beta A x^{k+1}) \\
z^{k+1} &= z^k - \alpha_k(z^k - x^{k+1} + \beta(A^\intercal u^{k+1} + \nabla h_{i(k)}(x^{k+1}))) \\
w^{k+1} &= w^k - \alpha_k(w^k + \gamma A x^{k+1}).
\end{aligned}$$

If total duality holds, $h_1,\dots,h_N$ are L-smooth, $\beta > 0$ and $\gamma > 0$ satisfy

$$\beta L/2 + \beta\gamma\lambda_{\max}(A^\intercal A) < 1,$$

and (7.2) holds, then $x^k \to x^\star$ and $u^k \to u^\star$ with probability 1.

BIBLIOGRAPHICAL NOTES

The proof of Theorem 4 follows what is now considered a standard argument, which was first presented in Bottou's 1991 thesis in French [Bot91, Section 3.3.1.4] and later in Bottou's 1998 paper in English [Bot99]. The proof of Theorem 5 follows the techniques presented in Andrieu, Moulines, and Priouret's 2005 paper [AMP05] and Bianchi's 2016 paper [Bia16]. Lemma 2 was first presented by Passty in 1979 [Pas79].

Stochastic Approximation The stochastic gradient method, also referred to as *stochastic approximation* in the classical literature, dates back to Robbins and Monro's 1951 paper [RM51]. Kushner and Yin's 2003 textbook [KY03] provides a comprehensive treatment of stochastic approximation. The technique of averaging was first proposed by Bruck in 1977 [Bru77] and Nemirovski and Yudin in 1978 [NY78] for the non-stochastic setup and by Ruppert in 1988 [Rup88] and Polyak in 1990 [Pol90] for the stochastic setup. The subgradient method for the non-stochastic setup was first proposed by Shor in the 1960s [Sho62, Sho64, Sho85]. With the rise of machine learning, literature on the stochastic subgradient method has exploded, as the method is used universally for training neural networks.

The stochastic proximal gradient method has been studied extensively in machine learning. Stochastic forward-backward splitting for the operator setup was studied by Rosasco, Villa, and Vũ in 2016 [RVV16]. Interestingly, the proximal *subgradient* method, stochastic or otherwise, has not been studied much; Bello-Cruz's 2017 paper seems to be the only prior work [BC17b]. Recently, the stochastic forward-backward method has received increased interest in machine learning due to minimax training of GANs [CGFL19, MLZ+19, RYY19, GBV+19, MKS+20]. For the more general problem of the form

$$\underset{x \in \mathbb{R}^n}{\text{find}} \quad 0 \in \left(\frac{1}{N} \sum_{i=1}^{N} (\mathbb{A}_i + \mathbb{B}_i) \right) x,$$

where $\mathbb{A}_i \colon \mathbb{R}^n \rightrightarrows \mathbb{R}^n$ is a maximal monotone operator on $\mathbb{R}^n$ with $\operatorname{dom} \mathbb{A}_i = \mathbb{R}^n$ and $\mathbb{B}_i \colon \mathbb{R}^n \rightrightarrows \mathbb{R}^n$ is a maximal monotone operator for $i = 1, \ldots, N$. The method

$$x^{k+1} \in \mathbb{J}_{\alpha_k \mathbb{B}_{i(k)}} (\mathbb{I} - \alpha_k \mathbb{A}_{i(k)}) x^k,$$

where $\alpha_k > 0$ and $i(k) \in \{1, \ldots, N\}$ is chosen independently uniformly at random, was studied by Bianchi and Hachem in 2016 [Bia16, BH16].

Expectation of Operators In this chapter, we focused on finite sums of operators to avoid measure-theoretic discussions. The *Aumann integral*, first presented by Aumann in 1965 [Aum65] and further studied by Rockafellar in 1969 [Roc69] and Bertsekas in 1973 [Ber73], generalizes the finite sum of multi-valued operators to general integrals and expectations. Theorems 4 and 5 and their proofs remain valid with minimal modification when the finite sum is replaced with the Aumann integral.

Demipositivity The demipositivity assumption was first formulated by Bruck [Bru75a]. The definition we use is equivalent to the definition formulated by Peypouquet and Sorin [PS10]. See Exercise 7.4. Paramonotone operators, defined in the bibliographical notes section of §9, are also demipositive.

EXERCISES

7.1 *Convergence rate of the projected stochastic subgradient method.* Consider the problem

$$\begin{array}{ll} \underset{x \in \mathbb{R}^n}{\text{minimize}} & \frac{1}{N} \sum_{i=1}^{N} f_i(x) \\ \text{subject to} & x \in C, \end{array}$$

where C is a nonempty closed convex set and f_i is a CCP function on $\mathbb{R}^n$ such that $C \subseteq \operatorname{dom} \partial f_i$ for $i = 1, \ldots, N$. Assume $\|\partial f_i(x)\| \leq G$ for any $x \in C$ for $i = 1, \ldots, N$. (This assumption is equivalent to assuming $f_1, \ldots, f_N$ are G-Lipschitz continuous on C.) Consider the projected stochastic subgradient method:

$$x^{k+1} \in \Pi_C \left(x^k - \alpha_k \partial f_{i(k)}(x^k) \right).$$

Show that

$$f(\bar{x}^k) - f(x^\star) \leq \frac{\|x^0 - x^\star\|^2 + G^2 \sum_{i=0}^k \alpha_k^2}{2 \sum_{i=0}^k \alpha_k},$$

where $\bar{x}^k$ is as defined in (7.3). If $\alpha_k = 1/(k+1)^p$, for what value of p do we have $f(\bar{x}^k) - f(x^\star) \to 0$?

Hint. Use the subgradient inequality to show

$$\|x^{i+1} - x^\star\|^2 \leq \|x^i - x^\star\|^2 - 2\alpha_i (f(x^i) - f(x^\star)) + \alpha^2 G^2.$$

Then sum both sides from $i = 0, \ldots, k$.

7.2 *Convergence rate under strong monotonicity.* In the setup for Theorem 4, furthermore assume $\mathbb{A} + \mathbb{B}$ is μ-strongly monotone. Show that if $\alpha_k = C/(k+1)$ for some large enough $C > 0$, then

$$\|x^k - x^\star\|^2 \leq O(1/k).$$

Hint. Let $U^0, U^1, \ldots$ be a nonnegative sequence and let $W^0, W^1, \ldots$ be a nonnegative *summable* sequence. Assume

$$U^{k+1} \leq \left(1 - \frac{\sigma}{k} + \frac{\tau}{k^2}\right) U^k + W^k - W^{k-1} + \frac{\rho}{k^2},$$

where $\sigma > 0$ and $\tau \geq 0$. Define

$$\tilde{U}^k = kU^k - \frac{\rho}{\sigma - 1},$$

and show

$$\tilde{U}^{k+1} \leq \left(1 - \frac{\sigma - 1}{k} + O\left(\frac{1}{k^2}\right)\right) \tilde{U}^k + O\left(\frac{1}{k^2}\right) + (k+1)(W^k - W^{k+1}).$$

Sum both sides to conclude

$$U^k \leq \frac{\rho}{(\sigma - 1)k} + o\left(\frac{1}{k}\right).$$

7.3 *Robust stochastic approximation.* Consider the problem

$$\begin{array}{ll} \underset{x \in \mathbb{R}^n}{\text{minimize}} & \dfrac{1}{N} \sum_{i=1}^N f_i(x) \\ \text{subject to} & x \in C, \end{array}$$

where f_i are CCP functions on $\mathbb{R}^n$ with $\operatorname{dom} f_i = \mathbb{R}^n$ for $i = 1, \ldots, N$, and C is a nonempty convex set. For notational simplicity, define

$$f(x) = \frac{1}{N} \sum_{i=1}^N f_i(x).$$

Assume there is a $G < \infty$ such that $\|\partial f_i(x)\| \leq G$ for any $x \in C$ for $i = 1,\ldots,N$. Consider the projected stochastic subgradient method:

$$x^{k+1} \in \Pi_C \left(x^k - \alpha_k \partial f_{i(k)}(x^k)\right).$$

Use Exercise 7.1 to show that, with $\alpha_k = \gamma/\sqrt{k+1}$ for any $\gamma > 0$, the averaged iterates $\bar{x}^k$ defined in (7.3) achieve the rate

$$f(\bar{x}^k) - f(x^\star) \leq O(1/\sqrt{k}).$$

Next, use Exercise 7.2 to show that, if f is μ-strongly convex and $\alpha_k = \gamma/(k+1)$ with large enough $\gamma > 0$, the iterates achieve the rate

$$f(x^k) - f(x^\star) \leq O(1/k).$$

Finally, consider the specific case $n = N = 1$, $f_1(x) = (\mu/2)x^2$, and $C = [-1,1]$. Show that with $x^0 = 1$, $\alpha_k = \gamma/(k+1)$, and $\gamma < 1/\mu$,

$$f(x^k) - f(x^\star) \geq O(1/k^{\mu\gamma}).$$

Remark. These results show that $\alpha_k = \gamma/(k+1)$ without averaging yields a rate of convergence highly sensitive to the choice of γ, while $\alpha_k = \gamma/\sqrt{k+1}$ with averaging more robustly provides the rate of $O(1/\sqrt{k})$.

7.4 *Equivalent definition of demipositivity.* Peypouquet and Sorin [PS10] define a maximal monotone operator $\mathbb{A}$ to be demipositive if there is an $x^\star \in \operatorname{Zer} \mathbb{A}$ such that for every sequence $(x^k, u^k) \in \mathbb{A}$ such that $x^k \to x^\infty$ and u^k is bounded:

$$\langle u^k, x^k - x^\star \rangle \to 0 \quad \Rightarrow \quad x^\infty \in \operatorname{Zer} \mathbb{A}.$$

Show that this definition of Peypouquet and Sorin is equivalent to our definition in §7.1.

7.5 Show that if $\mathbb{A}$ is maximal monotone and $\operatorname{int}(\operatorname{Zer} \mathbb{A}) \neq \emptyset$, then $\mathbb{A}$ is demipositive.
Hint. Let $x^\star \in \operatorname{int}(\operatorname{Zer} \mathbb{A})$. Assume for contradiction that there is a $x \notin \operatorname{Zer} \mathbb{A}$ such that $u \in \mathbb{A}x$ and $\langle u, x - x^\star \rangle = 0$. Then $x^\star + \varepsilon u \in \operatorname{Zer} \mathbb{A}$ for small enough $\varepsilon > 0$.

8 ADMM-Type Methods

In this chapter, we present the alternating direction method of multipliers (ADMM) and its variants, which we loosely refer to as *ADMM-type* methods. We first present FLiP-ADMM, a general and highly versatile variant, and establish its convergence directly without relying on the machinery of monotone operators. We then derive a wide range of ADMM-type methods from FLiP-ADMM.

8.1 FUNCTION-LINEARIZED PROXIMAL ADMM

Let f_1 and f_2 be CCP functions on $\mathbb{R}^p$ and g_1 and g_2 be CCP functions on $\mathbb{R}^q$. Let f_2 and g_2 be differentiable. For notational convenience, write

$$f = f_1 + f_2, \qquad g = g_1 + g_2.$$

Let $A \in \mathbb{R}^{n\times p}$, $B \in \mathbb{R}^{n\times q}$, and $c \in \mathbb{R}^n$. Consider the primal problem

$$\begin{array}{ll} \underset{x\in\mathbb{R}^p,\, y\in\mathbb{R}^q}{\text{minimize}} & f_1(x) + f_2(x) + g_1(y) + g_2(y) \\ \text{subject to} & Ax + By = c, \end{array} \tag{8.1}$$

generated by the Lagrangian

$$\mathbf{L}(x, y, u) = f(x) + g(y) + \langle u, Ax + By - c\rangle.$$

The method *function-linearized proximal alternating direction method of multipliers* (FLiP-ADMM) is

$$\begin{aligned} x^{k+1} &\in \underset{x\in\mathbb{R}^p}{\operatorname{argmin}} \left\{ f_1(x) + \langle \nabla f_2(x^k) + A^\intercal u^k, x\rangle + \frac{\rho}{2}\|Ax + By^k - c\|^2 + \frac{1}{2}\|x - x^k\|_P^2 \right\} \\ y^{k+1} &\in \underset{y\in\mathbb{R}^q}{\operatorname{argmin}} \left\{ g_1(y) + \langle \nabla g_2(y^k) + B^\intercal u^k, y\rangle + \frac{\rho}{2}\|Ax^{k+1} + By - c\|^2 + \frac{1}{2}\|y - y^k\|_Q^2 \right\} \\ u^{k+1} &= u^k + \varphi\rho(Ax^{k+1} + By^{k+1} - c), \end{aligned}$$

where $\rho > 0$, $\varphi > 0$, $P \in \mathbb{R}^{p\times p}$, $P \succeq 0$, $Q \in \mathbb{R}^{q\times q}$, and $Q \succeq 0$.

Theorem 6 Consider FLiP-ADMM. Assume total duality. Assume the x- and y-subproblems always have solutions. Assume f_2 is L_f-smooth and g_2 is L_g-smooth, where $L_f \geq 0$ and $L_g \geq 0$. Assume there is an $\varepsilon \in (0, 2-\varphi)$ such that

$$P \succeq L_f I, \qquad Q \succeq 0, \qquad \rho\left(1 - \frac{(1-\varphi)^2}{2-\varphi-\varepsilon}\right) B^\intercal B + Q \succeq 3L_g I.$$

Then

$$f(x^k) + g(y^k) \to f(x^\star) + g(y^\star), \qquad Ax^k + By^k - c \to 0,$$

where $(x^\star, y^\star)$ is a solution of the primal problem.

To clarify, when $f_2 = 0$ or $g_2 = 0$, we set $L_f = 0$ or $L_g = 0$, respectively.

8.1.1 Parameter Choices

FLiP-ADMM has several algorithmic parameters, φ, ρ, P, Q, f_2, and g_2. Their choices affect the number of iterations required to achieve a desired accuracy and the computational cost per iteration. The optimal choice for a given problem balances the number of iterations and cost per iteration.

The condition

$$\rho\left(1 - \frac{(1-\varphi)^2}{2-\varphi-\varepsilon}\right) B^\intercal B + Q \succeq 3L_g I \tag{8.2}$$

imposes constraints on φ and ρ. Note

$$1 - \frac{(1-\varphi)^2}{2-\varphi} \begin{cases} > 0, & \varphi \in (0, \frac{\sqrt{5}+1}{2}) \\ = 0, & \varphi = \frac{\sqrt{5}+1}{2} \\ < 0, & \varphi \in (\frac{\sqrt{5}+1}{2}, 2). \end{cases}$$

For each $\varphi \in (0, \frac{\sqrt{5}+1}{2})$, there is a small enough $\varepsilon > 0$ such that

$$1 - \frac{(1-\varphi)^2}{2-\varphi-\varepsilon} > 0,$$

so large ρ helps to satisfy (8.2). For each $\varphi \in [\frac{\sqrt{5}+1}{2}, 2)$ and all $\varepsilon \in (0, 2-\varphi)$,

$$1 - \frac{(1-\varphi)^2}{2-\varphi-\varepsilon} < 0,$$

so small ρ helps to satisfy (8.2).

The Dual Extrapolation Parameter φ

While the choice $\varphi = 1$ is most common, larger values for φ can provide a speedup. When $Q = 0$ and $L_g = 0$, condition (8.2) is satisfied with $\varphi \in (0, \frac{\sqrt{5}+1}{2})$. This is also the requirement for φ in the classical "golden ratio" ADMM setup, where $f_2 = 0$ ($L_f = 0$), $g_2 = 0$ ($L_g = 0$), $P = 0$, and $Q = 0$.

In the classical ADMM setup, where $f_2 = 0$, $g_2 = 0$, $P = 0$, and $Q = 0$, FLiP-ADMM reduces to

$$\begin{aligned} x^{k+1} &\in \operatorname*{argmin}_{x\in\mathbb{R}^p} \mathbf{L}_\rho(x, y^k, u^k) \\ y^{k+1} &\in \operatorname*{argmin}_{y\in\mathbb{R}^q} \mathbf{L}_\rho(x^{k+1}, y, u^k) \\ u^{k+1} &= u^k + \varphi\rho(Ax^{k+1} + By^{k+1} - c), \end{aligned}$$

where

$$\mathbf{L}_\rho(x, y, u) = f(x) + g(y) + \langle u, Ax + By - c\rangle + \frac{\rho}{2}\|Ax + By - c\|^2. \tag{8.3}$$

This is called the *golden ratio ADMM*, since the stepsize requirement is $0 < \varphi < (1 + \sqrt{5})/2$, and $(1 + \sqrt{5})/2 \approx 1.618$ is the golden ratio.

Penalty Parameter ρ

The parameter ρ controls the relative priority between primal and dual convergence. The Lyapunov function used to prove convergence of FLiP-ADMM contains the terms $\rho\|B(y^k - y^\star)\|^2$ (primal error) and $\frac{1}{\varphi\rho}\|u^k - u^\star\|^2$ (dual error), and large ρ prioritizes primal accuracy while small ρ prioritizes dual accuracy. When $0 < \varphi < \frac{\sqrt{5}+1}{2}$, we can use large ρ. When $\frac{\sqrt{5}+1}{2} \le \varphi < 2$, we can use small ρ.

Proximal Terms via P and Q

The letter "P" in FLi**P**-ADMM describes the presence of the *proximal terms*

$$\frac{1}{2}\|x - x^k\|_P^2, \qquad \frac{1}{2}\|y - y^k\|_Q^2.$$

Empirically, smaller P and Q lead to fewer required iterations. When $f_2 = 0$ and $g_2 = 0$, the choice $P = 0$ and $Q = 0$ is often optimal in the number of required iterations. In the cases considered in §8.2.1 and §8.2.3, however, we can use a nonzero P and Q to cancel out (linearize) unwieldy quadratic terms and thereby reduce the cost per iteration.

Linearizing Functions f_2 and g_2

When $f_2 = 0$, the x-update of FLiP-ADMM is

$$x^{k+1} \in \operatorname*{argmin}_{x\in\mathbb{R}^p} \left\{ \mathbf{L}_\rho(x, y^k, u^k) + \frac{1}{2}\|x - x^k\|_P^2 \right\},$$

where $\mathbf{L}_\rho$ is the augmented Lagrangian (8.3). When $f_2 \neq 0$,

$$\begin{aligned} x^{k+1} \in \operatorname*{argmin}_{x\in\mathbb{R}^p} \Big\{ & f_1(x) + f_2(x^k) + \langle \nabla f_2(x^k), x - x^k\rangle + g(y^k) \\ & + \langle u^k, Ax + By^k - c\rangle + \frac{\rho}{2}\|Ax + By^k - c\|^2 + \frac{1}{2}\|x - x^k\|_P^2 \Big\}, \end{aligned}$$

i.e., replace $f_2(x)$ with its first-order approximation $f_2(x^k) + \langle \nabla f_2(x^k), x - x^k\rangle$ in $\mathbf{L}_\rho(x, y^k, u^k)$ and minimize with respect to x. We call this *linearizing the function* or *function-linearization*. The same discussion holds for the y-update. The "FLi (Function-Linearized)" in **FLi**P-ADMM describes this feature of accessing f_2 and g_2 through their gradients.

FLiP-ADMM presents the choice of whether or not to use function linearization. Often, the choices $f_2 = 0$ and $g_2 = 0$ lead to fewer required iterations. In some cases, however, nonzero choices of f_2 and g_2 reduce the cost per iteration of solving the x- and y-subproblems.

8.1.2 Further Discussion

Solvability of Subproblems

We say the x- and y-subproblems are *solvable* if they have solutions (not necessarily unique). Solvability of the subproblems is not automatic, even when total duality holds and when f and g are CCP. In the derivation of ADMM in §3, solvability was ensured by assuming the additional regularity condition (3.6). In this section, we directly assume solvability instead. See the bibliographical notes of §3 for further discussion.

Notion of Convergence

The notion of convergence in Theorem 6 is different from what we had previously seen in Part I. The result establishes that the objective value converges to the optimal value and that the constraint violation converges to 0 rather than showing the iterates converge to a solution.

Relation to Method of Multipliers

The "AD (Alternating Direction)" in FLiP-**AD**MM describes the solving of the two x- and y-subproblems in an alternating fashion. The "MM" in FLiP-AD**MM** describes the method's similarity to the method of multipliers, which has only one primal subproblem.

When $q = 0$, the y-subproblem and B-matrix vanish, and the ADMM setup reduces to the method of multipliers setup, and we obtain methods similar to what we saw in Exercise 3.8. (Theorem 6 applies when $q = 0$.) In particular, when $q = 0, f_2 = 0, g_2 = 0$, $B = 0$, $P = 0$, and $Q = 0$, FLiP-ADMM reduces to the classical method of multipliers

$$
\begin{aligned}
x^{k+1} &\in \operatorname*{argmin}_{x} \left\{ f(x) + \langle u^k, Ax \rangle + \frac{\rho}{2}\|Ax - c\|^2 \right\} \\
u^{k+1} &= u^k + \varphi\rho(Ax^{k+1} - c),
\end{aligned}
$$

which converges for $\varphi \in (0, 2)$.

8.1.3 Scaled Form

With the substitution $v^k = (1/\rho)u^k$ for $k = 0, 1, \dots$, FLiP-ADMM becomes

$$
\begin{aligned}
x^{k+1} &\in \operatorname*{argmin}_{x \in \mathbb{R}^p} \left\{ f_1(x) + \langle \nabla f_2(x^k), x \rangle + \frac{\rho}{2}\|Ax + By^k - c + v^k\|^2 + \frac{1}{2}\|x - x^k\|_P^2 \right\} \\
y^{k+1} &\in \operatorname*{argmin}_{y \in \mathbb{R}^q} \left\{ g_1(y) + \langle \nabla g_2(y^k), y \rangle + \frac{\rho}{2}\|Ax^{k+1} + By - c + v^k\|^2 + \frac{1}{2}\|y - y^k\|_Q^2 \right\} \\
v^{k+1} &= v^k + \varphi(Ax^{k+1} + By^{k+1} - c).
\end{aligned}
$$

We call this the *scaled form* of FLiP-ADMM. In some cases, the scaled forms of ADMM-type methods are simpler than the original form and are therefore preferred. In this chapter, we use the unscaled form to make clear that the u^k-iterates represent (unscaled) dual variables.

8.1.4 Proof of Theorem 6

Proof. The assumption of total duality means $\mathbf{L}$ has a saddle point $(x^\star, y^\star, u^\star)$. Define

$$w^\star = \begin{bmatrix} x^\star \\ y^\star \\ u^\star \end{bmatrix}, \qquad w^k = \begin{bmatrix} x^k \\ y^k \\ u^k \end{bmatrix} \quad \text{for } k = 0, 1, \ldots.$$

Define $\eta = 2 - \varphi - \varepsilon$. Define the symmetric positive semidefinite matrices

$$M_0 = \frac{1}{2}\begin{bmatrix} P & 0 & 0 \\ 0 & \rho B^\intercal B + Q & 0 \\ 0 & 0 & \frac{1}{\varphi\rho} I \end{bmatrix}, \qquad M_1 = \frac{1}{2}\begin{bmatrix} 0 & 0 & 0 \\ 0 & Q + L_g I & 0 \\ 0 & 0 & \frac{\eta}{\varphi^2\rho} I \end{bmatrix},$$

$$M_2 = \frac{1}{2}\begin{bmatrix} P - L_f I & 0 & 0 \\ 0 & \rho\left(1 - \frac{(1-\varphi)^2}{\eta}\right) B^\intercal B + Q - 3L_g I & 0 \\ 0 & 0 & \frac{2-\varphi-\eta}{\varphi^2\rho} I \end{bmatrix}.$$

Define the Lyapunov function

$$V^k = \|w^k - w^\star\|_{M_0}^2 + \|w^k - w^{k-1}\|_{M_1}^2.$$

Proof Outline In stage 1, we use the definition of x^{k+1} and y^{k+1} as minimizers to obtain certain inequalities respectively relating x^{k+1} with $x^\star$ and y^{k+1} with $y^\star$. In stage 2, we use the definition of y^k and y^{k+1} as minimizers to obtain an inequality relating y^k with y^{k+1}. In stage 3, we use the inequalities of the previous stages to establish the key inequality

$$V^{k+1} \le V^k - \|w^{k+1} - w^k\|_{M_2}^2 - \left(\mathbf{L}(x^{k+1}, y^{k+1}, u^\star) - \mathbf{L}(x^\star, y^\star, u^\star)\right). \tag{8.4}$$

In stage 4, we use the summability argument to show convergence.

Stage 1 Generally, if $z^\star \in \operatorname{argmin}_z\{h_1(z) + h_2(z)\}$, where h_1 is convex and h_2 is differentiable convex, then $z^\star \in \operatorname{argmin}_z\{h_1(z) + \langle \nabla h_2(z^\star), z - z^\star\rangle\}$. This fact can be verified by considering the optimality conditions. In the x-subproblem defining x^{k+1}, we set $h_1 = f_1$ and h_2 to the remaining terms and get

$$\begin{aligned} 0 &\le f_1(x) - f_1(x^{k+1}) \\ &\quad + \left\langle \nabla f_2(x^k) + A^\intercal(u^k + \rho(Ax^{k+1} + By^k - c)) + P(x^{k+1} - x^k), x - x^{k+1}\right\rangle \end{aligned}$$

for any $x \in \mathbb{R}^p$. By convexity of f_2 and L_f-smoothness of f_2,

$$\begin{aligned} \langle \nabla f_2(x^k), x - x^{k+1}\rangle &= \langle \nabla f_2(x^k), x - x^k\rangle + \langle \nabla f_2(x^k), x^k - x^{k+1}\rangle \\ &\le f_2(x) - f_2(x^k) + f_2(x^k) - f_2(x^{k+1}) + \frac{L_f}{2}\|x^{k+1} - x^k\|^2 \\ &= f_2(x) - f_2(x^{k+1}) + \frac{L_f}{2}\|x^{k+1} - x^k\|^2. \end{aligned}$$

Adding the two inequalities, we get

$$\begin{aligned} 0 &\le f(x) - f(x^{k+1}) + \frac{L_f}{2}\|x^{k+1} - x^k\|^2 \\ &\quad + \left\langle A^\intercal(u^k + \rho(Ax^{k+1} + By^k - c)) + P(x^{k+1} - x^k), x - x^{k+1}\right\rangle. \end{aligned}$$

To simplify notation, define

$$\hat{u}^{k+1} = u^k + \rho(Ax^{k+1} + By^{k+1} - c)$$

and rewrite the inequality as

$$\begin{aligned} &f(x^{k+1}) - f(x) + \left\langle \hat{u}^{k+1}, A(x^{k+1} - x) \right\rangle \\ &\le \frac{L_f}{2}\|x^{k+1} - x^k\|^2 + \rho\left\langle B(y^{k+1} - y^k), A(x^{k+1} - x)\right\rangle - \left\langle x^{k+1} - x^k, x^{k+1} - x\right\rangle_P . \end{aligned} \tag{8.5}$$

Repeating analogous steps with the y-update, we get

$$\begin{aligned} &g(y^{k+1}) - g(y) + \left\langle \hat{u}^{k+1}, B(y^{k+1} - y)\right\rangle \\ &\qquad \le \frac{L_g}{2}\|y^{k+1} - y^k\|^2 - \left\langle y^{k+1} - y^k, y^{k+1} - y\right\rangle_Q \end{aligned} \tag{8.6}$$

for any $y \in \mathbb{R}^q$.

Set $x = x^\star$ in (8.5), set $y = y^\star$ in (8.6), add the two inequalities, add the identity

$$\langle u^\star - \hat{u}^{k+1}, Ax^{k+1} + By^{k+1} - c\rangle = \frac{1}{\rho}\langle u^\star - \hat{u}^{k+1}, \hat{u}^{k+1} - u^k\rangle,$$

and substitute

$$\begin{aligned} A(x^{k+1} - x^\star) &= \frac{1}{\varphi\rho}(u^{k+1} - u^k) - B(y^{k+1} - y^\star) \\ u^\star - \hat{u}^{k+1} &= \left(1 - \frac{1}{\varphi}\right)(u^{k+1} - u^k) - (u^{k+1} - u^\star) \\ \hat{u}^{k+1} - u^k &= \frac{1}{\varphi}\left(u^{k+1} - u^k\right) \end{aligned}$$

to get

$$\begin{aligned} &\mathbf{L}(x^{k+1}, y^{k+1}, u^\star) - \mathbf{L}(x^\star, y^\star, u^\star) \\ &\le \frac{L_f}{2}\|x^{k+1} - x^k\|^2 + \frac{L_g}{2}\|y^{k+1} - y^k\|^2 + \left(1 - \frac{1}{\varphi}\right)\frac{1}{\varphi\rho}\|u^{k+1} - u^k\|^2 \\ &\quad - 2\langle w^{k+1} - w^k, w^{k+1} - w^\star\rangle_{M_0} + \frac{1}{\varphi}\langle u^{k+1} - u^k, B(y^{k+1} - y^k)\rangle. \end{aligned} \tag{8.7}$$

Stage 2 Consider

$$\begin{aligned} &g(y^{k+1}) - g(y^k) + \left\langle \hat{u}^{k+1}, B(y^{k+1} - y^k)\right\rangle \\ &\qquad \le \frac{L_g}{2}\|y^{k+1} - y^k\|^2 - \left\langle y^{k+1} - y^k, y^{k+1} - y^k\right\rangle_Q, \end{aligned}$$

which follows from (8.6) with $y = y^k$, and

$$\begin{aligned} &g(y^k) - g(y^{k+1}) + \left\langle \hat{u}^k, B(y^k - y^{k+1})\right\rangle \\ &\qquad \le \frac{L_g}{2}\|y^k - y^{k-1}\|^2 - \left\langle y^k - y^{k-1}, y^k - y^{k+1}\right\rangle_Q, \end{aligned}$$

which follows from decrementing the indices $(k+1,k)$ to $(k,k-1)$ in (8.6) and using $y = y^{k+1}$. We add the two inequalities and reorganize to get

$$\begin{aligned}
&\frac{1}{\varphi}\langle u^{k+1} - u^k, B(y^{k+1} - y^k)\rangle \\
&\leq \frac{L_g}{2}\|y^{k+1} - y^k\|^2 + \frac{L_g}{2}\|y^k - y^{k-1}\|^2 - \|y^{k+1} - y^k\|_Q^2 \\
&\quad + \langle y^{k+1} - y^k, y^k - y^{k-1}\rangle_Q - \left(1 - \frac{1}{\varphi}\right)\langle u^k - u^{k-1}, B(y^{k+1} - y^k)\rangle.
\end{aligned}$$

We apply Young's inequality

$$\langle a, b\rangle \leq \frac{\zeta}{2}\|a\|^2 + \frac{1}{2\zeta}\|b\|^2, \qquad \forall\, a, b \in \mathbb{R}^n,\ \zeta > 0$$

to the two inner products on the right-hand side and reorganize to get

$$\begin{aligned}
&\frac{1}{\varphi}\langle u^{k+1} - u^k, B(y^{k+1} - y^k)\rangle \\
&\leq \frac{1}{2}\|y^{k+1} - y^k\|^2_{L_g I - Q + \frac{(1-\varphi)^2}{\eta}\rho B^\intercal B} + \frac{1}{2}\|y^k - y^{k-1}\|^2_{L_g I + Q} + \frac{\eta}{2\varphi^2\rho}\|u^k - u^{k-1}\|^2
\end{aligned} \tag{8.8}$$

for any $\eta > 0$. The left-hand side of this inequality is the last term on the right-hand side of (8.7).

Stage 3 Using

$$\|w^{k+1} - w^\star\|^2_{M_0} = \|w^k - w^\star\|^2_{M_0} - \|w^{k+1} - w^k\|^2_{M_0} + 2\langle w^{k+1} - w^k, w^{k+1} - w^\star\rangle_{M_0}$$

on the differences between V^{k+1} and V^k, we get

$$\begin{aligned}
V^{k+1} = V^k &- \|w^k - w^{k-1}\|^2_{M_1} + \|w^{k+1} - w^k\|^2_{M_1} - \|w^{k+1} - w^k\|^2_{M_0} \\
&+ 2\langle w^{k+1} - w^k, w^{k+1} - w^\star\rangle_{M_0}.
\end{aligned}$$

To this identity, we add the inequalities (8.8) and (8.7) to get

$$\begin{aligned}
V^{k+1} &\leq V^k - \frac{1}{2}\|x^{k+1} - x^k\|^2_{P - L_f I} - \frac{1}{2}\|y^{k+1} - y^k\|^2_{\left(1 - \frac{(1-\varphi)^2}{\eta}\right)\rho B^\intercal B + Q - 3L_g I} \\
&\quad - \frac{2 - \eta - \varphi}{2\varphi^2\rho}\|u^{k+1} - u^k\|^2 - \left(\mathbf{L}(x^{k+1}, y^{k+1}, u^\star) - \mathbf{L}(x^\star, y^\star, u^\star)\right) \\
&= V^k - \|w^{k+1} - w^k\|^2_{M_2} - \left(\mathbf{L}(x^{k+1}, y^{k+1}, u^\star) - \mathbf{L}(x^\star, y^\star, u^\star)\right)
\end{aligned}$$

for any $\eta > 0$, which is the key inequality (8.4). Note that

$$\mathbf{L}(x^{k+1}, y^{k+1}, u^\star) - \mathbf{L}(x^\star, y^\star, u^\star) \geq 0,$$

since $(x^\star, y^\star, u^\star)$ is a saddle point of $\mathbf{L}$.

Stage 4 Applying the summability argument on (8.4) tells us $\|w^{k+1} - w^k\|^2_{M_2} \to 0$ and $\mathbf{L}(x^{k+1}, y^{k+1}, u^\star) - \mathbf{L}(x^\star, y^\star, u^\star) \to 0$. Note that $\|w^{k+1} - w^k\|^2_{M_2} \to 0$ implies $u^{k+1} - u^k \to 0$ and thus $Ax^k + Bx^k - c \to 0$. Since

$$\begin{aligned}\mathbf{L}(x^{k+1}, y^{k+1}, u^\star) &= f(x^{k+1}) + g(y^{k+1}) + \underbrace{\langle u^\star, Ax^{k+1} + By^{k+1} - c\rangle}_{\to 0} \\ &\to \mathbf{L}(x^\star, y^\star, u^\star) = f(x^\star) + g(y^\star),\end{aligned}$$

we conclude $f(x^k) + g(y^k) \to f(x^\star) + g(y^\star)$. □

The inequalities we show in stages 1 and 2 are sometimes referred to as *variational inequalities* due to their connection to variational inequality problems. The key technical difficulty of the proof is the construction of the Lyapunov function V^k, which comes from the insights accumulated over the many papers studying various generalizations of ADMM.

8.2 DERIVED ADMM-TYPE METHODS

8.2.1 Linearized Methods

Consider the problem

$$\begin{array}{ll}\underset{x\in\mathbb{R}^p,\, y\in\mathbb{R}^q}{\text{minimize}} & f_1(x) + g_1(y) \\ \text{subject to} & Ax + By = c,\end{array}$$

where $f_2 = 0$ and $g_2 = 0$. We use the linearization technique of §3.5 with FLiP-ADMM. With $P = (1/\alpha)I - \rho A^\intercal A$ and $Q = (1/\beta)I - \rho B^\intercal B$, we get *linearized ADMM*:

$$\begin{aligned}x^{k+1} &= \mathrm{Prox}_{\alpha f}\left(x^k - \alpha A^\intercal(u^k + \rho(Ax^k + By^k - c))\right) \\ y^{k+1} &= \mathrm{Prox}_{\beta g}\left(y^k - \beta B^\intercal(u^k + \rho(Ax^{k+1} + By^k - c))\right) \\ u^{k+1} &= u^k + \varphi\rho(Ax^{k+1} + By^{k+1} - c).\end{aligned}$$

The stepsize requirement is satisfied with $1 \geq \alpha\rho\lambda_{\max}(A^\intercal A)$, $1 \geq \beta\rho\lambda_{\max}(B^\intercal B)$, and $0 < \varphi < (1+\sqrt{5})/2$. The linearized ADMM we had seen in §3.5 corresponds to the case $\varphi = 1$.

"Linearization" in the context of ADMM-type methods is an ambiguous term referring to more than one technique. While it most often refers to the technique of canceling out inconvenient quadratic terms, there are other "linearizations" as we will see in §8.2.2 and §8.2.6.

PDHG

Consider the problem

$$\begin{array}{ll}\underset{x\in\mathbb{R}^p,\, y\in\mathbb{R}^q}{\text{minimize}} & f_1(x) + g_1(y) \\ \text{subject to} & -Ix + By = 0.\end{array}$$

As discussed in §3.5, PDHG is an instance of FLiP-ADMM. With $\varphi = 1$, $P = 0$, and $Q = (1/\beta)I - \rho B^\intercal B$, we recover PDHG

$$\begin{aligned}\mu^{k+1} &= \mathrm{Prox}_{\rho f_1^*}\left(\mu^k + \rho B(2y^k - y^{k-1})\right)\\ y^{k+1} &= \mathrm{Prox}_{\beta g_1}\left(y^k - \beta B^\intercal \mu^{k+1}\right).\end{aligned}$$

The stepsize requirement is $1 \geq \beta\rho\lambda_{\max}(B^\intercal B)$.

8.2.2 Function-Linearized Methods

FLiP-ADMM linearizes the functions f_2 and g_2, i.e., it accesses f_2 and g_2 through their gradient evaluations rather than through minimization subproblems. This feature provides great flexibility.

Condat–Vũ

Consider the problem

$$\begin{array}{ll}\underset{x\in\mathbb{R}^p,\, y\in\mathbb{R}^q}{\text{minimize}} & f_1(x) + g_1(y) + g_2(y)\\ \text{subject to} & -Ix + By = 0.\end{array}$$

FLiP-ADMM with $\varphi = 1$, $P = 0$, and $Q = (1/\beta)I - \rho B^\intercal B$ is

$$\begin{aligned}x^{k+1} &= \mathrm{Prox}_{(1/\rho)f_1}\left((1/\rho)u^k + By^k\right)\\ y^{k+1} &= \mathrm{Prox}_{\beta g_1}\left(y^k - \beta\nabla g_2(y^k) - \beta B^\intercal(u^k - \rho(x^{k+1} - By^k))\right)\\ u^{k+1} &= u^k - \rho(x^{k+1} - By^{k+1}).\end{aligned}$$

Use the Moreau identity (2.12) to get

$$\begin{aligned}x^{k+1} &= (1/\rho)u^k + By^k - (1/\rho)\underbrace{\mathrm{Prox}_{\rho f_1^*}\left(u^k + \rho By^k\right)}_{=\mu^{k+1}}\\ y^{k+1} &= \mathrm{Prox}_{\beta g_1}\left(y^k - \beta\nabla g_2(x^k) - \beta B^\intercal \mu^{k+1}\right)\\ u^{k+1} &= \mu^{k+1} + \rho B(y^{k+1} - y^k),\end{aligned}$$

and we recover Condat–Vũ

$$\begin{aligned}\mu^{k+1} &= \mathrm{Prox}_{\rho f_1^*}\left(\mu^k + \rho B(2y^k - y^{k-1})\right)\\ y^{k+1} &= \mathrm{Prox}_{\beta g_1}\left(y^k - \beta\nabla g_2(y^k) - \beta B^\intercal \mu^{k+1}\right).\end{aligned}$$

Therefore, Condat–Vũ is a special case of FLiP-ADMM. The stepsize requirement of FLiP-ADMM translates to the requirement $1 \geq \beta\rho\lambda_{\max}(B^\intercal B) + 3\beta L_g$, which is worse than what we had seen in §3.3.

Doubly-Linearized ADMM
Consider the general problem

$$\begin{array}{ll} \underset{x\in\mathbb{R}^p,\, y\in\mathbb{R}^q}{\text{minimize}} & f_1(x) + f_2(x) + g_1(y) + g_2(y) \\ \text{subject to} & Ax + By = c. \end{array}$$

FLiP-ADMM with $P = (1/\alpha)I - \rho A^\intercal A$ and $Q = (1/\beta)I - \rho B^\intercal B$ is

$$\begin{aligned} x^{k+1} &= \mathrm{Prox}_{\alpha f_1}\left(x^k - \alpha\left(\nabla f_2(x^k) + A^\intercal u^k + \rho A^\intercal(Ax^k + By^k - c)\right)\right) \\ y^{k+1} &= \mathrm{Prox}_{\beta g_1}\left(y^k - \beta\left(\nabla g_2(y^k) + B^\intercal u^k + \rho B^\intercal(Ax^{k+1} + By^k - c)\right)\right) \\ u^{k+1} &= u^k + \varphi\rho(Ax^{k+1} + By^{k+1} - c). \end{aligned}$$

We call this method *doubly-linearized ADMM* as it linearizes both the quadratic terms and the functions f_2 and g_2. The stepsize requirement is satisfied with $1 \geq \alpha\rho\lambda_{\max}(A^\intercal A) + \alpha L_f$, $1 \geq \beta\rho\lambda_{\max}(B^\intercal B) + 3\beta L_g$, and $0 < \varphi < (1 + \sqrt{5})/2$. This method generalizes PDHG and Condat–Vũ.

Partial Linearization
Consider the problem

$$\begin{array}{ll} \underset{x\in\mathbb{R}^p,\, y\in\mathbb{R}^q}{\text{minimize}} & f_2(x) + g_1(y) + g_2(y) \\ \text{subject to} & Ax + By = c. \end{array}$$

Assume $\gamma I + \rho A^\intercal A$ is not easily invertible, but there is a $C \approx \rho A^\intercal A$ such that $\gamma I + C$ is easily invertible, for $\gamma > 0$. Choose $P = \gamma I + C - \rho A^\intercal A$, where $\gamma > \lambda_{\max}(\rho A^\intercal A - C)$. Since $C \approx \rho A^\intercal A$ is a close approximation, γ can be small.

In this case, the x-update of FLiP-ADMM is

$$x^{k+1} = x^k - (\gamma I + C)^{-1}(\nabla f_2(x^k) + A^\intercal u^k + \rho A^\intercal(Ax^k + By^k - c)).$$

The x-update is easy to compute, and we say it has been *partially linearized*. In contrast, the x-update of the doubly-linearized ADMM is

$$x^{k+1} = x^k - \frac{1}{\delta}(\nabla f_2(x^k) + A^\intercal u^k + \rho A^\intercal(Ax^k + By^k - c))$$

for some $\delta > \lambda_{\max}(\rho A^\intercal A)$. When $C \approx \rho A^\intercal A$, partial linearization reduces the number of required iterations compared to (full) linearization. This setup arises when $A^\intercal A$ is diagonally dominant in the regular basis or the discrete Fourier basis.

Example 8.1 *CT imaging with total variation regularization.* Consider the problem

$$\underset{x\in\mathbb{R}^p}{\text{minimize}} \quad \ell(Ax - b) + \lambda\|Dx\|_1,$$

where x represents an unknown 2D or 3D image reshaped into a vector, A is the discrete Radon transform operator, b represents the measurements, D is a finite difference operator, and ℓ is a CCP function. For simplicity, assume $A^\intercal A$ is invertible. The problem is equivalent to

$$\begin{array}{ll} \underset{x,y,z}{\text{minimize}} & \frac{1}{2}\ell(y) + \lambda\|z\|_1 \\ \text{subject to} & \begin{bmatrix} A \\ D \end{bmatrix} x - \begin{bmatrix} y \\ z \end{bmatrix} = \begin{bmatrix} b \\ 0 \end{bmatrix}. \end{array}$$

PDHG applied to the equivalent problem

$$\begin{aligned} u^{k+1} &= \mathrm{Prox}_{\rho\ell^*}\left(u^k + \rho A(2x^k - x^{k-1}) - \rho b\right) \\ v^{k+1} &= \Pi_{[-\lambda,\lambda]}\left(v^{k+1} + \rho D(2x^k - x^{k-1})\right) \\ x^{k+1} &= x^k - \frac{1}{\gamma}\left(A^\intercal u^{k+1} + D v^{k+1}\right) \end{aligned}$$

has a small cost per iteration, but sometimes requires too many iterations to converge. Classic ADMM with $\varphi = 1$ applied to the equivalent problem

$$\begin{aligned} u^{k+1} &= \mathrm{Prox}_{\rho\ell^*}\left(u^k + \rho A(2x^k - x^{k-1}) - \rho b\right) \\ v^{k+1} &= \Pi_{[-\lambda,\lambda]}\left(v^{k+1} + \rho D(2x^k - x^{k-1})\right) \\ x^{k+1} &= x^k - (\rho A^\intercal A + \rho D^\intercal D)^{-1}\left(A^\intercal u^{k+1} + D v^{k+1}\right) \end{aligned}$$

cannot be implemented as $(\rho A^\intercal A + \rho D^\intercal D)^{-1}$ is too expensive to compute. FLiP-ADMM, where (i) we update the y- and z-variables first, the x-variable second, and the dual variable last and (ii) we use the proximal term $P = \gamma I + C - \rho A^\intercal A - \rho D^\intercal D$ for the x-update and $\varphi = 1$ for the dual update, simplifies to

$$\begin{aligned} u^{k+1} &= \mathrm{Prox}_{\rho\ell^*}\left(u^k + \rho A(2x^k - x^{k-1}) - \rho b\right) \\ v^{k+1} &= \Pi_{[-\lambda,\lambda]}\left(v^{k+1} + \rho D(2x^k - x^{k-1})\right) \\ x^{k+1} &= x^k - (\gamma I + C)^{-1}\left(A^\intercal u^{k+1} + D v^{k+1}\right). \end{aligned}$$

See Exercise 8.2 for the derivation.

This partially linearized method can provide a significant speedup over PDHG. We can compute $(\gamma I + C)^{-1}$ efficiently with the fast Fourier transform; since $A^\intercal A$ and $D^\intercal D$ are discretizations of shift-invariant continuous operators, they are closely approximated by circulant matrices, which are diagonalizable by the discrete Fourier basis. A small $\gamma > \lambda_{\max}(\rho A^\intercal A + \rho D^\intercal D - C)$ makes P small and thereby minimizes the slowdown caused by the proximal term.

8.2.3 Block Splitting

Partition $x \in \mathbb{R}^p$ into m non-overlapping blocks of sizes $p_1, \ldots, p_m$. Write $x = (x_1, \ldots, x_m)$, so $x_i \in \mathbb{R}^{p_i}$ for $i = 1, \ldots, m$. Consider the problem

$$\begin{array}{ll} \underset{(x_1,\ldots,x_m)\in\mathbb{R}^p}{\text{minimize}} & \sum_{i=1}^m f_i(x_i) \\ \text{subject to} & Ax = c, \end{array} \tag{8.9}$$

where

$$A = \begin{bmatrix} A_{:,1} & A_{:,2} & \cdots & A_{:,m} \end{bmatrix}, \qquad Ax = A_{:,1}x_1 + A_{:,2}x_2 + \cdots + A_{:,m}x_m.$$

This problem is known as the multi-block ADMM problem or the extended monotropic program.

The objective function splits across the blocks $x_1,\dots,x_m$. However, the x-updates of FLiP-ADMM couple the m blocks in general; FLiP-ADMM with no function linearization and no y-block is

$$\begin{aligned} x^{k+1} &\in \operatorname*{argmin}_{x\in\mathbb{R}^p} \left\{ \sum_{i=1}^m f_i(x_i) + \langle A^\intercal u^k, x\rangle + \frac{\rho}{2}\|Ax-c\|^2 + \frac{1}{2}\|x-x^k\|_P^2 \right\} \\ u^{k+1} &= u^k + \rho(Ax-c), \end{aligned}$$

and the blocks $x_1^{k+1},\dots,x_m^{k+1}$ cannot be computed independently. In this section, we present techniques to obtain ADMM-type methods with split x-updates that can be computed independently in parallel.

Orthogonal Blocks

Consider problem (8.9). When the columns of A are block-wise orthogonal, i.e., $A_{:,i}^\intercal A_{:,j} = 0$ for all $i \neq j$, the x-updates of FLiP-ADMM with $P = 0$ split:

$$\begin{aligned} x_i^{k+1} &\in \operatorname*{argmin}_{x_i\in\mathbb{R}^{p_i}} \left\{ f_i(x_i) + \langle u^k - \rho c, A_{:,i}x_i\rangle + \frac{\rho}{2}\|A_{:,i}x_i\|^2 \right\} \qquad \text{for } i = 1,\dots,m \\ u^{k+1} &= u^k + \varphi\rho(Ax-c). \end{aligned}$$

Jacobi ADMM

Consider problem (8.9). Consider the matrix

$$P = \begin{bmatrix} \gamma I & -\rho A_{:,1}^\intercal A_{:,2} & \cdots & \cdots & -\rho A_{:,1}^\intercal A_{:,m} \\ -\rho A_{:,2}^\intercal A_{:,1} & \gamma I & \cdots & \cdots & -\rho A_{:,2}^\intercal A_{:,m} \\ \vdots & & \ddots & & \vdots \\ \vdots & & & \ddots & \vdots \\ -\rho A_{:,m}^\intercal A_{:,1} & -\rho A_{:,m}^\intercal A_{:,2} & \cdots & -\rho A_{:,m}^\intercal A_{:,(m-1)} & \gamma I \end{bmatrix},$$

which is positive semidefinite for $\gamma \geq \rho\lambda_{\max}(A^\intercal A)$. Let

$$\mathbf{L}_\rho(x,u) = \sum_{i=1}^m f_i(x_i) + \langle u, Ax-c\rangle + \frac{\rho}{2}\|Ax-c\|^2.$$

Let $x_{\neq i}^k$ denote all components of x^k excluding x_i^k. Then FLiP-ADMM is

$$\begin{aligned} x_i^{k+1} &= \operatorname*{argmin}_{x_i\in\mathbb{R}^{p_i}} \left\{ \mathbf{L}_\rho(x_i, x_{\neq i}^k, u^k) + \frac{\gamma}{2}\|x_i - x_i^k\|^2 \right\} \qquad \text{for } i = 1,\dots,m \\ u^{k+1} &= u^k + \varphi\rho\left(Ax^{k+1} - c\right). \end{aligned}$$

This method is called *Jacobi proximal ADMM* in analogy to the Jacobi method of numerical linear algebra; for $i = 1,\dots,m$, the update x_i^{k+1} is computed with the other blocks fixed to the older copies x_j^k for $j \neq i$. The stepsize requirement is satisfied with $\gamma \geq \rho\lambda_{\max}(A^\intercal A)$ and $\varphi \in (0,2)$.

The off-diagonal blocks of P remove the interaction between the x-blocks and thereby allow the x-update to split. Although we used γI for the diagonal blocks of P, other choices are possible; they just need to be, loosely speaking, sufficiently positive to ensure P is positive semidefinite. In Exercise 8.3, we use different diagonal blocks to perform linearizations with Jacobi ADMM.

Dummy Variables

Consider the problem

$$\begin{array}{ll} \underset{\substack{(x_1,\ldots,x_m)\in\mathbb{R}^p \\ y\in\mathbb{R}^n}}{\text{minimize}} & \sum_{i=1}^m f_i(x_i) + g(y) \\ \text{subject to} & Ax + y = c. \end{array}$$

Introduce dummy variables $z_1,\ldots,z_m$ and eliminate y to get the equivalent problem

$$\begin{array}{lll} \underset{\substack{(x_1,\ldots,x_m)\in\mathbb{R}^p \\ z_1,\ldots,z_m\in\mathbb{R}^n}}{\text{minimize}} & \sum_{i=1}^m f_i(x_i) + g\left(c - \sum_{i=1}^m z_i\right) & \\ \text{subject to} & A_{:,i}x_i - z_i = 0 & \text{for } i = 1,\ldots,m. \end{array}$$

We apply FLiP-ADMM with $P = 0$, $Q = 0$, no function linearization, and initial u-variables satisfying $u_1^0 = \cdots = u_m^0$. Then we can show $u_1^k = \cdots = u_m^k$ for $k = 1,\ldots,m$, and the iteration simplifies to

$$\begin{aligned} x_i^{k+1} &\in \underset{x_i\in\mathbb{R}^{p_i}}{\operatorname{argmin}} \left\{ f_i(x_i) + \left\langle u^k + \frac{\rho}{m}(Ax^k - z_{\text{sum}}^k), A_{:,i}x_i \right\rangle + \frac{\rho}{2}\left\|A_{:,i}(x_i - x_i^k)\right\|^2 \right\} \\ &\qquad\qquad\qquad\qquad\qquad\qquad\qquad\qquad\qquad\qquad \text{for } i = 1,\ldots,m \\ z_{\text{sum}}^{k+1} &= c - \operatorname{Prox}_{\frac{m}{\rho}g}\left(c - Ax^{k+1} - \frac{m}{\rho}u^k\right) \\ u^{k+1} &= u^k + \frac{\varphi\rho}{m}\left(Ax^{k+1} - z_{\text{sum}}^{k+1}\right). \end{aligned}$$

The stepsize requirement is $\varphi \in (0, (1+\sqrt{5})/2)$.

8.2.4 Consensus Technique

Consider the problem

$$\underset{x\in\mathbb{R}^p}{\text{minimize}} \quad \sum_{i=1}^n f_i(x).$$

Use the consensus technique to get the equivalent problem

$$\begin{array}{lll} \underset{x_1,\ldots,x_n,z\in\mathbb{R}^p}{\text{minimize}} & \sum_{i=1}^n f_i(x_i) & \\ \text{subject to} & x_i = z, & \text{for } i = 1,\ldots,n. \end{array}$$

Here, $x_i \in \mathbb{R}^p$ is a copy of $x \in \mathbb{R}^p$. Apply FLiP-ADMM with $P = 0$, $Q = 0$, no function linearization, and initial u-variables satisfying $u_1^0 + \cdots + u_n^0 = 0$ to get

$$\begin{aligned} x_i^{k+1} &= \operatorname*{argmin}_{x\in\mathbb{R}^p} \left\{f_i(x_i) + \langle u_i^k, x_i\rangle + \frac{\rho}{2}\|x_i - z^k\|^2\right\} && \text{for } i = 1,\dots,n \\ z^{k+1} &= \frac{1}{n}\sum_{i=1}^n x_i^{k+1} \\ u_i^{k+1} &= u_i^k + \varphi\rho(x_i^{k+1} - z^{k+1}) && \text{for } i = 1,\dots,n. \end{aligned}$$

The stepsize requirement is $\varphi \in (0,(1+\sqrt{5})/2)$. To clarify, each x_i represents a copy of the entire x and therefore has the same dimension. This contrasts with the block splitting of §8.2.3, where each x_i represented a single block of x.

In this version of the consensus technique, we constrain $x_1,\dots,x_n$ to equal a single z. In general, one can have multiple z-variables related through a graph structure. We explore this technique further in §11, in the context of decentralized optimization.

8.2.5 2-1-2 ADMM

Consider the problem

$$\begin{array}{ll} \underset{x\in\mathbb{R}^p,\, y\in\mathbb{R}^q}{\text{minimize}} & f(x) + g(y) \\ \text{subject to} & Ax + By = c. \end{array}$$

Assume g is a strongly convex quadratic function with affine constraints, i.e.,

$$g(y) = y^\intercal My + \mu^\intercal y + \delta_{\{y\in\mathbb{R}^q \mid Ny=v\}}(y)$$

for some positive definite $M \in \mathbb{R}^{q\times q}$, $N \in \mathbb{R}^{s\times q}$, and $v \in \mathcal{R}(N)$. If there is no affine constraint, we set $s = 0$. Define

$$\mathbf{L}_\rho(x,y,u) = f(x) + g(y) + \langle u, Ax + By - c\rangle + \frac{\rho}{2}\|Ax + By - c\|^2.$$

We call the method

$$\begin{aligned} y^{k+1/2} &= \operatorname*{argmin}_{y\in\mathbb{R}^q} \mathbf{L}_\rho(x^k, y, u^k) \\ x^{k+1} &\in \operatorname*{argmin}_{x\in\mathbb{R}^p} \mathbf{L}_\rho(x, y^{k+\frac{1}{2}}, u^k) \\ y^{k+1} &= \operatorname*{argmin}_{y\in\mathbb{R}^q} \mathbf{L}_\rho(x^{k+1}, y, u^k) \\ u^{k+1} &= u^k + \varphi\rho(Ax^{k+1} + By^{k+1} - c) \end{aligned}$$

2-1-2 ADMM. As we soon show, 2-1-2 ADMM is an instance of FLiP-ADMM and has the stepsize requirement of $\varphi \in (0,2)$.

Derivation

As we show in Exercise 8.10, the y update can be expressed as

$$y(x) = \operatorname*{argmin}_{y} \mathbf{L}_\rho(x, y, u^k) = -TB^\intercal Ax + t(u^k),$$

for a symmetric positive semidefinite $T \in \mathbb{R}^{q\times q}$ and a function t. Consider the FLiP-ADMM

$$\begin{aligned} (x^{k+1}, y^{k+1}) &\in \underset{x\in\mathbb{R}^p,\, y\in\mathbb{R}^q}{\operatorname{argmin}} \left\{ \mathbf{L}_\rho(x,y,u^k) + \frac{\rho}{2}\|x - x^k\|_P^2 \right\} \\ u^{k+1} &= u^k + \varphi\rho(Ax^{k+1} + By^{k+1} - c), \end{aligned} \tag{8.10}$$

with $P = A^\intercal BTB^\intercal A$. (In this instance of FLiP-ADMM, there is no second primal block, so there is no alternating update.) The stepsize requirement is $\varphi \in (0,2)$. The optimality condition is

$$\begin{aligned} 0 &\in \partial f(x^{k+1}) + A^\intercal \left(u^k + \rho(Ax^{k+1} + By^{k+1} - c)\right) + \rho A^\intercal BTB^\intercal A(x^{k+1} - x^k) \\ y^{k+1} &= -TB^\intercal Ax^{k+1} + t(u^k). \end{aligned}$$

On the other hand, the optimality condition of 2-1-2 ADMM is

$$\begin{aligned} y^{k+1/2} &= -TB^\intercal Ax^k + t(u^k) \\ 0 &\in \partial f(x^{k+1}) + A^\intercal \left(u^k + \rho(Ax^{k+1} + By^{k+1/2} - c)\right) \\ y^{k+1} &= -TB^\intercal Ax^{k+1} + t(u^k). \end{aligned}$$

Eliminating $y^{k+1/2}$ gives us the same optimality condition as that of (8.10). Therefore, 2-1-2 ADMM and FLiP-ADMM (8.10) are equivalent in the sense that they share the same set of iterates (x^k, y^k).

8.2.6 Trip-ADMM

Consider the more general problem

$$\begin{array}{ll} \underset{x\in\mathbb{R}^p,\, y\in\mathbb{R}^q}{\text{minimize}} & f_1(Cx) + f_2(x) + g_1(Dy) + g_2(y) \\ \text{subject to} & Ax + By = c, \end{array}$$

where $C \in \mathbb{R}^{r\times p}$ and $D \in \mathbb{R}^{s\times q}$. We can solve this problem with

$$\begin{aligned} &\left\{\begin{aligned} x^{k+1/2} &= x^k - \sigma\left(C^\intercal v^k + \nabla f_2(x^k) + A^\intercal u^k + \rho A^\intercal(Ax^k + By^k - c)\right) \\ v^{k+1} &= \operatorname{Prox}_{\tau f_1^*}\left(v^k + \tau C x^{k+1/2}\right) \\ x^{k+1} &= x^{k+1/2} - \sigma C^\intercal\left(v^{k+1} - v^k\right) \end{aligned}\right. \\ &\left\{\begin{aligned} y^{k+1/2} &= y^k - \sigma\left(D^\intercal w^k + \nabla g_2(y^k) + B^\intercal u^k + \rho B^\intercal(Ax^{k+1} + By^k - c)\right) \\ w^{k+1} &= \operatorname{Prox}_{\tau g_1^*}\left(w^k + \tau D y^{k+1/2}\right) \\ y^{k+1} &= y^{k+1/2} - \sigma D^\intercal\left(w^{k+1} - w^k\right) \end{aligned}\right. \\ &\quad u^{k+1} = u^k + \rho\left(Ax^{k+1} + By^{k+1} - c\right), \end{aligned}$$

which we call *Triple-linearized ADMM* (Trip-ADMM). The stepsize requirement is satisfied with $\rho > 0$, $\sigma > 0$, and $\tau > 0$,

$$\begin{aligned} 1 &\geq \sigma\rho\lambda_{\max}(A^\intercal A) + \sigma L_f, & 1 &\geq \sigma\rho\lambda_{\max}(B^\intercal B) + 3\sigma L_g, \\ 1 &\geq \sigma\tau\lambda_{\max}(CC^\intercal), & 1 &\geq \sigma\tau\lambda_{\max}(DD^\intercal), \end{aligned}$$

and under total duality, we have

$$\begin{aligned} f_1(Cx^k - \sigma\tilde{C}^\intercal\tilde{C}(v^{k+1}-v^k)) + f_2(x^k) + g_1(Dy^k - \sigma\tilde{D}^\intercal\tilde{D}(w^{k+1}-w^k)) + g_2(y^k) \\ \to f_1(Cx^\star) + f_2(x^\star) + g_1(Dy^\star) + g_2(y^\star) \\ \tilde{C}^\intercal\tilde{C}(v^{k+1}-v^k) \to 0, \qquad \tilde{D}^\intercal\tilde{D}(w^{k+1}-w^k) \to 0 \\ Ax^k + Bx^k - c \to 0. \end{aligned}$$

Derivation

First, we show that for any CCP function h on $\mathbb{R}^n$ and $A \in \mathbb{R}^{m\times n}$,

$$\begin{pmatrix} \mu^+ & \in \underset{\mu\in\mathbb{R}^m}{\operatorname{argmin}} \left\{ h^*(\mu) - \langle z_0, A^\intercal\mu\rangle + \frac{\alpha}{2}\|A^\intercal\mu\|_2^2 \right\} \\ x^+ & = z_0 - \alpha A^\intercal\mu^+ \end{pmatrix} \tag{8.11}$$

$$\Rightarrow \quad x^+ = \underset{x\in\mathbb{R}^n}{\operatorname{argmin}} \left\{ h(Ax) + \frac{1}{2\alpha}\|x - z_0\|_2^2 \right\},$$

provided that an argmin μ^+ exists. This follows from

$$\begin{aligned} x^+ =& z_0 - \alpha A^\intercal\mu^+,\ \mu^+ \text{ is an argmin} \\ \Leftrightarrow\quad & x^+ = z_0 - \alpha A^\intercal\mu^+,\ \partial h^*(\mu^+) - Az_0 + \alpha AA^\intercal\mu^+ \ni 0 \\ \Leftrightarrow\quad & x^+ = z_0 - \alpha A^\intercal\mu^+,\ \partial h^*(\mu^+) \ni Ax^+ \\ \Leftrightarrow\quad & A^\intercal\mu^+ + \frac{1}{\alpha}(x^+ - z_0) = 0,\ \mu^+ \in \partial h(Ax^+) \\ \Leftrightarrow\quad & A^\intercal\partial h(Ax^+) + \frac{1}{\alpha}(x^+ - z_0) \ni 0 \\ \Rightarrow\quad & x^+ \text{ is the argmin.} \end{aligned}$$

We now start the derivation. Consider the equivalent problem

$$\begin{array}{ll} \underset{\substack{x\in\mathbb{R}^p,\,\tilde{x}\in\mathbb{R}^r \\ y\in\mathbb{R}^q,\,\tilde{y}\in\mathbb{R}^s}}{\text{minimize}} & f_1(Cx + \tilde{C}\tilde{x}) + f_2(x) + g_1(Dy + \tilde{D}\tilde{y}) + g_2(y) \\ \text{subject to} & Ax + By = c \\ & \frac{1}{\sqrt{\rho\sigma}}\tilde{x} = 0 \\ & \frac{1}{\sqrt{\rho\sigma}}\tilde{y} = 0, \end{array}$$

where $\tilde{C} \in \mathbb{R}^{r\times r}$ and $\tilde{D} \in \mathbb{R}^{s\times s}$. FLiP-ADMM applied to this method is

$$\begin{aligned} (x^{k+1},\tilde{x}^{k+1}) \in \underset{x\in\mathbb{R}^p,\,\tilde{x}\in\mathbb{R}^r}{\operatorname{argmin}} & \Big\{ f_1(Cx + \tilde{C}\tilde{x}) + \langle \nabla f_2(x^k) + A^\intercal u^k, x\rangle + \frac{1}{\sqrt{\rho\sigma}}\langle \tilde{u}_x^k, \tilde{x}\rangle \\ & + \frac{\rho}{2}\|Ax + By^k - c\|^2 + \frac{1}{2\sigma}\|\tilde{x}\|^2 + \frac{1}{2}\|x - x^k\|_P^2 + \frac{1}{2}\|x - x^k\|_{\tilde{P}}^2 \Big\} \\ (y^{k+1},\tilde{y}^{k+1}) \in \underset{y\in\mathbb{R}^q,\,\tilde{y}\in\mathbb{R}^s}{\operatorname{argmin}} & \Big\{ g_1(Dy + \tilde{D}\tilde{y}) + \langle \nabla g_2(y^k) + B^\intercal u^k, y\rangle + \frac{1}{\sqrt{\rho\sigma}}\langle \tilde{u}_y^k, \tilde{y}\rangle \\ & + \frac{\rho}{2}\|Ax^{k+1} + By - c\|^2 + \frac{1}{2\sigma}\|\tilde{y}\|^2 + \frac{1}{2}\|y - y^k\|_Q^2 + \frac{1}{2}\|\tilde{y} - \tilde{y}^k\|_{\tilde{Q}}^2 \Big\} \\ u^{k+1} &= u^k + \varphi\rho(Ax^{k+1} + By^{k+1} - c) \\ \tilde{u}_x^{k+1} &= \tilde{u}_x^k + \varphi\sqrt{\rho/\sigma}\tilde{x}^{k+1} \\ \tilde{u}_y^{k+1} &= \tilde{u}_y^k + \varphi\sqrt{\rho/\sigma}\tilde{y}^{k+1}. \end{aligned}$$

Let

$$\varphi = 1, \qquad P = \frac{1}{\sigma}I - \rho A^\top A, \qquad \tilde{P} = 0, \qquad Q = \frac{1}{\sigma}I - \rho B^\top B, \qquad \tilde{Q} = 0.$$

Then we have

$$\begin{aligned}
(x^{k+1}, \tilde{x}^{k+1}) \in \underset{x\in\mathbb{R}^p,\, \tilde{x}\in\mathbb{R}^r}{\operatorname{argmin}} & \left\{ f_1(Cx + \tilde{C}\tilde{x}) + \frac{1}{2\sigma}\left\| \tilde{x} + \frac{\sqrt{\sigma}}{\sqrt{\rho}}\tilde{u}_x^k \right\|^2 \right. \\
& \left. + \frac{1}{2\sigma}\left\| x - x^k + \sigma\left(\nabla f_2(x^k) + A^\top u^k + \rho A^\top(Ax^k + By^k - c)\right)\right\|^2 \right\} \\
(y^{k+1}, \tilde{y}^{k+1}) \in \underset{y\in\mathbb{R}^q,\, \tilde{y}\in\mathbb{R}^s}{\operatorname{argmin}} & \left\{ g_1(Dy + \tilde{D}\tilde{y}) + \frac{1}{2\sigma}\left\| \tilde{y} + \frac{\sqrt{\sigma}}{\sqrt{\rho}}\tilde{u}_y^k \right\|^2 \right. \\
& \left. + \frac{1}{2\sigma}\left\| y - y^k + \sigma\left(\nabla g_2(y^k) + B^\top u^k + \rho B^\top(Ax^{k+1} + By^k - c)\right)\right\|^2 \right\} \\
u^{k+1} &= u^k + \rho(Ax^{k+1} + By^{k+1} - c) \\
\tilde{u}_x^{k+1} &= \tilde{u}_x^k + \sqrt{\rho/\sigma}\tilde{x}^{k+1} \\
\tilde{u}_y^{k+1} &= \tilde{u}_y^k + \sqrt{\rho/\sigma}\tilde{y}^{k+1}.
\end{aligned}$$

Using (8.11), we get

$$\begin{aligned}
v^{k+1} &= \underset{v\in\mathbb{R}^r}{\operatorname{argmin}} \left\{ f_1^*(v) + \left\langle \frac{\sqrt{\sigma}}{\sqrt{\rho}}\tilde{u}_x^k, \tilde{C}^\top v \right\rangle + \frac{\sigma}{2}\left(\|C^\top v\|^2 + \|\tilde{C}^\top v\|^2 \right) \right. \\
& \qquad \left. - \left\langle x^k - \sigma\left(\nabla f_2(x^k) + A^\top u^k + \rho A^\top(Ax^k + By^k - c)\right), C^\top v \right\rangle \right\} \\
x^{k+1} &= x^k - \sigma\left(\nabla f_2(x^k) + A^\top u^k + \rho A^\top(Ax^k + By^k - c)\right) - \sigma C^\top v^{k+1} \\
\tilde{x}^{k+1} &= -\frac{\sqrt{\sigma}}{\sqrt{\rho}}\tilde{u}_x^k - \sigma\tilde{C}^\top v^{k+1} \\
w^{k+1} &= \underset{w\in\mathbb{R}^s}{\operatorname{argmin}} \left\{ g_1^*(w) + \left\langle \frac{\sqrt{\sigma}}{\sqrt{\rho}}\tilde{u}_y^k, \tilde{D}^\top w \right\rangle + \frac{\sigma}{2}\left(\|D^\top w\|^2 + \|\tilde{D}^\top w\|^2 \right) \right. \\
& \qquad \left. - \left\langle y^k - \sigma\left(\nabla g_2(y^k) + B^\top u^k + \rho B^\top(Ax^{k+1} + By^k - c)\right), D^\top w \right\rangle \right\} \\
y^{k+1} &= y^k - \sigma\left(\nabla g_2(y^k) + B^\top u^k + \rho B^\top(Ax^{k+1} + By^k - c)\right) - \sigma D^\top w^{k+1} \\
\tilde{y}^{k+1} &= -\frac{\sqrt{\sigma}}{\sqrt{\rho}}\tilde{u}_y^k - \sigma\tilde{D}^\top w^{k+1} \\
u^{k+1} &= u^k + \rho(Ax^{k+1} + By^{k+1} - c) \\
\tilde{u}_x^{k+1} &= \tilde{u}_x^k + \sqrt{\rho/\sigma}\tilde{x}^{k+1} = -\sqrt{\rho\sigma}\tilde{C}^\top v^{k+1} \\
\tilde{u}_y^{k+1} &= \tilde{u}_y^k + \sqrt{\rho/\sigma}\tilde{y}^{k+1} = -\sqrt{\rho\sigma}\tilde{D}^\top w^{k+1}.
\end{aligned}$$

Eliminating some variables, we get

$$\begin{aligned}
v^{k+1} &= \underset{v\in\mathbb{R}^r}{\operatorname{argmin}} \left\{ f_1^*(v) - \sigma\left\langle v^k, \tilde{C}\tilde{C}^\top v \right\rangle + \frac{\sigma}{2}\left(\|C^\top v\|^2 + \|\tilde{C}^\top v\|^2 \right) \right. \\
& \qquad \left. - \left\langle x^k - \sigma\left(\nabla f_2(x^k) + A^\top u^k + \rho A^\top(Ax^k + By^k - c)\right), C^\top v \right\rangle \right\}
\end{aligned}$$

$$\begin{aligned}
x^{k+1} &= x^k - \sigma\left(\nabla f_2(x^k) + A^\intercal u^k + \rho A^\intercal(Ax^k + By^k - c) + C^\intercal v^{k+1}\right)\\
w^{k+1} &= \underset{w\in\mathbb{R}^s}{\operatorname{argmin}}\left\{g_1^*(w) - \sigma\left\langle w^k, \tilde{D}\tilde{D}^\intercal w\right\rangle + \frac{\sigma}{2}\left(\|D^\intercal w\|^2 + \|\tilde{D}^\intercal w\|^2\right)\right.\\
&\qquad \left. - \left\langle y^k - \sigma\left(\nabla g_2(y^k) + B^\intercal u^k + \rho B^\intercal(Ax^{k+1} + By^k - c)\right), D^\intercal w\right\rangle\right\}\\
y^{k+1} &= y^k - \sigma\left(\nabla g_2(y^k) + B^\intercal u^k + \rho B^\intercal(Ax^{k+1} + By^k - c) + D^\intercal w^{k+1}\right)\\
u^{k+1} &= u^k + \rho(Ax^{k+1} + By^{k+1} - c).
\end{aligned}$$

Next, we set

$$\tilde{C}\tilde{C}^\intercal = \frac{1}{\tau\sigma}I - CC^\intercal, \qquad \tilde{D}\tilde{D}^\intercal = \frac{1}{\tau\sigma}I - DD^\intercal$$

to get

$$\begin{aligned}
v^{k+1} &= \operatorname{Prox}_{\tau f_1^*}\\
&\left(v^k + \tau C\left(x^k - \sigma\left(\nabla f_2(x^k) + A^\intercal u^k + \rho A^\intercal(Ax^k + By^k - c) + C^\intercal v^k\right)\right)\right)\\
x^{k+1} &= x^k - \sigma\left(\nabla f_2(x^k) + A^\intercal u^k + \rho A^\intercal(Ax^k + By^k - c) + C^\intercal v^{k+1}\right)\\
w^{k+1} &= \operatorname{Prox}_{\tau g_1^*}\\
&\left(w^k + \tau D\left(y^k - \sigma\left(\nabla g_2(y^k) + B^\intercal u^k + \rho B^\intercal(Ax^{k+1} + By^k - c) + D^\intercal w^k\right)\right)\right)\\
y^{k+1} &= y^k - \sigma\left(\nabla g_2(y^k) + B^\intercal u^k + \rho B^\intercal(Ax^{k+1} + By^k - c) + D^\intercal w^{k+1}\right)\\
u^{k+1} &= u^k + \rho\left(Ax^{k+1} + By^{k+1} - c\right).
\end{aligned}$$

The minimization subproblems are solvable since they are evaluations of proximal operators of the CCP functions f_1 and g_1. We further simplify to get

$$\begin{aligned}
x^{k+1/2} &= x^k - \sigma\left(C^\intercal v^k + \nabla f_2(x^k) + A^\intercal u^k + \rho A^\intercal(Ax^k + By^k - c)\right)\\
v^{k+1} &= \operatorname{Prox}_{\tau f_1^*}\left(v^k + \tau C x^{k+1/2}\right)\\
x^{k+1} &= x^{k+1/2} - \sigma C^\intercal\left(v^{k+1} - v^k\right)\\
y^{k+1/2} &= y^k - \sigma\left(D^\intercal w^k + \nabla g_2(y^k) + B^\intercal u^k + \rho B^\intercal(Ax^{k+1} + By^k - c)\right)\\
w^{k+1} &= \operatorname{Prox}_{\tau g_1^*}\left(w^k + \tau D y^{k+1/2}\right)\\
y^{k+1} &= y^{k+1/2} - \sigma D^\intercal\left(w^{k+1} - w^k\right)\\
u^{k+1} &= u^k + \rho\left(Ax^{k+1} + By^{k+1} - c\right).
\end{aligned}$$

8.3 BREGMAN METHODS

A class of methods called the *Bregman* methods were developed and popularized in the image processing community. Later, it was discovered that the Bregman methods are related to the method of multipliers and ADMM. In this section, we briefly describe the relationship.

Bregman Distance Let f be a CCP function. When f is differentiable, we define the f-induced *Bregman distance* or *Bregman divergence* as

$$D_f(x,y) = f(x) - f(y) - \langle \nabla f(y), x - y \rangle.$$

When f is not differentiable, we use

$$D_f^v(x,y) = f(x) - f(y) - \langle v, x - y \rangle,$$

where $v \in \partial f(y)$.

D_f generalizes the squared Euclidean distance, since $D_f(x,y) = \|x - y\|^2$ when $f(x) = \|x\|^2$. $D_f(x,y) \geq 0$ follows from convexity of f. However, despite its name, the Bregman "distance" is not a mathematical distance (metric). In particular, $D_f(x,y)$ may not equal $D_f(y,x)$, and $D_f(x,y) = 0$ may hold for $x \neq y$.

Bregman Method and Method of Multipliers Consider the problem

$$\begin{array}{ll} \underset{x \in \mathbb{R}^n}{\text{minimize}} & f(x) \\ \text{subject to} & Ax = b, \end{array} \tag{8.12}$$

where f is CCP, $A \in \mathbb{R}^{m \times n}$, and $b \in \mathbb{R}^m$. Let $h(x) = \hat{h}(Ax - b)$, for some differentiable CCP function $\hat{h}$ such that $\hat{h}(0) = 0$ and $\hat{h}(u) > 0$ for $u \neq 0$. When f is differentiable, the *Bregman method* is

$$x^{k+1} = \underset{x \in \mathbb{R}^n}{\text{argmin}} \left\{ D_f(x, x^k) + \rho h(x) \right\},$$

where ρ is a scalar. When f is not differentiable, the Bregman method is

$$\begin{aligned} x^{k+1} &= \underset{x \in \mathbb{R}^n}{\text{argmin}} \left\{ D_f^{v^k}(x, x^k) + \rho h(x) \right\} \\ v^{k+1} &= v^k - \rho \nabla h(x^{k+1}), \end{aligned}$$

where $v^0 \in \partial f(x^0)$. The optimality condition of the first step ensures

$$v^{k+1} = v^k - \rho \nabla h(x^{k+1}) \in \partial f(x^{k+1}).$$

Since the argmin depends on x^k only through $v^k \in \partial f(x^k)$, we can pick any $v^0 \in \text{range}(\partial f)$ without explicitly specifying x^0. The method converges under certain mild conditions.

When $h(x) = \frac{1}{2}\|Ax - b\|^2$, the Bregman method with the change of variables $v^k = -A^\intercal u^k$ coincides with the method of multipliers

$$\begin{aligned} x^{k+1} &\in \underset{x \in \mathbb{R}^n}{\text{argmin}} \left\{ f(x) + \langle u^k, Ax \rangle + \frac{\rho}{2}\|Ax - b\|^2 \right\} \\ u^{k+1} &= u^k + \rho(Ax^{k+1} - b). \end{aligned}$$

Split Bregman Method and ADMM Consider the problem

$$\begin{array}{ll} \underset{x\in\mathbb{R}^p,\, y\in\mathbb{R}^q}{\text{minimize}} & f(x)+g(y) \\ \text{subject to} & Ax+y=c, \end{array}$$

where f and g are CCP functions, $A \in \mathbb{R}^{q\times p}$, and $c \in \mathbb{R}^q$. Let $h(x) = \frac{1}{2}\|Ax + y - c\|^2$ and apply the Bregman method to $f(x) + g(y)$ to get

$$\begin{aligned} (x^{k+1}, y^{k+1}) &\in \underset{x\in\mathbb{R}^p,\, y\in\mathbb{R}^q}{\operatorname{argmin}} \left\{ D_f^{v^k}(x, x^k) + D_g^{u^k}(y, y^k) + \frac{\rho}{2}\|Ax + y - b\|^2 \right\} \\ v^{k+1} &= v^k - \rho A^\intercal (Ax^{k+1} + y^{k+1} - c) \\ u^{k+1} &= u^k - \rho(Ax^{k+1} + y^{k+1} - c). \end{aligned}$$

Using $v^k = A^\intercal u^k$, we eliminate v^k to get

$$\begin{aligned} (x^{k+1}, y^{k+1}) &\in \underset{x\in\mathbb{R}^p,\, y\in\mathbb{R}^q}{\operatorname{argmin}} \left\{ f(x) + g(y) - \langle u^k, Ax + y\rangle + \frac{\rho}{2}\|Ax + y - c\|^2 \right\} \\ u^{k+1} &= u^k - \rho(Ax^{k+1} + y^{k+1} - c). \end{aligned}$$

The *split Bregman* method computes x^{k+1} and y^{k+1} approximately through alternating updates. When only one pass of sequential minimization of x and then y is performed, the split Bregman method coincides with ADMM.

8.4 CONCLUSION

In this section, we established the convergence of FLiP-ADMM and presented techniques for applying the method to a wide range of problems. The exercises further illustrate that the modular techniques can be combined to solve problems with complicated structures.

The analysis of FLiP-ADMM differs from that of Part I, where we derived various methods, including ADMM-type methods, as instances of monotone operator methods. There are two distinct approaches to analyze the classical ADMM. The first is to derive ADMM from DRS, as we did in §3. This approach leads to a possible intermediate dual variable update, as we discuss in Exercise 8.1. The second is to construct a Lyapunov function and analyze convergence directly. This approach leads to a possible dual extrapolation parameter φ. To the best of our knowledge, the fully general FLiP-ADMM cannot be reduced to a monotone operator splitting method and therefore must be analyzed directly with a Lyapunov function.

ADMM-type methods are "splitting methods" in that they decompose the optimization problem into smaller, simpler pieces and operate on them separately. They are intimately related to monotone operator methods, although, strictly speaking, they are not monotone operator methods themselves.

BIBLIOGRAPHICAL NOTES

FLiP-ADMM As individual components of ADMM-type methods, the dual extrapolation parameter, function linerization, and proximal terms are known techniques; FLiP-ADMM merely combines them. Among published ADMM-type methods, the randomized primal-dual block coordinate update method (RPDBU) of Gao, Xu, and Zhang [GXZ19] is most similar to FLiP-ADMM. RPDBU allows the x- and y-updates to be partially updated in a randomized coordinate-update fashion but does not incorporate the dual extrapolation parameter.

Early Development ADMM dates back to the 1970s. When Glowinski and Marroco was studying nonlinear Dirichlet problems in the form of

$$\underset{v \in V}{\text{minimize}} \quad f(Av) + g(v), \tag{8.13}$$

they reformulated it as

$$\begin{array}{ll} \underset{u \in AV,\, v \in V}{\text{minimize}} & f(u) + g(v) \\ \text{subject to} & u - Av = 0 \end{array} \tag{8.14}$$

and applied Hestenes and Powell's augmented Lagrangian method (ALM) [Hes69, Pow69]. In [GM75b], Glowinski and Marroco proposed to solve the ALM subproblem by updating u and v in an alternating manner while the Lagrange multipliers are fixed until a stopping criterion is met. Their approach hinted at ADMM. ADMM for (8.14) was first presented with a convergence proof by Gabay and Mercier [GM76, Algorithm 3.4]. They credited [GM75a] for numerical experiments of their algorithm without proof. Gabay and Mercier's proof assumes g is linear and establishes subsequence-convergence of the dual iterates for $\varphi \in (0, 2)$. When the objective functions are differentiable, they obtain convergence of the dual iterates. Then Glowinski and Fortin [FG83] took a large step forward to studying general convex f and g. They proved subsequence-convergence of the dual iterates for $\varphi \in (0, \frac{\sqrt{5}+1}{2})$. Later using [Opi67], Glowinski and Le Tallec obtained convergence of the dual iterates (weak convergence in infinite-dimensional Hilbert spaces) [GLT89]. Bertsekas and Tsitsiklis also presented a proof in [BT89, §3.4].

Relationship with DRS In [Roc76b] Rockafellar showed that ALM for a linearly constrained convex problem is the proximal point method applied to its dual problem, i.e., "ALM = PPM to the dual." Gabay [Gab83] extended this result to "ADMM = DRS to the dual" and "PRS-ADMM = PRS to the dual." (We discuss PRS-ADMM in Exercise 8.1.) Following Gabay's naming in [Gab83], many numerical analysts refer to ALM, ADMM, and PRS-ADMM as ALG1, ALG2, and ALG3, repsectively. Using this characterization, Eckstein generalized ADMM to allow the subproblems to be solved inexactly [Eck89, Proposition 4.7].

Although ADMM is equivalent to DRS applied to the dual, one can still directly apply ADMM to the dual. Eckstein [Eck89, Chapter 3.5] showed that ADMM is equivalent to DRS applied to the primal problem when $A = I$, Eckstein and Fukushima [EF94] showed the same for certain special problems, and Yan and Yin [YY16] extended their results to general ADMM.

Update Order Swapping the orders of the two subproblems of ADMM leads to different iterates in general, and Bauschke and Moursi studied the dependence on the iterates on this order [BM16]. Yan and Yin [YY16] showed that the two iterates generated by the two orders are in fact equivalent if one of the functions is quadratic. While either of the two orders leads to convergence, repeatedly switching the orders causes divergence as demonstrated by an example in [YY16]. Sun, Luo, and Ye [SLY20] showed that, for solving linear systems of equations, ADMM under randomly permuted orders converges in expectation.

Parameter Selection The general question of how to optimally choose the scalar parameters φ and ρ is open. Ghadimi, Teixeira, Shames, and Johansson characterized the optimal parameters for the specific problem of ℓ_2 regularized and constrained quadratic programming [GTSJ15]. There has been extensive work presenting adaptive methods for tuning ρ in various settings [HYW00, Woh17, XFG17, XFY$^+$17, XLLY17].

Dual Extrapolation Parameter The first appearance of $\varphi \in (0, (1 + \sqrt{5})/2)$, the golden ratio range of the dual extrapolation parameter, is due to Fortin and Glowinski [FG83, Glo84]. Xu showed that the same golden ratio range can be used for proximal ADMM [Xu07]. Tao and Yuan showed that when both f and g are quadratics, the parameter range extends to $\varphi \in (0, 2)$ [TY18].

Techniques Rockafellar presented the proximal method of multipliers [Roc76a]. Chen and Teboulle presented the *predictor corrector proximal multiplier method* [CT94]. Shefi and Teboulle [ST14] later identified Chen and Teboulle's method to be an instance of what we call the linearized method of multipliers. Eckstein presented proximal ADMM and showed it is Douglas–Rachford splitting applied to the saddle subdifferential [Eck94]. He, Liao, Han, and Yang further generalized proximal ADMM [HLHY02]. However, what we call the "linearization technique," where one chooses the proximal term carefully to cancel out and linearize quadratic terms, was likely unknown at the time of these publications. The first explicit description of the linearization technique is due to the concurrent work of Deng and Yin [DY16b] and Shefi and Teboulle [ST14].

Partial linearization in ADMM was first proposed by Deng and Yin [DY16b], and was applied to CT and PET imaging by Ryu, Ko, and Won [RKW20]. In particular, the method presented in Example 8.1 is the *near-circulant splitting* of [RKW20].

Function linearization in ADMM was first presented by Yang and Zhang [YZ11]; they applied function linearization to one quadratic subproblem. Lin, Ma, and Zhang [LMZ17] applied function linearization to general smooth convex functions in an ADMM-type method. Banert, Boţ, and Csetnek [BBC21] applied function linearization to one function with the ADMM with proximal terms and $B = I$. Gao, Xu, and Zhang applied function linearization to both x- and y-subproblems [GXZ19].

Jacobi-ADMM was first presented in [DLPY17, GHY14, HHY15, Tao14, HXY16, BK19]. Similar but different methods are proposed and analyzed in [TY12, WS17, WHML15].

The 2-1-2 technique was first presented by Sun, Toh, Yang, and Li in [STY15, LST16]. They also showed that the technique can be applied twice and that it can be combined with linearizations, as we do in Exercises 8.11 and 8.12. The 2-1-2 technique has also been used as the basis for the symmetric Gauss–Seidel ADMM of Li, Sun, and Toh [LST19].

3- and Multi-Block ADMM Since the early 2010s, there have been attempts to generalize ADMM to three or more blocks of primal variables. Chen, He, Ye, and Yuan [CHYY16] showed that the direct extension of ADMM to three blocks with sequential updates *does not* converge. On the other hand, the multi-block generalization *does* converge with additional assumptions or modifications [HTY12, HY12a, CSY13, LMZ15b, LST15, SLY20, LMZ16, HL17, CST17a, DY17b, HTY17, LST19, XXS19].

Using the reformulations of §8.2.3, the multi-block ADMM setup can be solved with existing ADMM-type methods, but an open question is finding a method that converges *fast* for the multi-block ADMM setup. The 2019 work of P. Xiao, Z. Xiao, and Sun [XXS19] provides excellent discussion on this subject and experimentally compares the three competitive methods: Gaussian back substitution ADMM by He, Tao, and Yuan [HTY12, HTY17], symmetric Gauss–Seidel ADMM by Li, Sun, Toh, and Chen [CST17a, LST19], and RP-ADMM by Sun, Luo, and Ye [SLY20].

Applications In image processing, Wang, Yang, Yin, and Zhang [WYYZ08] formulated a total variation deblurring problem with (8.14) such that both subproblems have closed-form solutions; however, their method is inexact ALM rather than ADMM. In compressed sensing, Yang and Zhang [YZ11] applied ADMM to ℓ_1-optimization. Wen, Goldfarb, and Yin [WGY10] used ADMM to solve semidefinite and conic programs, and O'Donoghue, Chu, Parikh, and Boyd [OCPB16] applied ADMM to the self-dual homogeneous embedding reformulation of conic programs. Lin, Ma, Ye, and Zhang [LMYZ21] used ADMM on a barrier formulation of linear programming. Yuan and Yang [YY13b, YY13a] used ADMM to recover sparse and low-rank components of a matrix. Yuan [Yua12] used ADMM for covariance matrix selection. Ma, Xue, and Zou [MXZ13] used ADMM for graphical model selection. Liu et al. [LMT+10] used ADMM for metric learning. Due to its popularity, ADMM has numerous other applications.

Relationship with Bregman Methods Bregman methods successively minimize a sequence of Bregman distances instead of updating Lagrange multipliers. Osher et al. [OBG+05] proposed the Bergman method in the context of image processing for finding an member of $\{x \mid \|Ax-b\| \le \varepsilon\}$ for some small $\varepsilon > 0$ that serves as a good reconstruction of the original image. This motivation is why the Bregman method looks like a method for minimizing h, rather than minimizing f. Osher et al. shows $h(x^k) \to \operatorname{argmin} h$ monotonically in [OBG+05]. Yin, Osher, Goldfard, and Darbon proved the Bregman method converges to a solution of the optimization problem [YOGD08, Section 5.1,5.2] and argued that the method is equivalent to the method of multipliers, and more generally the augmented Lagrangian method, under a change of variable [YOGD08, Section 3.4]. In [YOGD08, Section 5.3], they introduced linearized Bregman iteration. Both linearized Bregman and linearized method of multipliers have simpler subproblems, but they are not equivalent. Goldstein and Osher [GO09] introduced the Split Bregman method, equivalent to the method of multipliers under change of variables, and reported great empirical performance with a "single-pass," which is equivalent to ADMM for (8.14). Zhang, Burger, and Osher [ZBO11] linearized a split Bregman subproblem, obtaining a method that is equivalent to linearized ADMM for (8.14). These works helped to popularize ADMM-type methods in image processing. For other convergence results of the Bregman method, see [OBG+05] for the general setting, [YOGD08, YO13] for ℓ_1-norm and piece-wise linear functions, and [JZZ09].

Other ADMM-type Methods Wang and Banerjee [WB12] extended ADMM to the online setting. Suzuki [Suz13] and Ouyang, He, Tran, and Gray [OHTG13] concurrently extended ADMM to taking stochastic samples. Suzuki [Suz14], Zhong and Kwok [ZK14b], and Zheng and Kwok [ZK16] presented accelerated stochastic ADMM using variance reduction techniques. Yi and Pavel [YP19] applied ADMM to computing generalized Nash equilibrium. Chen, Chan, Ma, and Yang [CCMY15] and Boţ, Csetnek, and Hendrich [BCH15] presented ADMM and DRS with with inertial acceleration.

Convergence Rates Researchers have used different quantities to measure the rate of ADMM convergence. He and Yuan [HY12b] established an $O(1/k)$ rate that is applied to the violation of a variational-inequality optimality condition of ADMM. Monterio and Svaiter [MS13] showed another $O(1/k)$ rate that is applied to the sizes of some approximate subgradients and their approximate levels. Shefi and Teboulle [ST14] analyzed the convergence rates for the proximal and linearized ADMM. He and Yuan [HY15] showed that fixed-point residuals decay at an $O(1/k)$ rate. Davis and Yin [DY16a, DY17a] presented a comprehensive list of rates for function value and constraint violation corresponding to different smoothness and strong convexity assumptions. They also improved the $O(1/k)$ rate of some setups to $o(1/k)$ and establish tightness of the $o(1/k)$ rate by extending examples from [BBCN+14]. Lin, Ma, and Zhang [LMZ15b] presented conditions for this rate to hold for multi-block ADMM. Deng, Lai, Peng, and Yin established an $o(1/k)$ rate for Jacobi-ADMM [DLPY17].

With some additional assumptions, ADMM can be modified to achieve an accelerated sublinear rate. Assuming strong convexity on f and g, Goldstein, O'Donoghue, Setzer, and Baraniuk [GOSB14] used a modified ADMM to achieve an $O(1/k^2)$ rate. Ouyuang, Chen, Lan, and Pasiliao [OCLPJ15] introduced an accelerated linearized ADMM that also linearizes one objective function f. Assuming f is L_f Lipschitz differentiable, they established a rate in the form of $O(L_g/k^2 + C/k)$, where C includes quantities independent of L_g. Xu [Xu17] proposed a modified linearized ADMM that achieves a $O(1/k^2)$ rate when either f or $g_1 + g_2$ is strongly convex. His method linearizes the Lipschitz differentiable function g_2.

Under different combinations of assumptions, ADMM converges linearly. Deng and Yin [DY16b] proved linear convergence of ADMM for four combinations, and Davis and Yin [DY17a, §6] discovered more combinations. Giselsson [GB15], Giselsson and Boyd [GB17], Moursi and Vandenberghe [MV19], and Ryu, Taylor, Bergeling, and Giselsson [RTBG20] obtained tight linear rates for certain combinations. Hong and

Luo [HL17] established linear convergence for multi-block ADMM under an "error-bound" condition. Lin, Ma, and Zhang [LMZ15a] established linear convergence for multi-block ADMM under certain strong convexity, smoothness, and rank conditions.

Linear convergence of ADMM has also been studied for specific classes of problems. Eckstein and Bertsekas [EB90] proved it for linear programming. Bauschke et al. [BBCN+14] showed that, for finding an intersection point between two subspaces, the linear rate of ADMM is the cosine of the Friedrichs angle between the subspaces. Boley [Bol13] analyzed linear convergence of ADMM in different phases when applied to certain quadratic problems. Raghunathan and Di Cairano [RDC14] related the linear rate of ADMM applied to quadratic programming with linear equality and bound constraints to the spectrum of the Hessian and the Friedrichs angle. Aspelmeier, Charitha, and Luke [ACL16] established eventual linear convergence based on metric subregularity. Liang, Fadili, and Peyré [LFP17] studied local linear convergence and manifold identification.

Miscellaneous Eckstein showed that when total duality fails, ADMM diverges in the sense that it generates unbounded iterates [Eck89, Proposition 4.8]. There has been extensive work characterizing the manner in which the iterates diverge and using the divergent iterates to identify pathological problems [RDC14, SBG+20, BGSB19, LRY19, RLY19].

Given a Douglas–Rachford splitting operator T_{DRS}, there generally is no function f such that $T_{\mathrm{DRS}} = \mathrm{Prox}_f$ [DY16a], but, when it exists, it is called the Douglas–Rachford envelope. Patrinos, Stella, Bemporad, and Themelis showed that the Douglas–Rachford envelope exists under smoothness conditions and used it to accelerate DRS and ADMM [PSB14, TP20]. Liu and Yin [LY19] gave the conditions for the Davis–Yin envelope to exist.

Finally, the review papers by Boyd et al. [BPC+11], Eckstein [Eck12], and Glowinski [Glo14] serve as excellent tutorials on ADMM.

EXERCISES

8.1 *Peaceman–Rachford ADMM.* In §3, we derived ADMM as an instance of DRS. One may wonder: what if we used the FPI with

$$(1-\theta)I + \theta R_{\alpha\partial\tilde{f}} R_{\alpha\partial\tilde{g}},$$

where $\tilde{f}$ and $\tilde{g}$ are as defined in §3.1 or 3.2, with $\theta \in (0,1)$? Do we recover the golden ratio ADMM? The answer is no. Show that we instead get

$$\begin{aligned}
x^{k+1} &= \underset{x}{\operatorname{argmin}}\, \mathbf{L}_\alpha(x, y^k, u^k)\\
u^{k+1/2} &= u^k + \alpha(2\theta - 1)(Ax^{k+1} + By^k - c)\\
y^{k+1} &= \underset{y}{\operatorname{argmin}}\, \mathbf{L}_\alpha(x^{k+1}, y, u^{k+1/2})\\
u^{k+1} &= u^{k+1/2} + \alpha(Ax^{k+1} + By^{k+1} - c).
\end{aligned}$$

When $\theta = 1$, this method is called Peaceman–Rachford ADMM. Although Peaceman–Rachford ADMM does not converge in general, it can be faster than regular ADMM under additional assumptions [Gab83, HLWY14].

8.2 Provide the derivation of Example 8.1.
Hint. First, show that FLiP-ADMM can be written as

$$
\begin{aligned}
y^{k+1} &= \mathrm{Prox}_{\frac{1}{\rho}\ell}\left(Ax^k - b + \frac{1}{\rho}\xi^k\right)\\
z^{k+1} &= \mathrm{Prox}_{\frac{\lambda}{\rho}\|\cdot\|_1}\left(Dy^k + \frac{1}{\rho}\zeta^k\right)\\
x^{k+1} &= x^k - (\gamma I + C)^{-1}\left(A^\intercal(\xi^k + \rho(Ax^k - b) - \rho y^{k+1}) + D(\zeta^k + \rho Dx^k - \rho z^{k+1})\right)\\
\xi^{k+1} &= \xi^k + \rho(Ax^{k+1} - b - y^{k+1})\\
\zeta^{k+1} &= \zeta^k + \rho(Dx^{k+1} - z^{k+1}),
\end{aligned}
$$

where ξ^{k+1} and ζ^{k+1} are the dual variables. Then apply the Moreau identity (2.12).

8.3 *Jacobi doubly-linearized ADMM.* Consider the problem

$$
\begin{array}{ll}
\underset{x\in\mathbb{R}^p}{\text{minimize}} & \sum_{i=1}^{m}(f_i(x_i) + h_i(x_i))\\
\text{subject to} & Ax = c,
\end{array}
$$

where $(x_1,\dots,x_m) = x$, $f_1,\dots,f_m$ are proximable CCP functions, and $h_1,\dots,h_m$ are differentiable CCP functions. Find a method analogous to Jacobi ADMM with a split x-update utilizing the proximal operators of $f_1,\dots,f_m$. What are the stepsize requirements?

8.4 *Jacobi+1.* Consider the problem

$$
\begin{array}{ll}
\underset{x\in\mathbb{R}^p,\, y\in\mathbb{R}^q}{\text{minimize}} & \sum_{i=1}^{m} f_i(x_i) + g(y)\\
\text{subject to} & Ax + By = c.
\end{array}
$$

Find a method analogous to Jacobi ADMM that performs a split x-update and then the y-update. What are the stepsize requirements?

8.5 *More dummy variables.* Consider the problem

$$
\begin{array}{ll}
\underset{x\in\mathbb{R}^p,\, y\in\mathbb{R}^q}{\text{minimize}} & \sum_{i=1}^{m} f_i(x_i) + \sum_{j=1}^{\ell} g_j(y_j)\\
\text{subject to} & Ax + By = c,
\end{array}
$$

where $x = (x_1,\dots,x_m)$ with $x_i \in \mathbb{R}^{p_i}$ for $i = 1,\dots,m$ and $y = (y_1,\dots,y_\ell)$ with $y_j \in \mathbb{R}^{q_j}$ for $j = 1,\dots,\ell$. Introduce dummy variables $\xi_1,\dots,\xi_m$ and $\zeta_1,\dots,\zeta_\ell$ to get the equivalent problem

$$
\begin{array}{ll}
\underset{\substack{(x_1,\dots,x_m)\in\mathbb{R}^p\\ (y_1,\dots,y_\ell)\in\mathbb{R}^q\\ \xi_1,\dots,\xi_m,\zeta_1,\dots,\zeta_\ell\in\mathbb{R}^n}}{\text{minimize}} & \sum_{i=1}^{m} f_i(x_i) + \sum_{j=1}^{\ell} g_j(y_j)\\
\text{subject to} & A_{:,i}x_i = \xi_i \quad \text{for } i = 1,\dots,m\\
& B_{:,j}y_j = \zeta_j \quad \text{for } j = 1,\dots,\ell\\
& \sum_{i=1}^{m}\xi_j + \sum_{j=1}^{\ell}\zeta_j = c.
\end{array}
$$

Find an ADMM-type method with split x- and y-updates. In particular, apply FLiP-ADMM with the x- and ζ-variables updated first, y- and ξ-variables updated second, and the dual variables updated last. What are the stepsize requirements?

8.6 *The exchange problem.* Consider the problem

$$\begin{array}{ll}\underset{x_1,\ldots,x_m\in\mathbb{R}^n}{\text{minimize}} & \sum_{i=1}^{m} f_i(x_i)\\ \text{subject to} & x_1+\cdots+x_m=b,\\ & x_1,\ldots,x_m\geq 0.\end{array}$$

Assume f_i is L_f-smooth and CCP and evaluating ∇f_i is efficient, for $i=1,\ldots,m$. Provide an ADMM-type method to efficiently solve this problem. What are the stepsize requirements?

Remark. The economics interpretation of this problem is as follows. There are n goods with a total amount of $b_1,\ldots,b_n$. There are m agents exchanging these goods while preserving the total amount. Agents cannot have a negative amount of goods. The goal is to find the optimal exchange that minimizes the global cost of all agents.

8.7 *Canonical sharing problem.* Consider the problem

$$\underset{x_1,\ldots,x_m\in\mathbb{R}^n}{\text{minimize}} \quad f\left(\sum_{i=1}^{m} x_i\right)+\sum_{i=1}^{m} g_i(x_i).$$

Assume $f, g_1,\ldots,g_m$ are proximable CCP functions. Provide an ADMM-type method to efficiently solve this problem. What are the stepsize requirements?

Remark. The economics interpretation of this problem is as follows. There are m agents sharing n common resources. The common shared cost is represented by f and the individual costs by g_i. For example, f may represent the shared cost of pollution, while g_i represents the individual cost (or negative gain) incurred by performing actions producing pollutants. The goal is to minimize the sum of the global and all individual costs.

8.8 *Model parallelism vs. data parallelism.* Consider the problem

$$\underset{x\in\mathbb{R}^p}{\text{minimize}} \quad f(x)+g(Ax-b),$$

where $A\in\mathbb{R}^{n\times p}$ and $b\in\mathbb{R}^n$. Partition $x\in\mathbb{R}^p$ into m non-overlapping blocks of sizes $p_1,\ldots,p_m$, and write $x=(x_1,\ldots,x_m)$, so $x_i\in\mathbb{R}^{p_i}$ for $i=1,\ldots,m$. Partition $z\in\mathbb{R}^n$ into ℓ non-overlapping blocks of sizes $n_1,\ldots,n_\ell$, and write $z=(z_1,\ldots,z_\ell)$, so $z_j\in\mathbb{R}^{n_j}$ for $j=1,\ldots,\ell$. Assume $f(x)=f_1(x_1)+\cdots+f_m(x_m)$, where $f_1,\ldots,f_m$ are CCP functions. Assume $g(z)=g_1(z_1)+\cdots+g_\ell(z_\ell)$, where $g_1,\ldots,g_\ell$ are CCP functions.
The problem is equivalent to

$$\begin{array}{ll}\underset{\substack{(x_1,\ldots,x_m)\in\mathbb{R}^p\\ z\in\mathbb{R}^n}}{\text{minimize}} & \sum_{i=1}^{m} f_i(x_i)+g(z-b)\\ \text{subject to} & \sum_{i=1}^{m} A_{:,i}x_i=z,\end{array}$$

where

$$A=\begin{bmatrix}A_{:,1} & A_{:,2} & \cdots & A_{:,m}\end{bmatrix}, \qquad Ax=A_{:,1}x_1+A_{:,2}x_2+\cdots+A_{:,m}x_m.$$

Provide an ADMM-type method that updates the blocks $x_1,\ldots,x_m$ in parallel. The update for x_i should access the data $A_{:,i}$ for $i=1,\ldots,m$. What are the stepsize requirements?

Now, redefine $x_1,\dots,x_\ell$ and y as copies of x. The problem is also equivalent to

$$\begin{array}{ll}\underset{x_1,\dots,x_\ell,y\in\mathbb{R}^p}{\text{minimize}} & \sum_{j=1}^{\ell} g_j(A_{j,:}x_j - b_j) + f(y) \\ \text{subject to} & x_j = y, \qquad j = 1,\dots,\ell,\end{array}$$

where

$$A = \begin{bmatrix} A_{1,:} \\ A_{2,:} \\ \vdots \\ A_{\ell,:} \end{bmatrix}, \qquad Ax = \begin{bmatrix} A_{1,:}x \\ A_{2,:}x \\ \vdots \\ A_{\ell,:}x \end{bmatrix}.$$

Provide an ADMM-type method that updates the copies $x_1,\dots,x_\ell$ in parallel. The update for x_j should access the data $A_{j,:}$ and b_j for $j = 1,\dots,\ell$. What are the stepsize requirements?

Remark. In the context of machine learning, we say the first type of method utilizes *model parallelism* and the second type *data parallelism*. We view $(x_1,\dots,x_m) = x$ as a decomposition of the machine learning *model* represented by x into m blocks, and we view $(A_{1,:},b_1),\dots,(A_{\ell,:},b_\ell)$ as a decomposition of the *data* into ℓ blocks. In model parallelism, the ith parallel process updates the ith block using all data blocks (but only the data relevant to the ith block of the model). In data parallelism, the jth parallel process updates the entire model using $(A_{j,:},b_j)$, the jth block of data points.

8.9 *Consolidating blocks with graph coloring.* Consider the multi-block ADMM formulation

$$\begin{array}{ll}\underset{x\in\mathbb{R}^p,\, y\in\mathbb{R}^q}{\text{minimize}} & \sum_{i=1}^{m} f_i(x_i) \\ \text{subject to} & \sum_{i=1}^{m} A_{:,i}x_i = c.\end{array}$$

As discussed in §8.2.3, we can solve this problem with several ADMM-type methods utilizing block splitting. However, such methods can be slow; empirically, using proximal terms or introducing dummy variables slows down the iteration. On the other hand, we know that if the blocks are orthogonal, i.e., $A_{:,i}^\intercal A_{:,j} = 0$ for all $i \neq j$, then the classical ADMM has split x-updates.

Consider the graph $\mathcal{G} = (\mathcal{V},\mathcal{E})$ with the vertex set $\mathcal{V} = \{1,\dots,m\}$ representing the blocks. For the edge set $\mathcal{E}$, we have $\{i,j\} \in \mathcal{E}$ if and only if $i \neq j$ and $A_{:,i}^\intercal A_{:,j} \neq 0$.

Assume $\mathcal{G}$ has a 2-coloring, i.e., we can partition $\mathcal{V}$ into $\mathcal{V}_1$ and $\mathcal{V}_2$ such that for any $i,j \in \mathcal{V}_k$, $\{i,j\} \notin \mathcal{E}$ for $k = 1,2$. The problem is equivalent to

$$\begin{array}{ll}\underset{x\in\mathbb{R}^p,\, y\in\mathbb{R}^q}{\text{minimize}} & \sum_{i\in\mathcal{V}_1} f_i(x_i) + \sum_{i\in\mathcal{V}_2} f_i(x_i) \\ \text{subject to} & \sum_{i\in\mathcal{V}_1} A_{:,i}x_i + \sum_{i\in\mathcal{V}_2} A_{:,i}x_i = c.\end{array}$$

Provide an ADMM-type method that updates the primal blocks in $\mathcal{V}_1$ and $\mathcal{V}_2$ in an alternating manner. What are the stepsize requirements?

Next, assume $\mathcal{G}$ has a χ-coloring, i.e., we partition $\mathcal{V}$ into $\mathcal{V}_1,\dots,\mathcal{V}_\chi$ such that for any $i,j \in \mathcal{V}_k$, $\{i,j\} \notin \mathcal{E}$ for $k = 1,\dots,\chi$. The problem is equivalent to

$$\begin{array}{ll}\underset{x\in\mathbb{R}^p,\, y\in\mathbb{R}^q}{\text{minimize}} & \sum_{k=1}^{\chi} \sum_{i\in\mathcal{V}_k} f_i(x_i) \\ \text{subject to} & \sum_{k=1}^{\chi} \sum_{i\in\mathcal{V}_k} A_{:,i}x_i = c.\end{array}$$

Provide an ADMM-type method analogous to Jacobi ADMM that updates the primal blocks in $\mathcal{V}_1, \ldots, \mathcal{V}_\chi$ concurrently. What are the stepsize requirements?

8.10 *Quadratic subproblem with affine constraints.* Assume g_1 is a strongly convex quadratic function with affine constraints, i.e.,

$$g_1(y) = \frac{1}{2} y^\intercal C y + \langle d, y\rangle + \delta_{\{y \mid Ey=f\}}(y),$$

where $C \in \mathbb{R}^{q\times q}$, $C \succ 0$, $d \in \mathbb{R}^q$, $E \in \mathbb{R}^{s\times q}$ has linearly independent rows, and $f \in \mathbb{R}^s$. When $s = 0$, the affine constraints vanish. Show that the solution to the subproblem

$$y^{k+1} = \operatorname*{argmin}_{y\in\mathbb{R}^q} \left\{ g_1(y) + \langle \nabla g_2(y^k) + B^\intercal u^k, y\rangle + \frac{\rho}{2}\|Ax^k + By - c\|^2 + \frac{1}{2}\|y - y^k\|_Q^2 \right\}$$

is given by

$$y^{k+1} = -T\left(\frac{\rho}{2} B^\intercal A x^k + \nabla g_2(y^k) + B^\intercal u^k - Q y^k\right) - h$$

for some symmetric positive semidefinite matrix $T \in \mathbb{R}^{q\times q}$ and $h = T(d - \frac{\rho}{2}B^\intercal c) + M^{-1}E^\intercal(EM^{-1}E^\intercal)^{-1} f$ with $M = C + Q + \rho B^\intercal B$.

8.11 *Four-block ADMM with 2-1-2-4-3-4 updates.* Consider the problem

$$\begin{array}{ll} \underset{\substack{x_1\in\mathbb{R}^{p_1},\, x_2\in\mathbb{R}^{p_2}\\ x_3\in\mathbb{R}^{p_3},\, x_4\in\mathbb{R}^{p_4}}}{\text{minimize}} & f_1(x_1) + f_2(x_2) + f_3(x_3) + f_4(x_4) \\ \text{subject to} & A_{:,1}x_1 + A_{:,2}x_2 + A_{:,3}x_3 + A_{:,4}x_4 = c, \end{array}$$

where f_2 and f_4 are strongly convex quadratic functions with affine constraints as in Exercise 8.10. Find an ADMM-type method that updates the primal blocks in the order $x_2 \to x_1 \to x_2 \to x_4 \to x_3 \to x_4$, analogous to the 2-1-2 ADMM of §8.2.5. What are the stepsize requirements?

8.12 *2-1-2 ADMM with FLiP.* Consider the problem

$$\begin{array}{ll} \underset{x\in\mathbb{R}^p,\, y\in\mathbb{R}^q}{\text{minimize}} & f_1(x) + g_1(y) \\ \text{subject to} & Ax + By = c. \end{array}$$

Assume g is is a strongly convex quadratic functions with affine constraints as in Exercise 8.10. Consider the 2-1-2 ADMM method with function linearization and proximal terms:

$$\begin{aligned}
y^{k+1/2} &= \operatorname*{argmin}_{y\in\mathbb{R}^q} \left\{ g_1(y) + \langle \nabla g_2(y^k) + B^\intercal u^k, y\rangle + \frac{\rho}{2}\|Ax^k + By - c\|^2 + \frac{1}{2}\|y - y^k\|_Q^2 \right\} \\
x^{k+1} &\in \operatorname*{argmin}_{x\in\mathbb{R}^p} \left\{ f_1(x) + \langle \nabla f_2(x^k) + A^\intercal u^k, x\rangle + \frac{\rho}{2}\|Ax + By^{k+1/2} - c\|^2 + \frac{1}{2}\|x - x^k\|_P^2 \right\} \\
y^{k+1} &= \operatorname*{argmin}_{y\in\mathbb{R}^q} \left\{ g_1(y) + \langle \nabla g_2(y^k) + B^\intercal u^k, y\rangle + \frac{\rho}{2}\|Ax^{k+1} + By - c\|^2 + \frac{1}{2}\|y - y^k\|_Q^2 \right\} \\
u^{k+1} &= u^k + \varphi\rho(Ax^{k+1} + By^{k+1} - c).
\end{aligned}$$

To clarify, the function linearization and the proximal terms of the y-updates are centered at y^k for both the $y^{k+1/2}$- and y^{k+1}-updates. Show that this method reduces to an instance of FLiP-ADMM, analogous to the 2-1-2 ADMM of §8.2.5. What are the stepsize requirements?

8.13 *Alternate proof of Theorem 6.* In this exercise, we perform an alternate analysis for Stages 2 and 3 of the proof of Theorem 6 to obtain a different stepsize requirement. Bound the last term of (8.7) with

$$\frac{1}{\varphi}\langle u^{k+1} - u^k, B(y^{k+1} - y^k)\rangle \leq \frac{\eta}{2\varphi^2\rho}\|u^{k+1} - u^k\|^2 + \frac{1}{2}\|y^{k+1} - y^k\|^2_{\frac{\rho}{\eta}B^\intercal B}$$

to get

$$\|w^{k+1} - w^\star\|_{M_0} \leq \|w^k - w^\star\|_{M_0} - \|w^{k+1} - w^k\|_{M_3} - \left(\mathbf{L}(x^{k+1}, y^{k+1}, u^\star) - \mathbf{L}(x^\star, y^\star, u^\star)\right),$$

where

$$M_3 = \frac{1}{2}\begin{bmatrix} P - L_f I & 0 & 0 \\ 0 & \rho\left(1 - \frac{1}{\eta}\right)B^\intercal B + Q - L_g I & 0 \\ 0 & 0 & \frac{2-\varphi-\eta}{\varphi^2\rho}I \end{bmatrix}.$$

Show that FLiP-ADMM converges if

$$P \succeq L_f I, \qquad Q \succeq 0,$$

and there exist $\varepsilon \in (0, 2 - \varphi)$ such that

$$\rho\left(1 - \frac{1}{2 - \varphi - \varepsilon}\right)B^\intercal B + Q \succeq L_g I.$$

Remark. This new condition is useful when L_g is large, as it does not have a factor 3.

8.14 *Refinement of Theorem 6.* Unify the proof of Theorem 6 and the analysis of Exercise 8.13. For $\alpha \in [0, 1]$, define

$$V_\alpha^k = \|w^k - w^\star\|^2_{M_0} + \alpha\|w^k - w^{k-1}\|^2_{M_1}.$$

Show the key inequality

$$V_\alpha^{k+1} \leq V_\alpha^k - \|w^{k+1} - w^k\|^2_{\alpha M_2 + (1-\alpha)M_3} - \left(\mathbf{L}(x^{k+1}, y^{k+1}, u^\star) - \mathbf{L}(x^\star, y^\star, u^\star)\right),$$

where M_3 is as defined in Exercise 8.13. Show that FLiP-ADMM converges if

$$P \succeq L_f I, \qquad Q \succeq 0,$$

and there exist $\alpha \in [0, 1]$ and $\varepsilon \in (0, 2 - \varphi)$ such that

$$\rho\left(1 - \frac{\alpha(1 - \varphi)^2 + 1 - \alpha}{2 - \varphi - \varepsilon}\right)B^\intercal B + Q \succeq (2\alpha + 1)L_g I.$$

Remark. When $\varphi = 1$, the choice $\alpha = \varepsilon$ leads to the (sufficient) condition $Q \succ L_g I$.

9 Duality in Splitting Methods

In this chapter, we present Attouch–Théra duality, a duality framework for monotone inclusion problems that is analogous to, but *simpler* than convex duality. Convex duality has several distinct and complementary interpretations, and Attouch–Théra duality takes *one* and generalizes it to operators.

We discuss the intimate connection of Attouch–Théra duality with the base splitting methods of §2.7.1. This was a theoretical detail omitted in §2.7.2, and the duality provides a more complete understanding. Furthermore, Attouch–Théra duality has algorithmic utility, as dual solutions certify correctness of primal solutions.

9.1 FENCHEL DUALITY

In *Fenchel duality*, the primal problem is

$$\underset{x\in\mathbb{R}^n}{\text{minimize}} \quad f(x) + g(x), \tag{F-P}$$

where f and g are CCP functions, and the dual problem is

$$\underset{u\in\mathbb{R}^n}{\text{maximize}} \quad -f^*(-u) - g^*(u). \tag{F-D}$$

These primal-dual problem pairs are generated by the Lagrangian

$$\mathbf{L}(x,u) = f(x) + \langle x,u\rangle - g^*(u).$$

As we discussed in §1.3.9, the question of when total duality holds in Fenchel duality (and more generally in convex duality) is subtle.

We now discuss an interpretation of Fenchel duality that we later extend to operators. For simplicity, assume total duality holds and f, g, f^*, and g^* are differentiable. Under the assumptions, the primal problem is equivalent to

$$\underset{x\in\mathbb{R}^n}{\text{find}} \quad 0 = \nabla f(x) + \nabla g(x),$$

which we interpret as the problem of finding a point x such that the gradients of f and g at x sum to 0. Remember from §2.1 that $\nabla f^* = (\nabla f)^{-1}$. Under the assumptions, the dual problem is equivalent to

$$\underset{u\in\mathbb{R}^n}{\text{find}} \quad (\nabla f)^{-1}(-u) = (\nabla g)^{-1}(u),$$

which we interpret as the problem of finding the gradient u such that the point at which ∇f produces $-u$ and the point at which ∇g produces u agree. When f, g, f^*, and g^* are not differentiable, one can make a similar argument with subgradients.

This is one of the many viewpoints of convex duality; the primal viewpoint is to find the variable x, while the dual viewpoint is to find the subgradient u.

9.2 ATTOUCH–THÉRA DUALITY

Consider the monotone inclusion problem

$$\underset{x\in\mathbb{R}^n}{\text{find}} \quad 0 \in (\mathbb{A}+\mathbb{B})x,$$

where $\mathbb{A}$ and $\mathbb{B}$ are maximal monotone. Define $\mathbb{A}^{-\circledcirc} = (-\mathbb{I})\mathbb{A}^{-1}(-\mathbb{I})$, i.e., $\mathbb{A}^{-\circledcirc}(u) = -\mathbb{A}^{-1}(-u)$. The Attouch–Théra dual monotone inclusion problem is

$$\underset{u\in\mathbb{R}^n}{\text{find}} \quad 0 \in (\mathbb{A}^{-\circledcirc}+\mathbb{B}^{-1})u.$$

Attouch–Théra duality is, in a sense, easier than Fenchel duality since

$$\operatorname{Zer}(\mathbb{A}+\mathbb{B}) \neq \emptyset \quad \Leftrightarrow \quad \operatorname{Zer}(\mathbb{A}^{-\circledcirc}+\mathbb{B}^{-1}) \neq \emptyset.$$

This follows from

$$\begin{aligned} \exists x\,[0 \in (\mathbb{A}+\mathbb{B})x] \quad &\Leftrightarrow \quad \exists x,u\,[-u \in \mathbb{A}x,\ u \in \mathbb{B}x] \\ &\Leftrightarrow \quad \exists x,u\left[-x \in \mathbb{A}^{-\circledcirc}u,\ x \in \mathbb{B}^{-1}u\right] \\ &\Leftrightarrow \quad \exists u\left[0 \in (\mathbb{A}^{-\circledcirc}+\mathbb{B}^{-1})u\right]. \end{aligned}$$

In other words, a primal solution exists if and only if a dual solution exists. There is no notion of strong duality, as there are no function values.

In a certain sense, Attouch–Théra duality generalizes Fenchel duality, as monotone operators generalize subdifferential operators of convex functions. In a different sense, Attouch–Théra duality does not generalize Fenchel duality, as Attouch–Théra duality fails to capture and provide insight into the subtleties and difficulties of convex duality. In Fenchel duality, strong duality may fail, a primal solution may exist while a dual solution does not, or vice versa. There is no analog of such pathologies in Attouch–Théra duality.

Dual Solutions as Certificates

When solving a monotone inclusion problem with multi-valued operators, a dual solution certifies correctness of a primal solution. Therefore, it is desirable for a splitting method to produce solutions of both the primal and dual monotone inclusion problems. Given $(x^\star, u^\star)$, one can verify it is indeed a primal-dual solution pair by checking $-u^\star \in \mathbb{A}x^\star$ and $u^\star \in \mathbb{B}x^\star$. If only a primal solution is provided, we must verify that $0 \in \mathbb{A}x^\star + \mathbb{B}x^\star$. This can be difficult if there is no effective way to compute the Minkowski sum $\mathbb{A}x^\star + \mathbb{B}x^\star$. (On a computer, how would we represent the sets $\mathbb{A}x^\star$ and $\mathbb{B}x^\star$ and how would we compute the Minkowski sum?)

In practice, a method used to verify "correctness" of a primal-dual solution must be able to deal with inaccuracies, since an output of an iterative algorithm will be at most approximately correct. This issue relates how to design an effective termination criterion for the iterative methods. We avoid this discussion for the sake of simplicity.

9.3 DUALITY IN SPLITTING METHODS

We now present the intimate connection of the base splittings of §2.7.1 with Attouch–Théra duality. We also show that the splittings are primal-dual in the sense that they provide dual information.

9.3.1 FBS

The FPI with FBS

$$\begin{aligned} x^{k+1/2} &= x^k - \alpha \mathbb{A} x^k \\ x^{k+1} &= \mathbb{J}_{\alpha\mathbb{B}} x^{k+1/2} \end{aligned}$$

is often not considered a primal-dual method as there is no explicit reference to the dual problem or a dual variable. However, we can make the method primal-dual by writing

$$\begin{aligned} x^{k+1/2} &= x^k - \alpha \mathbb{A} x^k \\ u^{k+1/2} &= -\mathbb{A} x^k \\ x^{k+1} &= \mathbb{J}_{\alpha\mathbb{B}} x^{k+1/2} \\ u^{k+1} &= \alpha^{-1}(x^{k+1/2} - x^{k+1}). \end{aligned}$$

Note that $u^{k+1} \in \mathbb{B}x^{k+1}$. It is straightforward to verify that if $x^k \to x^\star$, then

$$u^{k+1/2} \to u^\star, \quad u^{k+1} \to u^\star, \quad u^\star \in \operatorname{Zer}(\mathbb{A}^{-\oslash} + \mathbb{B}^{-1}).$$

9.3.2 DRS

Characterization of Fixed Points

Using the Attouch–Théra dual, we can now characterize the fixed points of the PRS and DRS operators more concretely:

$$\operatorname{Fix}(\mathbb{R}_{\alpha\mathbb{A}}\mathbb{R}_{\alpha\mathbb{B}}) \subseteq \operatorname{Zer}(\mathbb{A} + \mathbb{B}) + \alpha \operatorname{Zer}(\mathbb{A}^{-\oslash} + \mathbb{B}^{-1}).$$

This follows from

$$\begin{aligned} z = \mathbb{R}_{\alpha\mathbb{A}}\mathbb{R}_{\alpha\mathbb{B}} z \quad &\Leftrightarrow \quad z + 2\mathbb{J}_{\alpha\mathbb{A}}(2\mathbb{J}_{\alpha\mathbb{B}} - \mathbb{I})z - 2\mathbb{J}_{\alpha\mathbb{B}} z = z,\ x = \mathbb{J}_{\alpha\mathbb{B}} z \\ &\Leftrightarrow \quad \mathbb{J}_{\alpha\mathbb{A}}(x - \alpha u) = x,\ z = x + \alpha u,\ u \in \mathbb{B}x \\ &\Leftrightarrow \quad x - \alpha u = x + \alpha v,\ v \in \mathbb{A}x,\ z = x + \alpha u,\ u \in \mathbb{B}x \\ &\Leftrightarrow \quad v = -u,\ v \in \mathbb{A}x,\ u \in \mathbb{B}x,\ z = x + \alpha u \\ &\Leftrightarrow \quad -u \in \mathbb{A}x,\ u \in \mathbb{B}x,\ z = x + \alpha u \\ &\Leftrightarrow \quad -u \in \mathbb{A}x,\ u \in \mathbb{B}x,\ -x \in \mathbb{A}^{-\oslash}u,\ x \in \mathbb{B}^{-1}u,\ z = x + \alpha u \\ &\Rightarrow \quad 0 \in (\mathbb{A} + \mathbb{B})x,\ 0 \in (\mathbb{A}^{-\oslash} + \mathbb{B}^{-1})u,\ z = x + \alpha u. \end{aligned}$$

Because the last step is not an equivalence, the characterization is an inclusion, not equality. See the bibliographical notes section for further discussion.

DRS in Primal-Dual Form

We can make the FPI with DRS more explicitly primal-dual by writing

$$
\begin{aligned}
x^{k+1/2} &= \mathbb{J}_{\alpha\mathbb{B}}(z^k) \\
u^{k+1/2} &= \frac{1}{\alpha}(z^k - x^{k+1/2}) \\
x^{k+1} &= \mathbb{J}_{\alpha\mathbb{A}}(2x^{k+1/2} - z^k) \\
u^{k+1} &= \frac{1}{\alpha}(x^{k+1} - x^{k+1/2} + \alpha u^{k+1/2}) \\
z^{k+1} &= z^k + x^{k+1} - x^{k+1/2}.
\end{aligned}
$$

Note that $u^{k+1/2} \in \mathbb{B}x^{k+1/2}$ and $-u^{k+1} \in \mathbb{A}x^{k+1}$. It is straightforward to verify that if $\operatorname{Zer}(\mathbb{A} + \mathbb{B}) \neq \emptyset$, then

$$
\begin{aligned}
x^{k+1/2} \to x^\star, \quad x^{k+1} \to x^\star, \quad x^\star \in \operatorname{Zer}(\mathbb{A} + \mathbb{B}) \\
u^{k+1/2} \to u^\star, \quad u^{k+1} \to u^\star, \quad u^\star \in \operatorname{Zer}(\mathbb{A}^{-\oslash} + \mathbb{B}^{-1}) \\
z^k \to x^\star + \alpha u^\star.
\end{aligned}
$$

Self-Dual Property of DRS

Interestingly, PRS and DRS are self-dual in the following sense:

$$
\mathbb{R}_\mathbb{A}\mathbb{R}_\mathbb{B} = \mathbb{R}_{\mathbb{A}^{-\oslash}}\mathbb{R}_{\mathbb{B}^{-1}}.
$$

This follows from using $\mathbb{J}_{\mathbb{A}^{-\oslash}} = \mathbb{I} + \mathbb{J}_\mathbb{A}(-\mathbb{I})$ (see Exercise 9.1) and the inverse resolvent identity $\mathbb{J}_{\mathbb{B}^{-1}} = \mathbb{I} - \mathbb{J}_\mathbb{B}$:

$$
\begin{aligned}
(2\mathbb{J}_{\mathbb{A}^{-\oslash}} - \mathbb{I})(2\mathbb{J}_{\mathbb{B}^{-1}} - \mathbb{I}) &= (2\mathbb{J}_\mathbb{A}(-\mathbb{I}) + \mathbb{I})(\mathbb{I} - 2\mathbb{J}_\mathbb{B}) \\
&= (2\mathbb{J}_\mathbb{A}(-\mathbb{I}) + \mathbb{I})(-\mathbb{I})(2\mathbb{J}_\mathbb{B} - \mathbb{I}) \\
&= (2\mathbb{J}_\mathbb{A} - \mathbb{I})(2\mathbb{J}_\mathbb{B} - \mathbb{I}).
\end{aligned}
$$

In fact, when $\alpha = 1$, we can write the FPI with DRS as

$$
\begin{aligned}
x^{k+1/2} &= \mathbb{J}_\mathbb{B}(z^k) \\
u^{k+1/2} &= \mathbb{J}_{\mathbb{B}^{-1}}(z^k) = z^k - x^{k+1/2} \\
x^{k+1} &= \mathbb{J}_\mathbb{A}(2x^{k+1/2} - z^k) \\
u^{k+1} &= \mathbb{J}_{\mathbb{A}^{-\oslash}}(2u^{k+1/2} - z^k) = x^{k+1} - x^{k+1/2} + u^{k+1/2} \\
z^{k+1} &= z^k + x^{k+1} - x^{k+1/2} = z^k + u^{k+1} - u^{k+1/2}.
\end{aligned}
$$

This form of the iteration nicely reveals the symmetry. (Algorithmically there is no need to use both the x- and u-variables.) When $\alpha \neq 1$, we have a similar, but slightly less elegant self-dual relationship.

As an aside, this self-dual property explains why the infimal postcomposition technique of §3.1 and the dualization technique of §3.2 yield the same method ADMM.

9.3.3 DYS

For the monotone inclusion problem

$$\underset{x\in\mathbb{R}^n}{\text{find}} \quad 0 \in (\mathbb{A}+\mathbb{B}+\mathbb{C})x,$$

where $\mathbb{A}$, $\mathbb{B}$, and $\mathbb{C}$ are maximal monotone and $\mathbb{C}$ is single-valued, we consider the Attouch–Théra dual

$$\underset{u\in\mathbb{R}^n}{\text{find}} \quad 0 \in ((\mathbb{A}+\mathbb{C})^{-\circledcirc}+\mathbb{B}^{-1})u.$$

Consider the DYS operator

$$\mathbb{I}-\mathbb{J}_{\alpha\mathbb{B}}+\mathbb{J}_{\alpha\mathbb{A}}(\mathbb{R}_{\alpha\mathbb{B}}-\alpha\mathbb{C}\mathbb{J}_{\alpha\mathbb{B}}).$$

With similar steps as before, we can characterize the fixed points with

$$\text{Fix}\,(\mathbb{I}-\mathbb{J}_{\alpha\mathbb{B}}+\mathbb{J}_{\alpha\mathbb{A}}(\mathbb{R}_{\alpha\mathbb{B}}-\alpha\mathbb{C}\mathbb{J}_{\alpha\mathbb{B}}))\subseteq \text{Zer}\,(\mathbb{A}+\mathbb{B}+\mathbb{C})+\alpha\text{Zer}\,((\mathbb{A}+\mathbb{C})^{-\circledcirc}+\mathbb{B}^{-1}).$$

We can make the FPI with DYS more explicitly primal-dual by writing

$$\begin{aligned}
x^{k+1/2} &= \mathbb{J}_{\alpha\mathbb{B}}(z^k)\\
u^{k+1/2} &= \frac{1}{\alpha}(z^k-x^{k+1/2})\\
x^{k+1} &= \mathbb{J}_{\alpha A}(2x^{k+1/2}-z^k-\alpha\mathbb{C}x^{k+1/2})\\
u^{k+1} &= \frac{1}{\alpha}(x^{k+1}-x^{k+1/2}+\alpha u^{k+1/2})\\
z^{k+1} &= z^k+x^{k+1}-x^{k+1/2}.
\end{aligned}$$

Note that $u^{k+1/2}\in\mathbb{B}x^{k+1/2}$ and $-u^{k+1}\in\mathbb{A}x^{k+1}+\mathbb{C}x^{k+1/2}$. It is straightforward to verify that if $z^k\to z^\star$, then

$$\begin{gathered}
x^{k+1/2}\to x^\star,\quad x^{k+1}\to x^\star,\quad x^\star\in\text{Zer}\,(\mathbb{A}+\mathbb{B}+\mathbb{C})\\
u^{k+1/2}\to u^\star,\quad u^{k+1}\to u^\star,\quad u^\star\in\text{Zer}\,((\mathbb{A}+\mathbb{C})^{-\circledcirc}+\mathbb{B}^{-1})\\
z^k\to x^\star+\alpha u^\star.
\end{gathered}$$

Finally, we note that DYS is not self-dual, as it uses an evaluation of $\mathbb{C}$, a primal operation.

BIBLIOGRAPHICAL NOTES

Fenchel duality was formalized by Fenchel in 1949 [Fen49]. Although Attouch–Théra duality is named after Attouch and Théra's 1996 paper [AT96], it was first formalized by Mercier in 1980 [Mer80, p. 40]. The self-dual property of DRS was first presented by Eckstein in 1989 [Eck89, Lemma 3.6 p. 133] and was further investigated in [BBHM12, YY16, BLM17, BM17].

In general, $\text{Fix}\,(\mathbb{R}_{\alpha\mathbb{A}}\mathbb{R}_{\alpha\mathbb{B}})\neq\text{Zer}\,(\mathbb{A}+\mathbb{B})+\alpha\text{Zer}\,(\mathbb{A}^{-\circledcirc}+\mathbb{B}^{-1})$. See Exercise 9.4 for a counterexample and see Exercise 9.6 for a complete characterization, $\text{Fix}\,(\mathbb{R}_{\alpha\mathbb{A}}\mathbb{R}_{\alpha\mathbb{B}})$.

A monotone operator is said to be *paramonotone* if

$$u\in\mathbb{A}x,\ v\in\mathbb{A}y,\ \langle u-v,x-y\rangle=0\quad\Rightarrow\quad v\in\mathbb{A}x,\ u\in\mathbb{A}y.$$

Bruck first presented the notion without naming the property [Bru75b]. Censor, Iusem, and Zenios named the property paramonotonicity [CIZ98]. Bauschke, Boţ, Hare, and Moursi [BBHM12] showed that if $\mathbb{A}$ and $\mathbb{B}$ are paramonotone, then we can characterize the fixed points of PRS and DRS with equality:

$$\operatorname{Fix}(\mathbb{R}_{\alpha A}\mathbb{R}_{\alpha B}) = \operatorname{Zer}(\mathbb{A}+\mathbb{B}) + \alpha \operatorname{Zer}(\mathbb{A}^{-\circledcirc} + \mathbb{B}^{-1}).$$

(Subdifferential operators of CCP functions are paramonotone. Cf. Exercise 9.5.)

EXERCISES

9.1 *Variation of the inverse resolvent identity.* Prove $\mathbb{J}_{\mathbb{A}^{-\circledcirc}} - \mathbb{J}_{\mathbb{A}}(-\mathbb{I}) = \mathbb{I}$.

9.2 Show that the the fixed points of the DYS operator satisfy

$$\operatorname{Fix}(\mathbb{I} - \mathbb{J}_{\alpha\mathbb{B}} + \mathbb{J}_{\alpha\mathbb{A}}(\mathbb{R}_{\alpha\mathbb{B}} - \alpha\mathbb{C}\mathbb{J}_{\alpha\mathbb{B}})) \subseteq \operatorname{Zer}(\mathbb{A}+\mathbb{B}+\mathbb{C}) + \alpha\operatorname{Zer}((\mathbb{A}+\mathbb{C})^{-\circledcirc} + \mathbb{B}^{-1}).$$

9.3 Let

$$f(x) = \begin{cases} -\sqrt{x} & \text{for } x \geq 0 \\ \infty & \text{otherwise,} \end{cases} \qquad g(x) = \delta_{\{0\}}(x),$$

where $x \in \mathbb{R}$. Show that while the primal problem

$$\underset{x\in\mathbb{R}}{\text{minimize}} \quad f(x) + g(x)$$

has a solution, its Fenchel dual

$$\underset{u\in\mathbb{R}}{\text{maximize}} \quad -f^*(-u) - g^*(u)$$

does not. Also show that $\operatorname{argmin}(f+g) \neq \operatorname{Zer}(\partial f + \partial g)$.

9.4 Consider the operators

$$\mathbb{A} = \mathbb{N}_{\mathbb{R}^2_+}, \qquad \mathbb{B} = \begin{bmatrix} 0 & -1 \\ 1 & 0 \end{bmatrix},$$

where $\mathbb{R}^2_+ = \{(x_1,x_2) \in \mathbb{R}^2 \mid x_1 \geq 0, x_2 \geq 0\}$. Show

(a) $\operatorname{Zer}(\mathbb{A}+\mathbb{B}) = \{(x_1,0) \in \mathbb{R}^2 \mid x_1 \geq 0\}$

(b) $\operatorname{Zer}(\mathbb{A}^{-\circledcirc} + \mathbb{B}^{-1}) = \{(0,u_2) \in \mathbb{R}^2 \mid u_2 \geq 0\}$

(c) $\operatorname{Fix}(\mathbb{R}_{\mathbb{A}}\mathbb{R}_{\mathbb{B}}) = \{(z,z) \in \mathbb{R}^2 \mid z \geq 0\}$

and conclude that

$$\operatorname{Fix}(\mathbb{R}_{\mathbb{A}}\mathbb{R}_{\mathbb{B}}) \neq \operatorname{Zer}(\mathbb{A}+\mathbb{B}) + \operatorname{Zer}(\mathbb{A}^{-\circledcirc} + \mathbb{B}^{-1}).$$

Hint. Use

$$\mathbb{B}^{-1} = -\mathbb{B}, \qquad \mathbb{J}_{\mathbb{B}} = \frac{1}{2}\begin{bmatrix} 1 & 1 \\ -1 & 1 \end{bmatrix}, \qquad \mathbb{R}_{\mathbb{B}} = -\mathbb{B}$$

and

$$\mathbb{N}_{\mathbb{R}^2_+}(x) = \begin{cases} \{y \in \mathbb{R}^2 \mid y_1 \leq 0, y_2 \leq 0, \langle x,y\rangle = 0\} & \text{if } x \in \mathbb{R}^2_+ \\ \emptyset & \text{if } x \notin \mathbb{R}^2_+. \end{cases}$$

Remark. This counterexample is due to Bauschke, Boţ, Hare, and Moursi [BBHM12].

9.5 *Fixed points of DRS with Fenchel duality.* Consider the Fenchel dual setup with primal and dual problems

$$\underset{x\in\mathbb{R}^n}{\text{minimize}} \quad f(x)+g(x), \qquad \underset{u\in\mathbb{R}^n}{\text{maximize}} \quad -f^*(-u)-g^*(u),$$

where f and g are CCP functions on $\mathbb{R}^n$, generated by

$$\mathbf{L}(x,u) = f(x) + \langle x,u\rangle - g^*(u).$$

Assume total duality holds. Write $X^\star$ and $U^\star$ for the sets of primal and dual solutions. Show that

$$\operatorname{Fix}\left(\mathbb{R}_{\alpha\partial f}\mathbb{R}_{\alpha\partial g}\right) = X^\star + \alpha U^\star.$$

Hint. Note that $[x^\star \in X^\star$ and $u^\star \in U^\star]$ if and only if $[(x^\star,u^\star)$ is a saddle point of $\mathbf{L}]$.

9.6 *Fixed points of DRS via primal-dual inclusion.* Consider the Attouch–Théra dual setup with primal and dual problems

$$\underset{x\in\mathbb{R}^n}{\text{find}} \quad 0\in(\mathbb{A}+\mathbb{B})x, \qquad \underset{u\in\mathbb{R}^n}{\text{find}} \quad 0\in(\mathbb{A}^{-\circledcirc}+\mathbb{B}^{-1})u,$$

where $\mathbb{A}$ and $\mathbb{B}$ are maximal monotone operators on $\mathbb{R}^n$. Write $X^\star$ and $U^\star$ for the sets of primal and dual solutions. Consider the *primal-dual* inclusion problem

$$\underset{x,u\in\mathbb{R}^n}{\text{find}} \quad 0\in\begin{bmatrix}\mathbb{A} & \mathbb{I}\\ -\mathbb{I} & \mathbb{B}^{-1}\end{bmatrix}\begin{bmatrix}x\\ u\end{bmatrix} = \begin{bmatrix}\mathbb{A}x+u\\ -x+\mathbb{B}^{-1}u\end{bmatrix}$$

and write $\Phi^\star \subseteq \mathbb{R}^n\times\mathbb{R}^n$ for its solution set. Show

(a) $\begin{bmatrix}I & 0\end{bmatrix}\Phi^\star = X^\star$,
(b) $\begin{bmatrix}0 & I\end{bmatrix}\Phi^\star = U^\star$, and
(c) $\begin{bmatrix}I & I\end{bmatrix}\Phi^\star = \operatorname{Fix}(\mathbb{R}_{\mathbb{A}}\mathbb{R}_{\mathbb{B}})$,

where $0, I\in\mathbb{R}^{n\times n}$ are the zero and identity matrices.
Clarification. For any $A,B\in\mathbb{R}^{n\times n}$, $\begin{bmatrix}A & B\end{bmatrix}\Phi^\star = \{\mathbb{A}x^\star + Bu^\star \mid (x^\star,u^\star)\in\Phi^\star\}$.
Hint. For (a), use the equivalences

$$\begin{aligned} 0\in\mathbb{A}x^\star+\mathbb{B}x^\star \quad &\Leftrightarrow \quad \exists\, u^\star \text{ such that } 0\in\mathbb{A}x^\star+u^\star,\ u^\star\in\mathbb{B}x^\star\\ &\Leftrightarrow \quad \exists\, u^\star \text{ such that } 0\in\mathbb{A}x^\star+u^\star,\ 0\in -x^\star+\mathbb{B}^{-1}u^\star. \end{aligned}$$

Remark. This result was first established by Bauschke, Boţ, Hare, and Moursi [BBHM12, Theorem 4.5]. The set $\Phi^\star$ is also referred to as the "extended solution set" and was first studied by Eckstein and Svaiter [ES08].

10 Maximality and Monotone Operator Theory

In this chapter, we digress and study monotone operator theory. Convex optimization theory, the main subject of study in this book, focuses on the derivation and analysis of convex optimization algorithms. In contrast, monotone operator theory views monotone operators as interesting objects in their own right and focuses on understanding them better.

One goal of this chapter is to provide theoretical completeness; we prove several results that were simply asserted in §2. Another goal is to provide a gentle exposure to the field of monotone operator theory. Readers who find this subject interesting can continue their study through standard references such as [Phe93, Sho97, FP03, BL06, BV10, Boţ10, BC17a].

Often, results in monotone operator theory are established in infinite-dimensional Banach or Hilbert spaces, where a new set of interesting challenges arise. Here, we limit our attention to finite-dimensional Euclidean spaces.

10.1 MAXIMALITY OF SUBDIFFERENTIAL

We say $\bar{\mathbb{A}}: \mathbb{R}^n \rightrightarrows \mathbb{R}^n$ is an *extension* of $\mathbb{A}: \mathbb{R}^n \rightrightarrows \mathbb{R}^n$ if $\operatorname{Gra} \bar{\mathbb{A}} \supseteq \operatorname{Gra} \mathbb{A}$. We say $\bar{\mathbb{A}}$ is a *proper* extension of $\mathbb{A}$ if the containment $\operatorname{Gra} \bar{\mathbb{A}} \supset \operatorname{Gra} \mathbb{A}$ is strict.

Recall that a monotone operator is maximal if it has no proper monotone extension. As we have discussed in §2, and as we will soon prove, if $\mathbb{A}: \mathbb{R}^n \rightrightarrows \mathbb{R}^n$ is maximal monotone, then $\operatorname{dom} \mathbb{J}_{\mathbb{A}} = \mathbb{R}^n$, which implies fixed-point iterations using $\mathbb{J}_{\mathbb{A}}$ are well defined.

Theorem 7 If $f: \mathbb{R}^n \to \mathbb{R}^n \cup \{\infty\}$ is CCP, then ∂f is maximal monotone.

Proof. We know ∂f is monotone. Assume for contradiction that there exists a $(\tilde{x}, \tilde{g}) \notin \partial f$ such that $\{(\tilde{x}, \tilde{g})\} \cup \partial f$ is monotone. Define $(x, g) \in \partial f$ with

$$x = \operatorname*{argmin}_{z} \left\{ f(z) + \frac{1}{2}\|z - (\tilde{x} + \tilde{g})\|^2 \right\} = \operatorname{Prox}_f(\tilde{x} + \tilde{g}), \qquad 0 = x - \tilde{x} + g - \tilde{g}.$$

We get $g \in \partial f(x)$ from the $0 \in x - \tilde{x} + \partial f(x) - \tilde{g}$, the optimality condition of the argmin. Since we assumed $(\tilde{x}, \tilde{g}) \notin \partial f$, either $x \neq \tilde{x}$ or $g \neq \tilde{g}$ (or both). Using $x - \tilde{x} = -g + \tilde{g}$, we have

$$\langle g - \tilde{g}, x - \tilde{x} \rangle = -\|x - \tilde{x}\|_2^2 = -\|g - \tilde{g}\|_2^2 < 0,$$

which contradicts the assumption that $\{(\tilde{x}, \tilde{g})\} \cup \partial f$ is monotone. □

The key idea of proof is that given $v \in \mathbb{R}^n$,

$$v \mapsto (\underbrace{\mathrm{Prox}_f(v)}_{=x}, \underbrace{v - \mathrm{Prox}_f(v)}_{=g}) \in \partial f$$

provides a unique decomposition $v = x + g$ such that $(x, g) \in \partial f$.

10.2 FITZPATRICK FUNCTION

For $\mathbb{A}\colon \mathbb{R}^n \rightrightarrows \mathbb{R}^n$, define the *Fitzpatrick function* $\mathbf{F}_{\mathbb{A}}\colon \mathbb{R}^n \times \mathbb{R}^n \to \mathbb{R} \cup \{\infty\}$ as

$$\mathbf{F}_{\mathbb{A}}(x, u) = \langle x, u \rangle - \inf_{(y,v) \in \mathbb{A}} \langle x - y, u - v \rangle = \sup_{(y,v) \in \mathbb{A}} \{\langle y, u \rangle + \langle x, v \rangle - \langle y, v \rangle\},$$

which is useful when $\mathbb{A}$ is maximal monotone. The equivalent definition with sup follows from expanding the inner product within the inf.

Lemma 3 *Assume $\mathbb{A}\colon \mathbb{R}^n \rightrightarrows \mathbb{R}^n$ is maximal monotone. Then*

- $\mathbf{F}_{\mathbb{A}}$ *is CCP,*
- $\mathbf{F}_{\mathbb{A}}(x, u) \geq \langle x, u \rangle$ *for all $x, u \in \mathbb{R}^n$, and*
- $\mathbf{F}_{\mathbb{A}}(x, u) = \langle x, u \rangle$ *if and only if $(x, u) \in \mathbb{A}$.*

We say $\mathbf{F}_{\mathbb{A}}$ is a *representative function* of $\mathbb{A}$, since $\mathbf{F}_{\mathbb{A}}$ is a convex extension of $\langle x, u \rangle$ from Gra $\mathbb{A}$ to $\mathbb{R}^n \times \mathbb{R}^n$ that furthermore satisfies $\mathbf{F}_{\mathbb{A}}(x, u) \geq \langle x, u \rangle$. The Fitzpatrick function is one of the several representative functions used in the monotone operator theory literature.

A common technique in monotone operator theory is to analyze a representative function to conclude results about the original operator. In our case specifically, analyzing $\mathbf{F}_{\mathbb{A}}$, a CCP function, is easier than directly analyzing $\mathbb{A}$, since we can rely on results from convex analysis.

Proof. If $(x, u) \in \mathbb{A}$, then $\langle x - y, u - v \rangle \geq 0$ for all $(y, v) \in \mathbb{A}$ by monotonicity, and the infimum

$$\inf_{(y,v) \in \mathbb{A}} \langle x - y, u - v \rangle = 0$$

is attained at (x, u). So $\mathbf{F}_{\mathbb{A}}(x, u) = \langle x, u \rangle$.

Assume $(x, u) \notin \mathbb{A}$. Then by maximality there exists a $(y, v) \in \mathbb{A}$ such that $\langle x - y, u - v \rangle < 0$. Therefore

$$\inf_{(y,v) \in \mathbb{A}} \langle x - y, u - v \rangle < 0$$

and $\mathbf{F}_{\mathbb{A}}(x, u) > \langle x, u \rangle$.

Define

$$f_{y,v}(x,u) = \langle y,u\rangle + \langle x,v\rangle - \langle y,v\rangle,$$

which is a closed convex function for all $(y,v) \in \mathbb{A}$. Then

$$\operatorname{epi} \mathbf{F}_{\mathbb{A}} = \bigcap_{(y,v)\in\mathbb{A}} \operatorname{epi} f_{y,v}$$

is a closed convex set as it is an intersection of closed convex sets.

Since $\mathbf{F}_{\mathbb{A}}(x,u) \geq f_{y,v}(x,u) > -\infty$ for any $(y,v) \in \mathbb{A}$, we have $\mathbf{F}_{\mathbb{A}} > -\infty$ always. On the other hand,

$$\mathbf{F}_{\mathbb{A}}(x,u) = \langle x,u\rangle < \infty$$

for any $(x,u) \in \mathbb{A}$. So $\mathbf{F}_{\mathbb{A}}$ is proper. □

Theorem 8, the Minty surjectivity theorem, is foundational to operator splitting methods as it ensures that methods using resolvents are well defined. We say the operator $\mathbb{I} + \mathbb{A}$ is *surjective* if range $(\mathbb{I} + \mathbb{A}) = \mathbb{R}^n$, i.e., for any $u \in \mathbb{R}^n$ there is an $x \in \mathbb{R}^n$ such that $u \in (\mathbb{I} + \mathbb{A})x$. If $\mathbb{I} + \mathbb{A}$ is surjective, then $\operatorname{dom} \mathbb{J}_{\mathbb{A}} = \mathbb{R}^n$.

Theorem 8 *(Minty surjectivity theorem)* If $\mathbb{A}\colon \mathbb{R}^n \rightrightarrows \mathbb{R}^n$ is maximal monotone, then range $(\mathbb{I} + \mathbb{A}) = \mathbb{R}^n$.

Proof. We want to show that $u \in$ range $(\mathbb{I} + \mathbb{A})$ for any $u \in \mathbb{R}^n$ and maximal monotone $\mathbb{A}$. To do so, we first establish $0 \in$ range $(\mathbb{I} + \mathbb{A})$ for any maximal monotone $\mathbb{A}$. Then the maximal monotone operator $\mathbb{B}(x) = \mathbb{A}(x) - u$ satisfies $0 \in$ range $(\mathbb{I} + \mathbb{B})$, which implies $u \in$ range $(\mathbb{I} + \mathbb{A})$ for any $u \in \mathbb{R}^d$.

We now complete the proof by showing $0 \in$ range $(\mathbb{I} + \mathbb{A})$. Define $(y,v) \in \mathbb{R}^n \times \mathbb{R}^n$ with

$$(y,v) = \operatorname*{argmin}_{(x,u)\in\mathbb{R}^n\times\mathbb{R}^n} \left\{ \mathbf{F}_{\mathbb{A}}(x,u) + \frac{1}{2}\|x\|^2 + \frac{1}{2}\|u\|^2 \right\} = \operatorname{Prox}_{\mathbf{F}_{\mathbb{A}}}(0,0).$$

This implies

$$\begin{bmatrix} -y \\ -v \end{bmatrix} \in \partial \mathbf{F}_{\mathbb{A}}(y,v).$$

Since $\mathbf{F}_{\mathbb{A}}$ is convex, the subgradient inequality tells us

$$\left\langle \begin{bmatrix} -y \\ -v \end{bmatrix}, \begin{bmatrix} x \\ u \end{bmatrix} - \begin{bmatrix} y \\ v \end{bmatrix} \right\rangle \leq \mathbf{F}_{\mathbb{A}}(x,u) - \mathbf{F}_{\mathbb{A}}(y,v) \qquad \forall\, (x,u) \in \mathbb{R}^n \times \mathbb{R}^n.$$

By Lemma 3,

$$\mathbf{F}_{\mathbb{A}}(x,u) - \mathbf{F}_{\mathbb{A}}(y,v) \leq \langle x,u\rangle - \langle y,v\rangle \qquad \forall\, (x,u) \in \mathbb{A}.$$

Combining the two inequalities and reorganize to get

$$\|y + v\|^2 \leq \langle x + v, u + y\rangle \qquad \forall\, (x,u) \in \mathbb{A}. \tag{10.1}$$

Since $0 \leq \|y + v\|^2$ and since $\mathbb{A}$ is maximal monotone, this implies $(-v,-y) \in \mathbb{A}$. By letting $(x,u) = (-v,-y)$ in (10.1), we get $v = -y$. Thus $(y,-y) \in \mathbb{A}$ and we have

$$0 \in (\mathbb{A} + \mathbb{I})(y).$$

□

The converse of Theorem 8 is true. As a consequence, we can show a monotone operator $\mathbb{A}: \mathbb{R}^n \rightrightarrows \mathbb{R}^n$ is maximal if $\operatorname{dom} \mathbb{J}_{\mathbb{A}} = \mathbb{R}^n$.

Theorem 9 If $\mathbb{A}: \mathbb{R}^n \rightrightarrows \mathbb{R}^n$ is monotone and range $(J + \mathbb{A}) = \mathbb{R}^n$ for a symmetric positive definite $J \in \mathbb{R}^{n\times n}$, then $\mathbb{A}$ is maximal monotone.

Proof. First consider the case $J = \mathbb{I}$. Assume $\{(x,u)\} \cup \mathbb{A}$ is monotone, i.e.,

$$0 \le \langle x - z, u - w\rangle \qquad \forall (z,w) \in \mathbb{A}.$$

To establish maximality, it is enough to show $(x,u) \in \mathbb{A}$. Since range $(\mathbb{I} + \mathbb{A}) = \mathbb{R}^n$, there is a y such that $x + u \in (\mathbb{I} + \mathbb{A})y$. Let

$$v = x + u - y \in \mathbb{A}y.$$

Then

$$0 \le \langle x - y, u - v\rangle = -\|x - y\|^2 = -\|u - v\|^2.$$

So $x = y$ and $u = v$, which implies $(x,u) \in \mathbb{A}$.

When $J \neq \mathbb{I}$. Then $J^{-1/2}\mathbb{A}J^{-1/2}$ is monotone and, because $J + \mathbb{A}$ is surjective,

$$\operatorname{range}(\mathbb{I} + J^{-1/2}\mathbb{A}J^{-1/2}) = \mathbb{R}^n.$$

This implies $J^{-1/2}\mathbb{A}J^{-1/2}$ is maximal and so is $\mathbb{A}$. □

Theorem 10 Let $\mathbb{A}: \mathbb{R}^n \rightrightarrows \mathbb{R}^n$ be maximal monotone and $f: \mathbb{R}^n \to \mathbb{R} \cup \{\infty\}$ be CCP. If $0 \in \operatorname{int}(\operatorname{dom} \mathbb{A} - \operatorname{dom} f)$, then $\mathbb{A} + \partial f$ is maximal monotone.

Theorem 10 is useful, because Theorems 11 and 12 easily follow from it. The proof of Theorem 10 is similar to the proof of Theorem 8 but somewhat more complicated. See Exercise 10.9.

Theorem 11 Let $\mathbb{A}: \mathbb{R}^n \rightrightarrows \mathbb{R}^n$ and $\mathbb{B}: \mathbb{R}^n \rightrightarrows \mathbb{R}^n$ be maximal monotone. If $\operatorname{dom} \mathbb{A} \cap \operatorname{int} \operatorname{dom} \mathbb{B} \neq \emptyset$, then $\mathbb{A} + \mathbb{B}$ is maximal monotone.

Proof. Define

$$C = \{(x,x) \mid x \in \mathbb{R}^n\}.$$

Then

$$\mathbb{N}_C(x,x) = \{(v,-v) \mid v \in \mathbb{R}^n\}$$

for any $x \in \mathbb{R}^n$. (Remember that $\mathbb{N}_C$ is the normal cone operator and $\mathbb{N}_C = \partial\delta_C$, where δ_C is the indicator function.)

Consider the operator $\mathbb{F}\colon \mathbb{R}^d \times \mathbb{R}^d \rightrightarrows \mathbb{R}^d \times \mathbb{R}^d$ defined as

$$\mathbb{F}(x,y) = \begin{bmatrix} x \\ y \end{bmatrix} + \underbrace{\begin{bmatrix} \mathbb{A}(x) \\ \mathbb{B}(y) \end{bmatrix}}_{=\mathbb{E}(x,y)} + \mathbb{N}_C(x,y).$$

Pick any $x_0 \in \operatorname{dom} A \cap \operatorname{int} \operatorname{dom} B$. Then there is an $\varepsilon > 0$ such that for any $\delta_1, \delta_2 \in \mathbb{R}^n$ satisfying $\|\delta_1\| < \varepsilon$ and $\|\delta_2\| < \varepsilon$, we have $x_0 + \delta_1 - \delta_2 \in \operatorname{dom} B$. Then

$$(x_0 - y, x_0 + \delta_1 - \delta_2 - y) \in \operatorname{dom} \mathbb{E} - \operatorname{dom} \mathbb{N}_C$$

for any $\|\delta_1\| < \varepsilon$, $\|\delta_2\| < \varepsilon$, and $y \in \mathbb{R}^n$. We let $y = x_0 - \delta_2$ to get

$$(\delta_2, \delta_1) \in \operatorname{dom} \mathbb{E} - \operatorname{dom} \mathbb{N}_C$$

for any $\|\delta_1\| < \varepsilon$ and $\|\delta_2\| < \varepsilon$. So $0 \in \operatorname{int}(\operatorname{dom} \mathbb{E} - \operatorname{dom} \mathbb{N}_C)$, and Theorems 10 and 8 tell us range $(\mathbb{F}) = \mathbb{R}^n \times \mathbb{R}^n$.

For any $u \in \mathbb{R}^n$, there is an $x \in \mathbb{R}^n$ such that

$$\begin{bmatrix} u \\ 0 \end{bmatrix} \in \begin{bmatrix} (\mathbb{A} + \mathbb{I})x \\ (\mathbb{B} + \mathbb{I})x \end{bmatrix} + \begin{bmatrix} v \\ -v \end{bmatrix}, \quad \begin{bmatrix} v \\ -v \end{bmatrix} \in \mathbb{N}_C(x).$$

Left-multiplying by

$$\begin{bmatrix} \mathbb{I} & \mathbb{I} \end{bmatrix}$$

gives us

$$u \in (\mathbb{A} + \mathbb{B} + 2\mathbb{I})x.$$

So range $(\mathbb{A} + \mathbb{B} + 2\mathbb{I}) = \mathbb{R}^n$, and $\mathbb{A} + \mathbb{B}$ is maximal by Theorem 9. □

Theorem 12 Let $\mathbb{A}\colon \mathbb{R}^n \rightrightarrows \mathbb{R}^n$ be maximal monotone and $M \in \mathbb{R}^{n \times m}$. If $\operatorname{int} \operatorname{dom} \mathbb{A} \cap \mathcal{R}(M) \neq \emptyset$. then $M^\intercal \mathbb{A} M\colon \mathbb{R}^m \rightrightarrows \mathbb{R}^m$ is maximal monotone.

The proof is similar to the proof of Theorem 11. See Exercise 10.8.

10.3 MAXIMALITY AND EXTENSION THEOREMS

Let P be a property of an operator such as monotonicity, θ-averagedness, or L-Lipschitz continuity. We say an operator $\mathbb{A}\colon \mathbb{R}^n \rightrightarrows \mathbb{R}^n$ is "maximal P" if there is no proper extension $\bar{\mathbb{A}}$ with property P. To clarify, if $\mathbb{A}$ is already maximal P, its maximal P "extension" $\bar{\mathbb{A}}$ is not proper, i.e., $\mathbb{A} = \bar{\mathbb{A}}$. In this section, we characterize maximal extensions of certain operator classes.

Whether a given operator can be extended while preserving certain properties is a classical question in analysis. Examples of classical extension results include the Hahn–Banach theorem, which states that a linear operator on a subspace $V \subseteq \mathbb{R}^n$ has an extension to all of $\mathbb{R}^n$ with the same norm, and the Kirszbraun–Valentine theorem, which states that an L-Lipschitz operator on a subset $S \subseteq \mathbb{R}^n$ has an L-Lipschitz extension to all of $\mathbb{R}^n$.

Theorem 13 A monotone operator has a maximal monotone extension.

Proof. Let $\mathbb{A}\colon \mathbb{R}^n \rightrightarrows \mathbb{R}^n$ be monotone and let

$$\mathcal{P} = \{\mathbb{B}\colon \mathbb{R}^n \rightrightarrows \mathbb{R}^n \mid \mathbb{B} \text{ is monotone and } \operatorname{Gra}\mathbb{A} \subseteq \operatorname{Gra}\mathbb{B}\},$$

which is nonempty. We impose the partial order on $\mathcal{P}$ with $\mathbb{B}_1 \preceq \mathbb{B}_2$ if and only if $\operatorname{Gra}\mathbb{B}_1 \subseteq \operatorname{Gra}\mathbb{B}_2$ for all $\mathbb{B}_1, \mathbb{B}_2 \in P$. Every chain C in $\mathcal{P}$ has the upper bound $\bar{\mathbb{B}} \in \mathcal{P}$ given by

$$\operatorname{Gra}\bar{\mathbb{B}} = \bigcup_{\mathbb{B}\in C} \operatorname{Gra}\mathbb{B}.$$

By Zorn's lemma, there is a maximal element $\bar{\mathbb{A}}$ in $\mathcal{P}$. This element $\bar{\mathbb{A}}$ extends $\mathbb{A}$ by the definition of $\mathcal{P}$ and cannot be properly extended as it is maximal in $\mathcal{P}$. □

Theorem 14 For $\mu > 0$, a μ-strongly monotone operator has a maximal μ-strongly monotone extension. Furthermore, if $\mathbb{A}\colon \mathbb{R}^n \rightrightarrows \mathbb{R}^n$ is μ-strongly monotone, then $\mathbb{A}$ is maximal μ-strongly monotone if and only if $\operatorname{range}(\mathbb{A}) = \mathbb{R}^n$.

Proof. Since μ-strong monotonicity of $\mathbb{A}$ is defined as

$$\langle \mathbb{A}x - \mathbb{A}y, x - y\rangle \geq \mu\|x - y\|^2 \qquad \forall x, y \in \mathbb{R}^n,$$

$\mathbb{A}$ is μ-strongly monotone if and only if $\mathbb{B} = \mathbb{A} - \mu\mathbb{I}$ is monotone.

Extending $\mathbb{A}$ and $\mathbb{B}$ are equivalent in the following sense. If $\bar{\mathbb{A}}$ is a μ-strongly monotone extension of $\mathbb{A}$, then $\bar{\mathbb{A}} - \mu\mathbb{I}$ is a monotone extension of $\mathbb{B}$. If $\bar{\mathbb{B}}$ is a monotone extension of $\mathbb{B}$, then $\bar{\mathbb{B}} + \mu\mathbb{I}$ is a μ-strongly monotone extension of $\mathbb{A}$. By Theorem 13, $\mathbb{B}$ has a maximal monotone extension $\bar{\mathbb{B}}$, and $\mathbb{A}$ has a maximal μ-strongly monotone extension $\bar{\mathbb{B}} + \mu\mathbb{I}$.

Moreover, $\mathbb{A}$ is maximal μ-strongly monotone if and only if $\mathbb{B}$ is maximal monotone. By Theorems 8 and 9, $\mathbb{B}$ is maximal monotone if and only if $\operatorname{range}(\mathbb{A}) = \operatorname{range}(\mathbb{B}+\mu\mathbb{I}) = \mathbb{R}^n$. Finally, chaining the equivalences provides the second stated result. □

Theorem 15 For $\beta > 0$, a β-cocoercive operator has a maximal β-cocoercive extension. Furthermore, if $\mathbb{A}\colon \mathbb{R}^n \rightrightarrows \mathbb{R}^n$ is β-cocoercive, then $\mathbb{A}$ is maximal β-cocoercive if and only if $\operatorname{dom}\mathbb{A} = \mathbb{R}^n$.

Proof. Note $\mathbb{A}$ is β-cocoercive if and only if $\mathbb{A}^{-1}$ is β-strongly monotone.

Extending $\mathbb{A}$ and $\mathbb{A}^{-1}$ are equivalent in the following sense. If $\bar{\mathbb{A}}$ is a β-cocoercive extension of $\mathbb{A}$, then $\bar{\mathbb{A}}^{-1}$ is a β-strongly monotone extension of $\mathbb{A}^{-1}$. If $\overline{\mathbb{A}^{-1}}$ is a β-strongly monotone extension of $\mathbb{A}^{-1}$, then $(\overline{\mathbb{A}^{-1}})^{-1}$ is a β-cocoercive extension of $\mathbb{A}$. By Theorem 14, $\mathbb{A}^{-1}$ has a maximal β-strongly monotone extension $\overline{\mathbb{A}^{-1}}$, and $\mathbb{A}$ has a maximal β-cocoercive extension $(\overline{\mathbb{A}^{-1}})^{-1}$.

Moreover, $\mathbb{A}$ is maximal β-cocoercive if and only if $\mathbb{A}^{-1}$ is maximal β-strongly monotone. By Theorem 14, $\mathbb{A}^{-1}$ is maximal β-strongly monotone if and only if $\operatorname{range}(\mathbb{A}^{-1}) =$

$\mathbb{R}^n$, which holds if and only if $\mathrm{dom}(\mathbb{A}) = \mathbb{R}^n$. Finally, chaining the equivalences provides the second stated result. □

Remember that a β-cocoercive operators must be single-valued. By Theorem 15, [$\mathbb{A}\colon \mathbb{R}^n \rightrightarrows \mathbb{R}^n$ is maximal β-cocoercive] is equivalent to [$\mathbb{A}\colon \mathbb{R}^n \to \mathbb{R}^n$ is β-cocoercive] since $\mathbb{A}\colon \mathbb{R}^n \to \mathbb{R}^n$ implies $\mathrm{dom}\,\mathbb{A} = \mathbb{R}^n$. For the sake of conciseness, we usually avoid the former expression.

Theorem 16 For $L > 0$, an L-Lipschitz operator has a maximal L-Lipschitz extension. Furthermore, if $\mathbb{A}\colon \mathbb{R}^n \rightrightarrows \mathbb{R}^n$ is L-Lipschitz, then $\mathbb{A}$ is maximal L-Lipschitz if and only if $\mathrm{dom}\,\mathbb{A} = \mathbb{R}^n$.

This result is known as the Kirszbraun–Valentine theorem. We defer the proof to Exercise 10.10.

BIBLIOGRAPHICAL NOTES

Minty's original proof of the surjectivity theorem [Min62] relied on the Kirszbraun–Valentine theorem [Kir34, Val43, Val45] rather than the Fitzpatrick function. We instead prove the Minty surjectivity theorem with the Fitzpatrick function and obtain the Kirszbraun–Valentine theorem as a consequence in Exercise 10.10.

Rockafellar first proved the sum of two maximal monotone operators is maximal under regularity conditions, but the proof was quite complicated [Roc70c]. The presented proof outlined in Exercise 10.9 is due to Simons, Zălinescu, and Borwein [SZ04, Zăl05, SZ05, Bor06]. Maximality of $M^\intercal AM$ under the regularity condition stated in Theorem 12 was first established by Robinson [Rob99].

The idea of representing maximal monotone operators with convex functions was first explored by Krauss [Kra85]. Fitzpatrick soon provided a different construction, which we now call the Fitzpatrick function [Fit88]. The usefulness of Fitzpatrick's construction, however, was discovered much later by Penot, Simons, and Zălinescu [Pen03, Pen04, SZ04, Zăl05, SZ05].

There is a large body of work in monotone operator theory studying extension theorems beyond the Hahn–Banach theorem or the Kirszbraun–Valentine theorem. The specific extension theorems of §10.3 were presented by Ryu, Taylor, Bergeling, and Giselsson [RTBG20], but the core ideas are present in prior work of Bauschke and Wang [BW10] and Minty [Min62].

One may wonder whether $\mathbf{F}_{\partial f}$ with CCP f has a simple form. Bauschke, McLaren, and Sendov characterizes $\mathbf{F}_{\partial f}$ in some special cases including

$$\mathbf{F}_{\partial\|\cdot\|}(x,u) = \begin{cases} \|x\| & \text{if } \|u\| \le 1 \\ \infty & \text{otherwise.} \end{cases}$$

$$\mathbf{F}_{\partial(1/2)\|\cdot\|^2}(x,u) = \frac{1}{4}\|x+u\|^2$$

$$\mathbf{F}_{\partial\delta_C}(x,u) = \delta_C(x) + \delta_C^*(u),$$

where C is a nonempty closed convex set, but $\mathbf{F}_{\partial f}$ seems to be a complicated object in general [BMS06].

EXERCISES

10.1 *Basic exercises on maximality.* Let $\mathbb{A}\colon \mathbb{R}^n \rightrightarrows \mathbb{R}^n$ be maximal monotone. Show:
(a) $\mathbb{A}^{-1}$ is maximal monotone,
(b) $M^\intercal \mathbb{A} M$ is maximal monotone when $M \in \mathbb{R}^{n\times n}$ be invertible,
(c) $[\langle u - v, x - y\rangle \geq 0$ for all $(x,u) \in \mathbb{A}]$ if and only if $[(y,v) \in \mathbb{A}]$,
(d) $\inf_{(x,u)\in\mathbb{A}} \langle u, x - y\rangle \leq 0$ for all $y \in \mathbb{R}^n$, and $\inf_{(x,u)\in\mathbb{A}}\langle u, x - y\rangle = 0$ if and only if $y \in \operatorname{Zer}\mathbb{A}$.
Remark. We already know that $M^\intercal \mathbb{T} M$ is maximal when $\mathcal{R}(M) \cap \operatorname{int}\operatorname{dom}\mathbb{T} \neq \emptyset$, and this immediately implies (b). However, provide a direct proof for (b) that does not rely on this result. This problem does not require any of the new tools from this chapter.

10.2 *Nonexpansiveness and monotonicity.* Show that if $\mathbb{T}$ is maximal nonexpansive (i.e., nonexpansive and $\operatorname{dom}\mathbb{T} = \mathbb{R}^n$ per Theorem 16) then $\mathbb{A} = \left(\frac{1}{2}\mathbb{T} + \frac{1}{2}\mathbb{I}\right)^{-1} - \mathbb{I}$ is maximal monotone.
Remark. Conversely, if $\mathbb{A}$ is a maximal monotone operator, then $2\mathbb{J}_\mathbb{A} - \mathbb{I}$ is maximal nonexpansive. Therefore, the transformation $\mathbb{A} \mapsto 2\mathbb{J}_\mathbb{A} - \mathbb{I}$ and its inverse $\mathbb{T} \mapsto \left(\frac{1}{2}\mathbb{T} + \frac{1}{2}\mathbb{I}\right)^{-1} - \mathbb{I}$ provide a one-to-one correspondence between maximal monotone operators and maximal nonexpansive operators.

10.3 *Closed graph theorem for maximal monotone operators.* Let $\mathbb{A}\colon \mathbb{R}^n \rightrightarrows \mathbb{R}^n$ be maximal monotone. Show $\mathbb{A}$ is *upper hemicontinuituous*, i.e., show that if $x^k \to x^\infty$, $u^k \to u^\infty$, and $u^k \in \mathbb{A}x^k$, then $u^\infty \in \mathbb{A}x^\infty$. (Upper hemicontinuity of $\mathbb{A}$ is equivalent to $\operatorname{Gra}\mathbb{A} \subset \mathbb{R}^n \times \mathbb{R}^n$ being a closed set.)
Hint. The proof can be done in one line using $\mathbf{F}_\mathbb{A}(x^\infty, u^\infty) \leq \liminf_{k\to\infty} \mathbf{F}_\mathbb{A}(x^k, u^k)$ and Lemma 3.

10.4 *Method of multipliers primal solution convergence without strict convexity.* Consider the method of multipliers under the stated conditions. Show that any accumulation point of $x^0, x^1, \ldots$ is a primal solution.
Hint. Use Exercise 10.3 and note Exercise 2.18.
Remark. The stated conditions are f is CCP, $\mathcal{R}(A^\intercal) \cap \operatorname{ri}\operatorname{dom} f^* \neq \emptyset$, a dual solution exists, $\alpha > 0$, and $\mathbf{L}_\alpha(x,u) = f(x) + \langle u, Ax - b\rangle + \frac{\alpha}{2}\|Ax - b\|^2$.

10.5 *Maximality by surjectivity.* Consider

$$\mathbf{L}(x,v) = f(x) + \langle v, Ax - b\rangle,$$

is the Lagrangian of (1.5). Assume f is CCP. Using Theorem 9, show that ∂L is maximal.
Hint. Use (2.7).

10.6 *Partial inverse.* Given an operator $\mathbb{A}\colon \mathbb{R}^{m+n} \rightrightarrows \mathbb{R}^{m+n}$, the *partial inverse* of $\mathbb{A}$ is the operator $\mathbb{A}^{1,-1}\colon \mathbb{R}^{m+n} \rightrightarrows \mathbb{R}^{m+n}$ defined with

$$\operatorname{Gra}\mathbb{A}^{1,-1} = \{((x,v),(u,y)) \in \mid (u,v) \in A(x,y)\}.$$

Note that $\mathbb{A}^{1,-1} = A$ if $n = 0$ and $\mathbb{A}^{1,-1} = \mathbb{A}^{-1}$ is $m = 0$. Show that if $\mathbb{A}$ is maximal monotone, then $\mathbb{A}^{1,-1}$ is maximal monotone.

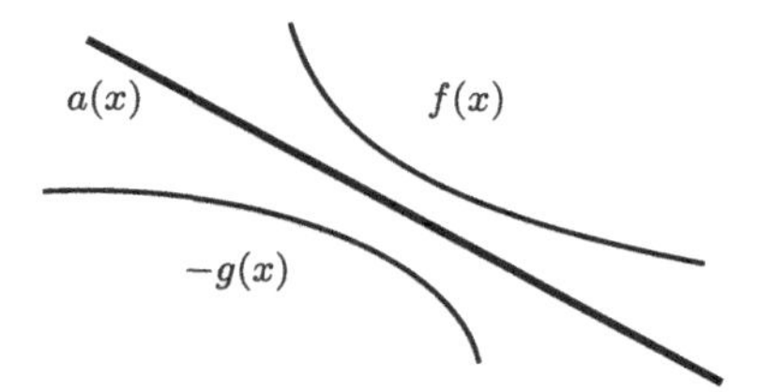

Figure 10.1 Illustration of the Hahn–Banach sandwich theorem

10.7 *Maximality of saddle subdifferential with partial inverse.* Let $L(x, v)$ be a convex-concave saddle function. Then

$$\mathbb{F}(x, y) = \sup_{v} \{\mathbf{L}(x, v) + \langle y, v \rangle\}$$

is called the partial conjugate of $\mathbf{L}$. Show that if $\mathbb{F}$ is CCP, then $\partial \mathbf{L}$ is maximal.
Hint. Show that $\partial \mathbf{L}$ is the partial inverse of $\partial \mathbb{F}$.

10.8 Prove Theorem 12.
Hint. Consider $C = \{(x, Mx) \mid x \in \mathbb{R}^n\}$ and $\mathbb{N}_C$. Left-multiply by $[I\ M^\intercal]$.

10.9 Prove Theorem 10.
Hint. The *Hahn–Banach sandwich theorem* states that if f and g are convex and $f \geq -g$, and $0 \in \operatorname{int}(\operatorname{dom} f - \operatorname{dom} g)$, then there is an affine function such that

$$f(x) \geq a(x) \geq -g(x).$$

See Figure 10.1 for an illustration. The *Fenchel–Young inequality* states that

$$\langle x, u \rangle \leq f(x) + f^*(u)$$

for any $x, u \in \mathbb{R}^n$ with equality if and only if $(x, u) \in \partial f$. Define $f_1(x) = f(x) + \frac{1}{2}\|x\|^2$. Show

$$\mathbf{F}_{\mathbb{A}}(x, u) \geq \langle x, u \rangle \geq -f_1(x) - f_1^*(-u).$$

By the Hahn–Banach sandwich theorem, there is $\mu, \nu \in \mathbb{R}^d$ such that

$$\mathbf{F}_{\mathbb{A}}(x, u) + f_1(y) + f_1^*(-v) \geq \langle \nu, x - y \rangle + \langle \mu, u - v \rangle$$

for any $x, u, y, v \in \mathbb{R}^n$. This implies

$$\langle x - \mu, u - \nu \rangle \geq -(\langle y, v \rangle + f_1(y) + f_1^*(-v)) + \langle y - \mu, v - \nu \rangle \tag{10.2}$$

for any $(x, u) \in \mathbb{A}$ and any $y, v \in \mathbb{R}^n$. The choice of (y, v) such that $v = \nu$ and $-v \in \partial f_1(y)$, possible by Theorems 7 and 8, shows

$$\langle x - \mu, u - \nu \rangle \geq 0$$

for all $(x, u) \in \mathbb{A}$. This tells us $(\mu, \nu) \in \mathbb{A}$. Plugging $x = \mu$ and $u = \nu$ into (10.2), we get

$$0 \geq \langle y, -v \rangle - f_1(y) + \langle \mu, -v \rangle - f_1^*(-v) + \langle \mu, \nu \rangle$$

for all $y, v \in \mathbb{R}^n$. Maximizing over y and v gives us

$$0 \geq f_1^*(-\nu) + f_1(\mu) + \langle \mu, \nu \rangle$$

This implies $(\mu, -\nu) \in \partial f_1$. This implies that $0 \in (\mathbb{I} + \mathbb{A} + \partial f)(\mu)$.

10.10 *Maximal Lipschitz operators.* Prove Theorem 16.
Hint. Use the fact that $\mathbb{A}$ is monotone if and only if $2\mathbb{J}_{\mathbb{A}} - I$ is nonexpansive.

10.11 *Maximal and strong monotone* $\Leftrightarrow$ *maximal strong monotone.* Let $\mu > 0$. Show that an operator is [(maximal monotone) and (μ-strongly monotone)] if and only if it is [maximal (μ-strongly monotone)].
Hint. The $\Rightarrow$ implication follows from the definitions (but you should explain why). For the $\Leftarrow$ implication, the question is whether it is possible for a maximal μ-strongly monotone operator $\mathbb{A}$ to have a proper extension $\bar{\mathbb{A}}$ that is monotone but not μ-strongly monotone. Use the fact that $\mathbb{A}^{-1}\colon \mathbb{R}^n \to \mathbb{R}^n$ is a continuous monotone operator by Theorem 15.
Remark. Fortunately [maximal and strong monotone] and [maximal strong monotone] mean the same thing, and there is no potential for confusion.

10.12 *Maximal and cocoercive* $\Leftrightarrow$ *maximal cocoercive.* Let $\beta > 0$. Show that an operator is [(maximal monotone) and (β-cocoercive)] if and only if it is [maximal (β-cocoercive)].
Hint. Use Exercise 10.11.

11 Distributed and Decentralized Optimization

In this chapter, we study distributed and decentralized methods that allow computational agents communicating over a network to collaboratively solve an optimization problem. Specifically, we solve

$$\underset{x\in\mathbb{R}^p}{\text{minimize}} \quad r(x) + \frac{1}{n}\sum_{i=1}^{n}(f_i(x) + h_i(x)), \tag{11.1}$$

where $r, f_1, \ldots, f_n$ are CCP (and proximable) and $h_1, \ldots, h_n$ are CCP and differentiable, in a computational setup where a server performs computation with r, agents $i = 1, \ldots, n$ each perform local computation with f_i and h_i, and the server and agents communicate over a network to find the (shared) solution $x^\star$. We distinguish *distributed* and *decentralized* methods as follows: distributed methods perform computation over a network (a broader class), while decentralized methods do so without central coordination (a subclass).

One application of distributed optimization is solving extremely large optimization problems that require the computing power of a cluster of computers communicating over a network. Another application is controlling a fleet of autonomous vehicles (such as drones) or a wireless sensor network, where individual agents make real-time decisions based on data gathered by itself and other agents. Decentralized methods are effective for these setups, as they reduce the high cost and latency of communication.

11.1 DISTRIBUTED OPTIMIZATION WITH CENTRALIZED CONSENSUS

In this section, we study distributed optimization methods based on the consensus technique of §2.7.4. We first present two base distributed methods and then present the primal and dual decomposition techniques, which allow us to transform problems into forms eligible for the base distributed methods. The relatively simple, centralized communication structure of these methods allows us to analyze them with the tools of §2.

Throughout this section, we write $C = \{(x_1, \ldots, x_n) \,|\, x_1 = \cdots = x_n \in \mathbb{R}^p\}$ for the consensus set, an unbound index i is assumed to range from $i = 1, \ldots, n$, and we write the mean over $i = 1,\dots,n$ with a bar notation as in $\bar{x}^k = (1/n)(x_1^k + \cdots + x_n^k)$ and $\bar{g}^k = (1/n)(g_1^k + \cdots + g_n^k)$.

11.1.1 Base Distributed Methods

Distributed Proximal Gradient Method Consider the problem

$$\underset{x\in\mathbb{R}^p}{\text{minimize}} \quad r(x) + \frac{1}{n}\sum_{i=1}^n h_i(x),$$

where r is a CCP function and $h_1,\dots,h_n$ are differentiable CCP functions. Using the consensus technique, we obtain the equivalent problem

$$\begin{array}{ll}\underset{x_1,\dots,x_n\in\mathbb{R}^p}{\text{minimize}} & r(x_1) + \frac{1}{n}\sum_{i=1}^n h_i(x_i)\\ \text{subject to} & (x_1,\dots,x_n)\in C.\end{array}$$

Apply FBS and use Exercise 2.29 to get

$$\begin{aligned} x_i^{k+1/2} &= x^k - \alpha\nabla h_i(x^k)\\ x^{k+1} &= \mathrm{Prox}_{\alpha r}\left(\frac{1}{n}\sum_{i=1}^n x_i^{k+1/2}\right),\end{aligned}$$

which is equivalent to

$$\begin{aligned} g_i^k &= \nabla h_i(x^k)\\ x^{k+1} &= \mathrm{Prox}_{\alpha r}\left(x^k - \alpha\bar{g}^k\right).\end{aligned}$$

We call this method the *distributed proximal gradient method*. Assume a solution exists, $h_1,\dots,h_n$ are L-smooth, and $\alpha\in(0,2/L)$. Then $x^k\to x^\star$.

This method is distributed, as it has a distributed implementation that alternates between local computation and centralized communication in a setup with n computational agents and a central node as in Figure 11.1: (i) each agent independently computes $g_i^k = \nabla h_i(x^k)$ and (ii) the agents send g_i^k to the central agent, the central agent computes their average and performs the proximal gradient step involving $\mathrm{Prox}_{\alpha r}$, and x^{k+1} is broadcast to all individual agents. The centralized communication and computation of the average of g_i^k in step (ii) is called a reduction operation in the parallel computing literature.

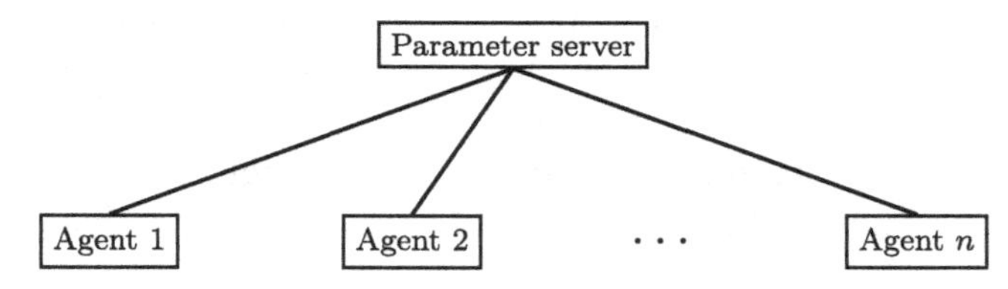

Figure 11.1 Depiction of a parameter-server network model. This network structure allows efficient distributed centralized optimization.

Distributed (Centralized) ADMM/DRS Consider the problem

$$\underset{x\in\mathbb{R}^p}{\text{minimize}} \quad \sum_{i=1}^n f_i(x),$$

where $f_1,\ldots,f_n$ are CCP functions. Using a variant of the consensus technique, we obtain the equivalent problem

$$\begin{array}{ll} \underset{\substack{x_1,\ldots,x_n\in\mathbb{R}^p\\ y\in\mathbb{R}^p}}{\text{minimize}} & \sum_{i=1}^n f_i(x_i) \\ \text{subject to} & x_i = y. \end{array}$$

Rewrite the constraints to fit ADMM's form

$$\begin{array}{ll} \underset{\substack{x_1,\ldots,x_n\in\mathbb{R}^p\\ y\in\mathbb{R}^p}}{\text{minimize}} & \sum_{i=1}^n f_i(x_i) \\ \text{subject to} & \begin{bmatrix} I & & \\ & I & \\ & & \ddots & \\ & & & I \end{bmatrix} \begin{bmatrix} x_1\\ x_2\\ \vdots\\ x_n \end{bmatrix} + \begin{bmatrix} -I\\ -I\\ \vdots\\ -I \end{bmatrix} y = 0, \end{array}$$

and apply ADMM to get

$$\begin{aligned} x_i^{k+1} &= \underset{x_i\in\mathbb{R}^p}{\text{argmin}} \left\{ f_i(x_i) + \langle u_i^k, x_i - y^k\rangle + \frac{\alpha}{2}\|x_i - y^k\|^2 \right\} \\ y^{k+1} &= \underset{y}{\text{argmin}} \left\{ \sum_{i=1}^n \langle u_i^k, x_i^{k+1} - y\rangle + \frac{\alpha}{2}\|x_i^{k+1} - y\|^2 \right\} = \frac{1}{n}\sum_{i=1}^n \left(x_i^{k+1} + \frac{1}{\alpha}u_i^k \right) \\ u_i^{k+1} &= u_i^k + \alpha(x_i^{k+1} - y^{k+1}). \end{aligned}$$

Simplify the iteration by noting that $u_1^k,\ldots,u_n^k$ has mean 0 after the initial iteration and eliminating y^k:

$$\begin{aligned} x_i^{k+1} &= \mathrm{Prox}_{(1/\alpha)f_i}\left(\bar{x}^k - (1/\alpha)u_i^k\right) \\ u_i^{k+1} &= u_i^k + \alpha(x_i^{k+1} - \bar{x}^{k+1}). \end{aligned}$$

We call this method distributed (centralized) ADMM. Convergence follows from the convergence of ADMM.

Distributed ADMM is also distributed, as it has a distributed implementation that alternates local computation and centralized communication: (i) each agent independently performs the u_i^k- and x_i^{k+1}-updates with local computation and (ii) the agents coordinate to compute $\bar{x}^{k+1}$ with a reduction.

Alternatively and equivalently, we can apply DRS to the problem obtained with the consensus technique

$$\underset{x_1,\ldots,x_n\in\mathbb{R}^p}{\text{minimize}} \quad \delta_C(x_1,\ldots,x_n) + \sum_{i=1}^n f_i(x_i)$$

to get

$$\begin{aligned} x_i^{k+1/2} &= \mathrm{Prox}_{(1/\alpha)f_i}(z_i) \\ z_i^{k+1} &= z_i^k - \bar{z}^k + 2\bar{x}^{k+1/2} - x_i^{k+1/2}. \end{aligned}$$

This is equivalent to the previously stated distributed ADMM. See Exercise 11.7.

11.1.2 Primal Decomposition Technique

The primal decomposition technique obtains a *master problem* through minimizing away local variables. This is a special case of the infimal postcomposition technique of §3.1.

Consider the problem

$$\underset{\substack{x_i \in \mathbb{R}^{p_i} \\ y \in \mathbb{R}^q}}{\text{minimize}} \quad r(y) + \frac{1}{n}\sum_{i=1}^{n} f_i(x_i, y),$$

where $r, f_1, \ldots, f_n$ are CCP. For a fixed $y \in \mathbb{R}^q$, the minimization over $x_1, \ldots, x_n$ decomposes into n embarrassingly parallel tasks. We call $x_1, \ldots, x_n$ *local variables* and y the *coupling variable*. With

$$\phi_i(y) = \inf_{x_i \in \mathbb{R}^{p_i}} f_i(x_i, y),$$

we obtain the equivalent master problem

$$\underset{y \in \mathbb{R}^q}{\text{minimize}} \quad r(y) + \frac{1}{n}\sum_{i=1}^{n} \phi_i(y),$$

which can be solved with methods chosen based on the properties of $r, \phi_1, \ldots, \phi_n$.

For example, when r is proximable and $\phi_1, \ldots, \phi_n$ are smooth, we can apply the proximal gradient method to solve the master problem:

$$y^{k+1} = \text{Prox}_{\alpha r}\left(y^k - \alpha \frac{1}{n}\sum_{i=1}^{n} \nabla \phi_i(y^k)\right).$$

Using Exercise 11.2, we express the method as

$$\begin{aligned}
x_i^\star(y^k) &\in \underset{x_i \in \mathbb{R}^{p_i}}{\text{argmax}} f_i(x_i, y^k) \\
(0, g_i^k) &\in \partial f_i(x_i^\star(y^k), y^k) \\
y^{k+1} &= \text{Prox}_{\alpha r}\left(y^k - \alpha \bar{g}^k\right),
\end{aligned}$$

provided that the argmins exist. This method has a distributed implementation, as the subproblems for computing g_i^k can be distributed.

When r is proximable but $\phi_1, \ldots, \phi_n$ are not smooth, we can apply the proximal subgradient method of §7. See Exercises 11.9 and 11.10 for using distributed ADMM/DRS.

Example 11.1 *Common bound problem.* Consider the setup where agents $i = 1, \ldots, n$ each reduce its cost $f_i(x_i)$ subject to the constraint $g_i(x_i) \preceq y$, where $\preceq$ denotes element-wise inequality, while paying a common cost $r(y)$:

$$\begin{array}{ll}
\underset{\substack{x_i \in \mathbb{R}^{p_i} \\ y \in \mathbb{R}^q}}{\text{minimize}} & r(y) + \sum_{i=1}^{n} f_i(x_i) \\
\text{subject to} & g_i(x_i) \preceq y.
\end{array}$$

This problem is equivalent to the master problem

$$\underset{y\in\mathbb{R}^q}{\text{minimize}} \quad r(y) + \frac{1}{n}\sum_{i=1}^{n} \phi_i(y),$$

where

$$\phi_i(y) = \inf_{x_i\in\mathbb{R}^{p_i}} \left\{ nf_i(x_i) + \delta_{\{(x_i,y)\,|\,g_i(x_i)\preceq y\}}(x_i,y) \right\}.$$

See Exercise 11.6 for evaluating the subdifferential $\partial\phi_i(y)$.

Example 11.2 *Resource sharing problem.* Consider the setup where agents $i = 1,\dots,n$ each reduces its cost $f_i(x_i)$ subject to a total resource constraint $\sum_{i=1}^n g_i(x_i) \preceq y$, where $\preceq$ denotes element-wise inequality, while paying a common cost $r(y)$:

$$\begin{array}{ll} \underset{\substack{x_i\in\mathbb{R}^{p_i}\\ y\in\mathbb{R}^q}}{\text{minimize}} & r(y) + \sum_{i=1}^{n} f_i(x_i) \\ \text{subject to} & \sum_{i=1}^{n} g_i(x_i) \preceq y. \end{array}$$

This problem is equivalent to the master problem

$$\underset{y_1,\dots,y_n\in\mathbb{R}^q}{\text{minimize}} \quad r(y_1 + \cdots + y_n) + \frac{1}{n}\sum_{i=1}^{n} \phi_i(y_i),$$

where $\phi_i(y_i) = \inf_{x_i\in\mathbb{R}^{p_i}} \{nf_i(x_i) + \delta_{\{(x_i,y_i)\,|\,g_i(x_i)\preceq y_i\}}(x_i,y_i)\}$. The solutions $y_1^\star,\dots,y_n^\star$ specify the optimal allocation of resources among the agents. By Exercise 1.8, if $r(y)$ is proximable, then so is $r(y_1 + \cdots + y_n)$. See Exercise 11.6 for evaluating $\partial\phi_i(y_i)$.

11.1.3 Dual Decomposition Technique

The dual decomposition technique obtains a master problem by taking the dual. This is essentially the same as the dualization technique of §3.2, but the focus is on obtaining a sum structure so that we can apply the base distributed methods.

Dual Decomposition with Coupling Variables Consider the problem

$$\underset{\substack{x_i\in\mathbb{R}^{p_i}\\ y\in\mathbb{R}^q}}{\text{minimize}} \quad \sum_{i=1}^{n} f_i(x_i,y),$$

where $f_1,\dots,f_n$ are CCP. This is the same problem as in the primal decomposition setup but with $r = 0$. The equivalent primal problem

$$\begin{array}{ll} \underset{\substack{x_1,\dots,x_n\in\mathbb{R}^p\\ z_1,\dots,z_n\in\mathbb{R}^q\\ y\in\mathbb{R}^q}}{\text{minimize}} & \sum_{i=1}^{n} f_i(x_i,z_i) \\ \text{subject to} & z_i = y \end{array}$$

is generated by the Lagrangian

$$\mathbf{L}(x_1,\ldots,x_n,y,z_1,\ldots,z_n,v_1,\ldots,v_n) = \sum_{i=1}^{n} \left(f_i(x_i,z_i) - \langle v_i, z_i - y \rangle \right).$$

With

$$\inf_{y\in\mathbb{R}^q} \sum_{i=1}^{n} \langle v_i, y\rangle = \begin{cases} 0, & \text{if } v_1 + \cdots + v_n = 0 \\ -\infty, & \text{otherwise} \end{cases}$$

and

$$\psi_i(v_i) = \sup_{\substack{x_i\in\mathbb{R}^p \\ z_i\in\mathbb{R}^q}} \left\{ -f_i(x_i,z_i) + \langle v_i, z_i\rangle \right\},$$

we obtain the master dual problem

$$\underset{v_1,\ldots,v_n\in\mathbb{R}^q}{\text{maximize}} \quad -\delta_{C^\perp}(v_1,\ldots,v_n) - \sum_{i=1}^{n} \psi_i(v_i),$$

where $C^\perp = \{(v_1,\ldots,v_n) \,|\, v_1 + \cdots + v_n = 0\}$. (See Exercise 11.8 for a discussion of $C^\perp$.) The master problem can be solved with methods chosen based on the properties of $\psi_1,\ldots,\psi_n$.

For example, when $\psi_1,\ldots,\psi_n$ are smooth, we can apply the projected gradient method

$$\begin{aligned} g_i^k &= \nabla\psi_i(v_i^k) \\ v_i^{k+1} &= v_i^k - \alpha(g_i^k - \bar{g}^k) \end{aligned}$$

provided that we initialize the iteration with $(v_1^0,\ldots,v_n^0) \in C^\perp$. Using Exercise 11.3, we express the method as

$$\begin{aligned} (x_i^\star(v_i^k), g_i^k) &\in \underset{\substack{x_i\in\mathbb{R}^p \\ g_i\in\mathbb{R}^q}}{\operatorname{argmin}} \left\{ -f_i(x_i,g_i) + \langle v_i^k, g_i\rangle \right\} \\ v_i^{k+1} &= v_i^k - \alpha(g_i^k - \bar{g}^k), \end{aligned}$$

provided that the argmins exist and we initialize the iteration with $(v_1^0,\ldots,v_n^0) \in C^\perp$. This method has a distributed implementation, as the subproblems for computing g_i^k can be distributed. When $\phi_1,\ldots,\phi_n$ are not smooth, we can apply the projected subgradient method of §7. See Exercises 11.9 and 11.10 for using distributed ADMM/DRS.

Dual Decomposition with Inequality Constraints Consider the problem of Example 11.2:

$$\begin{array}{ll} \underset{\substack{x_i\in\mathbb{R}^{p_i} \\ y\in\mathbb{R}^q}}{\text{minimize}} & r(y) + \dfrac{1}{n}\displaystyle\sum_{i=1}^{n} f_i(x_i) \\ \text{subject to} & \displaystyle\sum_{i=1}^{n} g_i(x_i) \preceq y, \end{array}$$

where r is a CCP function on $\mathbb{R}^q$, $f_1,\dots,f_n$ are respectively CCP functions on $\mathbb{R}^{p_1},\dots,\mathbb{R}^{p_n}$, and $\leq$ denotes element-wise inequality. Assume $g_i\colon \mathbb{R}^{p_i} \to \mathbb{R}^q$ has the form $g_i = (g_{i,1},\dots,g_{i,q})$ with scalar-valued CCP functions $g_{i,1},\dots,g_{i,q}$ for $i = 1,\dots,n$. Assume $g_{i,j}\colon \mathbb{R}^{p_i} \to \mathbb{R}$ (i.e., does not output ∞) for $i = 1,\dots,n$ and $j = 1,\dots,q$. This primal problem is generated by the Lagrangian

$$\mathbf{L}(x_1,\dots,x_n,y,u) = r(y) - \langle u,y\rangle + \frac{1}{n}\sum_{i=1}^{n}(f_i(x_i) + \langle u, g_i(x_i)\rangle) - \delta_{\mathbb{R}^n_+}(u),$$

where $\mathbb{R}^q_+$ denotes the nonnegative orthant. With

$$\psi_i(u) = \begin{cases} \sup_{x_i\in\mathbb{R}^{p_i}} (\langle -u, g_i(x_i)\rangle - f_i(x_i)) & \text{if } u \geq 0 \\ \infty & \text{otherwise,} \end{cases}$$

we obtain the master dual problem

$$\underset{u\in\mathbb{R}^q}{\text{maximize}} \quad -r^*(u) - \delta_{\mathbb{R}^q_+}(u) - \frac{1}{n}\sum_{i=1}^{n}\psi_i(u).$$

The master problem can be solved with methods chosen based on the properties of $\psi_1,\dots\psi_n$.

For example, when r^* is proximable and $\psi_1,\dots\psi_n$ are smooth, we can apply DYS and Exercise 11.3 to get

$$\begin{aligned} u^{k+1/2} &= \Pi_{\mathbb{R}^q_+}\left(\zeta^k\right) \\ x_i^{k+1} &\in \underset{x_i\in\mathbb{R}^{p_i}}{\operatorname{argmin}} \left\{f_i(x_i) + \langle u^{k+1/2}, g_i(x_i)\rangle\right\} \\ u^{k+1} &= \operatorname{Prox}_{\alpha r^*}\left(2u^{k+1/2} - \zeta^k + \frac{\alpha}{n}\sum_{i=1}^{n} g_i(x_i^{k+1})\right) \\ \zeta^{k+1} &= \zeta^k + u^{k+1} - u^{k+1/2}. \end{aligned}$$

When $r = \delta_{\{b\}}$ and $\psi_1,\dots\psi_n$ are smooth, then $r^*(u) = \langle u,b\rangle$ and we can apply the proximal gradient method and Exercise 11.3 to get

$$\begin{aligned} x_i^{k+1} &\in \underset{x_i\in\mathbb{R}^{p_i}}{\operatorname{argmin}} \left\{f_i(x_i) + \langle u^k, g_i(x_i)\rangle\right\} \\ u^{k+1} &= \Pi_{\mathbb{R}^q_+}\left(u^k + \frac{\alpha}{n}\sum_{i=1}^{n}\left(g_i(x_i^{k+1}) - b\right)\right). \end{aligned}$$

When $r = \delta_{\{b\}}$ but $\psi_1,\dots\psi_n$ are not smooth, we can apply the projected subgradient method of §7.

Recovering the Primal Solution The dual decomposition technique constructs a master dual problem, which can be naturally solved with distributed methods. Under certain strict convexity assumptions, the solution to the primal problem can be recovered from the dual problem. See Exercise 2.6.

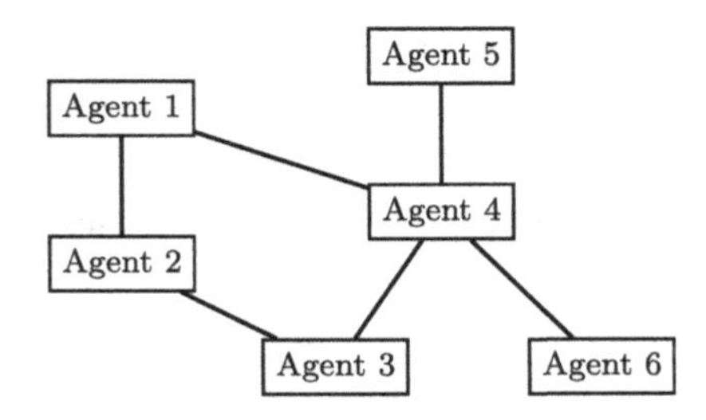

Figure 11.2 Depiction of a network without a central agent. Decentralized optimization is suitable for this network structure. We represent this network with the graph $G = (V, E)$, where $V = \{1, 2, 3, 4, 5, 6\}$ and $E = \{\{1, 2\}, \{1, 4\}, \{2, 3\}, \{3, 4\}, \{4, 5\}, \{4, 6\}\}$.

11.2 DECENTRALIZED OPTIMIZATION WITH GRAPH CONSENSUS

Decentralized optimization solves an optimization problem defined on a network without a central agent. In this section, we introduce the notion of graphs to represent the network and use them to derive and analyze decentralized optimization methods.

Networks and Graphs The word *graph* has two distinct meanings in mathematics. The first meaning, as in "we plot the graph $\sin(x)$ on a graphing calculator," concerns the relationship between the inputs and outputs of a function. The *graph* of an operator, which we denote as $\mathrm{Gra}\,\mathbb{A}$, and the scaled relative *graph* of §13 use this first meaning. In this chapter, we consider the second meaning, the use in discrete mathematics for representing networks.

A *graph* $G = (V, E)$, where V is the set of nodes and E is the set of edges, represents a network. Assume the network is finite and label the nodes 1 through n, i.e., $V = \{1, \ldots, n\}$. Assume the graph is undirected, i.e., an edge $\{i, j\} \in E$ is an unordered pair of distinct nodes i and j. Assume the graph has no self-loops, i.e., $\{i, i\} \notin E$ for all $i \in V$. Assume the graph is connected, i.e., for any $i, j \in V$ such that $i \neq j$, there is a sequence of edges

$$\{i, v_1\}, \{v_1, v_2\}, \ldots, \{v_{k-1}, v_k\}, \{v_k, j\} \in E$$

starting from i and ending at j. In this chapter, a node represents a computational agent that stores data and performs computation, and an edge $\{i, j\}$ represents a direct connection between i and j through which agents i and j can communicate. See Figure 11.2.

We use the words *network*, *agent*, and *connection* to refer to the physical infrastructure and *graph*, *node*, and *edge* to refer to their corresponding mathematical abstractions. If $\{i, j\} \in E$, then we say j is *adjacent* to i and that j is a *neighbor* of i (and vice versa). Write

$$N_i = \{j \in V \mid \{i, j\} \in E\}$$

for the set of neighbors of i and $|N_i|$ for the number of neighbors of i. N_i does not include i itself.

In the decentralized setup, we assume $r = 0$ in (11.1). Using the notation of graphs, we can recast problem (11.1) into

$$\begin{array}{ll} \underset{\{x_i\}_{i\in V}\subset\mathbb{R}^p}{\text{minimize}} & \sum_{i\in V} f_i(x_i) + h_i(x_i) \\ \text{subject to} & x_i = x_j \quad \forall\, \{i,j\} \in E. \end{array} \tag{11.2}$$

Throughout this chapter, we assume $G = (V, E)$ is a finite, undirected, connected graph without self-loops.

Because the network is connected, all agents can communicate with each other, just as any computer can communicate with any other computer over the internet. Any optimization method can be executed over the network by implementing communication between arbitrary nodes; if they are not directly connected, the communication is relayed over multiple edges. However, in distributed and decentralized optimization, communication, rather than computation, tends to be the bottleneck and relayed communication incurs a huge communication cost. Therefore, methods with reduced communication costs are preferred, and we consider algorithms that utilize communication across single edges, the most basic communication unit, without directly relying on long-range relayed communication.

11.2.1 Decentralized ADMM

Consider the setup (11.1) with $r = h_1 = \cdots = h_n = 0$. For every edge $e = \{i,j\}$, introduce a variable $y_e \in \mathbb{R}^p$ and replace the constraint $x_i = x_j$ of (11.2) with the two constraints $x_i = y_e$ and $x_j = y_e$ to obtain the equivalent problem

$$\begin{array}{ll} \underset{\substack{\{x_i\}_{i\in V}\\ \{y_e\}_{e\in E}}}{\text{minimize}} & \sum_{i\in V} f_i(x_i) \\ \text{subject to} & \begin{cases} x_i - y_e = 0 \\ x_j - y_e = 0 \end{cases} \quad \forall\, e = \{i,j\} \in E. \end{array} \tag{11.3}$$

For each $e = \{i,j\} \in E$, introduce the dual variables $u_{e,i}$ for $x_i - y_e = 0$ and $u_{e,j}$ for $x_j - y_e = 0$. The augmented Lagrangian is

$$\begin{aligned} \mathbf{L}_\alpha(x,y,u) = \sum_{i\in V} f_i(x_i) &+ \sum_{e=\{i,j\}} \left(\langle u_{e,i}, x_i - y_e\rangle + \langle u_{e,j}, x_j - y_e\rangle\right) \\ &+ \sum_{e=\{i,j\}} \frac{\alpha}{2}\left(\|x_i - y_e\|^2 + \|x_j - y_e\|^2\right), \end{aligned}$$

where $\sum_{e=\{i,j\}}$ is the summation over all edges $e = \{i,j\} \in E$. Express ADMM (3.8) applied to this setup as

$$\begin{aligned} x_i^{k+1} &= \underset{x_i\in\mathbb{R}^p}{\operatorname{argmin}} \left\{ f_i(x_i) + \sum_{j\in N_i} \left(\langle u^k_{\{i,j\},i}, x_i - y^k_{\{i,j\}}\rangle + \frac{\alpha}{2}\|x_i - y^k_{\{i,j\}}\|^2 \right) \right\} \quad \forall i \in V \\ y_e^{k+1} &= \underset{y_e\in\mathbb{R}^p}{\operatorname{argmin}} \left\{ \sum_{t=i,j} \left(\langle u^k_{e,t}, x_t^{k+1} - y_e\rangle + \frac{\alpha}{2}\|x_t^{k+1} - y_e\|^2 \right) \right\} \quad \forall e = \{i,j\} \in E \\ u_{e,t}^{k+1} &= u_{e,t}^k + \alpha(x_t^{k+1} - y_e^{k+1}) \qquad \forall e = \{i,j\} \in E,\ t = i,j. \end{aligned}$$

As is, this method can be implemented in a decentralized manner. However, we simplify further. Using the closed-form solution $y_e^{k+1} = \frac{1}{2}\sum_{t=i,j}(x_t^{k+1} + \frac{1}{\alpha}u_{e,t}^k)$, eliminate y_e^{k+1} in the u-update to get

$$\begin{aligned} u_{e,i}^{k+1} &= u_{e,i}^k + \alpha\left(x_i^{k+1} - \frac{1}{2}\sum_{t=i,j}\left(x_t^{k+1} + \frac{1}{\alpha}u_{e,t}^k\right)\right) \\ &= \frac{1}{2}(u_{e,i}^k - u_{e,j}^k) + \frac{\alpha}{2}(x_i^{k+1} - x_j^{k+1}), \quad \forall e = \{i,j\} \in E. \end{aligned}$$

Using $u_{e,i}^k + u_{e,j}^k = 0$ for all $e = \{i,j\}$ and $k = 1,2,\dots$, write $y_e^k = \frac{1}{2}(x_i^k + x_j^k)$, $u_{e,i}^{k+1} = u_{e,i}^k + \frac{\alpha}{2}(x_i^{k+1} - x_j^{k+1})$, and

$$\begin{aligned} x_i^{k+1} &= \underset{x_i \in \mathbb{R}^p}{\operatorname{argmin}}\left\{f_i(x_i) + \frac{\alpha}{2}\sum_{j\in N_i}\left\|x_i - \frac{1}{2}(x_i^k + x_j^k) + \frac{1}{\alpha}u_{\{i,j\},i}^k\right\|^2\right\} \\ &= \underset{x_i \in \mathbb{R}^p}{\operatorname{argmin}}\left\{f_i(x_i) + \frac{\alpha|N_i|}{2}\left\|x_i - \frac{1}{|N_i|}\sum_{j\in N_i}\left(\frac{1}{2}(x_i^k + x_j^k) - \frac{1}{\alpha}u_{\{i,j\},i}^k\right)\right\|^2\right\} \end{aligned}$$

for all $i \in V$. By defining $v_i^k = \frac{1}{|N_i|}\sum_{j\in N_i}\left(\frac{1}{2}(x_i^k + x_j^k) - \frac{1}{\alpha}u_{\{i,j\},i}^k\right)$ and $a_i^k = \frac{1}{|N_i|}\sum_{j\in N_i} x_j^k$, we obtain the simplified ADMM iteration:

$$x_i^{k+1} = \operatorname{Prox}_{(\alpha|N_i|)^{-1}f_i}(v_i^k) \qquad i \in V \tag{11.4a}$$

$$\begin{cases} a_i^{k+1} = \frac{1}{|N_i|}\sum_{j\in N_i} x_j^{k+1} \\ v_i^{k+1} = v_i^k + a_i^{k+1} - \frac{1}{2}a_i^k - \frac{1}{2}x_i^k \end{cases} \qquad i \in V. \tag{11.4b}$$

We call this method *decentralized ADMM*. Convergence follows from the convergence of ADMM. Step (11.4a) must be completed for all $i \in V$ before steps (11.4b) start for any i. The two steps in (11.4b) must be sequential at each i but can be out of sync across different i.

This method is decentralized, as it has a decentralized implementation that alternates local computation and communication with neighbors in a decentralized setup as in Figure 11.2: (i) each agent independently performs the v^k- and x^{k+1}-updates with local computation and (ii) the agents send x_i^{k+1} to its neighbors and each agent computes a_i^{k+1} by averaging the x_j^{k+1}'s received from its neighbors. The decentralized communication and computation of step (ii) is referred to as a *reduction operation in the neighborhood* or *neighborhood reduction*.

Decentralized FLiP-ADMM We can generalize decentralized ADMM to solve

$$\underset{\{x_i\}_{i\in V}}{\text{minimize}} \quad \sum_{i=1}^n f_i(x) + h_i(x),$$

the general formulation of (11.1), using FLiP-ADMM of §8 with $P = \beta I$, $Q = 0$, $\varphi = 1$, $\alpha > 0$, and $\beta > 0$:

$$x_i^{k+1} = \mathrm{Prox}_{(\alpha|N_i|+\beta)^{-1}f_i}\left((\alpha|N_i|+\beta)^{-1}\left(\alpha|N_i|v_i^k + \beta x_i^k - \nabla h_i(x_i^k)\right)\right) \qquad i \in V$$

$$\begin{cases} a_i^{k+1} = \frac{1}{|N_i|}\sum_{j\in N_i} x_j^{k+1} \\ v_i^{k+1} = v_i^k + a_i^{k+1} - \frac{1}{2}a_i^k - \frac{1}{2}x_i^k, \end{cases} \qquad i \in V$$

with initial points $v_i^0 = \frac{1}{2|N_i|}\sum_{j\in N_i}(x_i^0 + x_j^0)$ and arbitrary x_i^0, $i \in V$. See Exercise 11.11.

Decentralized Methods and Synchronization The decentralized methods of this chapter are synchronous in the sense that all agents must complete their computation and communication for the iteration to proceed. In some real-world decentralized networks, however, synchronization is a costly and unrealistic requirement. For such systems, one can use asynchronous decentralized methods, which combine the asynchrony studied in §6 with the decentralized methods of this chapter. See the bibliographical notes section.

11.3 DECENTRALIZED OPTIMIZATION WITH MIXING MATRICES

In this section, we introduce the notion of mixing matrices and use them to describe and analyze a broader class of decentralized optimization methods.

Decentralized Notation Define the *stack* operator and use boldface symbols to denote stacked variables

$$\mathbf{x} = \mathrm{stack}(x_1,\ldots,x_n) = \begin{bmatrix} - & x_1^\intercal & - \\ & \vdots & \\ - & x_n^\intercal & - \end{bmatrix} \in \mathbb{R}^{n\times p}. \tag{11.5}$$

Write $\mathbf{x}^k = \mathrm{stack}(x_1^k,\ldots,x_n^k)$ likewise to denote the iterates. Write both $x^\star \in \mathbb{R}^p$ and $\mathbf{x}^\star = \mathrm{stack}(x^\star,\ldots,x^\star) \in \mathbb{R}^{n\times p}$ to denote the solution of the optimization problem at hand. For any $\mathbf{x} = \mathrm{stack}(x_1,\ldots,x_n) \in \mathbb{R}^{n\times p}$ and $\mathbf{y} = \mathrm{stack}(y_1,\ldots,y_n) \in \mathbb{R}^{n\times p}$, define

$$\langle \mathbf{x}, \mathbf{y}\rangle = \sum_{i=1}^n \langle x_i, y_i\rangle.$$

For any symmetric positive semidefinite $A \in \mathbb{R}^{n\times n}$, define

$$\|\mathbf{x}\|_A^2 = \langle \mathbf{x}, A\mathbf{x}\rangle$$

and specifically define $\|\mathbf{x}\|^2 = \|\mathbf{x}\|_I^2 = \langle \mathbf{x}, \mathbf{x}\rangle$. Finally, define

$$f(\mathbf{x}) = \sum_{i=1}^n f_i(x_i), \qquad h(\mathbf{x}) = \sum_{i=1}^n h_i(x_i)$$

$$\mathrm{Prox}_{\alpha f}(\mathbf{x}) = \mathrm{stack}(\mathrm{Prox}_{\alpha f_1}(x_1),\ldots,\mathrm{Prox}_{\alpha f_n}(x_n))$$

$$\nabla h(\mathbf{x}) = \mathrm{stack}(\nabla h_1(x_1),\ldots,\nabla h_n(x_n)).$$

We say $\mathbf{x} = \mathrm{stack}(x_1,\ldots,x_n)$ is *in consensus* if $x_1 = \cdots = x_n$. A solution (or any feasible point) of (11.2) is in consensus. The methods of this chapter produce iterates that are in consensus in the limit.

11.3.1 Mixing Matrices

We informally say $W \in \mathbb{R}^{n\times n}$ is a *mixing matrix* when an application of W, within a distributed method, represents a round of communication and the aggregation of the communicated information. Throughout this chapter, let $\lambda_1,\dots,\lambda_n$ denote the eigenvalues of W.

We say W is a *decentralized* mixing matrix with respect to a graph $G=(V,E)$ if $W_{ij}=0$ when $i\neq j$ and $\{i,j\}\notin E$. (W_{ii} may be nonzero.) Consider the setup where agents $1,\dots,n$ each have access to the entries $y_1,\dots,y_n$ of the vector $y\in\mathbb{R}^n$, respectively, and we wish to evaluate the matrix-vector product Wy. When W is a decentralized mixing matrix, we can do so in a decentralized manner: since

$$(Wy)_i=\sum_{j=1}^{n}W_{ij}y_j=\sum_{j\in N_i\cup\{i\}}W_{ij}y_j,$$

agent i needs to communicate only with its neighbors.

Example 11.3 *Local averaging matrix.* With the mixing matrix $W\in\mathbb{R}^{n\times n}$ defined by

$$W_{ij}=\begin{cases}\frac{1}{|N_i|} & \text{if } \{i,j\}\in E\\ 0 & \text{otherwise}\end{cases}$$

for $i,j\in\{1,\dots,n\}$ and

$$\tilde{f}(\mathbf{x})=\sum_{i=1}^{n}\frac{1}{|N_i|}f_i(x_i),$$

we can express decentralized ADMM (11.4) as

$$\begin{aligned}\mathbf{x}^{k+1}&=\mathrm{Prox}_{\alpha\tilde{f}}(\mathbf{v}^k)\\ \mathbf{a}^{k+1}&=W\mathbf{x}^{k+1}\\ \mathbf{v}^{k+1}&=\mathbf{v}^k+\mathbf{a}^{k+1}-\frac{1}{2}\mathbf{a}^k-\frac{1}{2}\mathbf{x}^k.\end{aligned}$$

Decentralized Averaging As a motivating example for mixing matrices, consider decentralized averaging: each agent $i\in V$ has a vector $x_i\in\mathbb{R}^p$, and the goal is to compute the average $\bar{x}=\frac{1}{n}\sum_{i=1}^{n}x_i$ in a decentralized manner. This problem is a special case of (11.1) with $f_i(x)=\frac{1}{2}\|x-x_i\|^2$.

The method

$$\mathbf{x}^{k+1}=W\mathbf{x}^k \tag{11.6}$$

with the starting point $\mathbf{x}^0=\mathrm{stack}(x_1,\dots,x_n)$ and a decentralized mixing matrix $W\in\mathbb{R}^{n\times n}$ is called the *decentralized averaging* method. When $\mathbf{1}$ is applied to a matrix as a vector, it denotes the column vector whose every entry is 1. The method converges for all $\mathbf{x}^0$ if and only if $W\mathbf{1}=\mathbf{1}$, $\mathbf{1}^\intercal W=\mathbf{1}^\intercal$, and $1=|\lambda_1|>|\lambda_2|\geq\cdots\geq|\lambda_n|$. To clarify, $|\lambda_i|$ denotes the absolute value or modulus of the ith eigenvalue of W, sorted by absolute value. We leave the proof to Exercise 11.14.

Condition $W\mathbf{1} = \mathbf{1}$ implies that the set of $\mathbf{x}$-vectors in consensus (i.e., the components of $\mathbf{x}$ satisfy $x_1 = \cdots = x_n$) are fixed points of the iteration. Condition $\mathbf{1}^\intercal W = \mathbf{1}^\intercal$ implies the mean is preserved throughout the iteration. Finally, the eigenvalue condition implies the iteration converges. Note that λ_1 is real, i.e., $1 = \lambda_1$, since $W\mathbf{1} = \mathbf{1}$ and $\mathbf{1}^\intercal W = \mathbf{1}^\intercal$ imply that 1 is an eigenvalue of W.

Assumptions on Mixing Matrices A mixing matrix $W \in \mathbb{R}^{n\times n}$ used in decentralized optimization often satisfies some or all of the following assumptions:

$$W = W^\intercal \tag{11.7a}$$

$$\mathcal{N}(I - W) = \operatorname{span}(\mathbf{1}) \tag{11.7b}$$

$$1 = \lambda_1 > \max\{|\lambda_2|, \ldots, |\lambda_n|\}. \tag{11.7c}$$

Although assumption (11.7a) was not assumed in decentralized ADMM or decentralized averaging, it is common; methods with symmetric mixing matrices tend to be easier to analyze. Assumption (11.7b) implies $\mathbf{x}$ is in consensus if and only if $\mathbf{x} = W\mathbf{x}$ and is required for almost all decentralized optimization methods. For example, the mixing matrix of Example 11.3 satisfies assumption (11.7b) but not (11.7a) or (11.7c). Finally, assumption (11.7c) is assumed to establish the convergence of certain methods. Note that assumption (11.7a) implies the eigenvalues are real (but not necessarily nonnegative), and assumption (11.7b) implies $1 = \lambda_1$.

Example 11.4 *Laplacian-based mixing matrix.* Consider the symmetric mixing matrix $W \in \mathbb{R}^{n\times n}$ defined by

$$W = I - \frac{1}{\tau}L,$$

where L is the *graph Laplacian*

$$L_{ij} = \begin{cases} |N_i| & \text{if } i = j \\ -1 & \text{if } \{i,j\} \in E \\ 0 & \text{otherwise} \end{cases}$$

for $i,j \in \{1,\ldots,n\}$ and τ is a constant satisfying $\tau > \frac{1}{2}\lambda_{\max}(L)$. Using standard arguments with the graph Laplacian, one can show that $W\mathbf{1} = \mathbf{1}$ and $1 = \lambda_1 > \max\{|\lambda_2|, \ldots, |\lambda_n|\}$.

Example 11.5 *Metropolis mixing matrix.* Consider the symmetric mixing matrix $W \in \mathbb{R}^{n\times n}$ defined by

$$W_{ij} = \begin{cases} \frac{1}{\max\{|N_i|,|N_j|\}+\varepsilon} & \text{if } \{i,j\} \in E \\ 1 - \sum_{j\in N_i} W_{ij} & \text{if } i = j \\ 0 & \text{otherwise} \end{cases}$$

for $i,j \in \{1,\ldots,n\}$, where $\varepsilon > 0$. Using standard arguments with the Perron–Frobenius theory (W is a stochastic matrix for an irreducible and aperiodic Markov chain), one can show $W\mathbf{1} = \mathbf{1}$ and $1 = \lambda_1 > \max\{|\lambda_2|, \ldots, |\lambda_n|\}$.

Relationship with Stochastic Matrices Mixing matrices and stochastic matrices for Markov chains share some apparent similarities, but they do have some key differences. Given a Markov chain with states $1,\ldots,n$, its stochastic matrix $P \in \mathbb{R}^{n\times n}$ contains the transition probabilities as P_{ij} being the probability of transitioning from i to j, for all states i and j. Conversely, any matrix $P \in \mathbb{R}^{n\times n}$ satisfying $P_{ij} \geq 0$ for all i,j and $P\mathbf{1} = \mathbf{1}$ can be interpreted as a stochastic matrix of a Markov chain.

The first key difference between the two notions is that stochastic matrices have non-negative entries, while mixing matrices can have negative entries. See Exercise 11.16 for an example of a mixing matrix with negative entries. Another difference is in their primary use as linear operators. With a stochastic matrix $P \in \mathbb{R}^{n\times n}$ satisfying $P\mathbf{1} = \mathbf{1}$ (which means the total probability 1 is preserved), the key operation is the vector-matrix product

$$(\pi^{k+1})^\intercal = (\pi^k)^\intercal P,$$

and it represents the evolution of the state probabilities. With mixing matrix $W \in \mathbb{R}^{n\times n}$ satisfying $W\mathbf{1} = \mathbf{1}$ (which means a vector in consensus remains in consensus), the key operation is the matrix-(stacked vector) product

$$\mathbf{x}^{k+1} = W\mathbf{x}^k.$$

When a mixing matrix is a stochastic matrix, one can utilize the classical Markov chain theory based on the Perron–Frobenius theorem. For example, if $W \in \mathbb{R}^{n\times n}$ is a stochastic matrix for an irreducible Markov chain, then $\mathcal{N}(I-W) = \mathrm{span}(\mathbf{1})$ holds; if the Markov chain is irreducible and aperiodic, then $1 = \lambda_1 > \max\{|\lambda_2|,\ldots,|\lambda_n|\}$ holds. A Markov chain is *irreducible* if every state can be reached from every other state. A state of a Markov chain is *periodic* if the chain can return to the state only at multiples of some integer larger than 1. A Markov chain is *aperiodic* if none of its states is periodic. See the bibliographical notes section.

Dynamic Mixing Matrices For the sake of simplicity, we assumed the mixing matrices do not depend on the iteration. However, when the connectivity of the underlying graph is dynamic, one has to use a series of dynamic mixing matrices instead of a fixed one.

11.3.2 Inexact Decentralized Methods

Consider the setup with $r = f_1 = \cdots = f_n = 0$, and a symmetric mixing matrix $W = W^\intercal \in \mathbb{R}^{n\times n}$ satisfying $\mathcal{N}(I - W) = \mathrm{span}(\mathbf{1})$ and $1 = \lambda_1 > \max\{\lambda_2,\ldots,\lambda_n\}$. We write (11.1) equivalently as

$$\begin{array}{ll}\underset{\mathbf{x}\in\mathbb{R}^{n\times p}}{\text{minimize}} & h(\mathbf{x})\\ \text{subject to} & (I-W)\mathbf{x} = 0.\end{array} \tag{11.8}$$

We now consider inexact decentralized methods that solve a penalty formulations that approximate (11.8). When these inexact methods converge, they converge to an approximation of the original solution.

Decentralized Gradient Descent (DGD). Consider the penalty formulation

$$\underset{\mathbf{x}\in\mathbb{R}^{n\times p}}{\text{minimize}} \quad h(\mathbf{x}) + \frac{1}{2\alpha}\|\mathbf{x}\|^2_{I-W}. \tag{11.9}$$

Since the penalty term $\|\mathbf{x}\|^2_{I-W}$ equals 0 if and only if $\mathbf{x}$ is in consensus, we expect this formulation to approximate (11.8) well when $\alpha > 0$ is small. Gradient descent with stepsize α applied to this penalty formulation is

$$\begin{aligned}\mathbf{x}^{k+1} &= \mathbf{x}^k - \alpha\left(\nabla h(\mathbf{x}^k) + \frac{1}{\alpha}(I - W)\mathbf{x}^k\right)\\ &= W\mathbf{x}^k - \alpha\nabla h(\mathbf{x}^k).\end{aligned}$$

We call this method *decentralized gradient descent (DGD)* or the *combine-then-adapt* method. (The name combine-then-adapt is explained in the bibliographical notes.) DGD is decentralized when W is a decentralized mixing matrix: computing $W\mathbf{x}^k$ requires communication with neighbors, and all other operations require local computation. Assume the penalty formulation has a solution, $h_1,\dots,h_n$ are L-smooth, and $\alpha \in (0,(1 + \lambda_n(W))/L)$. Then $\mathbf{x}^k$ converges to a solution of the penalty formulation. The stepsize bound of $(1 + \lambda_n(W))/L$ follows from the stepsize requirement (stepsize) × (Lipschitz constant) < 2 of gradient descent.

Diffusion Further assume $\min\{\lambda_2,\dots,\lambda_n\} > 0$, i.e., assume W is positive definite and therefore invertible, and consider the penalty formulation

$$\underset{\mathbf{x}\in\mathbb{R}^{n\times p}}{\text{minimize}} \quad h(\mathbf{x}) + \frac{1}{2\alpha}\|\mathbf{x}\|^2_{W^{-1}-I}. \tag{11.10}$$

The forward-backward splitting FPI with $(\mathbb{I} + \alpha\mathbb{B})^{-1}(\mathbb{I} - \alpha\mathbb{A})$, where $\mathbb{A} = \nabla h$, $\mathbb{B} = \frac{1}{\alpha}(W^{-1} - I)$, is

$$\mathbf{x}^{k+1} = W(\mathbf{x}^k - \alpha\nabla h(\mathbf{x}^k)).$$

Note that W^{-1} appears in the analysis and formulation of the algorithm, but not within the iteration $\mathbf{x}^{k+1} = W(\mathbf{x}^k - \alpha\nabla h(\mathbf{x}^k))$.

This method is called the method of *diffusion* or the *adapt-then-combine* method. (The name adapt-then-combine is explained in the bibliographic notes.) Diffusion is also decentralized when W is a decentralized mixing matrix. Assume the penalty formulation has a solution, $h_1,\dots,h_n$ are L_h-smooth, and $\alpha \in (0,2/L_h)$. Then $\mathbf{x}^k$ converges to a solution of the penalty formulation.

Discussion The stepsize condition for diffusion $\alpha < 2/L$ is wider than the stepsize condition for DGD $\alpha < (1 + \lambda_n(W))/L$. Loosely speaking, use of a larger stepsize often leads to faster convergence.

When $\min\{\lambda_2,\dots,\lambda_n\} > 0$ does not hold, we can still use diffusion using the positive definite mixing matrix $(1-\theta)I + \theta W$ with $\theta \in (0, 1/(1 - \min\{\lambda_2,\dots,\lambda_n\}))$.

11.3.3 Exact Decentralized Methods

Consider the setup with $r = 0$ and a symmetric mixing matrix $W = W^\intercal \in \mathbb{R}^{n\times n}$ satisfying $\mathcal{N}(I - W) = \operatorname{span}(\mathbf{1})$ and $1 = \lambda_1 > \max\{\lambda_2,\dots,\lambda_n\}$. Since $I - W$ is symmetric positive semidefinite, there exists a symmetric $U \in \mathbb{R}^{n\times n}$ such that

$$U^2 = \frac{1}{2}(I - W).$$

Note, $\mathcal{N}(U) = \operatorname{span}(\mathbf{1})$.

The problem (11.1) is equivalent to

$$\underset{\mathbf{x}\in\mathbb{R}^{n\times p}}{\text{minimize}} \quad f(\mathbf{x}) + h(\mathbf{x}) + \delta_{\{0\}}(U\mathbf{x}), \tag{11.11}$$

where the indicator function $\delta_{\{0\}}(U\mathbf{x})$ encodes the constraint $U\mathbf{x} = 0$. In this section, we present decentralized methods based on primal-dual splitting methods that converge to an exact solution. The algorithms utilize W, while U is used only in the analysis.

PG-EXTRA Apply Condat–Vũ of Exercise 3.5 to (11.11) with $g = \delta_{\{0\}}$ (so $\text{Prox}_{\beta g^*} = \mathbb{I}$) to get

$$\begin{aligned}
\mathbf{u}^{k+1} &= \mathbf{u}^k + \beta U\mathbf{x}^k \\
\mathbf{x}^{k+1} &= \text{Prox}_{\alpha f}\left(\mathbf{x}^k - \alpha\nabla h(\mathbf{x}^k) - \alpha U(2\mathbf{u}^{k+1} - \mathbf{u}^k)\right).
\end{aligned}$$

To eliminate U, define $\mathbf{w}^k = \frac{1}{\beta}U\mathbf{u}^k = \frac{1}{2}(I - W)\sum_{j=0}^{k-1}\mathbf{x}^j$. Choose $\beta = \alpha^{-1}$ for simplicity and rearrange the terms to get

$$\begin{aligned}
\mathbf{x}^{k+1} &= \text{Prox}_{\alpha f}(W\mathbf{x}^k - \alpha\nabla h(\mathbf{x}^k) - \mathbf{w}^k) \\
\mathbf{w}^{k+1} &= \mathbf{w}^k + \frac{1}{2}(I - W)\mathbf{x}^k,
\end{aligned} \tag{11.12}$$

where we initialize $\mathbf{w}^0 = 0$, corresponding to $\mathbf{u}^0 = 0$ to avoid computing $U\mathbf{u}^0$, and set $\mathbf{x}^0$ arbitrarily.

This method is called *PG-EXTRA*. PG-EXTRA is decentralized when W is a decentralized mixing matrix. Assume total duality holds, $h_1,\ldots,h_n$ are L-smooth, and $0 < \alpha < (1 + \lambda_{\min}(W))/L$. Then, $\mathbf{x}^k \to \mathbf{x}^\star$. The stepsize bound follows from the stepsize requirement (3.13) of Condat–Vũ and $\lambda_{\max}(U^2) = 1/2 - 1/2\lambda_{\min}(W)$. The method *EXTRA* is the special case of PG-EXTRA with $f = 0$. See Exercise 11.17 for generalizations of PG-EXTRA.

NIDS Apply PD3O on (11.11) to get

$$\begin{aligned}
\mathbf{x}^{k+1} &= \text{Prox}_{\alpha f}(\mathbf{x}^k - \alpha U\mathbf{u}^k - \alpha\nabla h(\mathbf{x}^k)) \\
\mathbf{u}^{k+1} &= \mathbf{u}^k + \beta U\left(2\mathbf{x}^{k+1} - \mathbf{x}^k + \alpha\left(\nabla h(\mathbf{x}^k) - \nabla h(\mathbf{x}^{k+1})\right)\right).
\end{aligned}$$

We initialize $\mathbf{u}^0 = 0$ but set $\mathbf{x}^0$ arbitrarily. To eliminate U, define $\mathbf{z}^k = \mathbf{x}^k - \alpha U\mathbf{u}^k - \alpha\nabla h(\mathbf{x}^k)$. Choose $\beta = \alpha^{-1}$ for simplicity and rearrange the terms to get

$$\begin{aligned}
\mathbf{x}^{k+1} &= \text{Prox}_{\alpha f}(\mathbf{z}^k) \\
\mathbf{z}^{k+1} &= \mathbf{z}^k - \mathbf{x}^{k+1} + \frac{1}{2}(I + W)\left(2\mathbf{x}^{k+1} - \mathbf{x}^k + \alpha\left(\nabla h(\mathbf{x}^k) - \nabla h(\mathbf{x}^{k+1})\right)\right),
\end{aligned}$$

where we initialize $\mathbf{z}^0 = \mathbf{x}^0 - \alpha\nabla h(\mathbf{x}^0)$ but set $\mathbf{x}^0$ arbitrarily.

This method is called the *Network InDependent Stepsize* (NIDS) method. NIDS is decentralized when W is a decentralized mixing matrix. Assume total duality holds, $h_1,\ldots,h_n$ are L-smooth, and $\alpha \in (0, 2/L)$. Then $\mathbf{x}^k \to \mathbf{x}^\star$. Note that the choice of $\alpha \in (0, 2/L)$ is independent of the mixing matrix and, thus, the network topology.

Discussion of PG-EXTRA and NIDS The stepsize requirement of NIDS is more favorable than that of PG-EXTRA. A drawback of PG-EXTRA is that the stepsize α is affected by the eigenvalues of W, thus, also by the network structure. This not only limits the size of α but also makes the choice of α more difficult when the network is not fully known. In contrast, the stepsize α of NIDS can be chosen independently of W.

On the other hand, PG-EXTRA can compute $W\mathbf{x}^k$ and $\nabla h(\mathbf{x}^k)$ simultaneously, but NIDS must do its corresponding steps sequentially. Therefore, when those two steps cost similar amounts of time, PG-EXTRA can be implemented to run nearly twice as fast per iteration than NIDS.

When $f = 0$, we can simplify PG-EXTRA to one line. Apply $\mathrm{Prox}_{\alpha f} = \mathbb{I}$ to (11.12) and subtract $\mathbf{x}^k$ from $\mathbf{x}^{k+1}$ to get

$$\mathbf{x}^{k+1} - \mathbf{x}^k = W(\mathbf{x}^k - \mathbf{x}^{k-1}) - \alpha\left(\nabla h(\mathbf{x}^k) - \nabla h(\mathbf{x}^{k-1})\right) - (\mathbf{w}^k - \mathbf{w}^{k-1}).$$

Then use $\mathbf{w}^k - \mathbf{w}^{k-1} = \frac{1}{2}(I - W)\mathbf{x}^{k-1}$ to eliminate $\mathbf{w}^k - \mathbf{w}^{k-1}$ and obtain the one-line formula for $\mathbf{x}^{k+1}$ below. For comparison, also use $f = 0$ to simplify NIDS to one line:

$$\begin{aligned} \text{PG-EXTRA:} \quad \mathbf{x}^{k+1} &= \widetilde{W}(2\mathbf{x}^k - \mathbf{x}^{k-1}) + \alpha(\nabla h(\mathbf{x}^{k-1}) - \nabla h(\mathbf{x}^k)) \\ \text{NIDS:} \quad \mathbf{x}^{k+1} &= \widetilde{W}\left(2\mathbf{x}^k - \mathbf{x}^{k-1} + \alpha(\nabla h(\mathbf{x}^{k-1}) - \nabla h(\mathbf{x}^k))\right), \end{aligned}$$

where $\widetilde{W} = \frac{1}{2}(W + I)$. PG-EXTRA resembles DGD while NIDS resembles diffusion.

BIBLIOGRAPHICAL NOTES

Primal and Dual Decomposition Primal decomposition has its roots in the Dantzig–Wolfe decomposition [DW60] and Benders' decomposition [Ben62] for linear programming. Primal decomposition in the form we present was first presented by Geoffrion [Geo70], although the name "primal decomposition" was coined by Silverman [Sil72]. Dual decomposition, which is also called Lagrangian relaxation, is used widely not only in optimization with continuous variables [Eve63, SGJ11] but also in optimization with discrete variables [Lem01, Fis04]. For other overviews on decomposition methods, see [PC06, CLCD07].

Decentralized ADMM There has been a large body of work studying decentralized ADMM and its variants. Bertsekas and Tsitsiklis [BT89], Mateos, Bazerque, and Giannakis [MBG10], Schizas, Ribeiro, and Giannakis [SRG08], and Ling and Tian [LT10] studied various decentralized ADMM methods and Shi et al. [SLY+14], Chang, Hong, and Wang [CHW15], and Wei and Ozdaglar [WO13] further analyzed their convergence rates.

Mixing Matrices and Markov Chains Mixing matrices in decentralized optimization and Markov chains are closely related, as discussed in §11.3.1. We refer readers to the work of Boyd, Diaconis, and Xiao [XB04, BDX04] for further discussion on this connection, and in particular, the discussion on the mixing matrices of Examples 11.4 and 11.5.

Inexact Decentralized Methods Cattivelli, Lopes, and Sayed introduced DGD [CLS07, CS10]. We clarify that the original authors of DGD called it *distributed* gradient descent, but we have decided to call it *decentralized* gradient descent per our definitions of distributed and decentralized methods. Yuan, Ling, and Yin [YLY16] showed that, with a fixed stepsize α, DGD makes progress toward an $O(\alpha)$-size neighborhood of the solution to the original problem. Convergence to a solution (rather than a neighborhood) is possible with diminishing stepsizes as studied by Chen [Che12], Jakovetic, Xavier, and Moura [JXM14], and Zeng and Yin [ZY18].

In the community of bio-inspired network signal processing, subtracting $\alpha\nabla h$ is called adaptation, in analogy to organisms adapting to the environment, and applying W is called combination. This is why DGD is also called *combine-then-adapt*, while diffusion is called *adapt-then-combine*.

Exact Decentralized Methods Shi, Ling, Wu, and Yin presented EXTRA [SLWY15a] as the first method to achieve "exact convergence," i.e., the method converges to an exact solution (unlike DGD), while using gradients ∇h and a fixed stepsize. The name EXTRA is the abbreviation of **EX**act firs**T**-orde**R** **A**lgorithm. In a follow-up work, Shi, Ling, Wu, and Yin presented PG-EXTRA [SLWY15b] as a generalization of EXTRA that accommodates proximable functions. Li and Yan [LY17] show that PG-EXTRA, in fact, converges with parameters that are chosen more aggressively, specifically with $\alpha < (\frac{3}{4}(1+\lambda_n(W))+\frac{1}{2})/L$ and $\widetilde{W} = (W+I)/2$. Li, Shi, and Yan presented NIDS [LSY19] as an improvement upon PG-EXTRA that allows the stepsize α to be chosen independent of the network topology.

Directed Graph A graph is said to be a *directional graph* or *digraph* if its edges are directed from one node to another. If agent i can send information directly to agent j but not vice versa, we model this connection with a directed edge (i,j) and allow $W_{ji} \neq 0$ but require $W_{ij} = 0$ in the mixing matrix W. (So W is asymmetric.) Decentralized methods for digraphs often use the "push-sum" technique [KDG03, NO15], and it does not seem to be possible to obtain them via operator splitting.

Other Methods There is a large body of research on decentralized optimization methods not covered in this chapter. Terelius, Topcu, and Murray proposed a decentralized method [TTM11] based on dual decomposition. Zhu and Marinez [ZM10] introduced gradient tracking; see Exercises 11.18 and 11.19. The methods [NOSU17, XZSX15, LSY19, QL18] allow stepsizes chosen using agents' local information, similar to NIDS. Nedic, Olshevsky, and Shi [NOS17] introduced the method DIGing and showed it has linear convergence on certain time-varying graphs; see Exercise 11.19. Yuan, Ying, Zhao, and Sayed [YYZS19b, YYZS19a] proposed methods similar to NIDS and that also support left-stochastic matrices. Scaman et al. [SBB+17, SBB+18, SBB+19] established the lower bounds of gradient and communication complexities. Accelerated decentralized methods are another large body of work [QL20, SBB+18, LFYL20, SBB+17, ULGN20]. Lian et al. [LZZ+17, LZZL18] introduced decentralized variants of stochastic gradient descent for deep learning. Wu et al. [WYL+18] generalized PG-EXTRA to asynchronous communication with information delays.

EXERCISES

11.1 *Envelope theorem.* Let $f\colon X \times Y \to \mathbb{R}$ and $h(y) = \inf_{x\in X} f(x,y)$. Assume $X \subseteq \mathbb{R}^m$ is nonempty, $Y \subseteq \mathbb{R}^n$ is an open set, $h(y) > -\infty$ for all $y \in Y$, $f(x,y)$ differentiable in $y \in Y$ for all fixed $x \in X$, h is differentiable at $y \in Y$, and $x^\star(y) \in \operatorname{argmin}_{x\in X} f(x,y)$ exists. Show that

$$\nabla_y h(y) = (\nabla_y f)(x^\star(y), y).$$

Hint. Note that

$$f(x,z) - h(z) \geq 0$$

for all $x \in X$ and $z \in Y$. For a given $y \in Y$, show that

$$f(x^\star(y), z) - h(z),$$

as a function of z, is minimized at $z = y$.
Remark. When convexity is assumed, the differentiability assumptions can be dropped. These setups are explored in Exercises 11.2 and 11.3.
Remark. By the same reasoning, if $h(y) = \sup_{x\in X} f(x,y)$, then

$$x^\star(y) \in \operatorname*{argmax}_{x\in X} f(x,y)$$
$$\nabla_y h(y) = (\nabla_y f)(x^\star(y), y).$$

11.2 *Subgradients with partial minimization.* Let $f(x,y)$ with $x \in \mathbb{R}^m$ and $y \in \mathbb{R}^n$ be a convex function. (f is jointly convex in x and y.) Let $h(y) = \inf_{x\in\mathbb{R}^m} f(x,y)$. Show that
(a) $h\colon \mathbb{R}^n \to \mathbb{R} \cup \{\pm\infty\}$ is convex and
(b) for all $y \in \mathbb{R}^n$, if $x^\star(y) \in \operatorname{argmin}_{x\in\mathbb{R}^m} f(x,y)$ exists, then $(0,g) \in \partial f(x^\star(y), y)$ if and only if $g \in \partial h(y)$.

Remark. Even if f is CCP, h may not be proper.

11.3 *Subgradients with partial maximization.* Let $f(v,y)$ with $v \in \mathbb{R}^m$ and $y \in \mathbb{R}^n$ be convex in y for all fixed v. Let $h(y) = \sup_{v\in\mathbb{R}^m} f(v,y)$. Show that
(a) $h\colon \mathbb{R}^n \to \mathbb{R} \cup \{\pm\infty\}$ is convex and
(b) for all $y \in \mathbb{R}^n$, if $v^\star(y) \in \operatorname{argmax}_{v\in\mathbb{R}^m} f(v,y)$ exists, then $g \in (\partial_y f)(v^\star(y), y)$ implies $g \in \partial h(y)$. To clarify, $g \in (\partial_y f)(v^\star(y), y)$ means

$$f(v^\star(y), z) \geq f(v^\star(y), y) + \langle g, z - y\rangle \qquad \forall z \in \mathbb{R}^n.$$

11.4 *Primal and dual decomposition duality.* Let $f(x,y)$ with a function of $x \in \mathbb{R}^p$ and $y \in \mathbb{R}^q$. Define

$$\phi(y) = \inf_{x\in\mathbb{R}^p} f(x,y), \qquad \psi(v) = \sup_{\substack{x\in\mathbb{R}^p \\ y\in\mathbb{R}^q}} \{-f(x,y) + \langle v, y\rangle\}.$$

Show that $\psi = \phi^*$.

11.5 *Subgradients of indicator functions of linear constraints.* Prove the following statements.

(a) Let $A \in \mathbb{R}^{m\times n}$, $b \in \mathbb{R}^m$, and $C = \{x \in \mathbb{R}^n : Ax = b\}$. For $x \in C$, $\partial\delta_C(x) = \{A^\intercal u : u \in \mathbb{R}^m\}$. In particular, for $D = \{(x,y) \in \mathbb{R}^{n+m} : Ax - y = b\}$ and $(x,y) \in D$, $\partial\delta_D(x,y) = \{[A^\intercal u; -u] : u \in \mathbb{R}^m\}$.

(b) Let $A = [A_1; A_2] \in \mathbb{R}^{(m_1+m_2)\times n}$, $b = [b_1; b_2] \in \mathbb{R}^{(m_1+m_2)}$, and $C = \{x \in \mathbb{R}^n : Ax \leq b\}$. Suppose $x \in C$ satisfies $A_1x = b_1$ and $A_2x < b_2$. Then

$$\begin{aligned}\partial\delta_C(x) &= \{A^\intercal u : u = [u_1; u_2] \in \mathbb{R}^{(m_1+m_2)}, u_1 \geq 0, u_2 = 0\}\\ &= \{A_1^\intercal u_1 : u_1 \in \mathbb{R}^{m_1}, u_1 \geq 0\}.\end{aligned}$$

(To clarify, u_1 and u_2 correspond to A_1 and A_2, respectively.) In particular, for $D = \{(x,y) \in \mathbb{R}^{n+m} : Ax - y \leq b\}$ and $(x,y) \in C$ such that $A_1x = b_1$ and $A_2x < b_2$,

$$\partial\delta_D(x,y) = \{[A^\intercal u; -u] : u = [u_1; u_2] \in \mathbb{R}^m, u_1 \geq 0, u_2 = 0\}.$$

Vector u coincides with the Lagrange multipliers.

11.6 *Subgradients of minimization objective subject to linear constraints.* Consider

$$h(x,y) = \delta_{\{(x,y)\in\mathbb{R}^{p+q} \mid Ax\leq y\}}(x,y) + f(x),$$

where $A \in \mathbb{R}^{q\times p}$, f is CCP, and $\leq$ denotes element-wise inequality. Let $\phi(y) = \inf_{x\in\mathbb{R}^p} h(x,y)$, whose value is equal to the optimal value of

$$\begin{array}{ll}\underset{x\in\mathbb{R}^p}{\text{minimize}} & f(x)\\ \text{subject to} & Ax \leq y.\end{array}$$

Suppose $(x^\star(y), u^\star(y))$ is a primal-dual solution pair for which strong duality holds, i.e., $(x^\star(y), u^\star(y))$ is a saddle point of

$$\mathbf{L}(x,\mu) = f(x) + \langle \mu, Ax - y\rangle.$$

Show that $-u^\star(y) \in \partial\phi(y)$.

11.7 *Distributed ADMM = Distributed DRS.* Show that distributed ADMM

$$\begin{aligned}x_i^{k+1} &= \mathrm{Prox}_{(1/\alpha)f_i}\left(\bar{x}^k - (1/\alpha)u_i^k\right)\\ u_i^{k+1} &= u_i^k + \alpha(x_i^{k+1} - \bar{x}^{k+1}),\end{aligned}$$

and distributed DRS

$$\begin{aligned}x_i^{k+1/2} &= \mathrm{Prox}_{(1/\alpha)f_i}(z_i)\\ z_i^{k+1} &= z_i^k - \bar{z}^k + 2\bar{x}^{k+1/2} - x_i^{k+1/2}\end{aligned}$$

are equivalent in the sense that they generate an identical sequence of iterates after a change of variables.

Hint. Note that $u_1^k, \ldots, u_n^k$ has mean 0 after the initial iteration.

11.8 *Dual of consensus.* Consider the consensus set

$$C = (x_1, \ldots, x_n) \in \mathbb{R}^{pn} \in \{(x_1, \ldots, x_n) \mid x_1 = \cdots = x_n\}.$$

Show that the orthogonal complement of C (as defined in, say, Exercise 2.34) is

$$C^\perp = \{(v_1, \ldots, v_n) \in \mathbb{R}^{pn} \mid v_1 + \cdots + v_n = 0\}.$$

11.9 *DRS with primal decomposition.* Consider the primal decomposition formulation

$$\underset{z\in\mathbb{R}^q}{\text{minimize}} \quad \sum_{i=1}^{n} \phi_i(z),$$

with $\phi_i(z) = \inf_{x\in\mathbb{R}^p} f_i(x,z)$. Show that DRS applied to the consensus formulation

$$\underset{z_1,\dots,z_n\in\mathbb{R}^q}{\text{minimize}} \quad \sum_{i=1}^{n} \phi_i(z_i) + \delta_C(z_1,\dots,z_n)$$

is equivalent to distributed ADMM:

$$\begin{aligned} z_i^{k+1} &= \mathrm{Prox}_{(1/\alpha)\phi_i}\left(\bar{z}^k - (1/\alpha)u_i^k\right)\\ u_i^{k+1} &= u_i^k + \alpha(z_i^{k+1} - \bar{z}^{k+1}), \end{aligned}$$

for $i = 1,\dots,n$. For simplicity, assume all convex functions are CCP.
Remark. Note that $\mathrm{Prox}_{(1/\alpha)\phi_i}(z_0)$ can be computed by minimizing $f(x,z) + \frac{\alpha}{2}\|z - z_0\|^2$ with respect to x, z and returning z.

11.10 *DRS with dual decomposition.* Consider the dual decomposition formulation

$$\underset{v_1,\dots,v_n\in\mathbb{R}^q}{\text{minimize}} \quad \sum_{i=1}^{n} \psi_i(v_i) + \delta_{C^\perp}(v_1,\dots,v_n)$$

with $\psi_i(v_i) = \sup_{\substack{x_i\in\mathbb{R}^p\\ z_i\in\mathbb{R}^q}} \{-f_i(x_i,z_i) + \langle v_i, z_i\rangle\}$ and

$$C^\perp = \{(v_1,\dots,v_n) \in \mathbb{R}^{qn} \mid v_1 + \cdots + v_n = 0\}.$$

Show that DRS applied to this formulation is equivalent to the method of Exercise 11.9. For simplicity, assume all convex functions are CCP.
Hint. While it is possible to solve this problem by directly working out the application of DRS and establishing the equivalence, this alternative approach is more insightful. Use Exercises 11.8 and 2.34. Then use Exercise 11.4 and the self-dual property of DRS discussed in §9.3.

11.11 *FLiP-ADMM-based decentralized optimization.* Consider

$$\begin{array}{ll} \underset{\substack{\{x_i\}_{i\in V}\\ \{y_e\}_{e\in E}}}{\text{minimize}} & \sum_{i\in V} f_i(x_i) + h_i(x_i)\\ \text{subject to} & \begin{cases} x_i - y_e = 0\\ x_j - y_e = 0 \end{cases} \quad \forall\, e = \{i,j\} \in E, \end{array}$$

where f_i is CCP and h_i is CCP and L-smooth for $i \in V$. Derive *decentralized FLiP-ADMM* and use Theorem 6 to obtain convergence conditions.

11.12 *Another ADMM-based decentralized method.* Show that the formulation

$$\begin{array}{ll} \underset{\{x_i,y_i\}_{i\in V}}{\text{minimize}} & \sum_{i\in V} f_i(x_i)\\ \text{subject to} & x_i = y_i \quad \forall\, i \in V\\ & \begin{cases} x_i - y_j = 0\\ x_j - y_i = 0 \end{cases} \quad \forall\, \{i,j\} \in E \end{array}$$

is equivalent to (11.3). Apply ADMM to derive

$$x_i^{k+1} = \mathrm{Prox}_{(\alpha(|N_i|+1))^{-1}f_i}\left(\frac{1}{|N_i|+1}\sum_{j\in N_i\cup\{i\}} y_j^k - \frac{1}{\alpha}v_i^k\right)$$

$$y_i^{k+1} = \frac{1}{|N_i|+1} \sum_{j \in N_i \cup \{i\}} x_j^{k+1}$$

$$v_i^{k+1} = v_i^k + \alpha x_i^{k+1} - \frac{\alpha}{|N_i|+1} \sum_{j \in N_i \cup \{i\}} y_j^{k+1}$$

for all $i \in V$. Also, explain why the method is decentralized.

11.13 *Decentralized ADMM with a bipartite graph.* We say a graph $G = (V, E)$ is bipartite if there exists a partitioning V_l and V_r of V (i.e., $V_l \cup V_r = V$ and $V_l \cap V_r = \emptyset$) such that there are no edges within V_l and V_r (i.e., for all $\{i,j\} \in E$, $[i \in V_l$ and $j \in V_r]$ or $[j \in V_l$ and $i \in V_r]$). Assume $G = (V, E)$ is a bipartite graph. Show that ADMM directly applied to

$$\begin{array}{ll} \underset{\{x_i\}_{i \in V_l},\, \{y_j\}_{j \in V_r}}{\text{minimize}} & \sum_{i \in V_l} f_i(x_i) + \sum_{j \in V_r} f_j(y_j) \\ \text{subject to} & x_i = y_j \quad \forall \{i,j\} \in E \end{array}$$

(without introducing any new variables) is

$$x_i^{k+1} = \underset{x_i \in \mathbb{R}^p}{\operatorname{argmin}} \left\{ f_i(x_i) + \sum_{j \in N_i} \left(\langle u_{\{i,j\}}^k, x_i - y_j^k \rangle + \frac{\alpha}{2} \|x_i - y_j^k\|^2 \right) \right\} \qquad \forall i \in V$$

$$y_i^{k+1} = \underset{y_i \in \mathbb{R}^p}{\operatorname{argmin}} \left\{ f_j(y_j) + \sum_{i \in N_j} \left(\langle u_{\{i,j\}}^k, x_i^{k+1} - y_j \rangle + \frac{\alpha}{2} \|x_i^{k+1} - y_j\|^2 \right) \right\} \qquad \forall i \in V$$

$$u_e^{k+1} = u_e^k + \alpha (x_i^{k+1} - y_j^{k+1}) \qquad \forall e = \{i,j\} \in E.$$

Show that if $f_j = 0$ for all $j \in V_r$, then the method further simplifies to

$$x_i^{k+1} = \text{Prox}_{(\alpha |N_i|)^{-1} f_i}(v_i^k)$$

$$a_j^{k+1} = \frac{1}{|N_j|} \sum_{i \in N_j} x_j^{k+1}$$

$$v_i^{k+1} = v_i^k + \frac{1}{|N_i|} \sum_{j \in N_i} (2a_j^{k+1} - a_j^k) - x_i^{k+1}$$

for all $i \in V$.

Remark. Since ADMM updates the two blocks separately, the absence of edges within the partitions V_l and V_r leads to separable (in the sense of Example 5.2) ADMM subproblems. In fact, the formulations of (11.3) and Exercise 11.12 can be understood as constructing a bipartite graph and then applying ADMM: For each edge $\{i,j\} \in E$, introduce a new node k and replace $\{i,j\}$ with $\{i,k\}$ and $\{k,j\}$ (this operation is called the *subdivision of an edge* in graph theory) and place the original nodes in V_l and the new nodes (the y-nodes) in V_r.

11.14 Let $W \in \mathbb{R}^{n \times n}$. Consider the iteration

$$\mathbf{x}^{k+1} = W \mathbf{x}^k$$

for $k = 0, 1, \ldots$. Then $\mathbf{x}^k \to \mathbf{x}^\star$ holds for all starting points $\mathbf{x}^0 = \text{stack}(x_1^0, \ldots, x_n^0)$, where $x^\star = (1/n)(x_1^0 + \cdots + x_n^0)$ and $\mathbf{x}^\star = \text{stack}(x^\star, \ldots, x^\star)$, if and only if

(i) $W\mathbf{1} = \mathbf{1}$,
(ii) $\mathbf{1}^\intercal W = \mathbf{1}^\intercal$,
(iii) $1 = \lambda_1(W) > |\lambda_2(W)| \geq \cdots \geq |\lambda_n(W)|$.

For the sake of simplicity, assume W is diagonalizable, that is, assume there exists a decomposition $W = V\lambda V^{-1}$, where $\lambda \in \mathbb{C}^{n\times n}$ is a diagonal matrix with diagonal components $\lambda_1, \ldots, \lambda_n$ and $V \in \mathbb{C}^{n\times n}$ is invertible. (The Jordan canonical form can be used to prove the result without diagonalizability.)

Hint. If we let v_i be the ith column of V and $u_i^\intercal$ be the ith row of V^{-1}, then v_i is an eigenvector of W corresponding to λ_i, u_i is an eigenvector of $W^\intercal$ corresponding to λ_i, and

$$W^k = \sum_{i=1}^{n} \lambda_i^k v_i u_i^\intercal.$$

11.15 *Equivalence of consensus conditions.* Consider $\mathbf{x} = \text{stack}(x_1, \ldots, x_n) \in \mathbb{R}^{n\times p}$ as in (11.5) and a mixing matrix satisfying $W\mathbf{1} = \mathbf{1}$ and $\mathcal{N}(I - W) = \text{span}(\mathbf{1})$. Write $\lambda_1, \ldots, \lambda_n$ to denote the eigenvalues of W. Show that the following conditions are equivalent:

(i) $x_1 = \cdots = x_n$.

(ii) $(I - W)\mathbf{x} = 0$.

(iii) $\|\mathbf{x}\|_{I-W} = 0$, provided that $W = W^\intercal$ and $1 = \lambda_1 > \lambda_2 \geq \cdots \geq \lambda_n$.

(iv) $(W^{-1} - I)\mathbf{x} = 0$, provided that W is invertible.

(v) $\|\mathbf{x}\|_{W^{-1}-I} = 0$, provided that $W = W^\intercal$ and $1 = \lambda_1 > \lambda_2 \geq \cdots \geq \lambda_n > 0$.

(vi) $U\mathbf{x} = 0$, provided that $W = W^\intercal$, $1 = \lambda_1 > \lambda_2 \geq \cdots \geq \lambda_n$, and $U = U^\intercal$, and $U^2 = I - W$.

11.16 *Mixing matrices with negative entries.* Let $W \in \mathbb{R}^{n\times n}$ and consider the decentralized averaging method

$$\mathbf{x}^{k+1} = W\mathbf{x}^k.$$

Since

$$\|\mathbf{x}^k - \mathbf{x}^\star\| \sim (\rho(W - 11^\intercal/n))^k \, \|\mathbf{x}^0 - \mathbf{x}^\star\|, \tag{11.13}$$

where $\mathbf{1} \in \mathbb{R}^n$ is the vector with all entries 1 and ρ denotes the spectral radius, we interpret $\rho(W - \mathbf{1}\mathbf{1}^\intercal/n)$ as the asymptotic convergence rate. Next, consider a graph $G = (V, E)$ and assume W is decentralized with respect to G. Consider the problem of finding a decentralized mixing matrix with the fastest asymptotic convergence rate:

$$\begin{array}{ll} \underset{W\in\mathbb{R}^{n\times n}}{\text{minimize}} & \rho(W - \mathbf{1}\mathbf{1}^\intercal/n) \\ \text{subject to} & \mathbf{1}^\intercal W = \mathbf{1}^\intercal W,\ W\mathbf{1} = \mathbf{1} \\ & W_{ij} = 0,\ \{i,j\} \notin E,\ i \neq j. \end{array}$$

However, optimizing the spectral radius of asymmetric matrices is a difficult problem. So, we further assume W is symmetric:

$$\begin{array}{ll} \underset{W\in\mathbb{R}^{n\times n}}{\text{minimize}} & \sigma_{\max}(W - \mathbf{1}\mathbf{1}^\intercal/n) \\ \text{subject to} & W = W^\intercal,\ W\mathbf{1} = \mathbf{1} \\ & W_{ij} = 0,\ \{i,j\} \notin E,\ i \neq j, \end{array}$$

where $\sigma_{\max}$ denotes the maximum singular value. This problem is equivalent to

$$\begin{array}{ll} \underset{s\in\mathbb{R},\ W\in\mathbb{R}^{n\times n}}{\text{minimize}} & s \\ \text{subject to} & -sI \preceq W - \mathbf{1}\mathbf{1}^\intercal/n \preceq sI \\ & W = W^\intercal,\ W\mathbf{1} = \mathbf{1} \\ & W_{ij} = 0,\ \{i,j\} \notin E,\ i \neq j, \end{array}$$

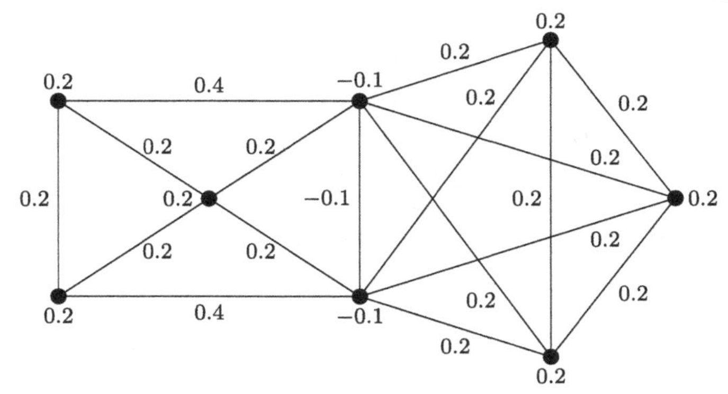

Figure 11.3 A small graph with 8 nodes and 17 edges. Each edge and node is labeled with the optimal symmetric weights, which give the minimum asymptotic convergence factor.

where $\preceq$ denotes the partial order in the sense of positive semidefinite matrices.

(a) Show (11.13).

(b) Numerically solve the problem instance depicted in Figure 11.3 and establish that the depicted solution is indeed optimal. (The solution is not unique.)

(c) The optimal mixing matrix of part (b) contains negative weights. Show that the negative weights are necessary to obtain the optimal mixing matrix by solving the optimization problem with the added constraint $W_{ij} \geq 0$ for all $i, j \in \{1, \dots, n\}$.

Remark. For optimization problems with somewhat complicated constraints, such as this one, it is often simpler to solve small problem instances with libraries such as CVX, CVXPY, or YALMIP. For large problem instances, it becomes necessary to use efficient splitting methods.

Remark. The approach of formulating the problem of finding the optimal symmetric mixing matrix and the particular example of Figure 11.3 was first presented by Xiao and Boyd [XB04].

11.17 *General form of PG-EXTRA.* Consider *two* symmetric mixing matrices $W, \widetilde{W} \in \mathbb{R}^{n\times n}$ satisfying

$$W \preceq \widetilde{W} \preceq \frac{1}{2}(W + I),$$
$$\mathcal{N}(\widetilde{W} - W) = \mathcal{N}(I - W) = \mathrm{span}(\mathbf{1}).$$

The choice $\widetilde{W} = \frac{1}{2}(W + I)$ is most common. Since $\widetilde{W} - W \succeq 0$, there exists a symmetric $U \in \mathbb{R}^{n\times n}$ such that $U^2 = \widetilde{W} - W$, and U satisfies $\mathcal{N}(U) = \mathrm{span}(\mathbf{1})$.

(a) Show that

$$\begin{array}{ll} \underset{\mathbf{x}\in\mathbb{R}^{n\times p}}{\text{minimize}} & f(\mathbf{x}) + h(\mathbf{x}) + \frac{1}{2\alpha}\|\mathbf{x}\|^2_{I-\widetilde{W}} \\ \text{subject to} & U\mathbf{x} = 0. \end{array}$$

is equivalent to problem (11.11) and that this primal problem is generated by the Lagrangian $\mathbf{L}(\mathbf{x}, \mathbf{u}) = f(\mathbf{x}) + h(\mathbf{x}) + \frac{1}{2\alpha}\|\mathbf{x}\|^2_{I-\widetilde{W}} + \langle \mathbf{u}, U\mathbf{x}\rangle$.

(b) Let

$$M = \begin{bmatrix} \alpha^{-1}I & -\frac{1}{2}U \\ -\frac{1}{2}U & \alpha I \end{bmatrix}.$$

Find a decomposition $\partial \mathbf{L} = \mathbb{F} + \mathbb{H}$ such that the FPI with $(M + \mathbb{F})^{-1}(M - \mathbb{H})$ is

$$\begin{aligned} \mathbf{x}^{k+1} &= \mathrm{Prox}_{\alpha f}(\widetilde{W}\mathbf{x}^k - \alpha \nabla h(\mathbf{x}^k) - \alpha U \mathbf{u}^k) \\ \mathbf{u}^{k+1} &= \mathbf{u}^k + \alpha^{-1} U \mathbf{x}^{k+1}, \end{aligned}$$

where $\mathbf{w}^0 = 0$.

(c) Show that the previous method with the initialization $\mathbf{u}^0 = \alpha^{-1} U \mathbf{x}^0$ and arbitrary $\mathbf{x}^0$ is equivalent to

$$\begin{aligned} \mathbf{x}^{k+1} &= \mathrm{Prox}_{\alpha f}\left(W\mathbf{x}^k - \alpha \nabla h(\mathbf{x}^k) - \mathbf{w}^k\right) \\ \mathbf{w}^{k+1} &= \mathbf{w}^k + (\widetilde{W} - W)\mathbf{x}^k. \end{aligned}$$

This method is called the *general form of PG-EXTRA*.
(*Hint*: substitute $\mathbf{w}^k = \sum_{j=0}^{k-1} (\widetilde{W} - W)\mathbf{x}^j$ and use $U^2 = \widetilde{W} - W$.)

11.18 *Consensus tracking in a network.* Consider a mixing matrix $W \in \mathbb{R}^{n \times n}$ satisfying (11.7a)–(11.7c), and let $\mathbf{y}^0, \mathbf{y}^1, \ldots \in \mathbb{R}^{n \times p}$ be a given sequence. We call $\mathbf{z}^0, \mathbf{z}^1, \ldots$ a *consensus tracking sequence* of $\mathbf{y}^0, \mathbf{y}^1, \ldots$ if $\mathbf{z}^0 = \mathbf{y}^0$ and

$$\mathbf{z}^{k+1} = W\mathbf{z}^k + \mathbf{y}^{k+1} - \mathbf{y}^k$$

for $k = 0, 1, \ldots$. Define $\mathbf{y}^k = \mathrm{stack}(y_1^k, \ldots, y_n^k)$ and $\mathbf{z}^k = \mathrm{stack}(z_1^k, \ldots, z_n^k)$. Show that

(a) $\frac{1}{n} \sum_{i=1}^n z_i^k = \frac{1}{n} \sum_{i=1}^n y_i^k$, and

(b) if $\lim_{k \to \infty} \mathbf{y}^k = \mathrm{stack}(y_1^\star, \ldots, y_n^\star)$, then $\lim_{k \to \infty} \mathbf{z}^k = \mathrm{stack}(\bar{y}^\star, \ldots, \bar{y}^\star)$, where $\bar{y}^\star = \frac{1}{n} \sum_{i=1}^n y_i^\star$.

Remark. The idea is that $\mathbf{z}^k$ "tracks" the mean $\frac{1}{n} \sum_{j \in V} y_j^k$ and is in consensus in the limit as $k \to \infty$. To clarify, $\lim_{k \to \infty} \mathbf{y}^k$ is not necessarily in consensus.

11.19 *DIGing.* Consider a mixing matrix $\overline{W} \in \mathbb{R}^{n \times n}$ satisfying (11.7a)–(11.7c) and the problem

$$\begin{array}{ll} \underset{\mathbf{x} \in \mathbb{R}^{n \times p}}{\text{minimize}} & h(\mathbf{x}) \\ \text{subject to} & (I - \overline{W})\mathbf{x} = 0, \end{array}$$

where h is differentiable. The method

$$\begin{aligned} \mathbf{x}^{k+1} &= \overline{W}\mathbf{x}^k - \alpha \mathbf{z}^k \\ \mathbf{z}^{k+1} &= \overline{W}\mathbf{z}^k + \nabla h(\mathbf{x}^{k+1}) - \nabla h(\mathbf{x}^k), \end{aligned}$$

where $\mathbf{x}^0$ is initialized to be in consensus and $\mathbf{z}^0 = \nabla h(\mathbf{x}^0)$, is called *Distributed Inexact Gradient method and a gradient tracking* (DIGing). The $\mathbf{x}^{k+1}$ update is similar to that of DGD but has $\alpha \nabla h(\mathbf{x}^k)$ replaced by $\mathbf{z}^k$. The $\mathbf{z}^k$ iterates track the average gradient $\frac{1}{n} \sum_{i=1}^n \nabla h_i(x_i^k)$ in the sense of Exercise 11.18. Show that DIGing is a special case of generalized PG-EXTRA of Exercise 11.17 with $f = 0$, $W = 2\overline{W} - I$, and $\widetilde{W} = \overline{W}\overline{W}$.

Remark. DIGing was introduced by Nedic, Olshevsky, and Shi [NOS17] for time-varying graphs, i.e., $\overline{W}$ is not fixed but changes with k. In this problem, we derive DIGing with a fixed $\overline{W}$ as an instance of PG-EXTRA.

11.20 *Local stepsizes of NIDS.* Consider a symmetric mixing matrix $W = W^\intercal \in \mathbb{R}^{n \times n}$ satisfying $\mathcal{N}(I - W) = \mathrm{span}(\mathbf{1})$ and $1 = \lambda_1 > \max\{\lambda_2, \ldots, \lambda_n\}$. Let $U \in \mathbb{R}^{n \times n}$ be a symmetric matrix satisfying $U^2 = \frac{1}{2}(I - W)$. Consider

$$\underset{\mathbf{x} \in \mathbb{R}^{n \times p}}{\text{minimize}} \quad f(\mathbf{x}) + h(\mathbf{x}) + \delta_{\{0\}}(U\mathbf{x}),$$

where the notation $f(\mathbf{x})$ and $h(\mathbf{x})$ is as defined for the NIDS setup. We assume $f_1,\dots,f_n$ and $h_1,\dots,h_n$ are CCP, but each h_i is L_i-smooth for $i=1,\dots,n$. Let us derive a variant of NIDS that allows each agent i to utilize its individual stepsize $\alpha_i < 2/L_i$.

Let $\gamma_i = \sqrt{2/L_i}$ for $i=1,\dots,n$, and define

$$\Gamma = \mathrm{diag}(\gamma_1,\dots,\gamma_n) = \begin{bmatrix} \gamma_1 & & \\ & \ddots & \\ & & \gamma_n \end{bmatrix}.$$

Define $\bar{\mathbf{x}} \in \mathbb{R}^{n\times p}$ with $\mathbf{x} = \Gamma\bar{\mathbf{x}}$. Define $\bar{f}(\bar{\mathbf{x}}) = f(\Gamma\bar{\mathbf{x}})$, $\bar{h}(\bar{\mathbf{x}}) = h(\Gamma\bar{\mathbf{x}})$, and $\overline{U} = U\Gamma$.

(a) Show that $\bar{h}$ is 2-smooth.

(b) Show that applying PD3O to the equivalent formulation

$$\underset{\bar{\mathbf{x}}\in\mathbb{R}^{n\times p}}{\text{minimize}} \quad \bar{f}(\bar{\mathbf{x}}) + \bar{h}(\bar{\mathbf{x}}) + \delta_{\{0\}}(\overline{U}\bar{\mathbf{x}})$$

yields the iteration

$$\begin{aligned} \mathbf{x}^{k+1} &= \mathrm{Prox}_F(\mathbf{z}^k) \\ \mathbf{z}^{k+1} &= \mathbf{z}^k - \mathbf{x}^{k+1} + M\left(2\mathbf{x}^{k+1} - \mathbf{x}^k + \mathrm{diag}(\alpha_1,\dots,\alpha_n)\left(\nabla h(\mathbf{x}^k) - \nabla h(\mathbf{x}^{k+1})\right)\right), \end{aligned}$$

where $\alpha_i = \alpha\gamma_i^2$ for $i=1,\dots,n$,

$$F(\mathbf{x}) = \sum_{i=1}^{n} \alpha_i f_i(x_i),$$

and $M = I - \frac{\beta}{2}\mathrm{diag}(\alpha_1,\dots,\alpha_n)(I-W)$.

(c) Show that the iteration converges if $\alpha_i \in (0, 2/L_i)$ for $i=1,\dots,n$ and $\alpha_1,\dots,\alpha_n,\beta > 0$ satisfy

$$\lambda_{\max}\left(\frac{\beta}{2}\mathrm{diag}(\sqrt{\alpha_1},\dots,\sqrt{\alpha_n})(I-W)\mathrm{diag}(\sqrt{\alpha_1},\dots,\sqrt{\alpha_n})\right) \le 1.$$

Remark. This version of NIDS was also presented in Li, Shi, and Yan's original NIDS paper [LSY19].

12 Acceleration

Theorem 1 establishes an $O(1/k)$ rate on the squared norm of the fixed-point residual, and a similar $O(1/k)$ rate can be established for the setups of Theorems 2, 3, and 6. It is natural to ask whether this rate can be improved. The answer is yes, at least in the worst-case rate.

In optimization, an *acceleration* is a modification of a base method that improves the convergence rate, and an improvement from $O(1/k)$ to $O(1/k^2)$ is most common for first-order algorithms. Acceleration is an active topic of research. In this chapter, we keep the discussion minimal and discuss Nesterov's AGM, which is the most well known, and APPM/OHM, which is most relevant to this book's content.

12.1 ACCELERATED GRADIENT METHOD

Consider the problem

$$\underset{x \in \mathbb{R}^n}{\text{minimize}} \quad f(x),$$

where f is convex and L-smooth. The method

$$\begin{aligned} x^{k+1} &= y^k - \frac{1}{L}\nabla f(y^k) \\ y^{k+1} &= x^{k+1} + \frac{k-1}{k+2}(x^{k+1} - x^k), \end{aligned}$$

where $x^0 = y^0$, is called Nesterov's accelerated gradient method (AGM).

Theorem 17 Assume the convex, L-smooth function f has a minimizer $x^\star$. Then AGM converges with the rate

$$f(x^k) - f(x^\star) \leq \frac{2L\|x^0 - x^\star\|^2}{k^2}$$

for $k = 1, 2, \ldots$.

We can equivalently write AGM as

$$\begin{aligned}
x^{k+1} &= y^k - \frac{1}{L}\nabla f(y^k) \\
z^{k+1} &= z^k - \frac{k+1}{2L}\nabla f(y^k) \\
y^{k+1} &= \left(1 - \frac{2}{k+2}\right)x^{k+1} + \frac{2}{k+2}z^{k+1},
\end{aligned}$$

where $x^0 = y^0 = z^0$; the two forms are equivalent in the sense that the generated x^k- and y^k-sequences are the same. See Exercise 12.1.

Proof of Theorem 17. We first make some preliminary observations. Define

$$\theta_k = \frac{k+1}{2}$$

for $k = -1, 0, 1, \ldots$. It is straightforward to verify

$$\theta_k^2 - \theta_k \le \theta_{k-1}^2 \tag{12.1}$$

for $k = 0, 1, \ldots$. We will use the inequalities

$$f(x^{k+1}) - f(y^k) + \frac{1}{2L}\|\nabla f(y^k)\|^2 \le 0 \tag{12.2}$$

$$f(y^k) - f(x^k) \le \langle \nabla f(y^k), y^k - x^k \rangle \tag{12.3}$$

$$f(y^k) - f(x^\star) \le \langle \nabla f(y^k), y^k - x^\star \rangle. \tag{12.4}$$

The first, (12.2), follows from L-smoothness, which implies $f(x) - \frac{L}{2}\|x - y^k\|^2$ is concave as a function of x, which in turn implies

$$f(x) - \frac{L}{2}\|x - y^k\|^2 \le f(y^k) + \langle \nabla f(y^k), x - y^k \rangle.$$

We plug in $x = x^{k+1} = y^k - \frac{1}{L}\nabla f(y^k)$ to get (12.2). The second and third inequalities, (12.3) and (12.4), follow from convexity of f.

$$V^k = \theta_{k-1}^2\left(f(x^k) - f(x^\star)\right) + \frac{L}{2}\|z^k - x^\star\|^2,$$

where z^k is as defined in the equivalent formulation of AGM. Then

$$\begin{aligned}
&V^{k+1} - V^k \\
&= \theta_k^2\left(f(x^{k+1}) - f(x^\star) + \frac{1}{2L}\|\nabla f(y^k)\|^2\right) - \theta_{k-1}^2(f(x^k) - f(x^\star)) \\
&\qquad - \theta_k\langle \nabla f(y^k), z^k - x^\star \rangle \\
&\le \theta_k^2\left(f(y^k) - f(x^\star)\right) - \theta_{k-1}^2(f(x^k) - f(x^\star)) - \theta_k\langle \nabla f(y^k), z^k - x^\star \rangle \\
&= (\theta_k^2 - \theta_k)(f(y^k) - f(x^k)) + \theta_k(f(y^k) - f(x^k)) + (\theta_k^2 - \theta_{k-1}^2)(f(x^k) - f(x^\star)) \\
&\qquad - \theta_k\langle \nabla f(y^k), z^k - x^\star \rangle \\
&\le (\theta_k^2 - \theta_k)(f(y^k) - f(x^k)) + \theta_k(f(y^k) - f(x^\star)) - \theta_k\langle \nabla f(y^k), z^k - x^\star \rangle \\
&\le (\theta_k^2 - \theta_k)\langle \nabla f(y^k), y^k - x^k \rangle + \theta_k\langle \nabla f(y^k), y^k - x^\star \rangle - \theta_k\langle \nabla f(y^k), z^k - x^\star \rangle \\
&= \theta_k\langle \nabla f(y^k), (1 - \theta_k)x^k + \theta_k y^k - z^k \rangle = 0,
\end{aligned}$$

where the first equality follows from

$$\frac{L}{2}\left\|z^k - x^\star - \frac{\theta_k}{L}\nabla f(y^k)\right\|^2 - \frac{L}{2}\|z^k - x^\star\|^2 = -\theta_k\langle \nabla f(y^k), z^k - x^\star\rangle + \frac{\theta_k^2}{2L}\|\nabla f(y^k)\|^2,$$

the first inequality follows from (12.2), the second inequality follows from (12.1), the third inequality follows from (12.3) and (12.4), and the final equality follows from the definition of z^k. This establishes $V^k \le V^{k-1} \le \cdots \le V^0$, and $V^k \le V^0$ implies

$$f(x^k) - f(x^\star) \le \frac{L}{2\theta_{k-1}^2}\|z^0 - x^\star\|^2 = \frac{2L}{k^2}\|z^0 - x^\star\|^2.$$

□

Comparison with Gradient Descent Gradient descent (GD) with stepsize $\alpha = 1/L$

$$x^{k+1} = x^k - \frac{1}{L}\nabla f(x^k)$$

converges in function value at the slower rate $O(1/k)$. To see this, define

$$V^k = k(f(x^k) - f(x^\star)) + \frac{L}{2}\|x^k - x^\star\|^2$$

and note

$$\begin{aligned}
&V^{k+1} - V^k \\
&= (k+1)(f(x^{k+1}) - f(x^k)) + f(x^k) - f(x^\star) + \frac{L}{2}\left(\frac{1}{L^2}\|\nabla f(x^k)\|^2 - \frac{2}{L}\langle \nabla f(x^k), x^k - x^\star\rangle\right) \\
&\le -\frac{k+1}{2L}\|\nabla f(x^k)\|^2 + \langle \nabla f(x^k), x^k - x^\star\rangle + \frac{1}{2L}\|\nabla f(x^k)\|^2 - \langle \nabla f(x^k), x^k - x^\star\rangle \\
&= -\frac{k}{2L}\|\nabla f(x^k)\|^2 \le 0,
\end{aligned}$$

where the inequality follows from analogues of (12.2) and (12.4). $V^k \le V^0$ implies

$$f(x^k) - f(x^\star) \le \frac{L}{2k}\|x^0 - x^\star\|^2 - \frac{L}{2k}\|x^k - x^\star\|^2 \le \frac{L}{2k}\|x^0 - x^\star\|^2.$$

Constructing the Lyapunov Function The nonincreasing quantity V^k in the proof of AGM and GD (and later in APPM) is called a Lyapunov function, energy function, or potential function, and the style of proof relying on such quantities is called a *Lyapunov analysis*. Not all convergence proofs in optimization use a Lyapunov function, but the ones that do tend to be more concise. Constructing a Lyapunov function is a highly nontrivial art, and we briefly outline the process for GD and AGM.

Imagine analyzing GD, and we suspect the convergence rate is $f(x^k) - f(x^\star) = O(1/k)$. We define $W^k = k(f(x^k) - f(x^\star))$ and, through some analysis, find

$$W^{k+1} - W^k \le \frac{L}{2}\|x^k - x^\star\|^2 - \frac{L}{2}\|x^{k+1} - x^\star\|^2.$$

So we define $V^k = k(f(x^k) - f(x^\star)) + \frac{L}{2}\|x^k - x^\star\|^2$ and present a Lyapunov analysis.

Encouraged by this success, one may try to prove a faster rate for GD by defining $W^k = t_k^2(f(x^k) - f(x^\star))$ with a yet unspecified t_k-sequence and analyzing $W^{k+1} - W^k$. If the t_k-sequence has a growth rate on the order of k, perhaps we can establish an $O(1/k^2)$ rate. However, such an effort does not lead to a rate faster than $O(1/k)$ for GD.

For AGM, we again define $W^k = t_{k-1}^2(f(x^k) - f(x^\star))$ and analyze $W^{k+1} - W^k$. With an analysis similar to what we have seen, we get

$$W^{k+1} - W^k \le \frac{L}{2}\|z^k - x^\star\|^2 - \frac{L}{2}\|z^{k+1} - x^\star\|^2.$$

for $t_k^2 - t_k \le t_{k-1}^2$ and $t_k \ge 0$. An admissible sequence is $t_k = (k+1)/2$. So we define $V^k = t_k^2(f(x^k) - f(x^\star)) + \frac{L}{2}\|z^k - x^\star\|^2$ and present a Lyapunov analysis. Instead of $k/2$, we can let $t_0 = 1$ and define $t_1, t_2, \ldots$ successively by $t_{k+1}^2 - t_{k+1} = t_k^2$, which gives a slightly better rate (see Exercise 12.3).

12.2 ACCELERATED PROXIMAL POINT AND OPTIMIZED HALPERN METHOD

Consider the problem

$$\underset{x\in\mathbb{R}^n}{\text{find}} \quad 0 \in \mathbb{A}x,$$

where $\mathbb{A}$ is maximal monotone. The method

$$\begin{aligned} y^{k+1} &= \mathbb{J}_{\mathbb{A}}x^k \\ x^{k+1} &= y^{k+1} + \frac{k}{k+2}(y^{k+1} - y^k) - \frac{k}{k+2}(y^k - x^{k-1}), \end{aligned}$$

where $y^0 = x^0$, is called the accelerated proximal point method (APPM).

Also consider the problem

$$\underset{x\in\mathbb{R}^n}{\text{find}} \quad x = \mathbb{T}x,$$

where $\mathbb{T}\colon \mathbb{R}^n \to \mathbb{R}^n$ is nonexpansive. We call

$$x^{k+1} = \frac{1}{k+2}x^0 + \frac{k+1}{k+2}\mathbb{T}x^k$$

the optimized Halpern method (OHM).

With $\mathbb{T} = \mathbb{R}_{\mathbb{A}}$, the two problems of finding elements of $\operatorname{Zer}\mathbb{A}$ and $\operatorname{Fix}\mathbb{T}$ are equivalent (cf. Exercise 10.2), and the two methods APPM and OHM are equivalent in the sense that the generated x^k- sequences are the same. (See Exercise 12.2.)

Theorem 18 Assume the maximal monotone operator $\mathbb{A}$ has a zero $x^\star$. Then APPM/OHM converges with the rate

$$\|x^{k-1} - \mathbb{J}_{\mathbb{A}}x^{k-1}\|^2 \le \frac{\|x^0 - x^\star\|^2}{k^2}$$

for $k = 1, 2, \ldots$.

Note that we can equivalently state this result as

$$\|\mathbb{T}x^{k-1} - x^{k-1}\|^2 \le \frac{4\|x^0 - x^\star\|^2}{k^2}.$$

Proof. Define $\tilde{\mathbb{A}}y^k = x^{k-1} - y^k$, which implies $\tilde{\mathbb{A}}y^k \in \mathbb{A}y^k$. Define

$$V^k = k^2\|\tilde{\mathbb{A}}y^k\|^2 + k\langle \tilde{\mathbb{A}}y^k, y^k - x^0\rangle + \frac{1}{2}\|x^0 - x^\star\|^2$$
$$= \frac{k^2}{2}\|\tilde{\mathbb{A}}y^k\|^2 + k\langle \tilde{\mathbb{A}}y^k, y^k - x^\star\rangle + \frac{1}{2}\|k\tilde{\mathbb{A}}y^k - (x^0 - x^\star)\|^2$$

for $k = 0,1,\ldots$. Note that $\langle \tilde{\mathbb{A}}y^k, y^k - x^\star\rangle \geq 0$ by monotonicity of $\mathbb{A}$, so $V^k \geq 0$ for $k = 0,1,\ldots$. From the equivalent OHM form, we have

$$y^{k+1} + \tilde{\mathbb{A}}y^{k+1} = x^k = \frac{1}{k+1}x^0 + \frac{k}{k+1}(2\mathbb{J}_{\mathbb{A}} - \mathbb{I})x^{k-1} = \frac{k}{k+1}(2y^k - (y^k + \tilde{\mathbb{A}}y^k)) + \frac{1}{k+1}x^0,$$

which we reorganized into

$$(k+1)\tilde{\mathbb{A}}y^{k+1} + (k+1)(y^{k+1} - x^0) + k\tilde{\mathbb{A}}y^k - k(y^k - x^0) = 0.$$

Then

$$\begin{aligned}
&V^{k+1} - V^k \\
&= (k+1)^2\|\tilde{\mathbb{A}}y^{k+1}\|^2 + (k+1)\langle \tilde{\mathbb{A}}y^{k+1}, y^{k+1} - x^0\rangle - k^2\|\tilde{\mathbb{A}}y^k\|^2 - k\langle \tilde{\mathbb{A}}y^k, y^k - x^0\rangle \\
&= \langle (k+1)\tilde{\mathbb{A}}y^{k+1}, \underbrace{(k+1)\tilde{\mathbb{A}}y^{k+1} + (y^{k+1} - x^0)}_{=-k(y^{k+1}-x^0)+k(y^k-x^0)-k\tilde{\mathbb{A}}y^k}\rangle - \langle k\tilde{\mathbb{A}}y^k, \underbrace{k\tilde{\mathbb{A}}y^k + (y^k - x^0)}_{=(k+1)(y^k-x^0)-(k+1)(y^{k+1}-x^0)-(k+1)\tilde{\mathbb{A}}y^{k+1}}\rangle \\
&= k(k+1)\langle \tilde{\mathbb{A}}y^{k+1}, y^k - y^{k+1} - \tilde{\mathbb{A}}y^k\rangle - k(k+1)\langle \tilde{\mathbb{A}}y^k, y^k - y^{k+1} - \tilde{\mathbb{A}}y^{k+1}\rangle \\
&= -k(k+1)\langle \tilde{\mathbb{A}}y^{k+1} - \tilde{\mathbb{A}}y^k, y^{k+1} - y^k\rangle \leq 0,
\end{aligned}$$

where the final inequality follows from monotonicity of $\mathbb{A}$. Finally, we conclude that

$$\frac{k^2}{2}\|\tilde{\mathbb{A}}y^k\|^2 \leq V^k \leq V^0 = \frac{1}{2}\|x^0 - x^\star\|^2. \qquad \square$$

12.3 WHEN DOES AN ACCELERATION ACCELERATE?

In optimization (and more generally in applied mathematics and computer science), convergence rates are usually established in the worst case, and the convergence observed in practice can be faster than this guarantee if the given problem instance does not represent the worst case. If an unaccelerated method actually converges at an $O(1/k)$ rate, then an $O(1/k^2)$ acceleration will accelerate the convergence. However, if the observed convergence is already faster than $O(1/k^2)$, the guarantee of the accelerated method, then the acceleration does not guarantee a speedup and may even slow down the convergence. See Exercise 12.9.

In practice, an acceleration is a technique that *sometimes* provides a speedup. When an "accelerated" variant of a method is available, one should try it out with the expectation that it may improve or worsen the convergence.

BIBLIOGRAPHICAL NOTES

Nesterov's accelerated gradient method was first presented in 1983 [Nes83]. Since then, there has been a large body of work studying accelerated first-order methods in optimization.

The accelerated $O(1/k^2)$ rate on the squared fixed-point residual for the problem of finding fixed points of nonexpansive mappings was first established by Sabach and Shtern in 2017 [SS17]. The "accelerated proximal point method" of this chapter has a better constant for the rate $O(1/k^2)$ and was independently discovered by Lieder [Lie21] and Kim [Kim21]. Specifically, the classical Halpern method [Hal67] has the form

$$x^{k+1} = \lambda_k x^0 + (1 - \lambda_k)\mathbb{T}x^k,$$

and Lieder showed that the specific choice of $\lambda_k = 1/(k+2)$ produces the stated $O(1/k^2)$ rate. Kim used the computer-assisted tool "performance estimation problem" [DT14, KF16, THG17, RTBG20] to generate the accelerated proximal point method as the method with the best theoretical guarantee among a certain class of methods.

Nesterov's original 1983 paper used a Lyapunov analysis [Nes83], but this proof is somewhat forgotten as Nesterov used a different "estimate sequence" technique in his later work [Nes88] for analyzing and constructing accelerated gradient methods. Tseng provided a unified Lyapunov analysis for a variety of accelerated methods [Tse08]. The particular Lyapunov analyses presented in this chapter were inspired by Bansal and Gupta [BG19] and Taylor and Bach [TB19].

In this chapter, we cover Nesterov's AGM, which is most well known, and APPM/OHM, which is most relevant to the content of this book. However, there are many other accelerations considered in the optimization literature: acceleration of the proximal point method applied to functions by Güler [Gül92], Chambolle–Pock and Davis–Yin accelerations applied to PDHG and DYS [CP11a, DY17b], generalization of Nesterov's acceleration to problems involving a closed convex set or a proximable function by Nesterov [Nes04, §2.2.3 and 2.2.4] and [Nes13] and by Beck and Teboulle [BT09], a different type of acceleration by Nesterov applied to structured convex-concave optimization problems [Nes05], Auslender and Teboulle's generalization of Nesterov's acceleration to the setup of Bregman divergences [AT06], Anderson acceleration applied to fixed-point iterations [And65, WN11], geometric descent as an alternative to Nesterov's acceleration by Bubeck, Lee, and Singh [BLS15], optimized gradient method as an improvement upon Nesterov's AGM by Drori, Teboulle, Kim, and Fessler [DT14, KF16, KF17], triple momemtum method as an improvement upon Nesterov's AGM in the strongly convex setup by Van Scoy, Freeman, and Lynch [VSFL18], catalyst acceleration as a meta-algorithm for accelerating unaccelerated methods by Lin, Mairal, and Harchaoui [LMH15, LMH18], and OGM-G for accelerating the reduction of the gradient magnitude by Kim and Fessler [KF21]. For the problem of minimizing smooth convex functions, one can establish a $O(1/k^4)$ rate on the squared gradient magnitude [NGGD20, Remark 2.1], which is significantly faster than the $O(1/k^2)$ rate of APPM/OHM.

EXERCISES

12.1 Show that the two forms of Nesterov AGM are equivalent.

12.2 Show that APPM and OHM are equivalent with $\mathbb{T} = 2\mathbb{J}_{\mathbb{A}} - \mathbb{I}$.

12.3 *AGM with optimal parameters.* Consider the problem

$$\underset{x\in\mathbb{R}^n}{\text{minimize}} \quad f(x),$$

where f is a convex, L-smooth function with a minimizer. Show that the method

$$\begin{aligned}
\varphi_{k+1} &= \frac{1+\sqrt{1+4\varphi_k^2}}{2}\\
x^{k+1} &= y^k - \frac{1}{L}\nabla f(y^k)\\
y^{k+1} &= x^{k+1} + \frac{\varphi_k - 1}{\varphi_{k+1}}(x^{k+1} - x^k),
\end{aligned}$$

where $k = 0, 1, \ldots$, $x^0 = y^0$, and $\varphi_0 = 1$, converges with the rate

$$f(x^k) - f(x^\star) \le \frac{L\|x^0 - x^\star\|^2}{2\varphi_{k-1}^2},$$

where $\varphi_{k-1} \ge \frac{k+1}{2}$. This method is also called Nesterov's accelerated gradient method. Show that the rate of this variant is slightly better than the rate shown in §12.1.

Hint. Use

$$V^k = \varphi_{k-1}^2\left(f(x^k) - f(x^\star)\right) + \frac{L}{2}\|z^k - x^\star\|^2.$$

12.4 *Backtracking linesearch.* Suppose the smoothness constant $L > 0$ exists but is not known. Consider the method

$$\begin{aligned}
x^{k+1} &= y^k - \frac{1}{2^i L_k}\nabla f(x^k),\\
&\qquad \text{try } i = 0, 1, \ldots \text{ until } f(x^{k+1}) - f(y^k) + \frac{1}{2^i L_k}\|\nabla f(x^k)\|^2 \le 0\\
L_{k+1} &= 2\cdot 2^i L_k\\
y^{k+1} &= x^{k+1} + \frac{k-1}{k+2}(x^{k+1} - x^k),
\end{aligned}$$

where $x^0 = y^0 \in \mathbb{R}^n$ and L_0 is an estimate of L. Show that if $L_0 \le L$, then

$$f(x^k) - f(x^\star) \le \frac{4L\|x^0 - x^\star\|^2}{k^2}.$$

Hint. The $L_k = 2^i L_{k-1}$ step represents a doubling of the estimate of L. This doubling happens finitely many times.

12.5 *Accelerated proximal gradient (FISTA).* Consider the problem

$$\underset{x\in\mathbb{R}^n}{\text{minimize}} \quad f(x) + g(x),$$

where f is convex and L-smooth and g is CCP. Assume $f+g$ has a minimizer. Consider the method

$$x^{k+1} = \mathrm{Prox}_{\frac{1}{L}g}(y^k - \frac{1}{L}\nabla f(y^k))$$
$$y^{k+1} = x^{k+1} + \frac{k-1}{k+2}(x^{k+1} - x^k),$$

where $x^0 = y^0$, is called the accelerated proximal gradient method and was proposed by Nesterov in 2004 [Nes04, §2.2.3 and 2.2.4] and Beck and Teboulle in 2009 [BT09]. The instance where $g(x) = \|x\|_1$ is more commonly known as the fast iterative shrinkage-thresholding algorithm (FISTA).
Show

$$f(x^k) + g(x^k) - f(x^\star) - g(x^\star) \leq \frac{2\|x^0 - x^\star\|^2}{\alpha(k+1)^2}.$$

Hint. Use

$$V^k = \theta_{k-1}^2 \left(f(x^k) + g(x^k) - f(x^\star) - g(x^\star) \right) + \frac{L}{2}\|z^k - x^\star\|^2.$$

12.6 *Strongly-convex accelerated gradient method.* Consider the problem

$$\underset{x\in\mathbb{R}^n}{\text{minimize}} \quad f(x),$$

where f is μ-strongly convex and L-smooth. The method

$$x^{k+1} = y^k - \frac{1}{L}\nabla f(y^k)$$
$$y^{k+1} = x^{k+1} + \frac{\sqrt{\kappa}-1}{\sqrt{\kappa}+1}(x^{k+1} - x^k),$$

where $k = 0, 1, \ldots$, $x^0 = y^0$, and $\kappa = L/\mu$, is called the strongly convex accelerated gradient method (SC-AGM). Show

$$f(x^k) - f(x^\star) \leq \frac{\mu + L}{2}\|x^0 - x^\star\|^2 e^{-k/\sqrt{\kappa}}.$$

Hint. Use

$$V^k = \left(1 + \frac{1}{\sqrt{\kappa}-1}\right)^k \left(f(x^k) - f(x^\star) + \frac{\mu}{2}\|z^k - x^\star\|^2 \right),$$

where $z^k = (1+\sqrt{\kappa})y^k - \sqrt{\kappa}x^k$.
Remark. The (unaccelerated) gradient method applied to this problem setup converges with the slower rate $\mathcal{O}(e^{-k/\kappa})$. See Exercise 13.5.

12.7 *Squared gradient norm of gradient descent.* Consider the problem

$$\underset{x\in\mathbb{R}^n}{\text{minimize}} \quad f(x),$$

where f is a convex, L-smooth function with a minimizer. Show that for GD

$$x^{k+1} = x^k - \frac{1}{L}\nabla f(x^k)$$

the quantity

$$V^k = (2k+1)L(f(x^k) - f(x^\star)) + k(k+2)\|\nabla f(x^k)\|^2 + L^2\|x^k - x^\star\|^2$$

is nonincreasing for $k = 0, 1, \ldots$.

Remark. This result implies the rate $\|\nabla f(x^k)\|^2 \le \frac{L}{k(k+2)}(L\|x^0 - x^\star\|^2 + f(x^0) - f(x^\star))$, which is on the same order as APPM. So when the the goal is to reduce the magnitude of the gradient of an L-smooth convex function, GD and APPM have the same rate. This result is due to Taylor and Bach [TB19].

12.8 *Gradient descent stepsize.* Consider the problem

$$\underset{x\in\mathbb{R}^n}{\text{minimize}} \quad f(x),$$

where f is a convex, L-smooth function with a minimizer. Show that gradient descent with stepsize α

$$x^{k+1} = x^k - \alpha \nabla f(x^k)$$

converges in function value at rate $O(1/k)$ for $\alpha \in (0, 2/L)$.

12.9 *Acceleration can slow you down.* Consider the specific problem instance

$$\underset{x\in\mathbb{R}^3}{\text{minimize}} \quad x_1^2 + 2x_2^2 + 3x_3^2.$$

Apply gradient descent, AGM, and SC-AGM of Exercise 12.6. Experimentally, what are the convergence rates on the objective value? Also apply the proximal point method and APPM. Experimentally, what are the convergence rates on the squared fixed-point residual?

12.10 *APPM/OHM for finding zeros of cocoercive operators.* Consider the problem

$$\underset{x\in\mathbb{R}^n}{\text{find}} \quad 0 \in \mathbb{A}x,$$

where $\mathbb{A}\colon \mathbb{R}^n \to \mathbb{R}^n$ is β-cocoercive with $\beta > 0$. Show that method

$$x^{k+1} = \frac{1}{k+2}x^0 + \frac{k+1}{k+2}(\mathbb{I} - 2\beta\mathbb{A})x^k$$

converges with the rate

$$\|\mathbb{A}x^k\|^2 \le \frac{\|x^0 - x^\star\|^2}{\beta^2(k+1)^2}$$

for $k = 0, 1, \ldots$. Also show that this method is equivalent to

$$\begin{aligned} y^{k+1} &= (\mathbb{I} - \beta\mathbb{A})x^k \\ x^{k+1} &= y^{k+1} + \frac{k}{k+2}(y^{k+1} - y^k) - \frac{k}{k+2}(y^k - x^{k-1}), \end{aligned}$$

for $k = 0, 1, \ldots$.

Hint. Show that $\mathbb{I} - 2\beta\mathbb{A}$ is nonexpansive and use Theorem 18.

13 Scaled Relative Graphs

In this chapter, we present a new notion called the scaled relative graph (SRG). The SRG provides a correspondence between algebraic operations on nonlinear operators and geometric operations on subsets of the 2D plane. We can think of the SRG as a signature of an operator analogous to how eigenvalues are a signature of a matrix. Using this machinery and elementary Euclidean geometry, we establish averagedness and contractiveness of certain operators and thereby establish convergence of corresponding fixed-point iterations. The geometric arguments constitute rigorous proofs and are not mere illustrations.

The geometric approach of this chapter contrasts with the analytical proofs based on inequalities, which are more common in the splitting methods literature. We make clear that the geometric proof techniques based on the SRG are meant to supplement, rather than replace the analytical techniques.

13.1 BASIC DEFINITIONS

Operator Classes We say $\mathcal{A}$ is a *class of operators* if $\mathcal{A}$ is a set of operators on $\mathbb{R}^n$ for all $n \in \mathbb{N}$. Note that $\mathbb{A}_1, \mathbb{A}_2 \in \mathcal{A}$ need not be defined on the same Euclidean spaces, i.e., $\mathbb{A}_1 \colon \mathbb{R}^n \rightrightarrows \mathbb{R}^n$, $\mathbb{A}_2 \colon \mathbb{R}^m \rightrightarrows \mathbb{R}^m$, and $n \neq m$ is possible.

Given classes of operators $\mathcal{A}$ and $\mathcal{B}$ and $\alpha > 0$, write

$$\begin{aligned}
\mathcal{A} + \mathcal{B} &= \{\mathbb{A} + \mathbb{B} \mid \mathbb{A} \in \mathcal{A},\ \mathbb{B} \in \mathcal{B},\ \mathbb{A} \colon \mathbb{R}^n \rightrightarrows \mathbb{R}^n,\ \mathbb{B} \colon \mathbb{R}^n \rightrightarrows \mathbb{R}^n\} \\
\mathcal{A}\mathcal{B} &= \{\mathbb{A}\mathbb{B} \mid \mathbb{A} \in \mathcal{A},\ \mathbb{B} \in \mathcal{B},\ \mathbb{A} \colon \mathbb{R}^n \rightrightarrows \mathbb{R}^n,\ \mathbb{B} \colon \mathbb{R}^n \rightrightarrows \mathbb{R}^n\} \\
\mathbb{J}_{\alpha\mathcal{A}} &= \{\mathbb{J}_{\alpha\mathbb{A}} \mid \mathbb{A} \in \mathcal{A},\ \mathbb{A} \colon \mathbb{R}^n \rightrightarrows \mathbb{R}^n\} \\
\mathbb{R}_{\alpha\mathcal{A}} &= 2\mathbb{J}_{\alpha\mathcal{A}} - \mathbb{I} = \{2\mathbb{J} - \mathbb{I} \mid \mathbb{J} \in \mathbb{J}_{\alpha\mathcal{A}},\ \mathbb{J} \colon \mathbb{R}^n \rightrightarrows \mathbb{R}^n,\ \mathbb{I} \colon \mathbb{R}^n \rightrightarrows \mathbb{R}^n\}
\end{aligned}$$

To clarify, these definitions require that $\mathbb{A}$ and $\mathbb{B}$ are operators on the same Euclidean space $\mathbb{R}^n$ (so n is shared), as otherwise the operations would not make sense. Also define

$$\mathcal{A}^{-1} = \{\mathbb{A}^{-1} \mid \mathbb{A} \in \mathcal{A}\}, \qquad \alpha\mathcal{A} = \{\alpha\mathbb{A} \mid \mathbb{A} \in \mathcal{A}\}.$$

For $L \in (0, \infty)$, define the class of L-Lipschitz operators as

$$\mathcal{L}_L = \left\{\mathbb{A} \colon \operatorname{dom} \mathbb{A} \to \mathbb{R}^n \mid \|\mathbb{A}x - \mathbb{A}y\|^2 \le L^2\|x - y\|^2,\ \forall x, y \in \operatorname{dom} \mathbb{A} \subseteq \mathbb{R}^n,\ n \in \mathbb{N}\right\}.$$

For $\beta \in (0, \infty)$, define the class of β-cocoercive operators as

$$\mathcal{C}_\beta = \left\{\mathbb{A}\colon \operatorname{dom}\mathbb{A} \to \mathbb{R}^n \mid \langle \mathbb{A}x - \mathbb{A}y, x - y\rangle \geq \beta\|\mathbb{A}x - \mathbb{A}y\|^2,\ \forall x, y \in \operatorname{dom}\mathbb{A} \subseteq \mathbb{R}^n,\ n \in \mathbb{N}\right\}.$$

Define the class of monotone operators as

$$\mathcal{M} = \left\{\mathbb{A}\colon \mathbb{R}^n \rightrightarrows \mathbb{R}^n \mid \langle \mathbb{A}x - \mathbb{A}y, x - y\rangle \geq 0,\ \forall x, y \in \operatorname{dom}\mathbb{A},\ n \in \mathbb{N}\right\}.$$

For $\mu \in (0, \infty)$, define the class of μ-strongly monotone operators as

$$\mathcal{M}_\mu = \{\mathbb{A}\colon \mathbb{R}^n \rightrightarrows \mathbb{R}^n \mid \langle \mathbb{A}x - \mathbb{A}y, x - y\rangle \geq \mu\|x - y\|^2,\ \forall x, y \in \operatorname{dom}\mathbb{A},\ n \in \mathbb{N}\}.$$

For $\theta \in (0, 1)$, define the class of θ-averaged operators as

$$\mathcal{N}_\theta = (1 - \theta)\mathbb{I} + \theta\mathcal{L}_1.$$

In these definitions, we do not impose any requirements on the domain or maximality of the operators. We define these classes to include operators on $\mathbb{R}^n$ for all $n \in \mathbb{N}$ in order to avoid discussing issues specific to the cases $n = 1$ and $n = 2$.

Write $\mathcal{F}_{\mu,L}$, $\mathcal{F}_{0,L}$, $\mathcal{F}_{\mu,\infty}$, and $\mathcal{F}_{0,\infty}$ for the sets of CCP functions on $\mathbb{R}^n$ for all $n \in \mathbb{N}$ that are respectively μ-strongly convex and L-smooth, convex and L-smooth, μ-strongly convex, and convex, for $0 < \mu < L < \infty$. Write

$$\partial\mathcal{F}_{\mu,L} = \{\partial f \mid f \in \mathcal{F}_{\mu,L}\},$$

where $0 \leq \mu < L \leq \infty$.

Basic Geometry For any $a, b \in \mathbb{R}^n$, let

$$\angle(a, b) = \begin{cases} \arccos\left(\frac{\langle a, b\rangle}{\|a\|\|b\|}\right) & \text{if } a \neq 0,\ b \neq 0 \\ 0 & \text{otherwise} \end{cases}$$

denote the angle between them. The *spherical triangle inequality* states that any nonzero $a, b, c \in \mathbb{R}^n$ satisfies

$$|\angle(a, b) - \angle(b, c)| \leq \angle(a, c) \leq \angle(a, b) + \angle(b, c).$$

Figure 13.1 illustrates the inequality. We use the spherical triangle inequality in Theorem 26 to argue that there is no need to consider a third dimension and that we can continue the analysis in 2D.

The *Stewart's theorem* states that for a triangle $\triangle ABC$ and Cevian $\overline{CD}$ to the side $\overline{AB}$,

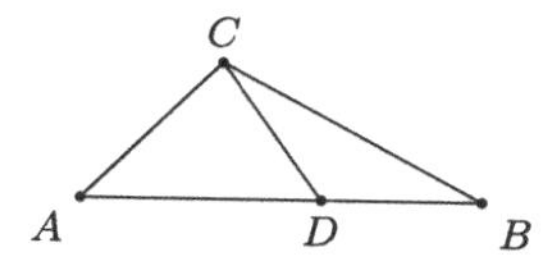

the lengths of the line segments satisfy

$$\overline{AD} \cdot \overline{CB}^2 + \overline{DB} \cdot \overline{AC}^2 = \overline{AB} \cdot \overline{CD}^2 + \overline{AD} \cdot \overline{DB}^2 + \overline{AD}^2 \cdot \overline{DB}.$$

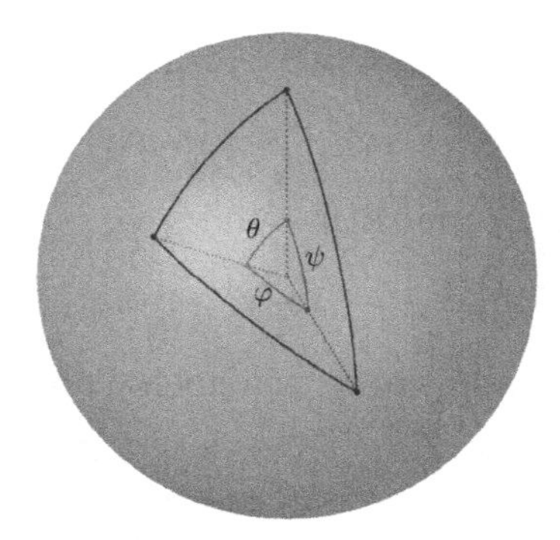

Figure 13.1 Spherical triangle inequality: $|\theta - \varphi| \leq \psi \leq \theta + \varphi$.

Extended Complex Plane and Inversive Geometry We use the *extended complex plane* $\overline{\mathbb{C}} = \mathbb{C} \cup \{\infty\}$ to represent the 2D plane and the point at infinity. Since complex numbers compactly represent rotations and scaling, this choice simplifies our notation compared to using $\mathbb{R}^2 \cup \{\infty\}$. We avoid the operations $\infty + \infty$, $0/0$, ∞/∞, and $0 \cdot \infty$. Otherwise, we adopt the convention of $z + \infty = \infty$, $z/\infty = 0$, $z/0 = \infty$, and $z \cdot \infty = \infty$.

We call $z \mapsto \bar{z}^{-1}$, a one-to-one map from $\overline{\mathbb{C}}$ to $\overline{\mathbb{C}}$, the *inversion* map. To clarify, $\bar{z}$ denotes the complex conjugate of z. In polar form, it is $re^{i\varphi} \mapsto (1/r)e^{i\varphi}$ for $0 \leq r \leq \infty$, i.e., inversion preserves the angle and inverts the magnitude.

Generalized circles consist of (finite) circles and lines with $\{\infty\}$. The interpretation is that a line is a circle with infinite radius. Inversion maps generalized circles to generalized circles, and we can perform it with the following semi-geometric procedure.

1. Draw a line L through the origin orthogonally intersecting the generalized circle. This means L intersects the boundary perpendicularly, which implies L goes through the circle's center when the generalized circle is finite.
2. Let $-\infty < x < y \leq \infty$ represent the signed distance of the intersecting points from the origin along this line. If the generalized circle is a line, then $y = \infty$.
3. Draw a generalized circle orthogonally intersecting L at $(1/x)$ and $(1/y)$.
4. When inverting a region with a generalized circle as the boundary, pick a point on L within the interior of the region to determine on which side of the boundary the inverted interior lies.

Examples 13.1 and 13.2 illustrate these steps.

Example 13.1 Illustration of inverting a disk. In step 1, we choose L to be the x-axis (although any line through the origin works). In steps 2 and 3, we identify x and y, invert them to x^{-1} and y^{-1}, and draw the generalized circle in the inverted plane to be the new boundary. In step 4, we determine that the interior of the disk is mapped to the exterior by noting that 1, a point invariant under the inversion map, is excluded in the original region and therefore is excluded in the inverted region.

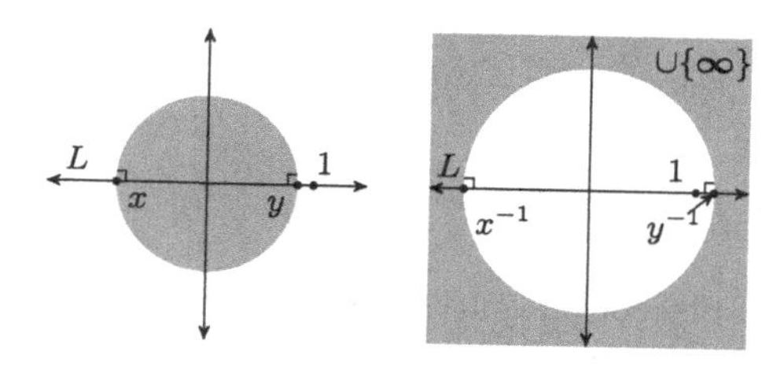

Example 13.2 The three vertical pairs illustrate inversion. In step 1, we choose L to be the x-axis. In step 4 we determine the interior by examining point 1: if 1 is included in the original region, it is included in the inverted region, and vice versa.

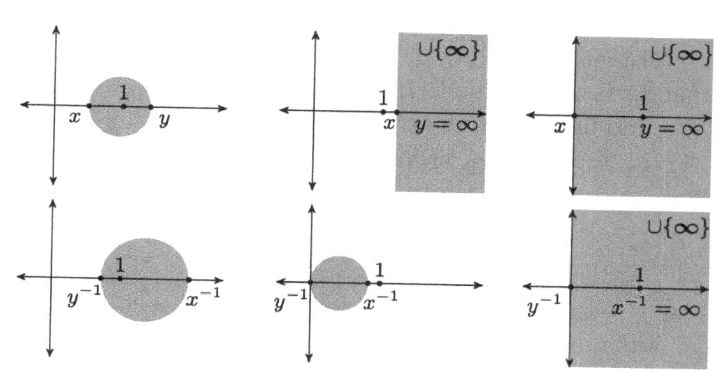

13.2 SCALED RELATIVE GRAPHS

In this section, we define the notion of *scaled relative graphs* (SRG). Loosely speaking, SRG maps the action of an operator to a set on the extended complex plane.

13.2.1 SRG of Operators

Consider an operator $\mathbb{A}: \mathbb{R}^n \rightrightarrows \mathbb{R}^n$. Let $x, y \in \mathbb{R}^n$ be a pair of inputs and let $u, v \in \mathbb{R}^n$ be their corresponding outputs, i.e., $u \in \mathbb{A}x$, and $v \in \mathbb{A}y$. The goal is to understand the change in output relative to the change in input.

First, consider the case $x \neq y$. Consider the complex conjugate pair

$$z = \frac{\|u - v\|}{\|x - y\|} \exp\left[\pm i\angle(u - v, x - y)\right].$$

The absolute value (magnitude) $|z| = \frac{\|u-v\|}{\|x-y\|}$ represents the size of the change in outputs relative to the size of the change in inputs. The argument (angle) $\angle(u-v, x-y)$ represents how much the change in outputs is aligned with the change in inputs. Equivalently, $\operatorname{Re} z$ and $\operatorname{Im} z$ respectively represent the components of $u - v$ aligned with and perpendicular to $x - y$:

$$\operatorname{Re} z = \operatorname{sgn}(\langle u - v, x - y\rangle)\frac{\|\Pi_{\operatorname{span}\{x-y\}}(u - v)\|}{\|x - y\|} = \frac{\langle u - v, x - y\rangle}{\|x - y\|^2}$$
$$\operatorname{Im} z = \pm\frac{\|\Pi_{\{x-y\}^\perp}(u - v)\|}{\|x - y\|}, \tag{13.1}$$

where $\Pi_{\operatorname{span}\{x-y\}}$ is the projection onto the span of $x - y$ and $\Pi_{\{x-y\}^\perp}$ is the projection onto the subspace orthogonal to $x - y$.

Define the SRG of an operator $\mathbb{A}\colon \mathbb{R}^n \rightrightarrows \mathbb{R}^n$ as

$$\mathcal{G}(\mathbb{A}) = \left\{ \frac{\|u - v\|}{\|x - y\|} \exp\left[\pm i\angle(u - v, x - y)\right] \,\middle|\, u \in \mathbb{A}x,\ v \in \mathbb{A}y,\ x \neq y \right\}$$
$$\Big(\cup \{\infty\} \text{ if } \mathbb{A} \text{ is multi-valued}\Big).$$

We clarify several points: (i) $\mathcal{G}(\mathbb{A}) \subseteq \overline{\mathbb{C}}$. (ii) $\infty \in \mathcal{G}(\mathbb{A})$ if and only if there is a point $x \in \mathbb{R}^n$ such that $\mathbb{A}x$ is multi-valued. (In this case, there exists $(x, u), (y, v) \in \mathbb{A}$ such that $x = y$ and $u \neq v$, and the idea is that $|z| = \|u - v\|/0 = \infty$, i.e., $u - v$ is infinitely larger than $x - y = 0$.) (iii) the $\pm$ makes $\mathcal{G}(\mathbb{A})$ symmetric about the real axis. (We include the $\pm$ because $\angle(u - v, x - y)$ always returns a nonnegative angle.)

Example 13.3 SRGs of the operators: $\Pi_L : \mathbb{R}^2 \to \mathbb{R}^2$ is the projection onto an arbitrary line L; $\mathbb{A}\colon \mathbb{R}^2 \to \mathbb{R}^2$ is defined as $\mathbb{A}(u, v) = (0, u)$; $\partial\|\cdot\|$ is the subdifferential of the Euclidean norm on $\mathbb{R}^n$ with $n \geq 2$; and $\mathbb{B}\colon \mathbb{R}^3 \to \mathbb{R}^3$ is defined as $\mathbb{B}(u, v, w) = (u, 2v, 3w)$. The shapes were obtained by plugging the operators into the definition of the SRG and performing direct calculations.

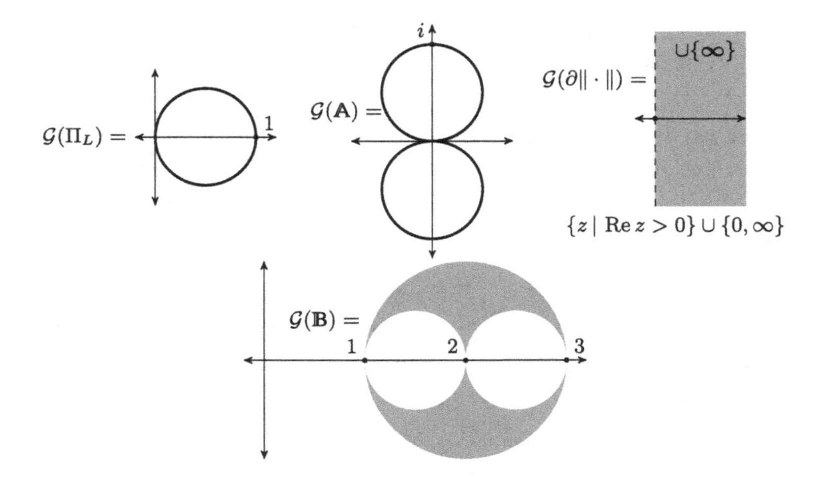

The SRG $\mathcal{G}(\mathbb{A})$ maps the action of the operator $\mathbb{A}$ to points in $\overline{\mathbb{C}}$. In the following sections, we will need to conversely take any point in $\overline{\mathbb{C}}$ and find an operator whose action maps to that point. Lemma 4 provides such constructions.

Lemma 4 *Take any $z = z_r + z_i i \in \mathbb{C}$. Define $\mathbb{A}_z : \mathbb{R}^2 \to \mathbb{R}^2$ and $\mathbb{A}_\infty : \mathbb{R}^2 \rightrightarrows \mathbb{R}^2$ as*

$$\mathbb{A}_z \begin{bmatrix} \zeta_1 \\ \zeta_2 \end{bmatrix} = \begin{bmatrix} z_r\zeta_1 - z_i\zeta_2 \\ z_r\zeta_2 + z_i\zeta_1 \end{bmatrix} \qquad \mathbb{A}_\infty(x) = \begin{cases} \mathbb{R}^2 & \text{if } x = 0 \\ \emptyset & \text{otherwise.} \end{cases}$$

Then,

$$\mathcal{G}(\mathbb{A}_z) = \{z, \bar{z}\}, \qquad \mathcal{G}(\mathbb{A}_\infty) = \{\infty\}.$$

If we write $\cong$ to identify an element of $\mathbb{R}^2$ with an element in $\mathbb{C}$ in that

$$\begin{bmatrix} x \\ y \end{bmatrix} \cong x + yi,$$

then we can view $\mathbb{A}_z$ as complex multiplication with z in the sense that

$$\mathbb{A}_z \begin{bmatrix} \zeta_1 \\ \zeta_2 \end{bmatrix} \cong z(\zeta_1 + \zeta_2 i).$$

Proof. Again, we write $\cong$ to identify an element of $\mathbb{R}^2$ with an element in $\mathbb{C}$. Write $z = r_z e^{i\theta_z}$. Consider any $x, y \in \mathbb{R}^2$ where $x \neq y$ and define $u = \mathbb{A}_z x$ and $v = \mathbb{A}_z y$. Then we can write

$$x - y = r_w \begin{bmatrix} \cos(\theta_w) \\ \sin(\theta_w) \end{bmatrix},$$

where $r_w > 0$, and

$$u - v = \mathbb{A}_z(x - y) \cong r_z r_w e^{i(\theta_z + \theta_w)}.$$

This gives us

$$\frac{\|u - v\|}{\|x - y\|} = r_z, \quad \angle(u - v, x - y) = |\theta_z|,$$

and

$$\mathcal{G}(\mathbb{A}_z) = \left\{ r_z e^{i\theta_z}, r_z e^{-i\theta_z} \right\}.$$

Now consider A_∞. By definition, $\infty \in \mathcal{G}(A_\infty)$. For any $u \in A_\infty x$ and $v \in A_\infty y$, we have $x = y = 0$, and therefore $\mathcal{G}(A_\infty)$ contains no finite $z \in \mathbb{C}$. We conclude $\mathcal{G}(A_\infty) = \{\infty\}$. □

13.2.2 SRG and Eigenvalues

For linear operators, the SRG generalizes eigenvalues: specifically, if $A \in \mathbb{R}^{n\times n}$ and $n = 1$ or $n \geq 3$, then $\lambda(A) \subseteq \mathcal{G}(A)$, where $\lambda(A)$ denotes the set of eigenvalues of A. It is also true (and not obvious to show) that $\mathcal{G}(A^\intercal) = \mathcal{G}(A)$ for any $A \in \mathbb{R}^{n\times n}$. See Examples 13.4 and 13.5. See the bibliographic notes section for further discussion.

Example 13.4 SRG of a 3×3 matrix. The three points denote the eigenvalues.

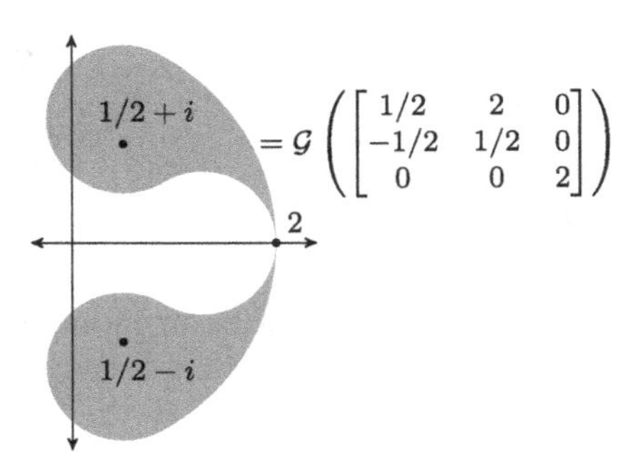

Example 13.5 For normal matrices, multiplicity of eigenvalues do not affect the SRG. (Left) SRG of an $n \times n$ normal matrix with one distinct real eigenvalue and three distinct complex conjugate eigenvalue pairs. (Right) SRG of an $n \times n$ symmetric matrix with distinct eigenvalues $\lambda_1 < \lambda_2 < \cdots < \lambda_6$.

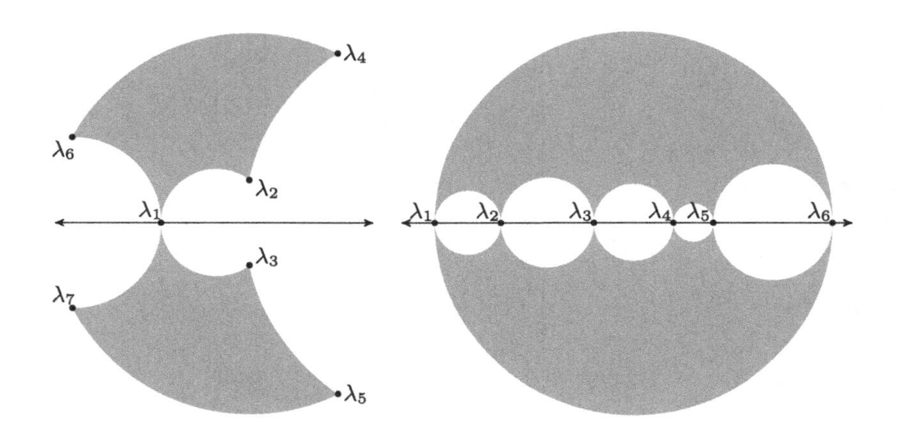

13.2.3 SRG of Operator Classes

Define the SRG of a collection of operators $\mathcal{A}$ as

$$\mathcal{G}(\mathcal{A}) = \bigcup_{\mathbb{A} \in \mathcal{A}} \mathcal{G}(\mathbb{A}).$$

We focus more on SRGs of operator classes, rather than individual operators, because theorems are usually stated with operator classes. For example, one might say "If $\mathbb{A}$ is $1/2$-cocoercive, i.e., if $\mathbb{A} \in \mathcal{C}_{1/2}$, then $\mathbb{I} - \mathbb{A}$ is nonexpansive."

Theorem 19 Let $\mu, \beta, L \in (0, \infty)$ and $\theta \in (0, 1)$. Then

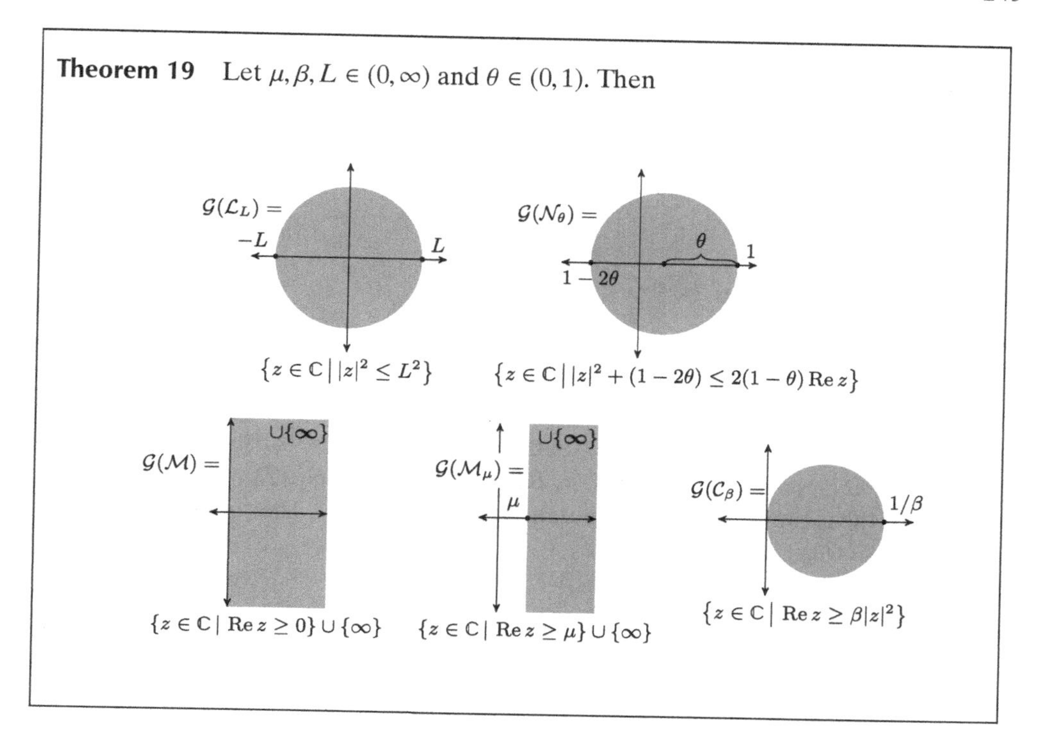

Proof. First, characterize $\mathcal{G}(\mathcal{L}_L)$. We have $\mathcal{G}(\mathcal{L}_L) \subseteq \{z \in \mathbb{C} \mid |z|^2 \leq L^2\}$ since

$$\mathbb{A} \in \mathcal{L}_L \quad \Rightarrow \quad \frac{\|\mathbb{A}x - \mathbb{A}y\|}{\|x - y\|} \leq L, \forall x, y \in \mathbb{R}^n, x \neq y \quad \Rightarrow \quad \mathcal{G}(\mathbb{A}) \subseteq \{z \in \mathbb{C} \mid |z|^2 \leq L^2\}.$$

Conversely, given any $z \in \mathbb{C}$ such that $|z| \leq L$, the operator $\mathbb{A}_z$ of Lemma 4 satisfies $\|\mathbb{A}_z x - \mathbb{A}_z y\| \leq L\|x - y\|$ for any $x, y \in \mathbb{R}^2$, i.e., $\mathbb{A}_z \in \mathcal{L}_L$, and $\mathcal{G}(\mathbb{A}_z) = \{z, \bar{z}\}$. Therefore, $\mathcal{G}(\mathcal{L}_L) \supseteq \{z \in \mathbb{C} \mid |z|^2 \leq L^2\}$.

Next, characterize $\mathcal{G}(\mathcal{M})$. For any $\mathbb{A} \in \mathcal{M}$, monotonicity implies

$$\frac{\langle u - v, x - y \rangle}{\|x - y\|^2} \geq 0, \quad \forall u \in \mathbb{A}x, v \in \mathbb{A}y, x \neq y.$$

Considering (13.1), we conclude $\mathcal{G}(\mathbb{A}) \backslash \{\infty\} \subseteq \{z \mid \operatorname{Re} z \geq 0\}$. On the other hand, given any $z \in \{z \mid \operatorname{Re} z \geq 0\}$, the operator $\mathbb{A}_z$ of Lemma 4 satisfies $\langle \mathbb{A}_z x - \mathbb{A}_z y, x - y \rangle \geq 0$ for any $x, y \in \mathbb{R}^2$, i.e., $\mathbb{A}_z \in \mathcal{M}$, and $\mathcal{G}(\mathbb{A}_z) = \{z, \bar{z}\}$. Therefore, $z \in \mathcal{G}(\mathbb{A}_z) \subset \mathcal{G}(\mathcal{M})$, and we conclude $\{z \mid \operatorname{Re} z \geq 0\} \subseteq \mathcal{G}(\mathcal{M})$. Finally, note that $\infty \in \mathcal{G}(\mathcal{M})$ is equivalent to saying that there exists a multi-valued operator in $\mathcal{M}$. The $\mathbb{A}_\infty$ of Lemma 4 is one such example.

We leave the characterization of $\mathcal{G}(\mathcal{M}_\mu)$, $\mathcal{G}(\mathcal{C}_\beta)$, and $\mathcal{G}(\mathcal{N}_\theta)$ to Exercise 13.12. □

As the operator classes $\mathcal{M}$, $\mathcal{M}_\mu$, $\mathcal{C}_\beta$, $\mathcal{L}_L$, and $\mathcal{N}_\theta$ are defined to include operators on $\mathbb{R}^n$ for all $n \in \mathbb{N}$, the use of Lemma 4 is sufficient, even though it only concerns the case $n = 2$.

Theorem 20 Let $0 < \mu < L < \infty$. Then

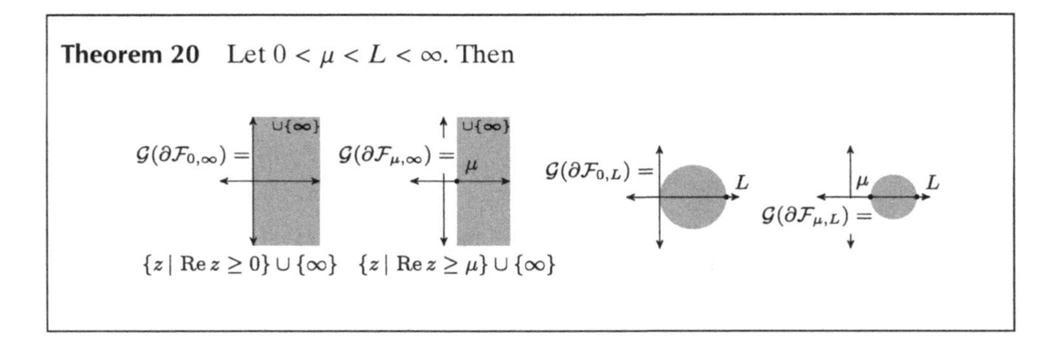

Proof. Since $\partial\mathcal{F}_{0,\infty} \subset \mathcal{M}$, we have $\mathcal{G}(\partial\mathcal{F}_{0,\infty}) \subseteq \mathcal{G}(\mathcal{M}) = \{z \in \mathbb{C} \mid \operatorname{Re} z \geq 0\} \cup \{\infty\}$ by Theorem 19. We claim $f\colon \mathbb{R}^2 \to \mathbb{R}$ defined by $f(x,y) = |x|$ satisfies $\mathcal{G}(\partial f) = \{z \in \mathbb{C} \mid \operatorname{Re} z \geq 0\} \cup \{\infty\}$. This tells us $\{z \in \mathbb{C} \mid \operatorname{Re} z \geq 0\} \cup \{\infty\} \subseteq \mathcal{G}(\partial\mathcal{F}_{0,\infty})$.

We prove the claim with basic computation. Let $f(x,y) = |x|$. The subgradient has the form $\partial f(x,y) = (h(x), 0)$ for h defined by

$$h(x) = \begin{cases} \{-1\} & \text{for } x < 0 \\ \{u \mid -1 \leq u \leq 1\} & \text{for } x = 0 \\ \{1\} & \text{for } x > 0. \end{cases}$$

Since ∂f is multi-valued at $(0,0)$, we have $\infty \in \mathcal{G}(\partial f)$. Since $\partial f(1,0) = \partial f(2,0)$, we have $0 \in \mathcal{G}(\partial f)$. The input-output pairs $(0,0) \in \partial f(0,0)$ and $(h(R\cos(\theta)),0) \in \partial f(R\cos(\theta), R\sin(\theta))$ map to the point $R^{-1}(|\cos(\theta)|, \pm\sin(\theta)) \in \mathbb{C}$. Clearly, the image of this map over the range $R \in (0,\infty)$, $\theta \in [0, 2\pi)$ is the right-hand plane except the origin. Hence $\mathcal{G}(\partial f) = \{z \in \mathbb{C} \mid \operatorname{Re} z \geq 0\} \cup \{\infty\}$.

We leave the characterization of $\mathcal{G}(\partial\mathcal{F}_{\mu,\infty})$ and $\mathcal{G}(\partial\mathcal{F}_{0,L})$ to Exercise 13.13. □

13.2.4 SRG-Full Classes

An operator defines its SRG. Conversely, can we examine the SRG and conclude something about the operator? To perform this type of reasoning, we need further conditions.

We say the class of operators $\mathcal{A}$ is *SRG-full* if

$$\mathbb{A} \in \mathcal{A} \quad \Leftrightarrow \quad \mathcal{G}(\mathbb{A}) \subseteq \mathcal{G}(\mathcal{A}).$$

Since $\mathbb{A} \in \mathcal{A} \Rightarrow \mathcal{G}(\mathbb{A}) \subseteq \mathcal{G}(\mathcal{A})$ already follows from the SRG's definition, the substance of this definition is $\mathcal{G}(\mathbb{A}) \subseteq \mathcal{G}(\mathcal{A}) \Rightarrow \mathbb{A} \in \mathcal{A}$. Essentially, a class is SRG-full if it can be fully characterized by its SRG; given an SRG-full class $\mathcal{A}$ and an operator A, we can check membership $\mathbb{A} \in \mathcal{A}$ by verifying (through geometric arguments) the containment $\mathcal{G}(\mathbb{A}) \subseteq \mathcal{G}(\mathcal{A})$ in the 2D plane.

SRG-fullness assumes the desirable property $\mathcal{G}(\mathbb{A}) \subseteq \mathcal{G}(\mathcal{A}) \Rightarrow \mathbb{A} \in \mathcal{A}$. We now discuss which classes possess this property.

Theorem 21 An operator class $\mathcal{A}$ is SRG-full if it is defined by

$$\mathbb{A} \in \mathcal{A} \quad \Leftrightarrow \quad h\left(\|u-v\|^2, \|x-y\|^2, \langle u-v, x-y\rangle\right) \leq 0, \quad \forall u \in \mathbb{A}x,\ v \in \mathbb{A}y$$

for some nonnegative homogeneous function $h\colon \mathbb{R}^3 \to \mathbb{R}$.

To clarify, h is nonnegative homogeneous if $\theta h(a,b,c) = h(\theta a, \theta b, \theta c)$ for all $\theta \geq 0$. (We do not assume h is smooth.) When a class $\mathcal{A}$ is defined by h as in Theorem 21, we say *h represents $\mathcal{A}$*. For example, the μ-strongly monotone class $\mathcal{M}_\mu$ is represented by $h(a,b,c) = \mu b - c$, since

$$\mathbb{A} \in \mathcal{M}_\mu \quad \Leftrightarrow \quad \mu\|x-y\|^2 \leq \langle u-v, x-y\rangle, \quad \forall u \in \mathbb{A}x,\ v \in \mathbb{A}y.$$

As another example, firmly nonexpansive class $\mathcal{N}_{1/2}$ is represented by $h(a,b,c) = a - c$, since

$$\mathbb{A} \in \mathcal{N}_{1/2} \quad \Leftrightarrow \quad \|u-v\|^2 \leq \langle u-v, x-y\rangle, \quad \forall u \in \mathbb{A}x,\ v \in \mathbb{A}y.$$

By Theorem 21, the classes $\mathcal{M}$, $\mathcal{M}_\mu$, $\mathcal{C}_\beta$, $\mathcal{L}_L$, and $\mathcal{N}_\theta$ are all SRG-full. See Exercise 13.14.

Proof. Since $\mathbb{A} \in \mathcal{A} \Rightarrow \mathcal{G}(\mathbb{A}) \subseteq \mathcal{G}(\mathcal{A})$ always holds, we show $\mathcal{G}(\mathbb{A}) \subseteq \mathcal{G}(\mathcal{A}) \Rightarrow \mathbb{A} \in \mathcal{A}$. Assume $\mathcal{A}$ is represented by h and an operator $\mathbb{A}\colon \mathbb{R}^n \rightrightarrows \mathbb{R}^n$ satisfies $\mathcal{G}(\mathbb{A}) \subseteq \mathcal{G}(\mathcal{A})$. Let $u_A \in \mathbb{A}x_A$ and $v_A \in \mathbb{A}y_A$ represent distinct evaluations, i.e., $x_A \neq y_A$ or $u_A \neq v_A$.

First, consider the case $x_A \neq y_A$. Then,

$$z = (\|u_A - v_A\|/\|x_A - y_A\|)\exp[i\angle(u_A - v_A, x_A - y_A)]$$

satisfies $z \in \mathcal{G}(\mathbb{A}) \subseteq \mathcal{G}(\mathcal{A})$. Since $z \in \mathcal{G}(\mathcal{A})$, there is an operator $\mathbb{B} \in \mathcal{A}$ such that $u_B \in \mathbb{B}x_B$ and $v_B \in \mathbb{B}y_B$ with

$$\frac{\|u_B - v_B\|^2}{\|x_B - y_B\|^2} = |z|^2, \qquad \frac{\langle u_B - v_B, x_B - y_B\rangle}{\|x_B - y_B\|^2} = \operatorname{Re} z.$$

Since h represents $\mathcal{A}$, we have

$$0 \geq h\left(\|u_B - v_B\|^2, \|x_B - y_B\|^2, \langle u_B - v_B, x_B - y_B\rangle\right),$$

and homogeneity gives us

$$\begin{aligned} 0 &\geq h\left(\frac{\|u_B - v_B\|^2}{\|x_B - y_B\|^2}, 1, \frac{\langle u_B - v_B, x_B - y_B\rangle}{\|x_B - y_B\|^2}\right) \\ &= h\left(|z|^2, 1, \operatorname{Re} z\right) = h\left(\frac{\|u_A - v_A\|^2}{\|x_A - y_A\|^2}, 1, \frac{\langle u_A - v_A, x_A - y_A\rangle}{\|x_A - y_A\|^2}\right). \end{aligned}$$

Finally, by homogeneity we have

$$h\left(\|u_A - v_A\|^2, \|x_A - y_A\|^2, \langle u_A - v_A, x_A - y_A\rangle\right) \leq 0.$$

Now consider the case $x_A = y_A$ and $u_A \neq v_B$. Then $\mathbb{A}$ is multi-valued and $\infty \in \mathcal{G}(\mathbb{A}) \subseteq \mathcal{G}(\mathcal{A})$. Since $\infty \in \mathcal{G}(\mathcal{A})$, there is a multi-valued operator $\mathbb{B} \in \mathcal{A}$ such that $u_B \in \mathbb{B}x_B$ and $v_B \in \mathbb{B}x_B$ with $u_B \neq v_B$. This implies $h(\|u_B - v_B\|^2, 0, 0) \leq 0$. Therefore, $h(\|u_A - v_A\|^2, 0, 0) \leq 0$.

In conclusion, (x_A, u_A) and (y_A, v_A), which represent arbitrary evaluations of $\mathbb{A}$, satisfy the inequality defined by h, and we conclude $\mathbb{A} \in \mathcal{A}$. □

Example 13.6 The classes $\partial\mathcal{F}_{0,\infty}$, $\partial\mathcal{F}_{\mu,\infty}$, $\partial\mathcal{F}_{0,L}$, and $\partial\mathcal{F}_{\mu,L}$ are *not* SRG-full. For example, the operator

$$\mathbb{A}(z_1, z_2) = \begin{bmatrix} 0 & -1 \\ 1 & 0 \end{bmatrix} \begin{bmatrix} z_2 \\ z_2 \end{bmatrix}$$

satisfies $\mathcal{G}(\mathbb{A}) = \{-i, i\} \subseteq \mathcal{G}(\partial\mathcal{F}_{0,\infty})$. However, $\mathbb{A} \notin \partial\mathcal{F}_{0,\infty}$ because there is no convex function f for which $\nabla f = D\mathbb{A}$.

For the sake of rigor and completeness, there is one degenerate case to keep in mind. The SRG-full class of operators $\mathcal{A}_{\text{null}}$ represented by $h(a,b,c) = a + b + |c|$ has $\mathcal{G}(\mathcal{A}_{\text{null}}) = \emptyset$. However, the class $\mathcal{A}_{\text{null}}$ is not itself empty; it contains operators whose graph contains zero or one pair, i.e., $A \in \mathcal{A}_{\text{null}}$ if and only if we have either (a) $\operatorname{dom} \mathbb{A} = \emptyset$ or (b) $\operatorname{dom} \mathbb{A} = x$ and $\mathbb{A}x = \{y\}$ for some $x, y \in \mathbb{R}^n$.

Role of Maximality

The notion of maximality is mostly orthogonal to the notion of the SRG. In particular, non-maximal operators have well-defined SRGs, and SRG-full classes contain non-maximal operators. By keeping the two notions separate, we avoid the geometric analyses via SRGs being entangled with the subtleties of maximality.

13.3 OPERATOR AND SRG TRANSFORMATIONS

In this section, we show how transformations of operators map to changes in their SRGs and analyze convergence of various fixed-point iterations.

13.3.1 Intersection

Theorem 22 If $\mathcal{A}$ and $\mathcal{B}$ are SRG-full classes, then $\mathcal{A} \cap \mathcal{B}$ is SRG-full, and

$$\mathcal{G}(\mathcal{A} \cap \mathcal{B}) = \mathcal{G}(\mathcal{A}) \cap \mathcal{G}(\mathcal{B}).$$

The substance of Theorem 22 is $\mathcal{G}(\mathcal{A} \cap \mathcal{B}) \supseteq \mathcal{G}(\mathcal{A}) \cap \mathcal{G}(\mathcal{B})$ since the containment $\mathcal{G}(\mathcal{A} \cap \mathcal{B}) \subseteq \mathcal{G}(\mathcal{A}) \cap \mathcal{G}(\mathcal{B})$ holds by definition, regardless of SRG-fullness.

Proof. Since $\mathcal{A}$ and $\mathcal{B}$ are SRG-full,

$$\begin{aligned} \mathcal{G}(\mathbb{C}) \subseteq \mathcal{G}(\mathcal{A} \cap \mathcal{B}) \subseteq \mathcal{G}(\mathcal{A}) \cap \mathcal{G}(\mathcal{B}) \quad &\Rightarrow \quad \mathcal{G}(\mathbb{C}) \subseteq \mathcal{G}(\mathcal{A}) \text{ and } \mathcal{G}(\mathbb{C}) \subseteq \mathcal{G}(\mathcal{B}) \\ &\Rightarrow \quad \mathbb{C} \in \mathcal{A} \text{ and } \mathbb{C} \in \mathcal{B} \\ &\Rightarrow \quad \mathbb{C} \in \mathcal{A} \cap \mathcal{B}, \end{aligned}$$

for an operator $\mathbb{C}$, and we conclude $\mathcal{A} \cap \mathcal{B}$ is SRG-full.

Assume $z \in \mathbb{C}$ satisfies $\{z, \bar{z}\} \subseteq \mathcal{G}(\mathcal{A}) \cap \mathcal{G}(\mathcal{B})$. Then, $\mathbb{A}_z$ of Lemma 4 satisfies $\mathcal{G}(\mathbb{A}_z) = \{z, \bar{z}\} \subseteq \mathcal{G}(\mathcal{A}) \cap \mathcal{G}(\mathcal{B})$. Since $\mathcal{A}$ and $\mathcal{B}$ are SRG-full, $\mathbb{A}_z \in \mathcal{A}$, $\mathbb{A}_z \in \mathcal{B}$, and $\{z, \bar{z}\} =$

$\mathcal{G}(\mathbb{A}_z) \subseteq \mathcal{G}(\mathcal{A} \cap \mathcal{B})$. If $\infty \in \mathcal{G}(\mathcal{A}) \cap \mathcal{G}(\mathcal{B})$, then a similar argument using $\mathbb{A}_\infty$ of Lemma 4 proves $\infty \in \mathcal{G}(\mathcal{A} \cap \mathcal{B})$. Therefore $\mathcal{G}(\mathcal{A}) \cap \mathcal{G}(\mathcal{B}) \subseteq \mathcal{G}(\mathcal{A} \cap \mathcal{B})$. Since the other containment, $\mathcal{G}(\mathcal{A} \cap \mathcal{B}) \subseteq \mathcal{G}(\mathcal{A}) \cap \mathcal{G}(\mathcal{B})$, holds by definition, we have the equality. □

Example 13.7 Theorem 22 does not apply when the operator classes are not SRG-full. For example, although

$$\partial \mathcal{F}_{\mu,L} = \partial \mathcal{F}_{\mu,\infty} \cap \partial \mathcal{F}_{0,L},$$

we have the strict containment:

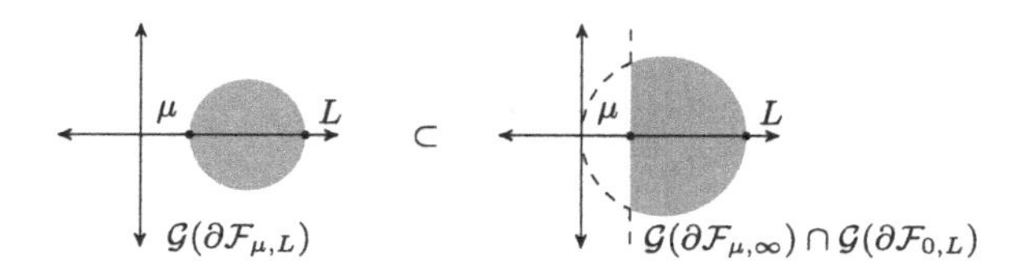

13.3.2 Scaling and Translation

Theorem 23 Let $\alpha \in \mathbb{R}$ and $\alpha \neq 0$. If $\mathcal{A}$ is a class of operators, then

$$\mathcal{G}(\alpha \mathcal{A}) = \alpha \mathcal{G}(\mathcal{A}), \qquad \mathcal{G}(\mathbb{I} + \mathcal{A}) = 1 + \mathcal{G}(\mathcal{A}).$$

If $\mathcal{A}$ is furthermore SRG-full, then $\alpha \mathcal{A}$ and $\mathbb{I} + \mathcal{A}$ are SRG-full.

Proof. $\mathcal{G}(\alpha \mathbb{A}) = \alpha \mathcal{G}(\mathbb{A})$ follows from the definition of the SRG, and $\mathcal{G}(\mathbb{I} + \mathbb{A}) = 1 + \mathcal{G}(\mathbb{A})$ follows from (13.1). The scaling and translation operations are reversible, and $\mathcal{G}((1/\alpha)\mathcal{A}) = (1/\alpha)\mathcal{G}(\mathcal{A})$ and $\mathcal{G}(\mathcal{A} - \mathbb{I}) = \mathcal{G}(A) - 1$. For any $\mathbb{B} : \mathbb{R}^n \rightrightarrows \mathbb{R}^n$,

$$\mathcal{G}(\mathbb{B}) \subseteq \mathcal{G}(\alpha \mathcal{A}) \quad \Rightarrow \quad \mathcal{G}((1/\alpha)\mathbb{B}) \subseteq \mathcal{G}(\mathcal{A}) \quad \Rightarrow \quad (1/\alpha)\mathbb{B} \in \mathcal{A} \quad \Rightarrow \quad \mathbb{B} \in \alpha \mathcal{A},$$

and we conclude $\alpha \mathcal{A}$ is SRG-full. By a similar reasoning, $\mathbb{I} + \mathcal{A}$ is SRG-full. □

Since a class of operators can consist of a single operator,

$$\mathcal{G}(\alpha \mathbb{A}) = \alpha \mathcal{G}(\mathbb{A}), \qquad \mathcal{G}(\mathbb{I} + \mathbb{A}) = 1 + \mathcal{G}(\mathbb{A})$$

holds for an individual operator $\mathbb{A}$.

Convergence Analysis: Gradient Descent

Consider the optimization problem

$$\underset{x \in \mathbb{R}^n}{\text{minimize}} \quad f(x),$$

where f is μ-strongly convex and L-smooth with $0 < \mu < L < \infty$. Gradient descent

$$x^{k+1} = x^k - \alpha \nabla f(x^k)$$

converges with rate

$$\|x^k - x^\star\| \leq (\max\{|1 - \alpha\mu|, |1 - \alpha L|\})^k \, \|x^0 - x^\star\|$$

for $\alpha \in (0, 2/L)$ by the following Proposition 2.

Proposition 2 *Let $0 < \mu < L < \infty$ and $\alpha \in (0, \infty)$. If $\mathcal{A} = \partial\mathcal{F}_{\mu,L}$, then $\mathbb{I} - \alpha\mathcal{A} \subseteq \mathcal{L}_R$ for*

$$R = \max\{|1 - \alpha\mu|, |1 - \alpha L|\}.$$

This result is tight in the sense that $\mathbb{I} - \alpha\mathcal{A} \not\subseteq \mathcal{L}_R$ for any smaller value of R.

Proof. By Theorems 20 and 23, we have the geometry

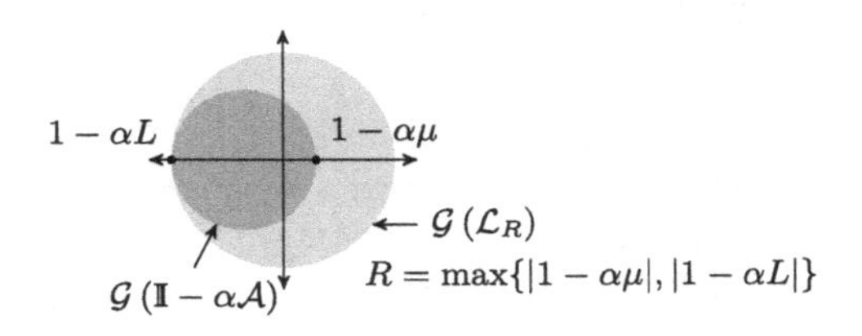

The containment of $\mathcal{G}(\mathbb{I} - \alpha\mathcal{A})$ holds for R and fails for smaller R. Since $\mathcal{L}_R$ is SRG-full by Theorem 21, the containment of the SRG in $\overline{\mathbb{C}}$ equivalent to the containment of the class. □

Convergence Analysis: Forward Step Method

Consider the monotone inclusion problem

$$\underset{x \in \mathbb{R}^n}{\text{find}} \quad 0 \in \mathbb{A}x,$$

where $\mathbb{A} \colon \mathbb{R}^n \to \mathbb{R}^n$. Consider the forward step method

$$x^{k+1} = x^k - \alpha\mathbb{A}x^k$$

under the following two setups.

Assume $\mathbb{A}$ is μ-strongly monotone and L-Lipschitz with $0 < \mu < L < \infty$. The forward step method converges with rate

$$\|x^k - x^\star\| \leq \left(1 - 2\alpha\mu + \alpha^2 L^2\right)^{k/2} \|x^0 - x^\star\|$$

for $\alpha \in (0, 2\mu/L^2)$ by the following Proposition 3.

Proposition 3 *Let $0 < \mu < L < \infty$ and $\alpha \in (0, \infty)$. If $\mathcal{A} = \mathcal{M}_\mu \cap \mathcal{L}_L$, then $\mathbb{I} - \alpha\mathcal{A} \subseteq \mathcal{L}_R$ for*

$$R = \sqrt{1 - 2\alpha\mu + \alpha^2 L^2}.$$

This result is tight in the sense that $\mathbb{I} - \alpha\mathcal{A} \not\subseteq \mathcal{L}_R$ for any smaller value of R.

Proof. First consider the case $\alpha\mu > 1$. By Theorems 19 and 23, we have the geometry

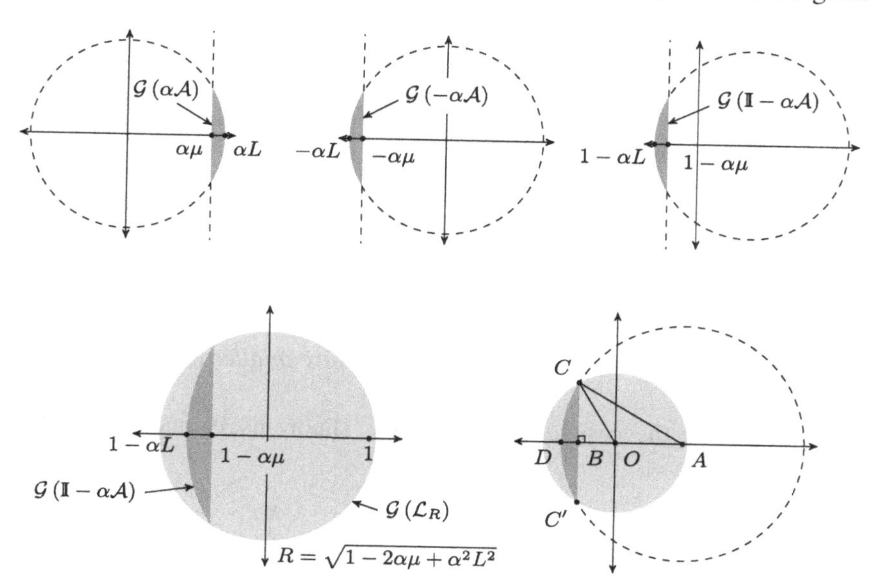

To clarify, O is the center of the circle with radius $\overline{OC}$ (lighter shade) and A is the center of the circle with radius $\overline{AC} = \overline{AD}$ defining the inner region (darker shade). With two applications of the Pythagorean theorem, we get

$$\begin{aligned}\overline{OC}^2 &= \overline{CB}^2 + \overline{BO}^2 = \overline{AC}^2 - \overline{BA}^2 + \overline{BO}^2 \\ &= (\alpha L)^2 - (\alpha\mu)^2 + (1-\alpha\mu)^2 = 1 - 2\alpha\mu + \alpha^2 L^2.\end{aligned}$$

Since $\overline{C'C}$ is a chord of circle O, it is within the circle. Since two nonidentical circles intersect at no more than two points, and since D is within circle O, arc $\widehat{CDC'}$ is within circle O. Finally, the region bounded by $\overline{C'C} \cup \widehat{CDC'}$ (darker shade) is within circle O (lighter shade).

The previous diagram illustrates the case $\alpha\mu > 1$. When $\alpha\mu = 1$ and $\alpha\mu < 1$, the geometries are slightly different, but the same arguments hold:

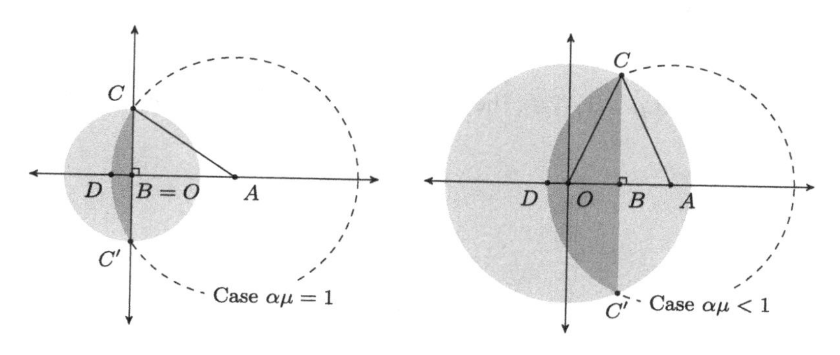

The containment holds for R and fails for smaller R. Since $\mathcal{L}_R$ is SRG-full by Theorem 21, the containment of the SRG in $\overline{\mathbb{C}}$ equivalent to the containment of the class. □

Assume $\mathbb{A}$ is μ-strongly monotone and β-cocoercive with $0 < \mu < 1/\beta < \infty$. The forward step method converges with rate

$$\|x^k - x^\star\| \leq \left(1 - 2\alpha\mu + \alpha^2\mu/\beta\right)^{k/2} \|x^0 - x^\star\|$$

for $\alpha \in (0, 2\beta)$ by the following Proposition 4.

Proposition 4 *Let $0 < \mu < 1/\beta < \infty$ and $\alpha \in (0, 2\beta)$. If $\mathcal{A} = \mathcal{M}_\mu \cap \mathcal{C}_\beta$, then $\mathbb{I} - \alpha\mathcal{A} \subseteq \mathcal{L}_R$ for*

$$R = \sqrt{1 - 2\alpha\mu + \alpha^2\mu/\beta}.$$

This result is tight in the sense that $\mathbb{I} - \alpha\mathcal{A} \nsubseteq \mathcal{L}_R$ for any smaller value of R.

Proof. First, consider the case $\mu < 1/(2\beta)$. By Theorems 19 and 23, we have the geometry

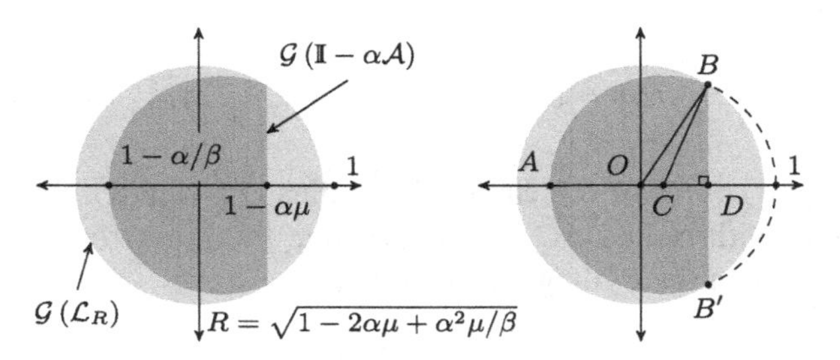

To clarify, O is the center of the circle with radius $\overline{OB}$ (lighter shade) and C is the center of the circle with radius $\overline{AC} = \overline{CB}$ defining the inner region (darker shade). With two applications of the Pythagorean theorem, we get

$$\begin{aligned}\overline{OB}^2 &= \overline{OD}^2 + \overline{DB}^2 = \overline{OD}^2 + \overline{BC}^2 - \overline{CD}^2 \\ &= (1 - \alpha\mu)^2 + (\alpha/(2\beta))^2 - (\alpha/(2\beta) - \alpha\mu)^2 = 1 - 2\alpha\mu + \alpha^2\mu/\beta.\end{aligned}$$

Since $\overline{B'B}$ is a chord of circle O, it is within the circle. Since two nonidentical circles intersect at at most two points, and since A is within circle O, arc $\widehat{BAB'}$ is within circle O. Finally, the region bounded by $\overline{B'B} \cup \widehat{BAB'}$ (darker shade) is within circle O (lighter shade).

When $\mu = 1/(2\beta)$ and $\mu > 1/(2\beta)$, the geometries are slightly different, but the same arguments hold:

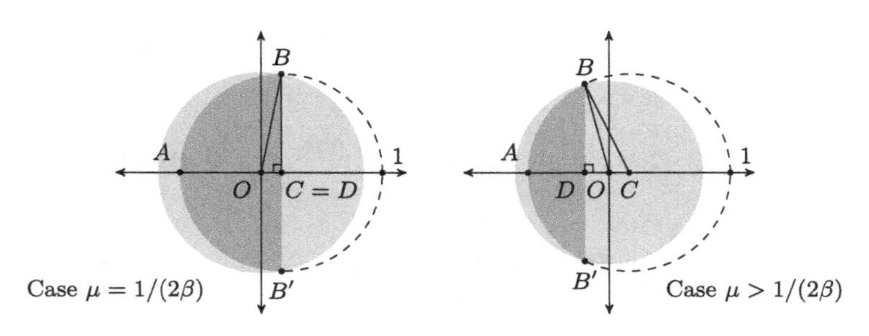

The containment holds for R and fails for smaller R. Since $\mathcal{L}_R$ is SRG-full by Theorem 21, the containment of the SRG in $\overline{\mathbb{C}}$ is equivalent to the containment of the class. □

13.3.3 Inversion

In this subsection, we relate inversion of operators with inversion (reciprocal) of complex numbers. This operation is intimately connected to inversive geometry.

Theorem 24 If $\mathcal{A}$ is a class of operators, then

$$\mathcal{G}(\mathcal{A}^{-1}) = (\mathcal{G}(\mathcal{A}))^{-1}.$$

If $\mathcal{A}$ is furthermore SRG-full, then $\mathcal{A}^{-1}$ is SRG-full.

To clarify, $(\mathcal{G}(\mathcal{A}))^{-1} = \{z^{-1} \mid z \in \mathcal{G}(\mathcal{A})\} \subseteq \overline{\mathbb{C}}$. Note that $(\mathcal{G}(\mathcal{A}))^{-1} = (\overline{\mathcal{G}(\mathcal{A})})^{-1}$, since $\mathcal{G}(\mathcal{A})$ is symmetric about the real axis, so we write the simpler $(\mathcal{G}(\mathcal{A}))^{-1}$ even though the inversion map we consider is $z \mapsto \bar{z}^{-1}$.

Proof. The equivalence of nonzero finite points, i.e.,

$$\mathcal{G}(\mathbb{A}^{-1})\backslash\{0,\infty\} = (\mathcal{G}(\mathbb{A})\backslash\{0,\infty\})^{-1},$$

follows from

$$\mathcal{G}(\mathbb{A})\backslash\{0,\infty\} = \left\{ \frac{\|u-v\|}{\|x-y\|} \exp\left[\pm i\angle(u-v,x-y)\right] \,\middle|\, (x,u),(y,v) \in \mathbb{A},\, x \neq y,\, u \neq v \right\}$$

and

$$\begin{aligned} &\mathcal{G}(\mathbb{A}^{-1})\backslash\{0,\infty\} \\ &\quad= \left\{ \frac{\|x-y\|}{\|u-v\|} \exp\left[\pm i\angle(x-y,u-v)\right] \,\middle|\, (u,x),(v,y) \in \mathbb{A}^{-1},\, x \neq y,\, u \neq v \right\} \\ &\quad= \left\{ \frac{\|x-y\|}{\|u-v\|} \exp\left[\pm i\angle(u-v,x-y)\right] \,\middle|\, (x,u),(y,v) \in \mathbb{A},\, x \neq y,\, u \neq v \right\} \\ &\quad= (\mathcal{G}(\mathbb{A})\backslash\{0,\infty\})^{-1}, \end{aligned}$$

where we use the fact that $\angle(a,b) = \angle(b,a)$.

The equivalence of the zero and infinite points follow from

$$\begin{aligned} \infty \in \mathcal{G}(\mathbb{A}) \quad &\Leftrightarrow \quad \exists\,(x,u),(x,v) \in \mathbb{A},\, u \neq v \\ &\Leftrightarrow \quad \exists\,(u,x),(v,x) \in \mathbb{A}^{-1},\, u \neq v \\ &\Leftrightarrow \quad 0 \in \mathcal{G}(\mathbb{A}^{-1}). \end{aligned}$$

With the same argument, we have $0 \in \mathcal{G}(\mathbb{A}) \Leftrightarrow \infty \in \mathcal{G}(\mathbb{A}^{-1})$.

The inversion operation is reversible. For any $\mathbb{B} : \mathbb{R}^n \rightrightarrows \mathbb{R}^n$,

$$\mathcal{G}(\mathbb{B}) \subseteq \mathcal{G}(\mathcal{A}^{-1}) \quad \Rightarrow \quad \mathcal{G}(\mathbb{B}^{-1}) \subseteq \mathcal{G}(\mathcal{A}) \quad \Rightarrow \quad \mathbb{B}^{-1} \in \mathcal{A} \quad \Rightarrow \quad \mathbb{B} \in \mathcal{A}^{-1},$$

and we conclude $\mathcal{A}^{-1}$ is SRG-full. □

Convergence Analysis: Proximal Point

Consider the monotone inclusion problem

$$\underset{x\in\mathbb{R}^n}{\text{find}} \quad 0 \in \mathbb{A}x,$$

where $\mathbb{A}$ is maximal μ-strongly monotone. Consider the proximal point method

$$x^{k+1} = \mathbb{J}_{\alpha\mathbb{A}}x^k.$$

By the following Proposition 5, the proximal point method converges exponentially with rate

$$\|x^k - x^\star\| \le \left(\frac{1}{1+\alpha\mu}\right)^k \|x^0 - x^\star\|$$

for $\alpha > 0$.

Proposition 5 *Let $\mu \in (0,\infty)$ and $\alpha \in (0,\infty)$. If $\mathcal{A} = \mathcal{M}_\mu$, then $\mathbb{J}_{\alpha\mathcal{A}} \subseteq \mathcal{L}_R$ for*

$$R = \frac{1}{1+\alpha\mu}.$$

This result is tight in the sense that $\mathbb{J}_{\alpha\mathcal{A}} \nsubseteq \mathcal{L}_R$ for any smaller value of R.

Proof. By Theorems 19, 23, and 24, we have the geometry

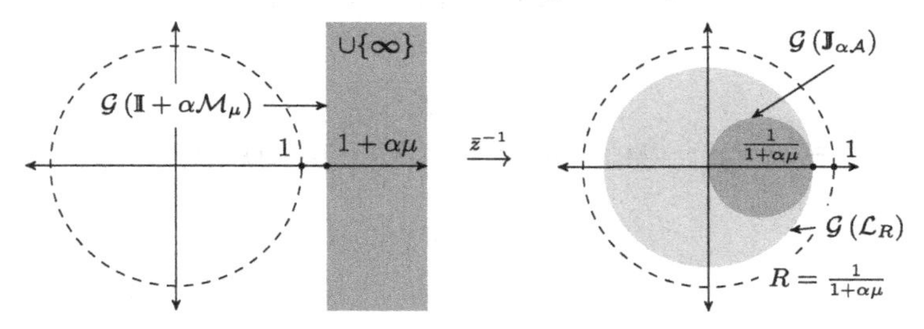

The containment holds for R and fails for smaller R. Since $\mathcal{L}_R$ is SRG-full by Theorem 21, the containment of the SRG in $\overline{\mathbb{C}}$ is equivalent to the containment of the class. □

Convergence Analysis: DRS

Consider the monotone inclusion problem

$$\underset{x\in\mathbb{R}^n}{\text{find}} \quad 0 \in (\mathbb{A}+\mathbb{B})x,$$

where $\mathbb{A}$ and $\mathbb{B}$ are maximal monotone. Assume $\mathbb{A}$ or $\mathbb{B}$ is μ-strongly monotone and β-cocoercive with $0 < \mu < 1/\beta < \infty$. Consider DRS:

$$z^{k+1} = \left(\tfrac{1}{2}\mathbb{I} + \tfrac{1}{2}\mathbb{R}_{\alpha\mathbb{A}}\mathbb{R}_{\alpha\mathbb{B}}\right)z^k.$$

DRS converges exponentially with rate

$$\|z^k - z^\star\| \le \left(\frac{1}{2} + \frac{1}{2}\sqrt{1 - \frac{4\alpha\mu}{1+2\alpha\mu+\alpha^2\mu/\beta}}\right)^k \|z^0 - z^\star\|$$

for $\alpha > 0$ by the following Proposition 6 and the argument of Exercise 13.11.

Proposition 6 *Let $0 < \mu < 1/\beta < \infty$ and $\alpha \in (0, \infty)$. If $\mathcal{A} = \mathcal{M}_\mu \cap \mathcal{C}_\beta$, then $\mathbb{R}_{\alpha\mathcal{A}} \subseteq \mathcal{L}_R$ for*

$$R = \sqrt{1 - \frac{4\alpha\mu}{1 + 2\alpha\mu + \alpha^2\mu/\beta}}.$$

This result is tight in the sense that $\mathbb{R}_{\alpha\mathcal{A}} \not\subseteq \mathcal{L}_R$ for any smaller value of R.

Proof. By Theorems 19, 23, and 24, we have the geometry

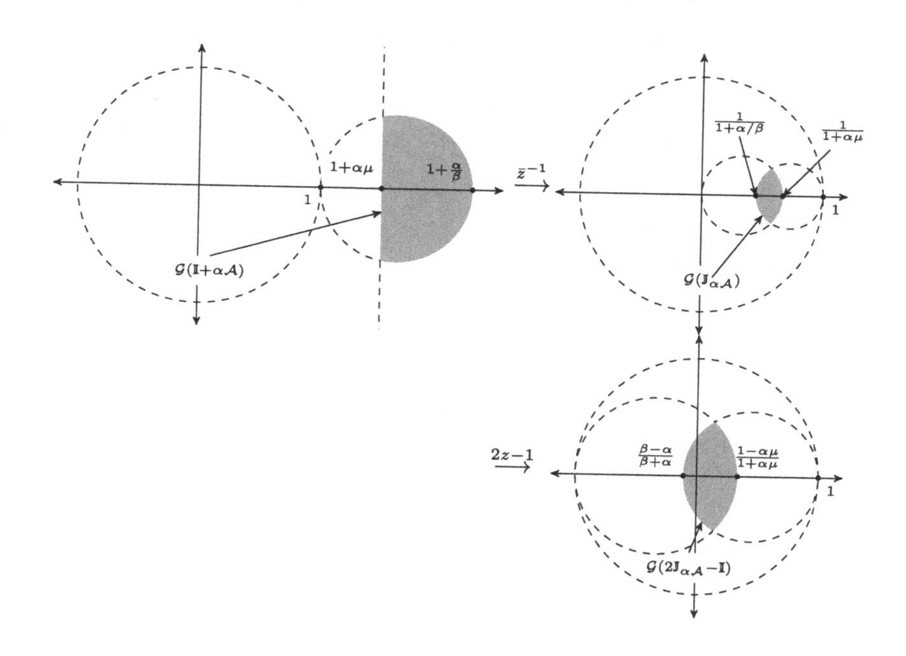

A closer look gives us

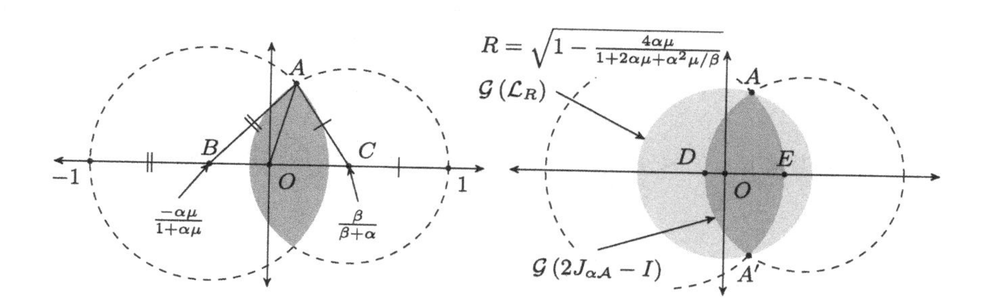

To clarify, B is the center of the circle with radius $\overline{BA}$, and C is the center of the circle with radius $\overline{CA}$. By Stewart's theorem, we have

$$
\begin{aligned}
\overline{OA}^2 &= \frac{\overline{OC}\cdot\overline{AB}^2+\overline{BO}\cdot\overline{CA}^2-\overline{BO}\cdot\overline{OC}\cdot\overline{BC}}{\overline{BC}} \\
&= \frac{\frac{\beta}{\alpha+\beta}\left(1-\frac{\alpha\mu}{1+\alpha\mu}\right)^2+\frac{\alpha\mu}{1+\alpha\mu}\left(1-\frac{\beta}{\alpha+\beta}\right)^2-\frac{\beta}{\alpha+\beta}\frac{\alpha\mu}{1+\alpha\mu}\left(\frac{\beta}{\alpha+\beta}+\frac{\alpha\mu}{1+\alpha\mu}\right)}{\frac{\beta}{\alpha+\beta}+\frac{\alpha\mu}{1+\alpha\mu}} \\
&= 1-\frac{4\alpha\mu}{1+2\alpha\mu+\alpha^2\mu/\beta}.
\end{aligned}
$$

Since two non-identical circles intersect at at most two points, and since D is within circle B, arc $\widehat{ADA'}$ is within circle O. By the same reasoning, arc $\widehat{A'EA}$ is within circle O. Finally, the region bounded by $\widehat{ADA'}\cup\widehat{A'EA}$ (darker shade) is within circle O (lighter shade).

The containment holds for R and fails for smaller R. Since $\mathcal{L}_R$ is SRG-full by Theorem 21, the containment of the SRG in $\overline{\mathbb{C}}$ is equivalent to the containment of the class. □

Consider the optimization problem

$$
\underset{x\in\mathbb{R}^n}{\text{minimize}} \quad f(x)+g(x),
$$

where f and g are CCP. Assume f or g is μ-strongly convex and L-smooth with $0<\mu<L<\infty$. Consider DRS:

$$
\begin{aligned}
x^{k+1/2} &= \mathrm{Prox}_{\alpha g}(z^k) \\
x^{k+1} &= \mathrm{Prox}_{\alpha f}(2x^{k+1/2}-z^k) \\
z^{k+1} &= z^k + x^{k+1} - x^{k+1/2}.
\end{aligned}
$$

DRS converges exponentially with rate

$$
\|z^k-z^\star\| \le \left(\frac{1}{2}+\frac{1}{2}\max\left\{\left|\frac{1-\alpha\mu}{1+\alpha\mu}\right|,\left|\frac{1-\alpha L}{1+\alpha L}\right|\right\}\right)^k \|z^0-z^\star\|
$$

by the following Proposition 7 and the argument of Exercise 13.11.

Proposition 7 *Let* $0<\mu<L<\infty$ *and* $\alpha\in(0,\infty)$. *If* $\mathcal{A}=\partial\mathcal{F}_{\mu,L}$, *then* $\mathbb{R}_{\alpha\mathcal{A}}\subseteq\mathcal{L}_R$ *for*

$$
R=\max\left\{\left|\frac{1-\alpha\mu}{1+\alpha\mu}\right|,\left|\frac{1-\alpha L}{1+\alpha L}\right|\right\}.
$$

This result is tight in the sense that $\mathbb{R}_{\alpha\mathcal{A}}\not\subseteq\mathcal{L}_R$ *for any smaller value of* R.

Proof. By Theorems 20, 23, and 24, we have the geometry

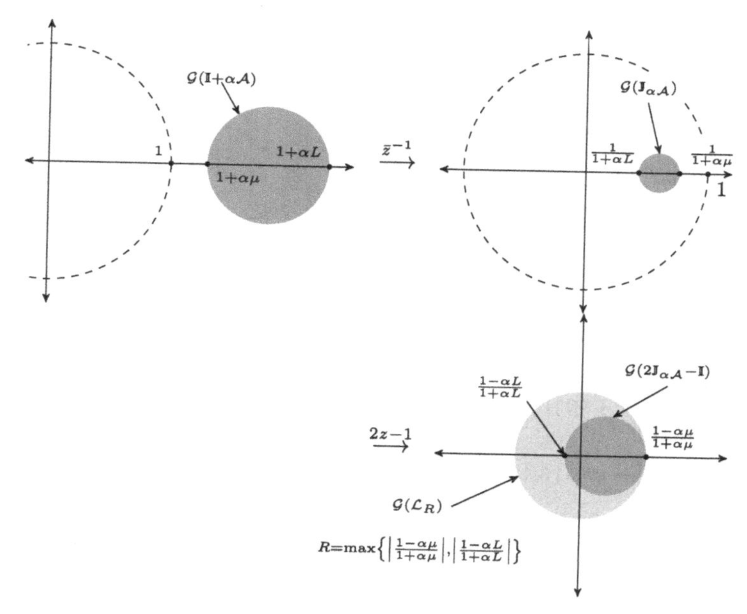

The containment holds for R and fails for smaller R. Since $\mathcal{L}_R$ is SRG-full by Theorem 21, the containment of the SRG in $\overline{\mathbb{C}}$ is equivalent to the containment of the class. □

13.3.4 Sum of Operators

Given $z, w \in \mathbb{C}$, define the *line segment* between z and w as

$$[z, w] = \{\theta z + (1-\theta)w \mid \theta \in [0,1]\}.$$

We say an SRG-full class $\mathcal{A}$ satisfies the *chord property* if $z \in \mathcal{G}(\mathcal{A})\backslash\{\infty\}$ implies $[z, \bar{z}] \subseteq \mathcal{G}(\mathcal{A})$. See Figure 13.2.

Theorem 25 Let $\mathcal{A}$ and $\mathcal{B}$ be SRG-full classes such that $\infty \notin \mathcal{G}(A)$ and $\infty \notin \mathcal{G}(B)$. Then

$$\mathcal{G}(\mathcal{A}+\mathcal{B}) \supseteq \mathcal{G}(\mathcal{A}) + \mathcal{G}(\mathcal{B}).$$

If $\mathcal{A}$ or $\mathcal{B}$ furthermore satisfies the chord property, then

$$\mathcal{G}(\mathcal{A}+\mathcal{B}) = \mathcal{G}(\mathcal{A}) + \mathcal{G}(\mathcal{B}).$$

Proof. We first show $\mathcal{G}(\mathcal{A}+\mathcal{B}) \supseteq \mathcal{G}(\mathcal{A}) + \mathcal{G}(\mathcal{B})$. Assume $\mathcal{G}(\mathcal{A}) \neq \emptyset$ and $\mathcal{G}(\mathcal{B}) \neq \emptyset$, as otherwise there is nothing to show. Let $z \in \mathcal{G}(\mathcal{A})$ and $w \in \mathcal{G}(\mathcal{B})$ and let $\mathbb{A}_z$ and $\mathbb{A}_w$ be their corresponding operators as defined in Lemma 4. Then it is straightforward to see that $\mathbb{A}_z + \mathbb{A}_w$ corresponds to complex multiplication with respect to $(z+w)$, and $z + w \in \mathcal{G}(\mathbb{A}_z + \mathbb{A}_w) \subseteq \mathcal{G}(\mathcal{A}+\mathcal{B})$.

Next, we show $\mathcal{G}(\mathcal{A}+\mathcal{B}) \subseteq \mathcal{G}(\mathcal{A}) + \mathcal{G}(\mathcal{B})$. Consider the case $\mathcal{G}(\mathcal{A}) \neq \emptyset$ and $\mathcal{G}(\mathcal{B}) \neq \emptyset$. Without loss of generality, assume it is $\mathcal{A}$ that satisfies the chord property. Consider

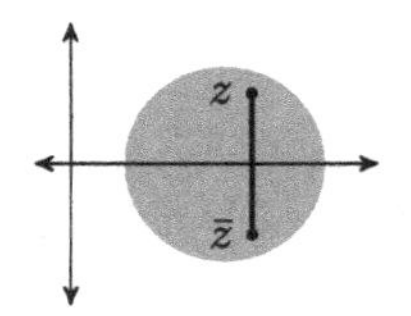

Figure 13.2 The chord property.

$\mathbb{A} + \mathbb{B} \in \mathcal{A} + \mathcal{B}$ such that $\mathbb{A} \in \mathcal{A}$ and $\mathbb{B} \in \mathcal{B}$. Consider $(x, u_A + u_B), (y, v_A + v_B) \in \mathbb{A} + \mathbb{B}$ such that $x \neq y$, $(x, u_A), (y, v_A) \in \mathbb{A}$, and $(x, u_B), (y, v_B) \in \mathbb{B}$. Define

$$z_A = \frac{\|u_A - v_A\|}{\|x - y\|} \exp\left[i\angle(u_A - v_A, x - y)\right] \in \mathcal{G}(\mathbb{A})$$
$$z_B = \frac{\|u_B - v_B\|}{\|x - y\|} \exp\left[i\angle(u_B - v_B, x - y)\right] \in \mathcal{G}(\mathbb{B})$$
$$z = \frac{\|u_A + u_B - v_A - v_B\|}{\|x - y\|} \exp\left[i\angle(u_A + u_B - v_A - v_B, x - y)\right] \in \mathcal{G}(\mathbb{A} + \mathbb{B}).$$

(Note that $\operatorname{Im} z_A, \operatorname{Im} z_B, \operatorname{Im} z \geq 0$.) Since

$$\operatorname{Re} z_A = \frac{\langle u_A - v_A, x - y\rangle}{\|x - y\|^2}, \qquad \operatorname{Re} z_B = \frac{\langle u_B - v_B, x - y\rangle}{\|x - y\|^2},$$
$$\operatorname{Re} z = \frac{\langle (u_A + u_B) - (v_A + v_B), x - y\rangle}{\|x - y\|^2},$$

we have $\operatorname{Re} z = \operatorname{Re} z_A + \operatorname{Re} z_B$. Using (13.1) and the triangle inequality, we have

$$\begin{aligned}\operatorname{Im} z &= \frac{\|\Pi_{\{x-y\}^\perp}(u_A + u_B - v_A - v_B)\|}{\|x - y\|} \\ &\leq \frac{\|\Pi_{\{x-y\}^\perp}(u_A - v_A)\| + \|\Pi_{\{x-y\}^\perp}(u_B - v_B)\|}{\|x - y\|} \\ &= \operatorname{Im} z_A + \operatorname{Im} z_B,\end{aligned}$$

and using the reverse triangle inequality, we have $\operatorname{Im} z \geq -\operatorname{Im} z_A + \operatorname{Im} z_B$. Together, we conclude

$$-\operatorname{Im} z_A + \operatorname{Im} z_B \leq \operatorname{Im} z \leq \operatorname{Im} z_A + \operatorname{Im} z_B$$

and

$$z \in [z_A, \overline{z_A}] + z_B, \qquad \overline{z} \in [z_A, \overline{z_A}] + \overline{z_B}.$$

This shows

$$\begin{aligned}\mathcal{G}(\mathcal{A} + \mathcal{B}) &\subseteq \{w_A + z_B \mid w_A \in [z_A, \overline{z_A}],\ z_A \in \mathcal{G}(\mathcal{A}),\ z_B \in \mathcal{G}(\mathcal{B})\} \\ &= \{w_A + z_B \mid w_A \in \mathcal{G}(\mathcal{A}),\ z_B \in \mathcal{G}(\mathcal{B})\} = \mathcal{G}(\mathcal{A}) + \mathcal{G}(\mathcal{B}),\end{aligned}$$

where the equality follows from the chord property.

Now, consider the case $\mathcal{G}(\mathcal{A}) = \emptyset$ or $\mathcal{G}(\mathcal{B}) = \emptyset$ (or both). Assume $\mathcal{G}(\mathcal{A}) = \emptyset$ without loss of generality, and let $\mathbb{A} \in \mathcal{A}$ and $\mathbb{B} \in \mathcal{B}$. Then, $\operatorname{dom} \mathbb{A}$ is empty or a singleton, and if $\{x\} = \operatorname{dom} \mathbb{A}$, then $\mathbb{A}x$ is a singleton. Therefore $\operatorname{dom}(\mathbb{A} + \mathbb{B}) \subseteq \operatorname{dom} \mathbb{A}$ is empty or a singleton, and if $\{x\} = \operatorname{dom} \mathbb{A}$, then $(\mathbb{A} + \mathbb{B})x$ is empty or a singleton, since $\mathbb{B}$ is single-valued. Therefore, $\mathcal{G}(\mathbb{A} + \mathbb{B}) = \emptyset$ and we conclude $\mathcal{G}(\mathcal{A} + \mathcal{B}) = \emptyset$. □

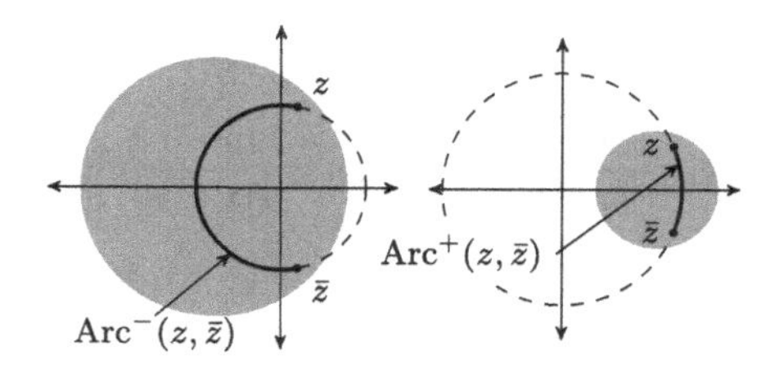

Figure 13.3 Left- and right-arc properties.

It is possible to generalize Theorem 25 to allow ∞ by excluding the following exception: if $\emptyset = \mathcal{G}(\mathcal{A})$ and $\infty \in \mathcal{G}(\mathcal{B})$, then $\{\infty\} = \mathcal{G}(\mathcal{A} + \mathcal{B})$.

13.3.5 Composition of Operators

Given $z \in \mathbb{C}$, define the *right-hand arc* between z and $\bar{z}$ as

$$\mathrm{Arc}^+(z, \bar{z}) = \left\{ re^{i(1-2\theta)\varphi} \,\middle|\, z = re^{i\varphi},\ \varphi \in (-\pi, \pi],\ \theta \in [0,1],\ r \geq 0 \right\}$$

and the *left-hand arc* as

$$\mathrm{Arc}^-(z, \bar{z}) = -\mathrm{Arc}^+(-z, -\bar{z}).$$

We say an SRG-full class $\mathcal{A}$ respectively satisfies the *left-arc property* and *right-arc property* if $z \in \mathcal{G}(\mathcal{A})\backslash\{\infty\}$ implies $\mathrm{Arc}^-(z, \bar{z}) \subseteq \mathcal{G}(\mathcal{A})$ and $\mathrm{Arc}^+(z, \bar{z}) \subseteq \mathcal{G}(\mathcal{A})$, respectively. We say $\mathcal{A}$ satisfies *an* arc property if the left- or right-arc property is satisfied. See Figure 13.3.

Theorem 26 Let $\mathcal{A}$ and $\mathcal{B}$ be SRG-full classes such that $\infty \notin \mathcal{G}(\mathcal{A})$, $\emptyset \neq \mathcal{G}(\mathcal{A})$, $\infty \notin \mathcal{G}(\mathcal{B})$, and $\emptyset \neq \mathcal{G}(\mathcal{B})$. Then,

$$\mathcal{G}(\mathcal{A}\mathcal{B}) \supseteq \mathcal{G}(\mathcal{A})\mathcal{G}(\mathcal{B}).$$

If $\mathcal{A}$ or $\mathcal{B}$ furthermore satisfies an arc property, then

$$\mathcal{G}(\mathcal{A}\mathcal{B}) = \mathcal{G}(\mathcal{B}\mathcal{A}) = \mathcal{G}(\mathcal{A})\mathcal{G}(\mathcal{B}).$$

Proof. We first show $\mathcal{G}(\mathcal{A}\mathcal{B}) \supseteq \mathcal{G}(\mathcal{A})\mathcal{G}(\mathcal{B})$. Assume $\mathcal{G}(\mathcal{A}) \neq \emptyset$ and $\mathcal{G}(\mathcal{B}) \neq \emptyset$, as otherwise there is nothing to show. Let $z \in \mathcal{G}(\mathcal{A})$ and $w \in \mathcal{G}(\mathcal{B})$ and let $\mathbb{A}_z$ and $\mathbb{A}_w$ be their corresponding operators as defined in Lemma 4. Then, it is straightforward to see that $\mathbb{A}_z\mathbb{A}_w$ corresponds to complex multiplication with respect to zw, and $zw \in \mathcal{G}(\mathbb{A}_z\mathbb{A}_w) \subseteq \mathcal{G}(\mathcal{A}\mathcal{B})$.

Next, we show $\mathcal{G}(\mathcal{A}\mathcal{B}) \subseteq \mathcal{G}(\mathcal{A})\mathcal{G}(\mathcal{B})$. Let $\mathbb{A} \in \mathcal{A}$ and $\mathbb{B} \in \mathcal{B}$. Consider $(u,s), (v,t) \in \mathbb{A}$ and $(x,u), (y,v) \in \mathbb{B}$, where $x \neq y$. This implies $(x,s), (y,t) \in \mathbb{A}\mathbb{B}$. Define

$$z = \frac{\|s - t\|}{\|x - y\|} \exp\left[i\angle(s - t, x - y)\right].$$

Consider the case $u = v$. Then $0 \in \mathcal{G}(\mathcal{B})$. Moreover, $s = t$, since $\mathbb{A}$ is single-valued (by the assumption $\infty \notin \mathcal{G}(\mathcal{A})$), and $z = 0$. Therefore, $z = 0 \in \mathcal{G}(\mathcal{A})\mathcal{G}(\mathcal{B})$.

Next, consider the case $u \neq v$. Define

$$z_A = \frac{\|s-t\|}{\|u-v\|}e^{i\varphi_A}, \quad z_B = \frac{\|u-v\|}{\|x-y\|}e^{i\varphi_B},$$

where $\varphi_A = \angle(s-t, u-v)$ and $\varphi_B = \angle(u-v, x-y)$. Consider the case where $\mathcal{A}$ satisfies the right-arc property. Using the spherical triangle inequality (further discussed in the appendix), we see that either $\varphi_A \geq \varphi_B$ and

$$\begin{aligned} z &\in \frac{\|s-t\|}{\|u-v\|}\frac{\|u-v\|}{\|x-y\|}\exp\left[i[\varphi_A - \varphi_B, \varphi_A + \varphi_B]\right] \\ &\subseteq \frac{\|s-t\|}{\|u-v\|}\frac{\|u-v\|}{\|x-y\|}\exp\left[i[\varphi_B - \varphi_A, \varphi_B + \varphi_A]\right] \\ &= z_B \mathrm{Arc}^+(z_A, \overline{z_A}) \end{aligned}$$

or $\varphi_A < \varphi_B$ and

$$\begin{aligned} z &\in \frac{\|s-t\|}{\|u-v\|}\frac{\|u-v\|}{\|x-y\|}\exp\left[i[\varphi_B - \varphi_A, \varphi_B + \varphi_A]\right] \\ &= z_B \mathrm{Arc}^+(z_A, \overline{z_A}). \end{aligned}$$

This gives us

$$z \in \underbrace{z_B}_{\in \mathcal{G}(\mathcal{B})} \underbrace{\mathrm{Arc}^+(z_A, \overline{z_A})}_{\subseteq \mathcal{G}(\mathcal{A})} \subseteq \mathcal{G}(\mathcal{A})\mathcal{G}(\mathcal{B}).$$

That $\bar{z} \in \mathcal{G}(\mathcal{A})\mathcal{G}(\mathcal{B})$ follows from the same argument. That $z, \bar{z} \in \mathcal{G}(\mathcal{A})\mathcal{G}(\mathcal{B})$ when instead $\mathcal{B}$ satisfies the right-arc property follows from the same argument.

Putting everything together, we conclude $\mathcal{G}(\mathcal{A}\mathcal{B}) = \mathcal{G}(\mathcal{A})\mathcal{G}(\mathcal{B})$ when $\mathcal{A}$ or $\mathcal{B}$ satisfies the right-arc property. When $\mathcal{A}$ satisfies the left-arc property, $-\mathcal{A}$ satisfies the right-arc property. So,

$$-\mathcal{G}(\mathcal{A}\mathcal{B}) = \mathcal{G}(-\mathcal{A}\mathcal{B}) = \mathcal{G}(-\mathcal{A})\mathcal{G}(\mathcal{B}) - \mathcal{G}(\mathcal{A})\mathcal{G}(\mathcal{B})$$

by Theorem 23, and we conclude $\mathcal{G}(\mathcal{A}\mathcal{B}) = \mathcal{G}(\mathcal{A})\mathcal{G}(\mathcal{B})$. When $\mathcal{B}$ satisfies the left-arc property, $\mathcal{B} \circ (-\mathbb{I})$ satisfies the right-arc property. So,

$$-\mathcal{G}(\mathcal{A}\mathcal{B}) = \mathcal{G}(\mathcal{A}\mathcal{B} \circ (-I)) = \mathcal{G}(\mathcal{A})\mathcal{G}(\mathcal{B} \circ (-\mathbb{I})) = -\mathcal{G}(\mathcal{A})\mathcal{G}(\mathcal{B})$$

by Theorem 23, and we conclude $\mathcal{G}(\mathcal{A}\mathcal{B}) = \mathcal{G}(\mathcal{A})\mathcal{G}(\mathcal{B})$. □

It is possible to generalize Theorem 26 to allow $\emptyset$ and ∞ by excluding the following exceptions: if $\emptyset = \mathcal{G}(\mathcal{A})$ and $\infty \in \mathcal{G}(\mathcal{B})$, then $\{\infty\} = \mathcal{G}(\mathcal{A}\mathcal{B})$; if $0 \in \mathcal{G}(\mathcal{A})$ and $\infty \in \mathcal{G}(\mathcal{B})$, then $\infty \in \mathcal{G}(\mathcal{A}\mathcal{B})$; if $\emptyset = \mathcal{G}(\mathcal{A})$ and $0 \in \mathcal{G}(\mathcal{B})$, then $\{0\} = \mathcal{G}(\mathcal{A}\mathcal{B})$ and $\emptyset = \mathcal{G}(\mathcal{B}\mathcal{A})$.

13.4 AVERAGEDNESS COEFFICIENTS

In this section, we establish the averagedness coefficients for the composition of averagedness operators and the DYS operator.

13.4.1 Composition of Averaged Operators

Theorem 27 Let $\mathbb{T}_1$ and $\mathbb{T}_2$ be θ_1- and θ_2-averaged operators on $\mathbb{R}^n$ with $\theta_1, \theta_2 \in (0,1)$. Then $\mathbb{T}_1\mathbb{T}_2$ is θ-averaged with

$$\theta = \frac{\theta_1 + \theta_2 - 2\theta_1\theta_2}{1 - \theta_1\theta_2}.$$

Proof. Note

$$z \in \mathcal{G}(\mathcal{N}_\theta) \quad \Leftrightarrow \quad |z - (1-\theta)|^2 \le \theta^2 \quad \Leftrightarrow \quad |z|^2 \le 1 - \frac{1-\theta}{\theta}|1-z|^2$$

by Theorem 19 and

$$\theta^2 - |z - (1-\theta)|^2 = \theta\left(1 - \frac{1-\theta}{\theta}|1-z|^2 - |z|^2\right).$$

Let $z_1 \in \mathcal{G}(\mathcal{N}_{\theta_1})$ and $z_2 \in \mathcal{G}(\mathcal{N}_{\theta_2})$. Then,

$$\begin{aligned}
|z_1z_2|^2 &\le |z_2|^2\left(1 - \frac{1-\theta_1}{\theta_1}|1-z_1|^2\right) \\
&\le 1 - \frac{1-\theta_2}{\theta_1}|1-z_2|^2 - \frac{1-\theta_1}{\theta_1}|1-z_1|^2|z_2|^2 \\
&= 1 - \frac{1-\theta}{\theta}|1-z_1z_2|^2 - \frac{\theta_1\theta_2}{\theta_1+\theta_2-2\theta_1\theta_2}\left|\frac{1-\theta_1}{\theta_1}(1-z_1)z_2 - \frac{1-\theta_2}{\theta_2}(1-z_2)\right|^2 \\
&\le 1 - \frac{1-\theta}{\theta}|1-z_1z_2|^2
\end{aligned}$$

and $z_1z_2 \in \mathcal{G}(\mathcal{N}_\theta)$. In other words, we have shown $\mathcal{G}(\mathcal{N}_{\theta_1})\mathcal{G}(\mathcal{N}_{\theta_2}) \subseteq \mathcal{G}(\mathcal{N}(\mathcal{G}_\theta))$. Since $\mathcal{N}_{\theta_1}$ satisfies an arc property, $\mathcal{G}(\mathcal{N}_{\theta_1})\mathcal{G}(\mathcal{N}_{\theta_2}) = \mathcal{G}(\mathcal{N}_{\theta_1}\mathcal{N}_{\theta_2})$ by Theorem 26. So,

$$\mathcal{G}(\mathcal{N}_{\theta_1}\mathcal{N}_{\theta_2}) = \mathcal{G}(\mathcal{N}_{\theta_1})\mathcal{G}(\mathcal{N}_{\theta_2}) \subseteq \mathcal{G}(\mathcal{N}_\theta)$$

implies $\mathcal{N}_{\theta_1}\mathcal{N}_{\theta_2} \subseteq \mathcal{N}_\theta$ by SRG-fullness of $\mathcal{N}_\theta$. □

13.4.2 Davis–Yin Splitting

Theorem 28 Assume $\mathbb{A}$, $\mathbb{B}$, and $\mathbb{C}$ are maximal monotone. Assume $\mathbb{C}$ is β-cocoercive and $\alpha \in (0, 2\beta)$. The DYS operator $\mathbb{I} - \mathbb{J}_{\alpha\mathbb{B}} + \mathbb{J}_{\alpha\mathbb{A}}(\mathbb{R}_{\alpha\mathbb{B}} - \alpha\mathbb{C}\mathbb{J}_{\alpha\mathbb{B}})$ is θ-averaged with

$$\theta = \frac{2\beta}{4\beta - \alpha}.$$

Lemma 5 *For $\theta \in (0,1)$, $\mathbb{T}$ is θ-averaged if and only if*

$$\|\mathbb{T}x - \mathbb{T}y\|^2 \le \|x-y\|^2 - \frac{1-\theta}{\theta}\|\mathbb{T}x - x - \mathbb{T}y + y\|^2 \qquad \forall, x, y \in \mathbb{R}^n.$$

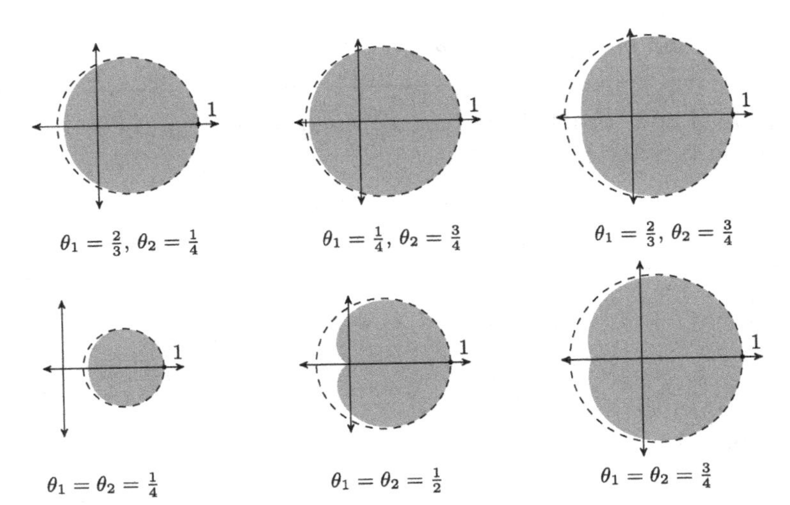

Figure 13.4 The shaded regions illustrate $\mathcal{G}(\mathcal{N}_{\theta_1}\mathcal{N}_{\theta_2})$ and the dashed circles illustrate $\mathcal{G}(\mathcal{N}_\theta)$ given by Theorem 27.

Proof. Note $\mathbb{T}$ is θ-averaged if and only if $\frac{1}{\theta}\mathbb{T} - \left(\frac{1}{\theta} - 1\right)\mathbb{I}$ is nonexpansive. The claim follows from

$$\begin{aligned} 0 &\geq \left\|\frac{1}{\theta}\mathbb{T}x - \left(\frac{1}{\theta} - 1\right)x - \frac{1}{\theta}\mathbb{T}y + \left(\frac{1}{\theta} - 1\right)y\right\|^2 - \|x - y\|^2 \\ &= \frac{1}{\theta}\left(\|\mathbb{T}x - \mathbb{T}y\|^2 + \frac{1-\theta}{\theta}\|\mathbb{T}x - x - \mathbb{T}y + y\|^2 - \|x - y\|^2\right). \end{aligned}$$

□

Proof of Theorem 28. For any $z^0, \hat{z}^0 \in \mathbb{R}^n$, let

$$\begin{aligned} x^{1/2} &= \mathbb{J}_{\alpha\mathbb{B}}(z^0) \\ x^1 &= \mathbb{J}_{\alpha\mathbb{A}}(2x^{1/2} - z^k - \alpha\mathbb{C}x^{1/2}) \\ z^1 &= z^0 + x^1 - x^{1/2} \end{aligned}$$

$$\begin{aligned} \hat{x}^{1/2} &= \mathbb{J}_{\alpha\mathbb{B}}(\hat{z}^0) \\ \hat{x}^1 &= \mathbb{J}_{\alpha\mathbb{A}}(2\hat{x}^{1/2} - \hat{z}^k - \alpha\mathbb{C}\hat{x}^{1/2}) \\ \hat{z}^1 &= \hat{z}^0 + \hat{x}^1 - \hat{x}^{1/2}. \end{aligned}$$

Define

$$\begin{aligned} \tilde{\mathbb{B}}x^{1/2} &= \frac{1}{\alpha}(z^0 - x^{1/2}) \\ \tilde{\mathbb{A}}x^1 &= \frac{1}{\alpha}(2x^{1/2} - z^k - \alpha\mathbb{C}x^{1/2} - x^1) \end{aligned}$$

$$\begin{aligned} \tilde{\mathbb{B}}\hat{x}^{1/2} &= \frac{1}{\alpha}(\hat{z}^0 - \hat{x}^{1/2}) \\ \tilde{\mathbb{A}}\hat{x}^1 &= \frac{1}{\alpha}(2\hat{x}^{1/2} - \hat{z}^k - \alpha\mathbb{C}\hat{x}^{1/2} - \hat{x}^1), \end{aligned}$$

which implies

$$\begin{aligned} \tilde{\mathbb{B}}x^{1/2} &\in \mathbb{B}x^{1/2} \\ \tilde{\mathbb{A}}x^1 &\in \mathbb{A}x^1 \end{aligned}$$

$$\begin{aligned} \tilde{\mathbb{B}}\hat{x}^{1/2} &\in \mathbb{B}\hat{x}^{1/2} \\ \tilde{\mathbb{A}}\hat{x}^1 &\in \mathbb{A}\hat{x}^1. \end{aligned}$$

Then,

$$\begin{aligned}\|z^1 - \hat{z}^1\|^2 &= \|z^0 - \hat{z}^0\|^2 - \frac{1-\theta}{\theta}\|z^1 - z^0 - \hat{z}^1 + \hat{z}^0\|^2 \\ &\quad - 2\alpha\langle \tilde{\mathbb{A}}x^1 - \tilde{\mathbb{A}}\hat{x}^1, x^1 - \hat{x}^1\rangle - 2\alpha\langle \tilde{\mathbb{B}}x^{1/2} - \tilde{\mathbb{B}}\hat{x}^{1/2}, x^{1/2} - \hat{x}^{1/2}\rangle \\ &\quad - 2\alpha\left(\langle \mathbb{C}x^{1/2} - \mathbb{C}\hat{x}^{1/2}, x^{1/2} - \hat{x}^{1/2}\rangle - \beta\|\mathbb{C}x^{1/2} - \mathbb{C}\hat{x}^{1/2}\|^2\right) \\ &\quad - \frac{\alpha^2}{2\beta}\left\|\tilde{\mathbb{A}}x^1 - \tilde{\mathbb{A}}\hat{x}^1 + \tilde{\mathbb{B}}x^{1/2} - \tilde{\mathbb{B}}\hat{x}^{1/2} - \frac{2\beta - \alpha}{\alpha}(\mathbb{C}x^{1/2} - \mathbb{C}\hat{x}^{1/2})\right\|^2 \\ &\le \|z^0 - \hat{z}^0\|^2 - \frac{1-\theta}{\theta}\|z^1 - z^0 - \hat{z}^1 + \hat{z}^0\|^2,\end{aligned}$$

where the inequality follows from monotonicity of $\mathbb{A}$ and $\mathbb{B}$ and β-cocoercivity of $\mathbb{C}$. Finally, the claim follows from Lemma 5. □

This proof does not use SRGs. Whether there is a simpler proof of Theorem 28 that relies on the SRG machinery is an open problem.

BIBLIOGRAPHICAL NOTES

Using circles or disks centered at the origin to illustrate contractive mappings is natural and likely common. Eckstein and Bertsekas's illustration of firm-nonexpansiveness via the disk with radius 1/2 centered at (1/2, 0) [Eck89, EB92] was, to the best of our knowledge, the first geometric illustration of notions from fixed-point theory other than nonexpansiveness and Lipschitz continuity. Since then, Giselsson and Boyd used similar illustrations in earlier versions of the paper [GB17] (the arXiv versions 1 through 3 have the geometric diagrams, but later versions do not) and more thoroughly in the lecture slides [Gis15]. Banjac and Goulart also utilize similar illustrations [BG18].

In complex analysis, the inversion map is known as the Möbius transformation [AF03, p. 366]. In classical Euclidean geometry, *inversive geometry* considers generally the inversion of the 2D plane about any circle [Ped70, p. 75]. Our inversion map $z \mapsto \bar{z}^{-1}$ is the inversion about the unit circle.

Stewart's theorem is due to [Ste46]. The proof of the spherical triangle inequality can be found in [RHY21]. The proof SRG generalizing eigenvalues relies on topological arguments and can be found in [RHY21]. To further understand Example 13.3, see [HRY20, HRY19] for follow-up work on drawing the SRG of linear operators. Pates identified a connection between the SRG of a linear operator and the numerical range of a related linear operator [Pat21]. This connection allows one to utilize existing machinery for the numerical range for drawing SRGs of linear operators. Furthermore, Pates used this connection and the Toeplitz–Hausdorff theorem to show that $\mathcal{G}(A^\top) = \mathcal{G}(A)$ for any $A \in \mathbb{R}^{n\times n}$. Finally, Chaffey, Forni, and Sepulchre has explored applications of the SRG in control theory [CFS21b, CFS21a].

Proposition 2 was first shown in [RHY21], Proposition 3 in [BC17a, Proposition 26.16], Proposition 4 in [RHY21, Fact 7], Proposition 5 in [BC17a, Proposition 23.13], Proposition 6 in [GB15, Theorem 7.2], and Proposition 7 in [GB17, Theorem 1].

The definition of the SRG, when restricted to linear operators, has a form similar to the pseudospectrum [TE05], but it is unclear if there is any meaningful connection. A connection between the SRG and the numerical range (field of values) [HJ91] was identified by Pates [Pat21].

In Example 13.3, the first, second, and third SRGs can be drawn by considering inputs of the form $[r\cos(\theta), r\sin(\theta)]$ for $r > 0$ and performing direct calculations. The fourth SRG (the one with $\mathbb{B}$) can be drawn with the results of [HRY19]. The SRG of Example 13.4 can be drawn with the ideas of [RHY21] and [HRY19]. The SRGs of Example 13.5 is discussed in [HRY19].

The averagedness coefficient of Theorem 27 was established by Ogura, Yamada, and Combettes [OY02, CY15]. Let

$$\mathcal{T}_{\alpha,\beta} = \{\mathbb{I} - \mathbb{J}_{\alpha\mathbb{B}} + \mathbb{J}_{\alpha\mathbb{A}}(\mathbb{R}_{\alpha\mathbb{B}} - \alpha\mathbb{C}\mathbb{J}_{\alpha\mathbb{B}}) \mid \mathbb{A}, \mathbb{B} \in \mathcal{M}, \mathbb{C} \in \mathcal{C}_\beta\}.$$

Theorem 28 states $\mathcal{G}(\mathcal{T}_{\alpha,\beta}) \subseteq \mathcal{G}\left(\mathcal{N}_{\frac{2\beta}{4\beta-\alpha}}\right)$ for $\alpha \in (0, 2\beta)$, but the stronger result

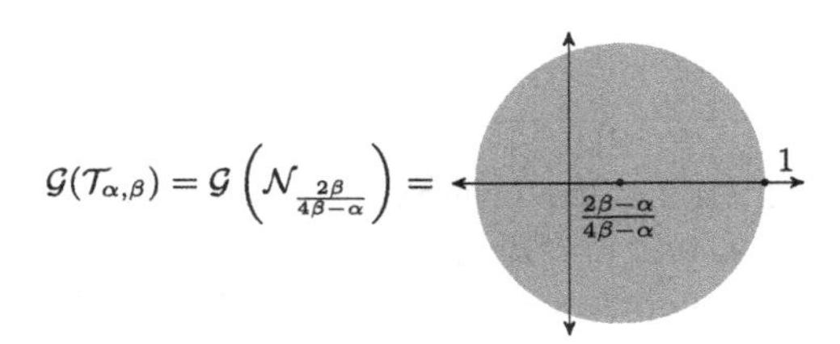

which implies tightness of the averagedness coefficient, was shown in [HRY20]. One may ask whether the averagedness coefficients established in Theorems 27 and 28 are tight (largest). In Figure 13.4, we can observe that the SRG touches the $\mathcal{G}(\mathcal{N}_\theta)$ at 1 and has matching "curvature." This observation has been formalized into a proof of tightness in [HRY20].

EXERCISES

13.1 *Contractive ⊂ averaged ⊂ nonexpansive.* Let $R < 1$. Show

$$\mathcal{L}_R \subset \mathcal{N}_{\frac{1+R}{2}} \subset \mathcal{L}_1.$$

13.2 Show $\mathbb{A} \in \mathcal{N}_{1/2} \;\Leftrightarrow\; \mathbb{A} \in \mathcal{C}_1 \;\Leftrightarrow\; \mathbb{I} - \mathbb{A} \in \mathcal{N}_{1/2} \;\Leftrightarrow\; 2\mathbb{A} - \mathbb{I} \in \mathcal{L}_1$.

13.3 Show that $\mathbb{A} \in \mathcal{N}_\theta$ if and only if $\mathbb{I} - \mathbb{A} \in \mathcal{C}_{1/(2\theta)}$.
Remark. We had proved this result in the proof of Theorem 2, but a proof using the SRG provides geometric intuition.

13.4 Show that if $\theta \in (0, 1/2)$, then $\mathcal{N}_\theta \subset \mathcal{M}_{1-2\theta}$, where the inclusion is strict.

13.5 *Optimal stepsize for gradient descent.* Let $0 < \mu < L < \infty$. We previously established that $\mathbb{I} - \alpha\partial\mathcal{F}_{\mu,L} \subseteq \mathcal{L}_R$ with

$$R = \max\{|1 - \alpha\mu|, |1 - \alpha L|\},$$

which provides an exponential rate of convergence for the gradient method

$$x^{k+1} = x^k - \alpha\nabla f(x^k).$$

What is the optimal choice of $\alpha > 0$ that minimizes the contraction factor? Describe $\mathcal{G}(\mathbb{I} - \alpha\partial\mathcal{F}_{\mu,L})$ with the optimal α.

13.6 Let $L > 0$. Consider

$$\underset{x\in\mathbb{R}^n}{\text{find}} \quad 0 = \mathbb{F}(x),$$

where $\mathbb{F}\colon \mathbb{R}^n \to \mathbb{R}^n$ is monotone and L-Lipschitz. Using $\mathcal{G}(\mathbb{I} - \alpha(\mathcal{M} \cap \mathcal{L}_L))$, explain why is it not possible to establish convergence of the forward step method

$$x^{k+1} = x^k - \alpha\mathbb{F}x^k$$

without further assumptions.

13.7 *Nonexpansive and inverse Lipschitz residual makes Krasnosel'skiĭ–Mann contractive.* Show that if $\mathbb{T}$ is nonexpansive and $(\mathbb{I}-\mathbb{T})^{-1}$ is γ-Lipschitz, with $\gamma \geq 1/2$ and $\theta \in (0,1)$, then

$$(1-\theta)\mathbb{I} + \theta\mathbb{T} \in \mathcal{L}\left(\sqrt{1 - \frac{\theta(1-\theta)}{\gamma^2}}\right).$$

Remark. This result was first shown in [LFP16].

13.8 *Proximal point with inverse Lipschitz operator.* Let $\alpha, \gamma \in (0,\infty)$. Show that if $\mathcal{A} = \mathcal{L}_\gamma^{-1} \cap \mathcal{M}$, then $J_{\alpha\mathcal{A}} \subseteq \mathcal{L}_R$ for

$$R = \frac{\gamma}{\sqrt{\alpha^2 + \gamma^2}}.$$

Also show that the result is tight in the sense that $J_{\alpha\mathcal{A}} \not\subseteq \mathcal{L}_R$ for any smaller value of R.

Remark. This result was first shown analytically in [Roc76b].

13.9 *Pseudononexpansive operators* An operator $\mathbb{T}$ on $\mathbb{R}^n$ is said to be *pseudononexpansive* if

$$\|\mathbb{T}x - \mathbb{T}y\|^2 \leq \|x-y\|^2 + \|(\mathbb{I}-\mathbb{T})x - (\mathbb{I}-\mathbb{T})y\|^2 \qquad \forall x, y \in \mathbb{R}^n.$$

Describe the SRG of the class of pseudononexpansive operators.

13.10 *Closedness of $\{C_\beta\}_{\beta>0}$ under addition.* Show that if $\beta_1, \beta_2 \in (0,\infty)$, then $C_{\beta_1} + C_{\beta_2} = C_{(\beta_1^{-1}+\beta_2^{-1})^{-1}}$.

13.11 *SRG of DRS.* Let $\mathcal{A}$ be an SRG-full operator class such that $\mathcal{G}(\mathcal{A}) \subseteq \mathcal{G}(\mathcal{L}_R)$ and R is tight in the sense that there exists a $z \in \mathcal{G}(\mathcal{A})$ such that $|z| = R$. Show that $\mathcal{G}(\mathcal{A}\mathcal{L}_1) = \mathcal{G}(\mathcal{L}_1\mathcal{A}) = \mathcal{G}(\mathcal{L}_R)$. Also show that

$$\mathcal{G}\left(\frac{1}{2}\mathbb{I} + \frac{1}{2}\mathcal{A}\mathcal{L}_1\right) = \mathcal{G}\left(\frac{1}{2}\mathbb{I} + \frac{1}{2}\mathcal{L}_1\mathcal{A}\right) \subseteq \mathcal{G}\left(\mathcal{L}_{\frac{1}{2}+\frac{1}{2}R}\right).$$

Remark. Since the SRG-full classes are defined to contain operators on $\mathbb{R}^n$ for all $n \geq 1$, it is sufficient to consider the case $n = 2$ and appeal to Lemma 4.

13.12 Complete the proof of Theorem 19. Specifically, given $\mathcal{G}(\mathcal{M}) = \{z \mid \operatorname{Re} z \geq 0\} \cup \{\infty\}$ and $\mathcal{G}(\mathcal{L}_L) = \{z \mid |z| \leq L\}$, prove the characterizations of $\mathcal{G}(\mathcal{N}_\theta)$, $\mathcal{G}(\mathcal{M}_\mu)$, and $\mathcal{G}(\mathcal{C}_\beta)$ asserted in Theorem 19.

Hint. Use $\mathcal{M}_\mu = \mu I + \mathcal{M}$, $(\mathcal{M}_\mu)^{-1} = \mathcal{C}_\mu$, and $(1-\theta)I + \theta\mathcal{L}_1 = \mathcal{N}_\theta$.

13.13 Complete the proof of Theorem 20. Specifically, given $\mathcal{G}(\mathcal{F}_{0,\infty}) = \{z \mid \operatorname{Re} z \geq 0\} \cup \{\infty\}$, prove the characterizations of $\mathcal{G}(\partial\mathcal{F}_{\mu,\infty})$, $\mathcal{G}(\partial\mathcal{F}_{0,L})$, and $\mathcal{G}(\partial\mathcal{F}_{\mu,L})$ asserted in Theorem 20.

Hint. Use $\partial\mathcal{F}_{\mu,\infty} = \mu\mathbb{I} + \partial\mathcal{F}_{0,\infty}$, $\partial\mathcal{F}_{0,L} = \left(\partial\mathcal{F}_{1/L,\infty}\right)^{-1}$, and $\partial\mathcal{F}_{\mu,L} = \mu\mathbb{I} + \partial\mathcal{F}_{0,L-\mu}$.

13.14 Describe nonnegative homogeneous functions representing $\mathcal{M}$, $\mathcal{C}_\beta$, $\mathcal{L}_L$, and $\mathcal{N}_\theta$.

13.15 *SRG of union.* Show that a result similar to Theorem 22 with a union, rather than an intersection, holds.

13.16 Show that if h and g represent SRG-full classes $\mathcal{A}$ and $\mathcal{B}$, then $\max\{h, g\}$ represents $\mathcal{A} \cap \mathcal{B}$. Also, which class does $\min\{h, g\}$ represent?

13.17 *SRG-invariant transformations.* Some operations do not change the SRG. Given $\mathbb{A}: \mathbb{R}^n \rightrightarrows \mathbb{R}^n$ and $w \in \mathbb{R}^n$, define the inner and outer shifts $\mathbb{A}_w: \mathbb{R}^n \rightrightarrows \mathbb{R}^n$ and ${}_w\mathbb{A}: \mathbb{R}^n \rightrightarrows \mathbb{R}^n$ as

$$\mathbb{A}_w(x) = \mathbb{A}(x - w), \qquad {}_w\mathbb{A}(x) = \mathbb{A}(x) - w.$$

Let $\mathbb{A}: \mathbb{R}^n \rightrightarrows \mathbb{R}^n$, $w \in \mathbb{R}^n$, and $U \in \mathbb{R}^{n \times n}$ be an orthogonal matrix. Show that

$$\mathcal{G}(\mathbb{A}) = \mathcal{G}(\mathbb{A}_w) = \mathcal{G}({}_w\mathbb{A}) = \mathcal{G}(U\mathbb{A}U^\intercal).$$

13.18 *Composition without an arc property.* Consider the SRG-full operator class $\mathcal{A}$ represented by $h(a, b, c) = |a - b| + |c|$. Show:

(a) $\mathcal{G}(\mathcal{A}) = \{\pm i\}$,

(b) linear operators on $\mathbb{R}^3$ representing 90 degrees rotations are in $\mathcal{A}$, and

(c) $\mathcal{G}(\mathcal{A}\mathcal{A}) = \{z \in \mathbb{C} \mid |z| = 1\}$.

Remark. Since $\mathcal{G}(\mathcal{A}\mathcal{A}) = \{z \in \mathbb{C} \mid |z| = 1\} \supset \mathcal{G}(\mathcal{A})\mathcal{G}(\mathcal{A})$ is a strict containment, the analysis shows we cannot fully drop the arc property in Theorem 26.

13.19 *Right-scalar multiplication.* Define $\mathbb{A}\alpha$, the *right-scalar multiplication* with an operator $\mathbb{A}$ and scalar α, as

$$(\mathbb{A}\alpha)(x) = \mathbb{A}(\alpha x).$$

Define $\mathcal{A}\alpha = \{\mathbb{A}\alpha \mid \mathbb{A} \in \mathcal{A}\}$. Show $\mathcal{G}(\mathcal{A}\alpha) = \alpha\mathcal{G}(\mathcal{A})$ for $\alpha \neq 0$.

13.20 *DRS with split strong monotonicity and cocoercivity is a contraction.* Assume $\mathcal{A} = \mathcal{M}_\mu$ and $\mathcal{B} = \mathcal{C}_\beta$. Show:

(a) $\mathbb{R}_\mathcal{B} = \mathcal{N}_{\frac{1}{1+\beta}}$,

(b) $-\mathbb{R}_\mathcal{A} = \mathcal{N}_{\frac{1}{1+\mu}}$,

(c) $-\mathbb{R}_\mathcal{A}\mathbb{R}_\mathcal{B} \subseteq \mathcal{N}_{\frac{\mu+\beta}{\mu+\beta+\mu\beta}}$,

(d) $\frac{1}{2}\mathbb{I} + \frac{1}{2}\mathbb{R}_\mathcal{A}\mathbb{R}_\mathcal{B} \subseteq \mathcal{L}_{\frac{\mu+\beta}{\mu+\beta+\mu\beta}}$, and

(e) $\mathbb{R}_\mathcal{A}\mathbb{R}_\mathcal{B} \not\subseteq \mathcal{L}_R$ for any $0 \leq R < 1$.

Hint. For part (c), use Theorem 27.

Remark. If $-\mathbb{A}$ is θ-averaged, we say $\mathbb{A}$ is *negatively averaged.* Giselsson defined the notion of negatively averaged operators and used it to prove fact (d) [GB15]. Ryu et al. used a similar argument to establish a contraction for the "plug-and-play" image reconstruction method in machine learning [RLW+19].

Appendix A Miscellaneous

Let $\mathcal{F}_0 \subseteq \mathcal{F}_1 \subseteq \cdots$ be a sequence of σ-algebras. Write $\mathbb{E}[X \mid \mathcal{F}_k]$ for the conditional expectation of a random variable X with respect to $\mathcal{F}_k$.

In this book, $\mathcal{F}_k$ represents the information before iteration k, and the Lyapunov function V^k is $\mathcal{F}_k$-measurable. Therefore, $\mathbb{E}\left[V^k \mid \mathcal{F}_k\right] = V^k$. To say this without using measure theoretic language, $\mathbb{E}[\cdot \mid \mathcal{F}_k]$ represents the expectation conditioned on the information before iteration k, and V^k is determined by the randomness of the iterations before k. For example, in Theorem 2 of §5.1, we (implicitly) use the Lyapunov function $V^k = \|x^k - x^\star\|^2$. Since x^k is determined by the (nonrandom) starting point x^0 and the random indices $i(0), i(1), \ldots, i(k-1)$, we have $\mathbb{E}\left[V^k \mid \mathcal{F}_k\right] = V^k$, since there is no randomness in V^k once we condition on the information before iteration k.

Theorem 29 *Supermartingale convergence theorem.* Let V^k and S^k be $\mathcal{F}_k$-measurable random variables satisfying $V^k \geq 0$ and $S^k \geq 0$ almost surely for $k = 0, 1, \ldots$. Assume

$$\mathbb{E}\left[V^{k+1} \mid \mathcal{F}_k\right] \leq \mathbb{E}_k V^k - S^k$$

holds for $k = 0, 1, \ldots$. Then

1. $V^k \to V^\infty$
2. $\sum_{k=0}^{\infty} S^k < \infty$

almost surely. (Note that the limit V^∞ is a random variable.)

This supermartingale convergence theorem is due to Doob [Doo53] and its proof can be found in many standard textbooks on probability theory. (The standard supermartingale convergence theorem is slightly more general.)

Theorem 30 Let V^k, S^k, and U^k be $\mathcal{F}_k$-measurable random variables satisfying $V^k \geq 0$, $S^k \geq 0$, and $U^k \geq 0$ almost surely for $k = 0, 1, \ldots$. Let $\beta_0, \beta_1, \ldots$ be nonnegative (nonrandom) scalars satisfying

$$\sum_{k=0}^{\infty} \beta_k < \infty.$$

Assume

$$\mathbb{E}\left[V^{k+1} \mid \mathcal{F}_k\right] \leq (1+\beta_k)V^k - S^k + U^k$$

and

$$\sum_{i=1}^{\infty} U^i < \infty$$

almost surely. Then

1. $V^k \to V^\infty$
2. $\sum_{k=0}^{\infty} S^k < \infty$

almost surely. (Note that the limit V^∞ is a random variable.)

This "almost supermartingale" convergence theorem is due to Robbins and Siegmund [RS85]

Theorem 31 Let M^k be a $\mathcal{F}_k$-measurable random variable for $k = 0, 1, \ldots$. Assume

$$\mathbb{E}\left[M^{k+1} \mid \mathcal{F}_k\right] = M^k$$

almost surely,

$$\mathbb{E}[\|M_k\|^2] < \infty \qquad \forall\, k = 0, 1, \ldots,$$

and

$$\sum_{k=0}^{\infty} \mathbb{E}\left[\|M_{k+1} - M_k\|^2 \mid \mathcal{F}_k\right] < \infty$$

almost surely. Then

$$M^k \to M^\infty$$

almost surely. (Note that the limit M^∞ is a random variable.)

The proof of this result can be found as Theorem 5.3.33 of Dembo's lecture notes [Dem19].

References

[AAC18] F. J. Aragón Artacho and R. Campoy. A new projection method for finding the closest point in the intersection of convex sets. *Computational Optimization and Applications*, 69(1):99–132, 2018.

[AAC19] F. J. Aragón Artacho and R. Campoy. Computing the resolvent of the sum of maximally monotone operators with the averaged alternating modified reflections algorithm. *Journal of Optimization Theory and Applications*, 181(3):709–726, 2019.

[ACL16] T. Aspelmeier, C. Charitha, and D. R. Luke. Local linear convergence of the ADMM/Douglas–Rachford algorithms without strong convexity and application to statistical imaging. *SIAM Journal on Imaging Sciences*, 9(2):842–868, 2016.

[AF03] M. J. Ablowitz and A. S. Fokas. *Complex Variables: Introduction and Applications*. Cambridge Texts in Applied Mathematics. Cambridge University Press, Cambridge, second ed., 2003.

[AHU58] K. J. Arrow, L. Hurwicz, and H. Uzawa. *Studies in Linear and Non-Linear Programming*. Stanford University Press, 1958.

[Amd67] G. M. Amdahl. Validity of the single processor approach to achieving large-scale computing capabilities. In *AFIPS Conference Proceedings*, 1967.

[AMP05] C. Andrieu, É. Moulines, and P. Priouret. Stability of stochastic approximation under verifiable conditions. *SIAM Journal on Control and Optimization*, 44(1):283–312, 2005.

[And65] D. G. Anderson. Iterative procedures for nonlinear integral equations. *Journal of The ACM*, 12(4):547–560, 1965.

[AT96] H. Attouch and M. Théra. A general duality principle for the sum of two operators. *Journal of Convex Analysis*, 3(1):1–24, 1996.

[AT06] A. Auslender and M. Teboulle. Interior gradient and proximal methods for convex and conic optimization. *SIAM Journal on Optimization*, 16(3):697–725, 2006.

[Att77] H. Attouch. Convergence of convex functions, subs-differentials and associated semi-groups. *Comptes Rendus Hebdomadaires des Séances de l'Académie des sciences, Série A*, 284(10):539–542, 1977.

[Att84] H. Attouch. *Variational Convergence for Functions and Operators*. Applicable Mathematics Series, vol. 1. Pitman Advanced Publishing Program, 1984.

[Aum65] R. J. Aumann. Integrals of set-valued functions. *Journal of Mathematical Analysis and Applications*, 12(1):1–12, 1965.

[Aus92] A. Auslender. Asymptotic properties of the Fenchel dual functional and applications to decomposition problems. *Journal of Optimization Theory and Applications*, 73(3):427–449, 1992.

[Ban22] S. Banach. Sur les opérations dans les ensembles abstraits et leur application aux équations intégrales. *Fundamenta Mathematicae*, 3(1):133–181, 1922.

[BB96] H. H. Bauschke and J. M. Borwein. On projection algorithms for solving convex feasibility problems. *SIAM Review*, 38(3):367–426, 1996.

[BBC21] S. Banert, R. I. Boţ, and E. R. Csetnek. Fixing and extending some recent results on the ADMM algorithm. *Numerical Algorithms*, 86(3):1303–1325, 2021.

[BBCN+14] H. H. Bauschke, J. Y. Bello Cruz, T. T. A. Nghia, H. M. Phan, and X. Wang. The rate of linear convergence of the Douglas–Rachford algorithm for subspaces is the cosine of the Friedrichs angle. *Journal of Approximation Theory*, 185:63–79, 2014.

[BBHM12] H. H. Bauschke, R. I. Boţ, W. L. Hare, and W. M. Moursi. Attouch–Théra duality revisited: Paramonotonicity and operator splitting. *Journal of Approximation Theory*, 164(8):1065–1084, 2012.

[BBL97] H. H. Bauschke, J. M. Borwein, and A. S. Lewis. The method of cyclic projections for closed convex sets in Hilbert space. *Contemporary Mathematics*, 204:1–38, 1997.

[BBR78] J. B. Baillon, R. E. Bruck, and S. Reich. On the asymptotic behavior of nonexpansive mappings and semigroups in Banach spaces. *Houston Journal of Mathematics*, 4(1):1–9, 1978.

[BC10] H. H. Bauschke and P. L. Combettes. The Baillon–Haddad theorem revisited. *Journal of Convex Analysis*, 17(3–4):781–787, 2010.

[BC17a] H. H. Bauschke and P. L. Combettes. *Convex Analysis and Monotone Operator Theory in Hilbert Spaces*. Springer International Publishing, second ed., 2017.

[BC17b] J. Y. Bello Cruz. On proximal subgradient splitting method for minimizing the sum of two nonsmooth convex functions. *Set-Valued and Variational Analysis*, 25(2):245–263, 2017.

[BCH15] R. I. Boţ, E. R. Csetnek, and C. Hendrich. Inertial Douglas–Rachford splitting for monotone inclusion problems. *Applied Mathematics and Computation*, 256:472–487, 2015.

[BCHH15] R. I. Boţ, E. R. Csetnek, A. Heinrich, and C. Hendrich. On the convergence rate improvement of a primal-dual splitting algorithm for solving monotone inclusion problems. *Mathematical Programming*, 150(2):251–279, 2015.

[BCR05] H. H. Bauschke, P. L. Combettes, and S. Reich. The asymptotic behavior of the composition of two resolvents. *Nonlinear Analysis: Theory, Methods and Applications*, 60(2):283–301, 2005.

[BD15] S. Boyd and J. Duchi. Exercises for EE364b. Homework Exercises EE364b, Stanford University, 2015.

[BD18] L. M. Briceño-Arias and D. Davis. Forward-backward-half forward algorithm for solving monotone inclusions. *SIAM Journal on Optimization*, 28(4):2839–2871, 2018.

[BDV18] S. Boyd, J. Duchi, and L. Vandenberghe. Subgradients. Lecture Note EE364b, Stanford University, 2018.

[BDX04] S. Boyd, P. Diaconis, and L. Xiao. Fastest mixing Markov chain on a graph. *SIAM Review*, 46(4):667–689, 2004.

[Bec19] S. Becker. The Chen-Teboulle algorithm is the proximal point algorithm. *arXiv:1908.03633*, 2019.

[Ben62] J. F. Benders. Partitioning procedures for solving mixed-variables programming problems. *Numerische Mathematik*, 4(1):238–252, 1962.

[Ber73] D. P. Bertsekas. Stochastic optimization problems with nondifferentiable cost functionals. *Journal of Optimization Theory and Applications*, 12(2):218–231, 1973.

[Ber83] D. P. Bertsekas. Distributed asynchronous computation of fixed points. *Mathematical Programming*, 27(1):107–120, 1983.

[Ber09] D. P. Bertsekas. *Convex Optimization Theory*. Athena Scientific, 2009.

[Ber16] D. P. Bertsekas. *Nonlinear Programming*. Athena Scientific, third ed., 2016.

[BG18] G. Banjac and P. J. Goulart. Tight global linear convergence rate bounds for operator splitting methods. *IEEE Transactions on Automatic Control*, 63(12):4126–4139, 2018.

[BG19] N. Bansal and A. Gupta. Potential-function proofs for gradient methods. *Theory of Computing*, 15(4):1–32, 2019.

[BGLS95] J. F. Bonnans, J. C. Gilbert, C. Lemaréchal, and C. A. Sagastizábal. A family of variable metric proximal methods. *Mathematical Programming*, 68(1):15–47, 1995.

[BGMS20] H. H. Bauschke, S. Gretchko, W. M. Moursi, and M. Saurette. Edelstein's astonishing affine isometry. https://www.tandfonline.com/doi/abs/10.1080/00029890.2021.1962151

[BGSB19] G. Banjac, P. Goulart, B. Stellato, and S. Boyd. Infeasibility detection in the alternating direction method of multipliers for convex optimization. *Journal of Optimization Theory and Applications*, 183(2):490–519, 2019.

[BH77] J.-B. Baillon and G. Haddad. Quelques propriétés des opérateurs angle-bornés etn-cycliquement monotones. *Israel Journal of Mathematics*, 26(2):137–150, 1977.

[BH16] P. Bianchi and W. Hachem. Dynamical behavior of a stochastic Forward–Backward algorithm using random monotone operators. *Journal of Optimization Theory and Applications*, 171(1):90–120, 2016.

[BI98] R. S. Burachik and A. N. Iusem. A generalized proximal point algorithm for the variational inequality problem in a Hilbert space. *SIAM Journal on Optimization*, 8(1):197–216, 1998.

[Bia16] P. Bianchi. Ergodic convergence of a stochastic proximal point algorithm. *SIAM Journal on Optimization*, 26(4):2235–2260, 2016.

[BJ76] C. A. Botsaris and D. H. Jacobson. A Newton-type curvilinear search method for optimization. *Journal of Mathematical Analysis and Applications*, 54(1):217–229, 1976.

[BK19] E. Börgens and C. Kanzow. Regularized Jacobi-type ADMM-methods for a class of separable convex optimization problems in Hilbert spaces. *Computational Optimization and Applications*, 73(3):755–790, 2019.

[BL78] H. Brezis and P. L. Lions. Produits infinis de resolvantes. *Israel Journal of Mathematics*, 29(4):329–345, 1978.

[BL06] J. Borwein and A. S. Lewis. *Convex Analysis and Nonlinear Optimization*. Springer, second ed., 2006.

[BLM17] H. H. Bauschke, B. Lukens, and W. M. Moursi. Affine nonexpansive operators, Attouch–Théra duality and the Douglas–Rachford algorithm. *Set-Valued and Variational Analysis*, 25(3):481–505, 2017.

[BLS15] S. Bubeck, Y. T. Lee, and M. Singh. A geometric alternative to Nesterov's accelerated gradient descent. *arXiv:1506.08187*, 2015.

[BM16] H. H. Bauschke and W. M. Moursi. On the order of the operators in the Douglas–Rachford algorithm. *Optimization Letters*, 10(3):447–455, 2016.

[BM17] H. H. Bauschke and W. M. Moursi. On the Douglas–Rachford algorithm. *Mathematical Programming*, 164(1):263–284, 2017.

[BMS06] H. H. Bauschke, D. A. McLaren, and H. S. Sendov. Fitzpatrick functions: Inequalities, examples, and remarks on a problem by S. Fitzpatrick. *Journal of Convex Analysis*, 13(3):499–523, 2006.

[Bol13] D. Boley. Local linear convergence of the alternating direction method of multipliers on quadratic or linear programs. *SIAM Journal on Optimization*, 23(4):2183–2207, 2013.

[Bon11] S. Bonettini. Inexact block coordinate descent methods with application to non-negative matrix factorization. *IMA Journal of Numerical Analysis*, 31(4):1431–1452, 2011.

[Bor06] J. M. Borwein. Maximal monotonicity via convex analysis. *Journal of Convex Analysis*, 13(3–4):561–586, 2006.

[Bot91] L. Bottou. *Une Approche Théorique de l'Apprentissage Connexionniste: Applications à La Reconnaissance de La Parole*. PhD Thesis, Université de Paris XI, Orsay, France, 1991.

[Bot99] L. Bottou. On-line learning and stochastic approximations. In D. Saad, ed., *On-Line Learning in Neural Networks*, Publications of the Newton Institute, pages 9–42. Cambridge University Press, 1999.

[Boţ10] R. I. Boţ. *Conjugate Duality in Convex Optimization*, vol. 637 of Lecture Notes in Economics and Mathematical Systems. Springer-Verlag, 2010.

[BP66] F. E. Browder and W. V. Petryshyn. The solution by iteration of nonlinear functional equations in Banach spaces. *Bulletin of the American Mathematical Society*, 72(3):571–575, 1966.

[BP67] F. E. Browder and W. V. Petryshyn. Construction of fixed points of nonlinear mappings in Hilbert space. *Journal of Mathematical Analysis and Applications*, 20(2):197–228, 1967.

[BPC+11] S. Boyd, N. Parikh, E. Chu, B. Peleato, and J. Eckstein. Distributed optimization and statistical learning via the alternating direction method of multipliers. *Foundations and Trends in Machine Learning*, 3(1):1–122, 2011.

[BQ99] J. V. Burke and M. Qian. A variable metric proximal point algorithm for monotone operators. *SIAM Journal on Control and Optimization*, 37(2):353–375, 1999.

[Bre71] H. Brezis. On a problem of T. Kato. *Communications on Pure and Applied Mathematics*, 24(1):1–6, 1971.

[Bri15] L. M. Briceño-Arias. Forward-Douglas–Rachford splitting and forward-partial inverse method for solving monotone inclusions. *Optimization*, 64(5):1239–1261, 2015.

[Bru75a] R. E. Bruck. Asymptotic convergence of nonlinear contraction semigroups in Hilbert space. *Journal of Functional Analysis*, 18(1):15–26, 1975.

[Bru75b] R. E. Bruck. An iterative solution of a variational inequality for certain monotone operators in Hilbert space. *Bulletin of the American Mathematical Society*, 81(5):890–892, 1975.

[Bru77] R. E. Bruck. On the weak convergence of an ergodic iteration for the solution of variational inequalities for monotone operators in Hilbert space. *Journal of Mathematical Analysis and Applications*, 61(1):159–164, 1977.

[BSS16] M. Burger, A. Sawatzky, and G. Steidl. First order algorithms in variational image processing. In R. Glowinski, S. J. Osher, and W. Yin, eds., *Splitting Methods in Communication, Imaging, Science, and Engineering*, Scientific Computation, pages 345–407. Springer, 2016.

[BT89] D. P. Bertsekas and J. N. Tsitsiklis. *Parallel and Distributed Computation: Numerical Methods*. Prentice Hall, 1989.

[BT09] A. Beck and M. Teboulle. A fast iterative shrinkage-thresholding algorithm for linear inverse problems. *SIAM Journal on Imaging Sciences*, 2(1):183–202, 2009.

[BT13] A. Beck and L. Tetruashvili. On the convergence of block coordinate descent type methods. *SIAM Journal on Optimization*, 23(4):2037–2060, 2013.

[BV04] S. Boyd and L. Vandenberghe. *Convex Optimization*. Cambridge University Press, 2004.

[BV10] J. M. Borwein and J. D. Vanderwerff. *Convex Functions: Constructions, Characterizations and Counterexamples*. Encyclopedia of Mathematics and Its Applications. Cambridge University Press, 2010.

[BW10] H. H. Bauschke and X. Wang. Firmly nonexpansive and Kirszbraun–Valentine extensions: A constructive approach via monotone operator theory. In A. Leizarowitz, B. S. Mordukhovich, I. Shafrir, and A. J. Zaslavski, eds., *Nonlinear Analysis and Optimization I: Nonlinear Analysis*, pages 55–64. American Mathematics Society, 2010.

[Cau47] A.-L. Cauchy. Méthode générale pour la résolution des systémes d'équations simultanées. *Comptes Rendus Hebdomadaires des Séances de l'Académie des Sciences*, 25:536–538, 1847.

[CCCP] G. Chierchia, E. Chouzenoux, P. L. Combettes, and J.-C. Pesquet. The proximity operator repository. User's guide. http://proximity-operator.net/download/guide.pdf.

[CCMY15] C. Chen, R. H. Chan, S. Ma, and J. Yang. Inertial proximal ADMM for linearly constrained separable convex optimization. *SIAM Journal on Imaging Sciences*, 8(4):2239–2267, 2015.

[CCPV14] P. L. Combettes, L. Condat, J.-C. Pesquet, and B. C. Vũ. A forward-backward view of some primal-dual optimization methods in image recovery. In *IEEE International Conference on Image Processing*, 2014.

[CDR15] S. Chaturapruek, J. C. Duchi, and C. Ré. Asynchronous stochastic convex optimization: the noise is in the noise and SGD don't care. In *Neural Information Processing Systems*, 2015.

[CFKS17a] L. Cannelli, F. Facchinei, V. Kungurtsev, and G. Scutari. Asynchronous parallel nonconvex large-scale optimization. In *International Conference on Acoustics, Speech and Signal Processing*, 2017.

[CFKS17b] L. Cannelli, F. Facchinei, V. Kungurtsev, and G. Scutari. Asynchronous parallel algorithms for nonconvex big-data optimization. Part II: Complexity and numerical results. *arXiv:1701.04900*, 2017.

[CFS21a] T. Chaffey, F. Forni, and R. Sepulchre. Graphical nonlinear system analysis. *arXiv:2107.11272*, 2021.

[CFS21b] T. Chaffey, F. Forni, and R. Sepulchre. Scaled relative graphs for system analysis. https://ieeexplore.ieee.org/abstract/document/9683092.

[CGFL19] T. Chavdarova, G. Gidel, F. Fleuret, and S. Lacoste-Julien. Reducing noise in GAN training with variance reduced extragradient. In *Neural Information Processing Systems*, 2019.

[Che12] A. I.-A. Chen. *Fast Distributed First-Order Methods*. PhD Thesis, Massachusetts Institute of Technology, Department of Electrical Engineering and Computer Science, 2012.

[CHLZ12] B. Chen, S. He, Z. Li, and S. Zhang. Maximum block improvement and polynomial optimization. *SIAM Journal on Optimization*, 22(1):87–107, 2012.

[CHW15] T.-H. Chang, M. Hong, and X. Wang. Multi-agent distributed optimization via inexact consensus ADMM. *IEEE Transactions on Signal Processing*, 63(2):482–497, 2015.

[CHYY16] C. Chen, B. He, Y. Ye, and X. Yuan. The direct extension of ADMM for multi-block convex minimization problems is not necessarily convergent. *Mathematical Programming*, 155(1):57–79, 2016.

[CIZ98] Y. Censor, A. N. Iusem, and S. A. Zenios. An interior point method with Bregman functions for the variational inequality problem with paramonotone operators. *Mathematical Programming*, 81(3):373–400, 1998.

[CKCH22] L. Condat, D. Kitahara, A. Contreras, and A. Hirabayashi. Proximal splitting algorithms for convex optimization: A tour of recent advances, with new twists. arXiv:1912.00137, 2022.

[CLCD07] M. Chiang, S. H. Low, A. R. Calderbank, and J. C. Doyle. Layering as optimization decomposition: A mathematical theory of network architectures. *Proceedings of the IEEE*, 95(1):255–312, 2007.

[CLS07] F. S. Cattivelli, C. G. Lopes, and A. H. Sayed. A diffusion rls scheme for distributed estimation over adaptive networks. In *Signal Processing Advances in Wireless Communications*, 2007.

[CM69] D. Chazan and W. Miranker. Chaotic relaxation. *Linear Algebra and Its Applications*, 2(2):199–222, 1969.

[Com04] P. L. Combettes. Solving monotone inclusions via compositions of nonexpansive averaged operators. *Optimization*, 53(5–6):475–504, 2004.

[Con13] L. Condat. A primal-dual splitting method for convex optimization involving Lipschitzian, proximable and linear composite terms. *Journal of Optimization Theory and Applications*, 158(2):460–479, 2013.

[CP08] P. L. Combettes and J.-C. Pesquet. A proximal decomposition method for solving convex variational inverse problems. *Inverse Problems*, 24(6):065014, 2008.

[CP11a] A. Chambolle and T. Pock. A first-order primal-dual algorithm for convex problems with applications to imaging. *Journal of Mathematical Imaging and Vision*, 40(1):120–145, 2011.

[CP11b] P. L. Combettes and J.-C. Pesquet. Proximal splitting methods in signal processing. In H. H. Bauschke, R. S. Burachik, P. L. Combettes, V. Elser, D. R. Luke, and H. Wolkowicz, eds., *Fixed-Point Algorithms for Inverse Problems in Science and Engineering*, pages 185–212. Springer, 2011.

[CP15] P. L. Combettes and J.-C. Pesquet. Stochastic quasi-Fejér block-coordinate fixed point iterations with random sweeping. *SIAM Journal on Optimization*, 25(2):1221–1248, 2015.

[CP16a] A. Chambolle and T. Pock. An introduction to continuous optimization for imaging. *Acta Numerica*, 25:161–319, 2016.

[CP16b] A. Chambolle and T. Pock. On the ergodic convergence rates of a first-order primal-dual algorithm. *Mathematical Programming*, 159(1):253–287, 2016.

[CP19] P. L. Combettes and J.-C. Pesquet. Stochastic quasi-Fejér block-coordinate fixed point iterations with random sweeping II: Mean-square and linear convergence. *Mathematical Programming*, 174(1–2):433–451, 2019.

[CPR16] E. Chouzenoux, J.-C. Pesquet, and A. Repetti. A block coordinate variable metric forward–backward algorithm. *Journal of Global Optimization*, 66(3):457–485, 2016.

[CS10] F. S. Cattivelli and A. H. Sayed. Diffusion LMS strategies for distributed estimation. *IEEE Transactions on Signal Processing*, 58(3):1035–1048, 2010.

[CSFK16] L. Cannelli, G. Scutari, F. Facchinei, and V. Kungurtsev. Parallel asynchronous lock-free algorithms for nonconvex big-data optimization. In *Asilomar Conference on Signals, Systems and Computers*, 2016.

[CST17a] L. Chen, D. Sun, and K.-C. Toh. An efficient inexact symmetric Gauss–Seidel based majorized ADMM for high-dimensional convex composite conic programming. *Mathematical Programming*, 161(1):237–270, 2017.

[CST17b] L. Chen, D. Sun, and K.-C. Toh. A note on the convergence of ADMM for linearly constrained convex optimization problems. *Computational Optimization and Applications*, 66(2):327–343, 2017.

[CSY13] C. Chen, Y. Shen, and Y. You. On the convergence analysis of the alternating direction method of multipliers with three blocks. *Abstract and Applied Analysis*, 2013:183961, 2013.

[CT93] G. Chen and M. Teboulle. Convergence analysis of a proximal-like minimization algorithm using Bregman functions. *SIAM Journal on Optimization*, 3(3):538–543, 1993.

[CT94] G. Chen and M. Teboulle. A proximal-based decomposition method for convex minimization problems. *Mathematical Programming*, 64(1):81–101, 1994.

[CT05] E. J. Candes and T. Tao. Decoding by linear programming. *IEEE Transactions on Information Theory*, 51(12):4203–4215, 2005.

[CT06] E. J. Candes and T. Tao. Near-optimal signal recovery from random projections: Universal encoding strategies? *IEEE Transactions on Information Theory*, 52(12):5406–5425, 2006.

[CV95] C. Cortes and V. Vapnik. Support-vector networks. *Machine Learning*, 20(3):273–297, 1995.

[CV14] P. L. Combettes and B. C. Vũ. Variable metric forward–backward splitting with applications to monotone inclusions in duality. *Optimization*, 63(9):1289–1318, 2014.

[CV20] C. Clason and T. Valkonen. *Introduction to Nonsmooth Analysis and Optimization. arXiv:2001.00216*, 2020.

[CW90] D. Coppersmith and S. Winograd. Matrix multiplication via arithmetic progressions. *Computational Algebraic Complexity Editorial*, 9(3):251–280, 1990.

[CW05] P. L. Combettes and V. R. Wajs. Signal recovery by proximal forward-backward splitting. *Multiscale Modeling and Simulation*, 4(4):1168–1200, 2005.

[CY15] P. L. Combettes and I. Yamada. Compositions and convex combinations of averaged nonexpansive operators. *Journal of Mathematical Analysis and Applications*, 425(1):55–70, 2015.

[CZ92] Y. Censor and S. A. Zenios. Proximal minimization algorithm with D-functions. *Journal of Optimization Theory and Applications*, 73(3):451–464, 1992.

[Dav16] D. Davis. The asynchronous PALM algorithm for nonsmooth nonconvex problems. *arXiv:1604.00526*, 2016.

[DDC14] A. Defazio, J. Domke, and T. S. Caetano. Finito: A faster, permutable incremental gradient method for big data problems. In *International Conference on Machine Learning*, 2014.

[D'E59] D. A. D'Esopo. A convex programming procedure. *Naval Research Logistics Quarterly*, 6(1):33–42, 1959.

[DEC65] G. B. Dantzig, E. Eisenberg, and R. W. Cottle. Symmetric dual nonlinear programs. *Pacific Journal of Mathematics*, 15(3):809–812, 1965.

[Dem19] A. Dembo. Lecture notes on probability theory: Stanford statistics 310. Lecture Note STAT310/MATH230, Department of Mathematics, Stanford University, 2019.

[DLPY17] W. Deng, M.-J. Lai, Z. Peng, and W. Yin. Parallel multi-block ADMM with $o(1/k)$ convergence. *Journal of Scientific Computing*, 71(2):712–736, 2017.

[Don06] D. L. Donoho. Compressed sensing. *IEEE Transactions on Information Theory*, 52(4):1289–1306, 2006.

[Doo53] J. L. Doob. *Stochastic Processes*. Wiley, 1953.

[DR56] J. Douglas and H. H. Rachford. On the numerical solution of heat conduction problems in two and three space variables. *Transactions of the American Mathematical Society*, 82(2):421–439, 1956.

[DRT11] I. S. Dhillon, P. Ravikumar, and A. Tewari. Nearest neighbor based greedy coordinate descent. In *Neural Information Processing Systems*, 2011.

[DST15] Y. Drori, S. Sabach, and M. Teboulle. A simple algorithm for a class of nonsmooth convex–concave saddle-point problems. *Operations Research Letters*, 43(2):209–214, 2015.

[DT14] Y. Drori and M. Teboulle. Performance of first-order methods for smooth convex minimization: A novel approach. *Mathematical Programming*, 145(1):451–482, 2014.

[Dur10] R. Durrett. *Probability: Theory and Examples*. Cambridge Series in Statistical and Probabilistic Mathematics. Cambridge University Press, fourth ed., 2010.

[DW60] G. B. Dantzig and P. Wolfe. Decomposition Principle for Linear Programs. *Operations Research*, 8(1):101–111, 1960.

[DY16a] D. Davis and W. Yin. Convergence rate analysis of several splitting schemes. In R. Glowinski, S. J. Osher, and W. Yin, eds., *Splitting Methods in Communication, Imaging, Science and Engineering*, Chapter 4, pages 115–163. Springer, 2016.

[DY16b] W. Deng and W. Yin. On the global and linear convergence of the generalized alternating direction method of multipliers. *Journal of Scientific Computing*, 66(3):889–916, 2016.

[DY17a] D. Davis and W. Yin. Faster convergence rates of relaxed Peaceman–Rachford and ADMM under regularity assumptions. *Mathematics of Operations Research*, 42(3):783–805, 2017.

[DY17b] D. Davis and W. Yin. A three-operator splitting scheme and its optimization applications. *Set-Valued and Variational Analysis*, 25(4):829–858, 2017.

[EB90] J. Eckstein and D. Bertsekas. An alternating direction method for linear programming. LIDS Technical Reports LIDS-P 1967, Laboratory for Information and Decision Systems, Massachusetts Institute of Technology, 1990.

[EB92] J. Eckstein and D. P. Bertsekas. On the Douglas–Rachford splitting method and the proximal point algorithm for maximal monotone operators. *Mathematical Programming*, 55(1–3):293–318, 1992.

[Eck89] J. Eckstein. *Splitting Methods for Monotone Operators with Applications to Parallel Optimization*. PhD Thesis, Massachusetts Institute of Technology, Department of Civil Engineering, 1989.

[Eck94] J. Eckstein. Some saddle-function splitting methods for convex programming. *Optimization Methods and Software*, 4(1):75–83, 1994.

[Eck12] J. Eckstein. Augmented Lagrangian and alternating direction methods for convex optimization: A tutorial and some illustrative computational results. RUTCOR Research Report RRR RRR 32-2012, RUTCOR Rutgers Center for Operations Research Rutgers University, 2012.

[Ede64] M. Edelstein. On non-expansive mappings of Banach spaces. *Mathematical Proceedings of the Cambridge Philosophical Society*, 60(3):439–447, 1964.

[EF94] J. Eckstein and M. Fukushima. Some reformulations and applications of the alternating direction method of multipliers. In W. W. Hager, D. W. Hearn, and P. M. Pardalos, eds., *Large Scale Optimization*, pages 115–134. Springer, 1994.

[EHJT04] B. Efron, T. Hastie, I. Johnstone, and R. Tibshirani. Least angle regression. *The Annals of Statistics*, 32(2):407–499, 2004.

[ER11] R. Escalante and M. Raydan. *Alternating Projection Methods*. Fundamentals of Algorithms. Society for Industrial and Applied Mathematics, 2011.

[ES08] J. Eckstein and B. F. Svaiter. A family of projective splitting methods for the sum of two maximal monotone operators. *Mathematical Programming*, 111(1):173–199, 2008.

[Eve63] H. Everett. Generalized Lagrange multiplier method for solving problems of optimum allocation of resources. *Operations Research*, 11(3):399–417, 1963.

[EZC10] E. Esser, X. Zhang, and T. F. Chan. A general framework for a class of first order primal-dual algorithms for convex optimization in imaging science. *SIAM Journal on Imaging Sciences*, 3(4):1015–1046, 2010.

[Fen49] W. Fenchel. On conjugate convex functions. *Canadian Journal of Mathematics*, 1(1):73–77, 1949.

[Fen53] W. Fenchel. Convex Cones, Sets, and Functions. Lecture Note from notes by D. W. Blackett of lectures, Princeton University Department of Mathematics, 1953.

[FG83] M. Fortin and R. Glowinski. On decomposition-coordination methods using an augmented Lagrangian. In M. Fortin and R. Glowinski, eds., *Studies in Mathematics and Its Applications*, volume 15, pages 97–146. Elsevier, 1983.

[FGH21] M. P. Friedlander, A. Goodwin, and T. Hoheisel. From perspective maps to epigraphical projections. *arXiv:2102.06809*, 2021.

[Fis04] M. L. Fisher. The Lagrangian relaxation method for solving integer programming problems. *Management Science*, 50(12_supplement):1861–1871, 2004.

[Fit88] S. Fitzpatrick. Representing monotone operators by convex functions. In S. Fitzpatrick and J. Giles, eds., *Workshop/Miniconference on Functional Analysis and Optimization*, pages 59–65. Centre for Mathematics and its Applications, Mathematical Sciences Institute, The Australian National University, Canberra AUS, 1988.

[FNW07] M. A. T. Figueiredo, R. D. Nowak, and S. J. Wright. Gradient projection for sparse reconstruction: Application to compressed sensing and other inverse problems. *IEEE Journal of Selected Topics in Signal Processing*, 1(4):586–597, 2007.

[FP03] F. Facchinei and J.-S. Pang. *Finite-Dimensional Variational Inequalities and Complementarity Problems*. Springer-Verlag, 2003.

[FR15] O. Fercoq and P. Richtárik. Accelerated, parallel, and proximal coordinate descent. *SIAM Journal on Optimization*, 25(4):1997–2023, 2015.

[FS00] A. Frommer and D. B. Szyld. On asynchronous iterations. *Journal of Computational and Applied Mathematics*, 123(1–2):201–216, 2000.

[Gab83] D. Gabay. Application of the methods of multipliers to variational inequalities. In M. Fortin and R. Glowinski, eds., *Augmented Lagrangians: Application to the Numerical Solution of Boundary Value Problems*, pages 299–331. North-Holland, 1983.

[GB15] P. Giselsson and S. Boyd. Metric selection in fast dual forward–backward splitting. *Automatica*, 62:1–10, 2015.

[GB17] P. Giselsson and S. Boyd. Linear convergence and metric selection for Douglas–Rachford splitting and ADMM. *IEEE Transactions on Automatic Control*, 62(2):532–544, 2017.

[GBV+19] G. Gidel, H. Berard, G. Vignoud, P. Vincent, and S. Lacoste-Julien. A variational inequality perspective on generative adversarial networks. In *International Conference on Learning Representation*, 2019.

[Geo70] A. M. Geoffrion. Primal resource-directive approaches for optimizing nonlinear decomposable systems. *Operations Research*, 18(3):375–403, 1970.

[GHY14] G. Gu, B. He, and X. Yuan. Customized proximal point algorithms for linearly constrained convex minimization and saddle-point problems: A unified approach. *Computational Optimization and Applications*, 59(1):135–161, 2014.

[Gis15] P. Giselsson. Lunds universitet, lecture notes: Large-scale convex optimization. Lecture Note, Lunds Universitet, Department of Automatic Control, 2015.

[Glo84] R. Glowinski. *Numerical Methods for Nonlinear Variational Problems*. Springer-Verlag, 1984.

[Glo14] R. Glowinski. On alternating direction methods of multipliers: A historical perspective. In W. Fitzgibbon, Y. A. Kuznetsov, P. Neittaanmäki, and O. Pironneau, eds., *Modeling, Simulation and Optimization for Science and Technology*, volume 34, pages 59–82. Springer, 2014.

[GLT89] R. Glowinski and P. Le Tallec. *Augmented Lagrangian and Operator-Splitting Methods in Nonlinear Mechanics*. Society for Industrial and Applied Mathematics, 1989.

[GM75a] R. Glowinski and A. Marrocco. Sur l'approximation, par éléments finis d'ordre un, et la résolution, par pénalisation-dualité d'une classe de problèmes de Dirichlet non linéaires. *Revue Française d'Automatique, Informatique, Recherche Opérationnelle. Analyse Numérique*, 9(2):41–76, 1975.

[GM75b] R. Glowinski and A. Marroco. Sur l'approximation, par éléments finis d'ordre un, et la résolution, par pénalisation-dualité d'une classe de problèmes de Dirichlet non linéaires. *Revue Française d'Automatique, Informatique, Recherche Opérationnelle. Analyse Numérique*, 9(2):41–76, 1975.

[GM76] D. Gabay and B. Mercier. A dual algorithm for the solution of nonlinear variational problems via finite element approximation. *Computers and Mathematics with Applications*, 2(1):17–40, 1976.

[GO09] T. Goldstein and S. Osher. The split Bregman method for L1-regularized problems. *SIAM Journal on Imaging Sciences*, 2(2):323–343, 2009.

[Gol64] A. A. Goldstein. Convex programming in Hilbert space. *Bulletin of the American Mathematical Society*, 70(5):709–710, 1964.

[GOSB14] T. Goldstein, B. O'Donoghue, S. Setzer, and R. Baraniuk. Fast alternating direction optimization methods. *SIAM Journal on Imaging Sciences*, 7(3):1588–1623, 2014.

[GR84] K. Goebel and S. Reich. *Uniform Convexity, Hyperbolic Geometry, and Nonexpansive Mappings*. Marcel Dekker, 1984.

[GS00] L. Grippo and M. Sciandrone. On the convergence of the block nonlinear Gauss–Seidel method under convex constraints. *Operations Research Letters*, 26(3):127–136, 2000.

[GTSJ15] E. Ghadimi, A. Teixeira, I. Shames, and M. Johansson. Optimal parameter selection for the alternating direction method of multipliers (ADMM): Quadratic problems. *IEEE Transactions on Automatic Control*, 60(3):644–658, 2015.

[Gül91] O. Güler. On the convergence of the proximal point algorithm for convex minimization. *SIAM Journal on Control and Optimization*, 29(2):403–419, 1991.

[Gül92] O. Güler. New proximal point algorithms for convex minimization. *SIAM Journal on Optimization*, 2(4):649–664, 1992.

[GXZ19] X. Gao, Y.-Y. Xu, and S.-Z. Zhang. Randomized primal-dual proximal block coordinate updates. *Journal of the Operations Research Society of China*, 7(2):205–250, 2019.

[Hal67] B. Halpern. Fixed points of nonexpanding maps. *Bulletin of the American Mathematical Society*, 73(6):957–961, 1967.

[Hes69] M. R. Hestenes. Multiplier and gradient methods. *Journal of Optimization Theory and Applications*, 4(5):303–320, 1969.

[HHY15] B. He, L. Hou, and X. Yuan. On full Jacobian decomposition of the augmented lagrangian method for separable convex programming. *SIAM Journal on Optimization*, 25(4):2274–2312, 2015.

[Hil57] C. Hildreth. A quadratic programming procedure. *Naval Research Logistics Quarterly*, 4(1):79–85, 1957.

[HJ91] R. A. Horn and C. R. Johnson. *Topics in Matrix Analysis*. Cambridge University Press, 1991.

[HL93] J.-B. Hiriart-Urruty and C. Lemaréchal. *Convex Analysis and Minimization Algorithms I*, volume 2. Springer, 1993.

[HL01] J.-B. Hiriart-Urruty and C. Lemaréchal. *Fundamentals of Convex Analysis*. Springer-Verlag, 2001.

[HL17] M. Hong and Z.-Q. Luo. On the linear convergence of the alternating direction method of multipliers. *Mathematical Programming*, 162(1):165–199, 2017.

[HLHY02] B. He, L.-Z. Liao, D. Han, and H. Yang. A new inexact alternating directions method for monotone variational inequalities. *Mathematical Programming*, 92(1):103–118, 2002.

[HLWY14] B. He, H. Liu, Z. Wang, and X. Yuan. A strictly contractive Peaceman–Rachford splitting method for convex programming. *SIAM Journal on Optimization*, 24(3):1011–1040, 2014.

[HRY19] X. Huang, E. K. Ryu, and W. Yin. Scaled relative graph of normal matrices. *arXiv:2001.02061*, 2019.

[HRY20] X. Huang, E. K. Ryu, and W. Yin. Tight coefficients of averaged operators via scaled relative graph. *Journal of Mathematical Analysis and Applications*, 490(1):124211, 2020.

[HS19] J. Haochen and S. Sra. Random shuffling beats SGD after finite epochs. In *International Conference on Machine Learning*, 2019.

[HTY12] B. He, M. Tao, and X. Yuan. Alternating direction method with gaussian back substitution for separable convex programming. *SIAM Journal on Optimization*, 22(2):313–340, 2012.

[HTY17] B. He, M. Tao, and X. Yuan. Convergence rate analysis for the alternating direction method of multipliers with a substitution procedure for separable convex programming. *Mathematics of Operations Research*, 42(3):662–691, 2017.

[HWRL17] M. Hong, X. Wang, M. Razaviyayn, and Z.-Q. Luo. Iteration complexity analysis of block coordinate descent methods. *Mathematical Programming*, 163(1):85–114, 2017.

[HXY16] B. He, H.-K. Xu, and X. Yuan. On the proximal Jacobian decomposition of ALM for multiple-block separable convex minimization problems and its relationship to ADMM. *Journal of Scientific Computing*, 66(3):1204–1217, 2016.

[HY12a] D. Han and X. Yuan. A note on the alternating direction method of multipliers. *Journal of Optimization Theory and Applications*, 155(1):227–238, 2012.

[HY12b] B. He and X. Yuan. On the $O(1/n)$ convergence rate of the Douglas–Rachford alternating direction method. *SIAM Journal on Numerical Analysis*, 50(2):700–709, 2012.

[HY15] B. He and X. Yuan. On non-ergodic convergence rate of Douglas–Rachford alternating direction method of multipliers. *Numerische Mathematik*, 130(3):567–577, 2015.

[HY17] R. Hannah and W. Yin. More iterations per second, same quality — Why asynchronous algorithms may drastically outperform traditional ones. *arXiv:1708.05136*, 2017.

[HY18] R. Hannah and W. Yin. On unbounded delays in asynchronous parallel fixed-point algorithms. *Journal of Scientific Computing*, 76(1):299–326, 2018.

[HYD15] C.-J. Hsieh, H.-F. Yu, and I. S. Dhillon. PASSCoDe: Parallel ASynchronous Stochastic dual Co-Ordinate Descent. In *International Conference on Machine Learning*, Lille, France, 2015.

[HYW00] B. S. He, H. Yang, and S. L. Wang. Alternating direction method with self-adaptive penalty parameters for monotone variational inequalities. *Journal of Optimization Theory and Applications*, 106(2):337–356, 2000.

[HYZ08] E. T. Hale, W. Yin, and Y. Zhang. Fixed-point continuation for ℓ_1-minimization: Methodology and convergence. *SIAM Journal on Optimization*, 19(3):1107–1130, 2008.

[Ius99] A. N. Iusem. Augmented Lagrangian methods and proximal point methods for convex optimization. *Investigación Operativa*, 8(1–3):11–49, 1999.

[Jen06] J. L. W. V. Jensen. Sur les fonctions convexes et les inégalités entre les valeurs moyennes. *Acta Mathematica*, 30(1):175–193, 1906.

[JXM14] D. Jakovetić, J. Xavier, and J. M. F. Moura. Fast distributed gradient methods. *IEEE Transactions on Automatic Control*, 59(5):1131–1146, 2014.

[JZZ09] R.-Q. Jia, H. Zhao, and W. Zhao. Convergence analysis of the Bregman method for the variational model of image denoising. *Applied and Computational Harmonic Analysis*, 27(3):367–379, 2009.

[Kac60] R. I. Kachurovskii. Monotone operators and convex functionals. *Uspekhi Matematicheskikh Nauk*, 15(4):213–215, 1960.

[KDG03] D. Kempe, A. Dobra, and J. Gehrke. Gossip-based computation of aggregate information. In *IEEE Symposium on Foundations of Computer Science*, 2003.

[KF16] D. Kim and J. A. Fessler. Optimized first-order methods for smooth convex minimization. *Mathematical Programming*, 159(1–2):81–107, 2016.

[KF17] D. Kim and J. A. Fessler. On the convergence analysis of the optimized gradient method. *Journal of Optimization Theory and Applications*, 172(1):187–205, 2017.

[KF21] D. Kim and J. A. Fessler. Optimizing the efficiency of first-order methods for decreasing the gradient of smooth convex functions. *Journal of Optimization Theory and Applications*, 188(1):192–219, 2021.

[Kim21] D. Kim. Accelerated proximal point method for maximally monotone operators. *Mathematical Programming*, 190, 57–87, 2021. https://link.springer.com/article/10.1007/s10107-021-01643-0

[Kir34] M. D. Kirszbraun. Über die zusammenziehende und Lipschitzsche Transformationen. *Fundamenta Mathematicae*, 22(1):77–108, 1934.

[KP15] N. Komodakis and J.-C. Pesquet. Playing with duality: An overview of recent primal-dual approaches for solving large-scale optimization problems. *IEEE Signal Processing Magazine*, 32(6):31–54, 2015.

[Kra55] M. A. Krasnosel'skii. Two remarks on the method of successive approximations. *Uspekhi Matematicheskikh Nauk*, 10(1):123–127, 1955.

[Kra85] E. Krauss. A representation of maximal monotone operators by saddle functions. *Revue Roumaine de Mathématique Pures et Appliquées*, 30(10):823–837, 1985.

[KY03] H. J. Kushner and G. G. Yin. *Stochastic Approximation and Recursive Algorithms and Applications*. Springer, second ed., 2003.

[KYW19] S. Ko, D. Yu, and J.-H. Won. Easily parallelizable and distributable class of algorithms for structured sparsity, with optimal acceleration. *Journal of Computational and Graphical Statistics*, 28(4):821–833, 2019.

[LAP+14] M. Li, D. G. Andersen, J. W. Park, A. J. Smola, A. Ahmed, V. Josifovski, J. Long, E. J. Shekita, and B.-Y. Su. Scaling distributed machine learning with the parameter server. In *USENIX Conference on Operating Systems Design and Implementation*, 2014.

[LCZ14] K. Lange, E. C. Chi, and H. Zhou. A brief survey of modern optimization for statisticians. *International Statistical Review*, 82(1):46–70, 2014.

[Lem92] B. Lemaire. About the convergence of the proximal method. In W. Oettli and D. Pallaschke, eds., *Advances in Optimization*, pages 39–51. Springer, Berlin, Heidelberg, 1992.

[Lem01] C. Lemaréchal. Lagrangian relaxation. In M. Jünger and D. Naddef, eds., *Computational Combinatorial Optimization: Optimal or Provably Near-Optimal Solutions*, Lecture Notes in Computer Science, pages 112–156. Springer, 2001.

[LFP16] J. Liang, J. Fadili, and G. Peyré. Convergence rates with inexact non-expansive operators. *Mathematical Programming*, 159(1):403–434, 2016.

[LFP17] J. Liang, J. Fadili, and G. Peyré. Local convergence properties of Douglas–Rachford and alternating direction method of multipliers. *Journal of Optimization Theory and Applications*, 172(3):874–913, 2017.

[LFYL20] H. Li, C. Fang, W. Yin, and Z. Lin. Decentralized accelerated gradient methods with increasing penalty parameters. *IEEE Transactions on Signal Processing*, 68:4855–4870, 2020.

[LHLL15] X. Lian, Y. Huang, Y. Li, and J. Liu. Asynchronous parallel stochastic gradient for nonconvex optimization. In *Neural Information Processing Systems*, 2015.

[Lie21] F. Lieder. On the convergence rate of the Halpern-iteration. *Optimization Letters*, 15(2):405–418, 2021.

[Lin94] E. Lindelöf. Sur l'applications de la méthode des approximations successives aux équations différentielles ordinaires du premier ordre. *Comptes Rendus Hebdomadaires des Séances de l'Académie des Sciences*, 116(3):454–456, 1894.

[LL10] D. Leventhal and A. S. Lewis. Randomized methods for linear constraints: Convergence rates and conditioning. *Mathematics of Operations Research*, 35(3):641–654, 2010.

[LM79] P. L. Lions and B. Mercier. Splitting algorithms for the sum of two nonlinear operators. *SIAM Journal on Numerical Analysis*, 16(6):964–979, 1979.

[LMH15] H. Lin, J. Mairal, and Z. Harchaoui. A universal catalyst for first-order optimization. In *Neural Information Processing Systems*, 2015.

[LMH18] H. Lin, J. Mairal, and Z. Harchaoui. Catalyst acceleration for first-order convex optimization: From theory to practice. *Journal of Machine Learning Research*, 18(212):1–54, 2018.

[LMT+10] W. Liu, S. Ma, D. Tao, J. Liu, and P. Liu. Semi-supervised sparse metric learning using alternating linearization optimization. In *SIGKDD International Conference on Knowledge Discovery and Data Mining*, 2010.

[LMYZ21] T. Lin, S. Ma, Y. Ye, and S. Zhang. An ADMM-based interior-point method for large-scale linear programming. *Optimization Methods and Software*, 36(2-3):389–424, 2021.

[LMZ15a] T. Lin, S. Ma, and S. Zhang. On the global linear convergence of the ADMM with multiblock variables. *SIAM Journal on Optimization*, 25(3):1478–1497, 2015.

[LMZ15b] T.-Y. Lin, S.-Q. Ma, and S.-Z. Zhang. On the sublinear convergence rate of multi-block ADMM. *Journal of the Operations Research Society of China*, 3(3):251–274, 2015.

[LMZ16] T. Lin, S. Ma, and S. Zhang. Iteration complexity analysis of multi-block ADMM for a family of convex minimization without strong convexity. *Journal of Scientific Computing*, 69(1):52–81, 2016.

[LMZ17] T. Lin, S. Ma, and S. Zhang. An extragradient-based alternating direction method for convex minimization. *Foundations of Computational Mathematics*, 17(1):35–59, 2017.

[LO09] Y. Li and S. Osher. Coordinate descent optimization for ℓ^1 minimization with application to compressed sensing; a greedy algorithm. *Inverse Problems and Imaging*, 3(3):487–503, 2009.

[LP66] E. S. Levitin and B. T. Polyak. Constrained minimization methods. *Zhurnal Vychislitel'noi Matematiki i Matematicheskoi Fiziki*, 6(5):787–823, 1966.

[LP89] B. Lemaire and J.-P. Penot. The proximal algorithm. In *New Methods in Optimization and Their Industrial Uses*, volume 87, pages 73–87. Birkhäuser, 1989.

[LPL17] R. Leblond, F. Pedregosa, and S. Lacoste-Julien. ASAGA: Asynchronous parallel SAGA. In *International Conference on Artificial Intelligence and Statistics*, 2017.

[LRY19] Y. Liu, E. K. Ryu, and W. Yin. A new use of Douglas–Rachford splitting for identifying infeasible, unbounded, and pathological conic programs. *Mathematical Programming*, 177(1):225–253, 2019.

[LS87] J. Lawrence and J. E. Spingarn. On fixed points of non-expansive piecewise isometric mappings. *Proceedings of the London Mathematical Society*, s3-55(3):605–624, 1987.

[LST15] M. Li, D. Sun, and K.-C. Toh. A convergent 3-block semi-proximal ADMM for convex minimization problems with one strongly convex block. *Asia-Pacific Journal of Operational Research*, 32(04):1550024, 2015.

[LST16] X. Li, D. Sun, and K.-C. Toh. A Schur complement based semi-proximal ADMM for convex quadratic conic programming and extensions. *Mathematical Programming*, 155(1):333–373, 2016.

[LST19] X. Li, D. Sun, and K.-C. Toh. A block symmetric Gauss–Seidel decomposition theorem for convex composite quadratic programming and its applications. *Mathematical Programming*, 175(1):395–418, 2019.

[LSY19] Z. Li, W. Shi, and M. Yan. A decentralized proximal-gradient method with network independent step-sizes and separated convergence rates. *IEEE Transactions on Signal Processing*, 67(17):4494–4506, 1 2019.

[LT92] Z.-Q. Luo and P. Tseng. On the convergence of the coordinate descent method for convex differentiable minimization. *Journal of Optimization Theory and Applications*, 72(1):7–35, 1992.

[LT10] Q. Ling and Z. Tian. Decentralized sparse signal recovery for compressive sleeping wireless sensor networks. *IEEE Transactions on Signal Processing*, 58(7):3816–3827, 2010.

[LUZ15] Z. Li, A. Uschmajew, and S. Zhang. On convergence of the maximum block improvement method. *SIAM Journal on Optimization*, 25(1):210–233, 2015.

[LV11] I. Loris and C. Verhoeven. On a generalization of the iterative soft-thresholding algorithm for the case of non-separable penalty. *Inverse Problems*, 27(12):125007, 2011.

[LW15] J. Liu and S. J. Wright. Asynchronous stochastic coordinate descent: Parallelism and convergence properties. *SIAM Journal on Optimization*, 25(1):351–376, 2015.

[LW19] C.-P. Lee and S. J. Wright. Random permutations fix a worst case for cyclic coordinate descent. *IMA Journal of Numerical Analysis*, 39(3):1246–1275, 2019.

[LWR+14] J. Liu, S. Wright, C. Re, V. Bittorf, and S. Sridhar. An asynchronous parallel stochastic coordinate descent algorithm. *International Conference on Machine Learning*, 2014.

[LWR+15] J. Liu, S. J. Wright, C. Ré, V. Bittorf, and S. Sridhar. An asynchronous parallel stochastic coordinate descent algorithm. *Journal of Machine Learning Research*, 16(1):285–322, 2015.

[LY17] Z. Li and M. Yan. A primal-dual algorithm with optimal stepsizes and its application in decentralized consensus optimization. *arXiv:1711.06785*, 2017.

[LY19] Y. Liu and W. Yin. An envelope for Davis–Yin splitting and strict saddle point avoidance. *Journal of Optimization Theory and Applications*, 181(2):567–587, 2019.

[LZZ+17] X. Lian, C. Zhang, H. Zhang, C.-J. Hsieh, W. Zhang, and J. Liu. Can decentralized algorithms outperform centralized algorithms? A case study for decentralized parallel stochastic gradient descent. In *Neural Information Processing Systems*, 2017.

[LZZL18] X. Lian, W. Zhang, C. Zhang, and J. Liu. Asynchronous decentralized parallel stochastic gradient descent. In *International Conference on Machine Learning*, 2018.

[Ma20] F. Ma. A revisit of Chen–Teboulle's proximal-based decomposition method. *arXiv:2006.11255*, 2020.

[Mai13] J. Mairal. Optimization with first-order surrogate functions. In *International Conference on Machine Learning*, 2013.

[Mar70] B. Martinet. Régularisation d'inéquations variationnelles par approximations successives. *Revue Française d'Informatique et de Recherche Opérationnelle, Série Rouge*, 4(3):154–158, 1970.

[Mar72a] B. Martinet. *Algorithmes Pour La Résolution de Problèmes d'optimisation et de Minimax*. PhD Thesis, Université Scientifique et Médicale de Grenoble, 1972.

[Mar72b] B. Martinet. Determination approchée d'un point fixe d'une application pseudo-contractante. *Comptes Rendus de l'Académie des Sciences, Série A*, 274(2):163–165, 1972.

[MBG10] G. Mateos, J. A. Bazerque, and G. B. Giannakis. Distributed sparse linear regression. *IEEE Transactions on Signal Processing*, 58(10):5262–5276, 2010.

[Mer80] B. Mercier. *Inéquations Variationnelles de La Mécanique*. Publications Mathématiques d'Orsay. Université de Paris-Sud, Département de mathématique, 1980.

[Min62] G. J. Minty. Monotone (nonlinear) operators in Hilbert space. *Duke Mathematical Journal*, 29(3):341–346, 1962.

[Min64] G. J. Minty. On the monotonicity of the gradient of a convex function. *Pacific Journal of Mathematics*, 14(1):243–247, 1964.

[MKS+20] K. Mishchenko, D. Kovalev, E. Shulgin, P. Richtarik, and Y. Malitsky. Revisiting stochastic extragradient. In *International Conference on Artificial Intelligence and Statistics*, 2020.

[MLZ+19] P. Mertikopoulos, B. Lecouat, H. Zenati, C.-S. Foo, V. Chandrasekhar, and G. Piliouras. Optimistic mirror descent in saddle-point problems: Going the extra (gradient) mile. In *International Conference on Learning Representations*, 2019.

[Mor62] J. J. Moreau. Fonctions convexes duales et points proximaux dans un espace hilbertien. *Comptes rendus hebdomadaires des séances de l'Académie des sciences*, 255:2897–2899, 1962.

[Mor65] J. J. Moreau. Proximité et dualité dans un espace hilbertien. *Bulletin de la Société Mathématique de France*, 93:273–299, 1965.

[MP65] O. L. Mangasarian and J. Ponstein. Minmax and duality in nonlinear programming. *Journal of Mathematical Analysis and Applications*, 11:504–518, 1965.

[MS13] R. D. C. Monteiro and B. F. Svaiter. Iteration-complexity of block-decomposition algorithms and the alternating direction method of multipliers. *SIAM Journal on Optimization*, 23(1):475–507, 2013.

[MV19] W. M. Moursi and L. Vandenberghe. Douglas–Rachford splitting for the sum of a Lipschitz continuous and a strongly monotone operator. *Journal of Optimization Theory and Applications*, 183(1):179–198, 2019.

[MXZ13] S. Ma, L. Xue, and H. Zou. Alternating direction methods for latent variable Gaussian graphical model selection. *Neural Computation*, 25(8):2172–2198, 2013.

[MZ19] W. M. Moursi and Y. Zinchenko. A note on the equivalence of operator splitting methods. In H. H. Bauschke, R. S. Burachik, and D. R. Luke, eds., *Splitting Algorithms, Modern Operator Theory, and Applications*, pages 331–349. Springer, 2019.

[Nes83] Y. Nesterov. A method of solving a convex programming problem with convergence rate $O(1/k^2)$. *Doklady Akademii Nauk SSSR*, 269(3):543–547, 1983. https://vsokolov.org/courses/750/2018/files/nesterov.pdf

[Nes88] Y. Nesterov. On an approach to the construction of optimal methods of minimization of smooth convex functions. *Ekonomika i Mateaticheskie Metody*, 24(3):509–517, 1988.

[Nes04] Y. Nesterov. *Introductory Lectures on Convex Optimization: A Basic Course*. Springer, 2004.

[Nes05] Y. Nesterov. Smooth minimization of non-smooth functions. *Mathematical Programming*, 103(1):127–152, 2005.

[Nes12] Y. Nesterov. Efficiency of coordinate descent methods on huge-scale optimization problems. *SIAM Journal on Optimization*, 22(2):341–362, 2012.

[Nes13] Y. Nesterov. Gradient methods for minimizing composite functions. *Mathematical Programming*, 140(1):125–161, 2013.

[NGGD20] Y. Nesterov, A. Gasnikov, S. Guminov, and P. Dvurechensky. Primal-dual accelerated gradient methods with small-dimensional relaxation oracle. *Optimization Methods and Software*, pages 1–38, 2020.

[NJN19] D. Nagaraj, P. Jain, and P. Netrapalli. SGD without replacement: Sharper rates for general smooth convex functions. In *International Conference on Machine Learning*, 2019.

[NO15] A. Nedic and A. Olshevsky. Distributed optimization over time-varying directed graphs. *IEEE Transactions on Automatic Control*, 60(3):601–615, 2015.

[NOS17] A. Nedić, A. Olshevsky, and W. Shi. Achieving geometric convergence for distributed optimization over time-varying graphs. *SIAM Journal on Optimization*, 27(4):2597–2633, 2017.

[NOSU17] A. Nedić, A. Olshevsky, W. Shi, and C. A. Uribe. Geometrically convergent distributed optimization with uncoordinated step-sizes. In *American Control Conference*, 2017.

[NP06] C. P. Niculescu and L.-E. Persson. *Convex Functions and Their Applications*. CMS Books in Mathematics. Springer, New York, NY, 2006.

[NSL+15] J. Nutini, M. Schmidt, I. H. Laradji, M. Friedlander, and H. Koepke. Coordinate descent converges faster with the gauss-southwell rule than random selection. In *International Conference on Machine Learning*, 2015.

[NY78] A. Nemirovski and D. B. Yudin. Cezari convergence of the gradient method of approximating saddle point of convex-concave functions. *Doklady Akademii Nauk SSSR*, 239(5):1056–1059, 1978.

[OBG+05] S. Osher, M. Burger, D. Goldfarb, J. Xu, and W. Yin. An iterative regularization method for total variation-based image restoration. *Multiscale Modeling and Simulation*, 4(2):460–489, 2005.

[OCLPJ15] Y. Ouyang, Y. Chen, G. Lan, and E. Pasiliao Jr. An accelerated linearized alternating direction method of multipliers. *SIAM Journal on Imaging Sciences*, 8(1):644–681, 2015.

[OCPB16] B. O'Donoghue, E. Chu, N. Parikh, and S. Boyd. Conic optimization via operator splitting and homogeneous self-dual embedding. *Journal of Optimization Theory and Applications*, 169(3):1042–1068, 2016.

[OHTG13] H. Ouyang, N. He, L. Q. Tran, and A. Gray. Stochastic alternating direction method of multipliers. In *International Conference on Machine Learning*, 2013.

[Opi67] Z. Opial. Weak convergence of the sequence of successive approximations for nonexpansive mappings. *Bulletin of the American Mathematical Society*, 73(4):591–597, 1967.

[OV20] D. O'Connor and L. Vandenberghe. On the equivalence of the primal-dual hybrid gradient method and Douglas–Rachford splitting. *Mathematical Programming*, 179(1–2):85–108, 2020.

[OY02] N. Ogura and I. Yamada. Non-strictly convex minimization over the fixed point set of an asymptotically shrinking nonexpansive mapping. *Numerical Functional Analysis and Optimization*, 23(1–2):113–137, 2002.

[Pas79] G. B. Passty. Ergodic convergence to a zero of the sum of monotone operators in Hilbert space. *Journal of Mathematical Analysis and Applications*, 72(2):383–390, 1979.

[Pat21] R. Pates. The scaled relative graph of a linear operator. *arXiv:2106.05650*, 2021.

[PB14a] N. Parikh and S. Boyd. Block splitting for distributed optimization. *Mathematical Programming Computation*, 6(1):77–102, 2014.

[PB14b] N. Parikh and S. Boyd. Proximal algorithms. *Foundations and Trends in Optimization*, 1(3):127–239, 2014.

[PC06] D. P. Palomar and M. Chiang. A tutorial on decomposition methods for network utility maximization. *IEEE Journal on Selected Areas in Communications*, 24(8):1439–1451, 2006.

[PC11] T. Pock and A. Chambolle. Diagonal preconditioning for first order primal-dual algorithms in convex optimization. In *International Conference on Computer Vision*, 2011.

[PCBC09] T. Pock, D. Cremers, H. Bischof, and A. Chambolle. An algorithm for minimizing the Mumford-Shah functional. In *International Conference on Computer Vision*, 2009.

[Ped70] D. Pedoe. *A Course Geometry for Colleges and Universities*. Cambridge University Press, 1970.

[Pen03] J.-P. Penot. Is convexity useful for the study of monotonicity? In R. P. Agarwal and D. O'Regan, eds., *Nonlinear Analysis and Applications: To V. Lakshmikantham on His 80th Birthday*, pages 807–822. Kluwer Academic Publishers, 2003.

[Pen04] J.-P. Penot. The relevance of convex analysis for the study of monotonicity. *Nonlinear Analysis: Theory, Methods and Applications*, 58(7):855–871, 2004.

[Phe93] R. R. Phelps. *Convex Functions, Monotone Operators and Differentiability*. Springer-Verlag, second ed., 1993.

[Pic90] É. Picard. Mémoire sur la théorie des équations aux dérivées partielles et la méthode des approximations successives. *Journal de Mathématiques Pures et Appliquées 4éme Série*, 6(2):145–210, 1890.

[Pie84] G. Pierra. Decomposition through formalization in a product space. *Mathematical Programming*, 28(1):96–115, 1984.

[PLL17] F. Pedregosa, R. Leblond, and S. Lacoste-Julien. Breaking the nonsmooth barrier: A scalable parallel method for composite optimization. In *Neural Information Processing Systems*, 2017.

[PLS08] L. A. Parente, P. A. Lotito, and M. V. Solodov. A class of inexact variable metric proximal point algorithms. *SIAM Journal on Optimization*, 19(1):240–260, 2008.

[PN15] A. Patrascu and I. Necoara. Efficient random coordinate descent algorithms for large-scale structured nonconvex optimization. *Journal of Global Optimization*, 61(1):19–46, 2015.

[Pol87] B. T. Polyak. *Introduction to Optimization*. Optimization Software, 1987.

[Pol90] B. T. Polyak. New method of stochastic approximation type. *Automat. i Telemekh*, 51(7):98–107, 1990.

[Pow69] M. J. D. Powell. A method for nonlinear constraints in minimization problems. In R. Fletcher, ed., *Optimization: Symposium of the Institute of Mathematics and Its Applications, University of Keele, England, 1968*, pages 283–298. Academic Press, 1969.

[PR55] D. W. Peaceman and J. Rachford, H. H. The numerical solution of parabolic and elliptic differential equations. *Journal of the Society for Industrial and Applied Mathematics*, 3(1):28–41, 1955.

[PR15] J.-C. Pesquet and A. Repetti. A class of randomized primal-dual algorithms for distributed optimization. *Journal of Nonlinear and Convex Analysis*, 16(12):2453–2490, 2015.

[PS10] J. Peypouquet and S. Sorin. Evolution equations for maximal monotone operators: Asymptotic analysis in continuous and discrete time. *Journal of Convex Analysis*, 17(3–4):1113–1163, 2010.

[PSB14] P. Patrinos, L. Stella, and A. Bemporad. Douglas–Rachford splitting: Complexity estimates and accelerated variants. In *IEEE Conference on Decision and Control*, 2014.

[PWX+16] Z. Peng, T. Wu, Y. Xu, M. Yan, and W. Yin. Coordinate-friendly structures, algorithms and applications. *Annals of Mathematical Sciences and Applications*, 1(1):57–119, 2016.

[PXYY16] Z. Peng, Y. Xu, M. Yan, and W. Yin. ARock: An algorithmic framework for asynchronous parallel coordinate updates. *SIAM Journal on Scientific Computing*, 38(5):A2851–A2879, 2016.

[PXYY19] Z. Peng, Y. Xu, M. Yan, and W. Yin. On the convergence of asynchronous parallel iteration with unbounded delays. *Journal of the Operations Research Society of China*, 7(1):5–42, 2019.

[PYY13] Z. Peng, M. Yan, and W. Yin. Parallel and distributed sparse optimization. In *Asilomar Conference on Signals, Systems and Computers*. IEEE, 2013.

[QL18] G. Qu and N. Li. Harnessing smoothness to accelerate distributed optimization. *IEEE Transactions on Control of Network Systems*, 5(3):1245–1260, 2018.

[QL20] G. Qu and N. Li. Accelerated distributed Nesterov gradient descent. *IEEE Transactions on Automatic Control*, 65(6):2566–2581, 2020.

[QR16a] Z. Qu and P. Richtárik. Coordinate descent with arbitrary sampling I: Algorithms and complexity. *Optimization Methods and Software*, 31(5):829–857, 2016.

[QR16b] Z. Qu and P. Richtárik. Coordinate descent with arbitrary sampling II: Expected separable overapproximation. *Optimization Methods and Software*, 31(5):858–884, 2016.

[QSMR19] X. Qian, A. Sailanbayev, K. Mishchenko, and P. Richtárik. MISO is making a comeback with better proofs and rates. *arXiv:1906.01474*, 2019.

[Rag19] H. Raguet. A note on the forward-Douglas–Rachford splitting for monotone inclusion and convex optimization. *Optimization Letters*, 13(4):717–740, 2019.

[Ray13] M. Raynal. *Concurrent Programming: Algorithms, Principles, and Foundations: Algorithms, Principles, and Foundations*. Springer-Verlag, 2013.

[RB16] E. K. Ryu and S. Boyd. Primer on monotone operator methods. *Applied and Computational Mathematics*, 15(1):3–43, 2016.

[RDC14] A. U. Raghunathan and S. Di Cairano. ADMM for convex quadratic programs: Q-linear convergence and infeasibility detection. *arXiv:1411.7288*, 2014.

[Rei79] S. Reich. Weak convergence theorems for nonexpansive mappings in Banach spaces. *Journal of Mathematical Analysis and Applications*, 67(2):274–276, 1979.

[Rei85] S. Reich. Averaged mappings in the Hilbert ball. *Journal of Mathematical Analysis and Applications*, 109(1):199–206, 1985.

[RFP13] H. Raguet, J. Fadili, and G. Peyré. A generalized forward-backward splitting. *SIAM Journal on Imaging Sciences*, 6(3):1199–1226, 2013.

[RGP20] S. Rajput, A. Gupta, and D. Papailiopoulos. Closing the convergence gap of SGD without replacement. In *International Conference on Machine Learning*, 2020.

[RHL13] M. Razaviyayn, M. Hong, and Z.-Q. Luo. A unified convergence analysis of block successive minimization methods for nonsmooth optimization. *SIAM Journal on Optimization*, 23(2):1126–1153, 2013.

[RHY21] E. K. Ryu, R. Hannah, and W. Yin. Scaled relative graphs: Nonexpansive operators via 2D Euclidean geometry. *Mathematical Programming*, 2021.

[RKW20] E. K. Ryu, S. Ko, and J.-H. Won. Splitting with near-circulant linear systems: Applications to total variation CT and PET. *SIAM Journal on Scientific Computing*, 42(1):B185–B206, 2020.

[RLW+19] E. Ryu, J. Liu, S. Wang, X. Chen, Z. Wang, and W. Yin. Plug-and-play methods provably converge with properly trained denoisers. In *International Conference on Machine Learning*, 2019.

[RLY19] E. K. Ryu, Y. Liu, and W. Yin. Douglas–Rachford splitting and ADMM for pathological convex optimization. *Computational Optimization and Applications*, 74(3):747–778, 2019.

[RM51] H. Robbins and S. Monro. A stochastic approximation method. *The Annals of Mathematical Statistics*, 22(3):400–407, 1951.

[Rob99] S. M. Robinson. Composition duality and maximal monotonicity. *Mathematical Programming*, 85(1):1–13, 1999.

[Roc64] R. T. Rockafellar. Minimax theorems and conjugate saddle-functions. *Mathematica Scandinavica*, 14(2):151–173, 1964.

[Roc66] R. T. Rockafellar. Characterization of the subdifferentials of convex functions. *Pacific Journal of Mathematics*, 17(3):497–510, 1966.

[Roc68] R. T. Rockafellar. A general correspondence between dual minimax problems and convex programs. *Pacific Journal of Mathematics*, 25(3):597–611, 1968.

[Roc69] R. T. Rockafellar. Measurable dependence of convex sets and functions on parameters. *Journal of Mathematical Analysis and Applications*, 28(1):4–25, 1969.

[Roc70a] R. T. Rockafellar. Monotone operators associated with saddle-functions and minimax problems. In F. E. Browder, ed., *Nonlinear Functional Analysis, Part 1*, volume 18 of *Proceedings of Symposia in Pure Mathematics*, pages 241–250. American Mathematical Society, 1970.

[Roc70b] R. T. Rockafellar. On the maximal monotonicity of subdifferential mappings. *Pacific Journal of Mathematics*, 33(1):209–216, 1970.

[Roc70c] R. T. Rockafellar. On the maximality of sums of nonlinear monotone operators. *Transactions of the American Mathematical Society*, 149(1):75–88, 1970.

[Roc70d] R. T. Rockafellar. *Convex Analysis*. Princeton University Press, 1970.

[Roc73] R. T. Rockafellar. The multiplier method of Hestenes and Powell applied to convex programming. *Journal of Optimization Theory and Applications*, 12(6):555–562, 1973.

[Roc74] R. T. Rockafellar. *Conjugate Duality and Optimization*. CBMS-NSF Regional Conference Series in Applied Mathematics. Society for Industrial and Applied Mathematics, 1974.

[Roc76a] R. T. Rockafellar. Augmented Lagrangians and applications of the proximal point algorithm in convex programming. *Mathematics of Operations Research*, 1(2):97–116, 1976.

[Roc76b] R. T. Rockafellar. Monotone operators and the proximal point algorithm. *SIAM Journal on Control and Optimization*, 14(5):877–898, 1976.

[Roc78] R. T. Rockafellar. Monotone operators and augmented Lagrangian methods in nonlinear programming. In O. L. Mangasarian, R. R. Meyer, and S. M. Robinson, eds., *Nonlinear Programming 3*, pages 1–25. Academic Press, 1978.

[Ros69] J. L. Rosenfeld. A case study in programming for parallel-processors. *Communications of The ACM*, 12(12):645–655, 1969.

[RRWN11] B. Recht, C. Re, S. Wright, and F. Niu. Hogwild!: A lock-free approach to parallelizing stochastic gradient descent. In *Neural Information Processing Systems*, 2011.

[RS81] S. F. Roehrig and R. C. Sine. The structure of ω-limit sets of nonexpansive maps. *Proceedings of the American Mathematical Society*, 81(3):398–400, 1981.

[RS85] H. Robbins and D. Siegmund. A convergence theorem for non negative almost supermartingales and some applications. In T. L. Lai and D. Siegmund, eds., *Herbert Robbins Selected Papers*, pages 111–135. Springer, 1985.

[RS98] S. Reich and D. Shoikhet. Averages of holomorphic mappings and holomorphic retractions on convex hyperbolic domains. *Studia Mathematica*, 130(3):231–244, 1998.

[RT14] P. Richtárik and M. Takáč. Iteration complexity of randomized block-coordinate descent methods for minimizing a composite function. *Mathematical Programming*, 144(1):1–38, 2014.

[RT16] P. Richtárik and M. Takáč. On optimal probabilities in stochastic coordinate descent methods. *Optimization Letters*, 10(6):1233–1243, 2016.

[RTBG20] E. K. Ryu, A. B. Taylor, C. Bergeling, and P. Giselsson. Operator splitting performance estimation: Tight contraction factors and optimal parameter selection. *SIAM Journal on Optimization*, 30(3):2251–2271, 2020.

[Rup88] D. Ruppert. Efficient estimations from a slowly convergent Robbins–Monro process. ORIE Technical Reports 781, Cornell University Operations Research and Industrial Engineering, 1988.

[RVV16] L. Rosasco, S. Villa, and B. C. Vũ. Stochastic forward-backward splitting for monotone inclusions. *Journal of Optimization Theory and Applications*, 169(2):388–406, 2016.

[RYY19] E. K. Ryu, K. Yuan, and W. Yin. ODE analysis of stochastic gradient methods with optimism and anchoring for minimax problems and GANs. *arXiv:1905.10899*, 2019.

[SBB+17] K. Seaman, F. Bach, S. Bubeck, Y. T. Lee, and L. Massoulié. Optimal algorithms for smooth and strongly convex distributed optimization in networks. In *International Conference on Machine Learning*, 2017.

[SBB+18] K. Scaman, F. Bach, S. Bubeck, L. Massoulié, and Y. T. Lee. Optimal algorithms for non-smooth distributed optimization in networks. In *Neural Information Processing Systems*, 2018.

[SBB+19] K. Scaman, F. Bach, S. Bubeck, Y. T. Lee, and L. Massoulié. Optimal convergence rates for convex distributed optimization in networks. *Journal of Machine Learning Research*, 20(159):1–31, 2019.

[SBG+20] B. Stellato, G. Banjac, P. Goulart, A. Bemporad, and S. Boyd. OSQP: An operator splitting solver for quadratic programs. *Mathematical Programming Computation*, 12(4):637–672, 2020.

[Sch57] H. Schaefer. Über die methode sukzessiver approximationen. *Jahresbericht der Deutschen Mathematiker-Vereinigung*, 59:131–140, 1957.

[SCMR20] A. Salim, L. Condat, K. Mishchenko, and P. Richtárik. Dualize, split, randomize: Fast nonsmooth optimization algorithms. *NeurIPS 2020 Workshop OPT2020: Optimization for Machine Learning*, 2020.

[SGJ11] D. Sontag, A. Globerson, and T. Jaakkola. Introduction to dual decomposition for inference. In S. Sra, S. Nowozin, and S. J. Wright, eds., *Optimization for Machine Learning*, pages 219–254. Massachusetts Institute of Technology Press, 2011.

[SH15] R. Sun and M. Hong. Improved iteration complexity bounds of cyclic block coordinate descent for convex problems. In *Neural Information Processing Systems*, 2015.

[Sho62] N. Z. Shor. An application of the method of gradient descent to the solution of the network transportation problem. *Materialy Naucnovo Seminara po Teoret i Priklad. Voprosam Kibernet. i Issted. Operacii, Nucnyi Sov. po Kibernet, Akad. Nauk Ukrain. SSSR, vyp*, 1:9–17, 1962.

[Sho64] N. Z. Shor. *On the Structure of Algorithms for Numerical Solution of Problems of Optimal Planning and Design*. PhD Thesis, Institute of Mathematics of National Academy of Sciences of Ukraine, 1964.

[Sho85] N. Z. Shor. *Minimization Methods for Non-Differentiable Functions*. Springer-Verlag, 1985.

[Sho97] R. E. Showalter. *Montone Operators in Banach Space and Nonlinear Partial Differential Equations*. American Mathematical Society, 1997.

[SHY17] T. Sun, R. Hannah, and W. Yin. Asynchronous coordinate descent under more realistic assumptions. In *Neural Information Processing Systems*, 2017.

[Sil72] G. J. Silverman. Primal decomposition of mathematical programs by resource allocation: I. Basic theory and a direction-finding procedure. *Operations Research*, 20(1):58–74, 1972.

[SK03] S. K. Shevade and S. S. Keerthi. A simple and efficient algorithm for gene selection using sparse logistic regression. *Bioinformatics*, 19(17):2246–2253, 2003.

[Sla50] M. Slater. Lagrange multipliers revisited: A contribution to non-linear programming. Cowles Commision Discussion Papers Math 403, Cowles Foundation for Research in Economics, 1950.

[SLWY15a] W. Shi, Q. Ling, G. Wu, and W. Yin. EXTRA: An exact first-order algorithm for decentralized consensus optimization. *SIAM Journal on Optimization*, 25(2):944–966, 2015.

[SLWY15b] W. Shi, Q. Ling, G. Wu, and W. Yin. A proximal gradient algorithm for decentralized composite optimization. *IEEE Transactions on Signal Processing*, 63(22):6013–6023, 2015.

[SLY+14] W. Shi, Q. Ling, K. Yuan, G. Wu, and W. Yin. On the linear convergence of the ADMM in decentralized consensus optimization. *IEEE Transactions on Signal Processing*, 62(7):1750–1761, 2014.

[SLY20] R. Sun, Z.-Q. Luo, and Y. Ye. On the efficiency of random permutation for ADMM and coordinate descent. *Mathematics of Operations Research*, 45(1):233–271, 2020.

[SN10] A. Smola and S. Narayanamurthy. An architecture for parallel topic models. *Proceedings of the VLDB Endowment*, 3(1–2):703–710, 2010.

[Spi83] J. E. Spingarn. Partial inverse of a monotone operator. *Applied Mathematics and Optimization*, 10(1):247–265, 1983.

[Spi85] J. E. Spingarn. Applications of the method of partial inverses to convex programming: Decomposition. *Mathematical Programming*, 32(2):199–223, 1985.

[SRG08] I. D. Schizas, A. Ribeiro, and G. B. Giannakis. Consensus in ad hoc WSNs with noisy links – Part I: Distributed estimation of deterministic signals. *IEEE Transactions on Signal Processing*, 56(1):350–364, 2008.

[SS86] F. Santosa and W. W. Symes. Linear inversion of band-limited reflection seismograms. *SIAM Journal on Scientific and Statistical Computing*, 7(4):1307–1330, 1986.

[SS17] S. Sabach and S. Shtern. A first order method for solving convex bilevel optimization problems. *SIAM Journal on Optimization*, 27(2):640–660, 2017.

[ST09] S. Shalev-Shwartz and A. Tewari. Stochastic methods for ℓ_1 regularized loss minimization. In *International Conference on Machine Learning*, Montreal, Quebec, Canada, 2009.

[ST11] S. Shalev-Shwartz and A. Tewari. Stochastic methods for ℓ_1 regularized loss minimization. *The Journal of Machine Learning Research*, 12(52):1865–1892, 2011.

[ST13] A. Saha and A. Tewari. On the nonasymptotic convergence of cyclic coordinate descent methods. *SIAM Journal on Optimization*, 23(1):576–601, 2013.

[ST14] R. Shefi and M. Teboulle. Rate of convergence analysis of decomposition methods based on the proximal method of multipliers for convex minimization. *SIAM Journal on Optimization*, 24(1):269–297, 2014.

[Ste46] M. Stewart. *Some General Theorems of Considerable Use in the Higher Parts of Mathematics*. W. Sands, A. Murray, and J. Cochran, 1746.

[Sto63] J. Stoer. Duality in nonlinear programming and the minimax theorem. *Numerische Mathematik*, 5(1):371–379, 1963.

[Sto64] J. Stoer. Über einen Dualitätssatz der nichtlinearen programmierung. *Numerische Mathematik*, 6(1):55–58, 1964.

[Str69] V. Strassen. Gaussian elimination is not optimal. *Numerische Mathematik*, 13(4):354–356, 1969.

[STXY16] H.-J. M. Shi, S. Tu, Y. Xu, and W. Yin. A primer on coordinate descent algorithms. arXiv:1610.00040, 2017.

[STY15] D. Sun, K.-C. Toh, and L. Yang. A convergent 3-block SemiProximal alternating direction method of multipliers for conic programming with 4-Type constraints. *SIAM Journal on Optimization*, 25(2):882–915, 2015.

[Suz13] T. Suzuki. Dual averaging and proximal gradient descent for online alternating direction multiplier method. In *International Conference on Machine Learning*, 2013.

[Suz14] T. Suzuki. Stochastic dual coordinate ascent with alternating direction method of multipliers. In *International Conference on Machine Learning*, 2014.

[SXB14] J. V. Shi, Y. Xu, and R. G. Baraniuk. Sparse bilinear logistic regression. UCLA CAM Reports 14-12, University of California, Los Angeles, 2014.

[SY21] R. Sun and Y. Ye. Worst-case complexity of cyclic coordinate descent: $O(n^2)$ gap with randomized version. *Mathematical Programming*, 185(1):487–520, 2021.

[SZ04] S. Simons and C. Zălinescu. A new proof for Rockafellar's characterization of maximal monotone operators. *Proceedings of the American Mathematical Society*, 132(10):2969–2972, 2004.

[SZ05] S. Simons and C. Zălinescu. Fenchel duality, Fitzpatrick functions and maximal monotonicity. *Journal of Nonlinear and Convex Analysis*, 6(1):1–22, 2005.

[SZ13] S. Shalev-Shwartz and T. Zhang. Stochastic dual coordinate ascent methods for regularized loss. *Journal of Machine Learning Research*, 14(1):567–599, 2013.

[Tao14] M. Tao. Some parallel splitting methods for separable convex programming with the $o(1/t)$ convergence rate. *Pacific Journal on Optimization*, 10(2):359–384, 2014.

[TB87] P. Tseng and D. P. Bertsekas. Relaxation methods for problems with strictly convex separable costs and linear constraints. *Mathematical Programming*, 38(3):303–321, 1987.

[TB19] A. Taylor and F. Bach. Stochastic first-order methods: Non-asymptotic and computer-aided analyses via potential functions. In *Conference on Learning Theory*, 2019.

[TBT90] P. Tseng, D. P. Bertsekas, and J. N. Tsitsiklis. Partially asynchronous, parallel algorithms for network flow and other problems. *SIAM Journal on Control and Optimization*, 28(3):678–710, 1990.

[TE05] L. N. Trefethen and M. Embree. *Spectra and Pseudospectra: The Behavior of Nonnormal Matrices and Operators*. Princeton University Press, 2005.

[THG17] A. B. Taylor, J. M. Hendrickx, and F. Glineur. Smooth strongly convex interpolation and exact worst-case performance of first-order methods. *Mathematical Programming*, 161(1):307–345, 2017.

[Tib96] R. Tibshirani. Regression shrinkage and selection via the Lasso. *Journal of the Royal Statistical Society: Series B (Methodological)*, 58(1):267–288, 1996.

[TP20] A. Themelis and P. Patrinos. Douglas–Rachford splitting and ADMM for nonconvex optimization: Tight convergence results. *SIAM Journal on Optimization*, 30(1):149–181, 2020.

[Tse90a] P. Tseng. Dual ascent methods for problems with strictly convex costs and linear constraints: A unified approach. *SIAM Journal on Control and Optimization*, 28(1):214–242, 1990.

[Tse90b] P. Tseng. Further applications of a splitting algorithm to decomposition in variational inequalities and convex programming. *Mathematical Programming*, 48(1):249–263, 1990.

[Tse91] P. Tseng. Applications of a splitting algorithm to decomposition in convex programming and variational inequalities. *SIAM Journal on Control and Optimization*, 29(1):119–138, 1991.

[Tse01] P. Tseng. Convergence of a block coordinate descent method for nondifferentiable minimization. *Journal of Optimization Theory and Applications*, 109(3):475–494, 2001.

[Tse08] P. Tseng. On accelerated proximal gradient methods for convex-concave optimization. *submitted to SIAM Journal on Optimization*, 2008.

[TTM11] H. Terelius, U. Topcu, and R. M. Murray. Decentralized multi-agent optimization via dual decomposition. *IFAC Proceedings Volumes*, 44(1):11245–11251, 2011.

[TY09] P. Tseng and S. Yun. A coordinate gradient descent method for nonsmooth separable minimization. *Mathematical Programming*, 117(1–2):387–423, 2009.

[TY12] M. Tao and X. Yuan. An inexact parallel splitting augmented Lagrangian method for monotone variational inequalities with separable structures. *Computational Optimization and Applications*, 52(2):439–461, 2012.

[TY18] M. Tao and X. Yuan. On Glowinski's open question on the alternating direction method of multipliers. *Journal of Optimization Theory and Applications*, 179(1):163–196, 2018.

[ULGN20] C. A. Uribe, S. Lee, A. Gasnikov, and A. Nedić. A dual approach for optimal algorithms in distributed optimization over networks. In *Information Theory and Applications Workshop*, 2020.

[Val43] F. A. Valentine. On the extension of a vector function so as to preserve a Lipschitz condition. *Bulletin of the American Mathematical Society*, 49(2):100–108, 1943.

[Val45] F. A. Valentine. A Lipschitz condition preserving extension for a vector function. *American Journal of Mathematics*, 67(1):83–93, 1945.

[Ver96] M. Verkama. Random relaxation of fixed-point iteration. *SIAM Journal on Scientific Computing*, 17(4):906–912, 1996.

[VSFL18] B. Van Scoy, R. A. Freeman, and K. M. Lynch. The fastest known globally convergent first-order method for minimizing strongly convex functions. *IEEE Control Systems Letters*, 2(1):49–54, 2018.

[Vũ13a] B. C. Vũ. A splitting algorithm for dual monotone inclusions involving cocoercive operators. *Advances in Computational Mathematics*, 38(3):667–681, 2013.

[Vũ13b] B. C. Vũ. A variable metric extension of the forward-backward-forward algorithm for monotone operators. *Numerical Functional Analysis and Optimization*, 34(9):1050–1065, 2013.

[War63] J. Warga. Minimizing certain convex functions. *Journal of the Society for Industrial and Applied Mathematics*, 11(3):588–593, 1963.

[WB12] H. Wang and A. Banerjee. Online alternating direction method. In *International Conference on Machine Learning*, Edinburgh, Scotland, UK, 2012.

[WGY10] Z. Wen, D. Goldfarb, and W. Yin. Alternating direction augmented Lagrangian methods for semidefinite programming. *Mathematical Programming Computation*, 2(3–4):203–230, 2010.

[WHML15] X. Wang, M. Hong, S. Ma, and Z.-Q. Luo. Solving multiple-block separable convex minimization problems using two-block alternating direction method of multipliers. *Pacific Journal of Optimization*, 11(4):645–667, 2015.

[WL08] T. T. Wu and K. Lange. Coordinate descent algorithms for Lasso penalized regression. *The Annals of Applied Statistics*, 2(1):224–244, 2008.

[WL20] S. J. Wright and C.-P. Lee. Analyzing random permutations for cyclic coordinate descent. *Mathematics of Computation*, 89(325):2217–2248, 2020.

[WN11] H. F. Walker and P. Ni. Anderson acceleration for fixed-point iterations. *SIAM Journal on Numerical Analysis*, 49(4):1715–1735, 2011.

[WO13] E. Wei and A. Ozdaglar. On the $O(1/k)$ convergence of asynchronous distributed alternating direction method of multipliers. In *Global Conference on Signal and Information Processing*, 2013.

[Woh17] B. Wohlberg. ADMM penalty parameter selection by residual balancing. *arXiv:1704.06209*, 2017.

[Wri15] S. J. Wright. Coordinate descent algorithms. *Mathematical Programming*, 151(1):3–34, 2015.

[WS17] J. J. Wang and W. Song. An algorithm twisted from generalized ADMM for multi-block separable convex minimization models. *Journal of Computational and Applied Mathematics*, 309:342–358, 2017.

[WYL+18] T. Wu, K. Yuan, Q. Ling, W. Yin, and A. H. Sayed. Decentralized consensus optimization with asynchrony and delays. *IEEE Transactions on Signal and Information Processing over Networks*, 4(2):293–307, 2018.

[WYYZ08] Y. Wang, J. Yang, W. Yin, and Y. Zhang. A new alternating minimization algorithm for total variation image reconstruction. *SIAM Journal on Imaging Sciences*, 1(3):248–272, 2008.

[XB04] L. Xiao and S. Boyd. Fast linear iterations for distributed averaging. *Systems and Control Letters*, 53(1):65–78, 2004.

[XFG17] Z. Xu, M. Figueiredo, and T. Goldstein. Adaptive ADMM with spectral penalty parameter selection. In *International Conference on Artificial Intelligence and Statistics*, 2017.

[XFY+17] Z. Xu, M. A. T. Figueiredo, X. Yuan, C. Studer, and T. Goldstein. Adaptive relaxed ADMM: Convergence theory and practical implementation. In *Computer Vision and Pattern Recognition*, 2017.

[XLLY17] Y. Xu, M. Liu, Q. Lin, and T. Yang. ADMM without a fixed penalty parameter: Faster convergence with new adaptive penalization. In *Neural Information Processing Systems*, 2017.

[Xu07] M. H. Xu. Proximal alternating directions method for structured variational inequalities. *Journal of Optimization Theory and Applications*, 134(1):107–117, 2007.

[Xu15] Y. Xu. Alternating proximal gradient method for sparse nonnegative Tucker decomposition. *Mathematical Programming Computation*, 7(1):39–70, 2015.

[Xu17] Y. Xu. Accelerated first-order primal-dual proximal methods for linearly constrained composite convex programming. *SIAM Journal on Optimization*, 27(3):1459–1484, 2017.

[XXS19] P. Xiao, Z. Xiao, and R. Sun. Understanding limitation of two symmetrized orders by worst-case complexity. *arXiv:1910.04366*, 2019.

[XY13] Y. Xu and W. Yin. A block coordinate descent method for regularized multiconvex optimization with applications to nonnegative tensor factorization and completion. *SIAM Journal on Imaging Sciences*, 6(3):1758–1789, 2013.

[XY17] Y. Xu and W. Yin. A globally convergent algorithm for nonconvex optimization based on block coordinate update. *Journal of Scientific Computing*, 72(2):700–734, 2017.

[XYLC19] L. Xiao, A. W. Yu, Q. Lin, and W. Chen. DSCOVR: Randomized primal-dual block coordinate algorithms for asynchronous distributed optimization. *Journal of Machine Learning Research*, 20(43):1–58, 2019.

[XZSX15] J. Xu, S. Zhu, Y. C. Soh, and L. Xie. Augmented distributed gradient methods for multi-agent optimization under uncoordinated constant stepsizes. In *IEEE Conference on Decision and Control*, 2015.

[Yan18a] M. Yan. A new primal-dual algorithm for minimizing the sum of three functions with a linear operator. *Journal of Scientific Computing*, 76(3):1698–1717, 2018.

[Yan18b] M. Yan. Primal-dual algorithms for the sum of two and three functions. Lecture Slides, `https://mingyan08.github.io/Slides/PD30.pdf`, 2018.

[YLY16] K. Yuan, Q. Ling, and W. Yin. On the convergence of decentralized gradient descent. *SIAM Journal on Optimization*, 26(3):1835–1854, 2016.

[YO13] W. Yin and S. Osher. Error forgetting of Bregman iteration. *Journal of Scientific Computing*, 54(2–3):684–695, 2013.

[YOGD08] W. Yin, S. Osher, D. Goldfarb, and J. Darbon. Bregman iterative algorithms for ℓ_1-minimization with applications to compressed sensing. *SIAM Journal on Imaging Sciences*, 1(1):143–168, 2008.

[YP19] P. Yi and L. Pavel. Distributed generalized Nash equilibria computation of monotone games via double-layer preconditioned proximal-point algorithms. *IEEE Transactions on Control of Network Systems*, 6(1):299–311, 2019.

[YT11] S. Yun and K.-C. Toh. A coordinate gradient descent method for ℓ_1-regularized convex minimization. *Computational Optimization and Applications*, 48(2):273–307, 2011.

[YTT11] S. Yun, P. Tseng, and K.-C. Toh. A block coordinate gradient descent method for regularized convex separable optimization and covariance selection. *Mathematical Programming*, 129(2):331–355, 2011.

[Yua12] X. Yuan. Alternating direction method for covariance selection models. *Journal of Scientific Computing*, 51(2):261–273, 2012.

[YY13a] J. Yang and X. Yuan. Linearized augmented Lagrangian and alternating direction methods for nuclear norm minimization. *Mathematics of Computation*, 82(281):301–329, 2013.

[YY13b] X. Yuan and J. Yang. Sparse and low-rank matrix decomposition via alternating direction methods. *Pacific Journal on Optimization*, 9(1):167–180, 2013.

[YY16] M. Yan and W. Yin. Self equivalence of the alternating direction method of multipliers. In R. Glowinski, S. J. Osher, and W. Yin, eds., *Splitting Methods in Communication, Imaging, Science and Engineering*, pages 165–194. Springer, 2016.

[YYZS19a] K. Yuan, B. Ying, X. Zhao, and A. H. Sayed. Exact diffusion for distributed optimization and learning – Part I: Algorithm development. *IEEE Transactions on Signal Processing*, 67(3):708–723, 2019.

[YYZS19b] K. Yuan, B. Ying, X. Zhao, and A. H. Sayed. Exact diffusion for distributed optimization and learning – Part II: Convergence analysis. *IEEE Transactions on Signal Processing*, 67(3):724–739, 2019.

[YZ11] J. Yang and Y. Zhang. Alternating direction algorithms for ℓ_1-problems in compressive sensing. *SIAM Journal on Scientific Computing*, 33(1):250–278, 2011.

[Zad70] N. Zadeh. Note – A note on the cyclic coordinate ascent method. *Management Science*, 16(9):642–644, 1970.

[Zăl05] C. Zălinescu. A new proof of the maximal monotonicity of the sum using the Fitzpatrick function. In F. Giannessi and A. Maugeri, eds., *Variational Analysis and Applications*, pages 1159–1172. Springer, 2005.

[Zar60] E. H. Zarantonello. Solving functional equations by contractive averaging. Technical Report 160, Mathematics Research Center, United States Army, University of Wisconsin, 1960.

[ZBO11] X. Zhang, M. Burger, and S. Osher. A unified primal-dual algorithm framework based on Bregman iteration. *Journal of Scientific Computing*, 46(1):20–46, 2011.

[ZC08] M. Zhu and T. Chan. An efficient primal-dual hybrid gradient algorithm for total variation image restoration. UCLA CAM Reports 08-34, University of California, Los Angeles, 2008.

[Zha04] T. Zhang. Solving large scale linear prediction problems using stochastic gradient descent algorithms. In *International Conference on Machine Learning*, 2004.

[ZK14a] R. Zhang and J. Kwok. Asynchronous distributed ADMM for consensus optimization. In *International Conference on Machine Learning*, 2014.

[ZK14b] W. Zhong and J. T. Kwok. Fast stochastic alternating direction method of multipliers. In *International Conference on Machine Learning*, 2014.

[ZK16] S. Zheng and J. T. Kwok. Fast-and-light stochastic ADMM. In *International Joint Conference on Artificial Intelligence*, 2016.

[ZM10] M. Zhu and S. Martínez. Discrete-time dynamic average consensus. *Automatica*, 46(2):322–329, 2010.

[ZMB+18] Z. Zhou, P. Mertikopoulos, N. Bambos, P. Glynn, Y. Ye, L.-J. Li, and L. Fei-Fei. Distributed asynchronous optimization with unbounded delays: How slow can you go? In *International Conference on Machine Learning*, 2018.

[ZX15] Y. Zhang and L. Xiao. Stochastic primal-dual coordinate method for regularized empirical risk minimization. In *International Conference on Machine Learning*, 2015.

[ZX17] Y. Zhang and L. Xiao. Stochastic primal-dual coordinate method for regularized empirical risk minimization. *Journal of Machine Learning Research*, 18(84):1–42, 2017.

[ZXC+16] N. Zhou, Y. Xu, H. Cheng, J. Fang, and W. Pedrycz. Global and local structure preserving sparse subspace learning: An iterative approach to unsupervised feature selection. *Pattern Recognition*, 53:87–101, 2016.

[ZY18] J. Zeng and W. Yin. On nonconvex decentralized gradient descent. *IEEE Transactions on Signal Processing*, 66(11):2834–2848, 2018.

Index